Numerical Optimization and Algorithms: 2nd Edition

Numerical Optimization and Algorithms: 2nd Edition

Guest Editors

Shuai Li

Dunhui Xiao

MDPI

Basel • Beijing • Wuhan • Barcelona • Belgrade • Novi Sad • Cluj • Manchester

Guest Editors
Shuai Li
Faculty of Information Technology and Electrical Engineering
University of Oulu
Oulu
Finland

Dunhui Xiao
School of Mathematical Sciences
Tongji University
Shanghai
China

Editorial Office
MDPI AG
Grosspeteranlage 5
4052 Basel, Switzerland

This is a reprint of the Special Issue, published open access by the journal *Algorithms* (ISSN 1999-4893), freely accessible at: https://www.mdpi.com/journal/algorithms/special_issues/1XMDC2A8II.

For citation purposes, cite each article independently as indicated on the article page online and as indicated below:

Lastname, A.A.; Lastname, B.B. Article Title. *Journal Name* **Year**, *Volume Number*, Page Range.

ISBN 978-3-7258-7358-6 (Hbk)
ISBN 978-3-7258-7359-3 (PDF)
https://doi.org/10.3390/books978-3-7258-7359-3

Contents

About the Editors

Shuai Li

Shuai Li is currently a Full Professor with the Faculty of Information Technology and Electrical Engineering (ITEE), University of Oulu, and also an Adjunct Professor with the VTT-Technology Research Center of Finland. Shuai Li's main research interests are nonlinear optimization and intelligent control with their applications to robotics. He has published over 200 SCI-indexed journal papers (including more than 90 on IEEE transactions) in peer-reviewed journals. Shuai Li is available for supervising both master and PhD students. PhD graduates from his group are now working in leading universities in Hong Kong, India, and China as professors. Shuai Li is a Fellow of IET (Institution of Engineering and Technology), a Fellow of BCS (British Computer Society) and a Fellow of IMA (Institute of Mathematics and its Applications).

Dunhui Xiao

Dunhui Xiao is a Professor at the School of Mathematical Sciences, Tongji University, and a recipient of the National High-Level Overseas Young Talent Award. He serves as the Vice Director of the Information Office (concurrent) and Head of the Computational Mathematics Division at Tongji University. Additionally, he is an Executive Member of the 11th Standing Committee of the Computational Mathematics Division of the Chinese Mathematical Society, a Standing Member of the AI Practical Applications Society, a Committee Member of Shanghai CSIAM, and the Vice Director of the Tongji Branch of the Zhangjiang Institute for Mathematical Research. He obtained his PhD from Imperial College London, where he conducted his post-doc research as well. Then, he worked as a lecturer at Swansea University's Zienkiewicz Centre for Computational Engineering, one of the birthplaces of the finite element method. He has led multiple national and international research projects, including those funded by the UK's EPSRC, Royal Society, and China's national funding agencies. He serves as a reviewer for multiple academic journals and funding agencies, including the UK Research Councils, ERC in Europe and the Ministry of Science and Technology in China. His research focuses on model order reduction, data-driven models, physics-data hybrid computational models, data assimilation, domain decomposition, and computational fluid dynamics.

algorithms

MDPI

Article

Hybrid Newton-like Inverse Free Algorithms for Solving Nonlinear Equations

Ioannis K. Argyros [1,*], Santhosh George [2], Samundra Regmi [3] and Christopher I. Argyros [4]

1 Department of Computing and Mathematical Sciences, Cameron University, Lawton, OK 73505, USA
2 Department of Mathematical & Computational Science, National Institute of Technology Karnataka, Surathkal 575025, India; sgeorge@nitk.edu.in
3 Department of Mathematics, University of Houston, Houston, TX 77205, USA; sregmi5@uh.edu
4 School of Computational Science and Engineering, Georgia Institute of Technology, North Avenue Atlanta, Atlanta, GA 30332, USA; cargyros3@gatech.edu
* Correspondence: iargyros@cameron.edu

Abstract: Iterative algorithms requiring the computationally expensive in general inversion of linear operators are difficult to implement. This is the reason why hybrid Newton-like algorithms without inverses are developed in this paper to solve Banach space-valued nonlinear equations. The inverses of the linear operator are exchanged by a finite sum of fixed linear operators. Two types of convergence analysis are presented for these algorithms: the semilocal and the local. The Fréchet derivative of the operator on the equation is controlled by a majorant function. The semi-local analysis also relies on majorizing sequences. The celebrated contraction mapping principle is utilized to study the convergence of the Krasnoselskij-like algorithm. The numerical experimentation demonstrates that the new algorithms are essentially as effective but less expensive to implement. Although the new approach is demonstrated for Newton-like algorithms, it can be applied to other single-step, multistep, or multipoint algorithms using inverses of linear operators along the same lines.

Keywords: hybrid-Newton-like algorithm; fixed sum of operators; Banach space; Fréchet derivative; convergence

MSC: 65G99; 65H10; 49H17; 49M15

1. Introduction

Let T_1, T_2 stand for Banach spaces; $\Omega \subset T_1$ be an open and convex set; and $G : \Omega \longrightarrow T_2$ denote a differentiable operator according to Fréchet [1–4].

A plethora of applications from diverse areas of research such as Optimization and Computational Sciences are reduced using mathematical modeling [5–14] to locate a solution x^* of a nonlinear equation like

$$G(x) = 0 \tag{1}$$

The closed version of a solution $x^* \in \Omega$ is possible only in special cases. Consequently, most solution approaches utilized by researchers and practitioners are iterative when the sequence is generated approximating the solution x^*.

The algorithm of successive substitutions or the algorithm of iteration or the Picard algorithm is a simple and important algorithm for solving linear as well as nonlinear equations. This algorithm originated in antiquity, appearing in the writings of Heron of Alexandria in the second century B.C. in relation to root extraction. Later, Cauchy, as well as Picard, employed this algorithm to assure the existence of solutions of differential equations. It is defined for $P : T_1 \longrightarrow T_1$ and each $n = 0, 1, 2, \ldots$ by

$$z_0 \in \Omega,\ z_{n+1} = P(z_n).$$

Banach inaugurated the abstract formulation of this algorithm followed by Cacciopoli and Weisinger. We refer the readers to the reference by Berinde [15], Krasnoselskij [16], and Kantorovich et al. [3] for further information on the convergence conditions (see also [17–23]). Concerning the convergence order of this algorithm, it is only linear. Thus, there is a need for introducing algorithms of convergence order higher than one.

Newton's algorithm is without a doubt the most well-known algorithm of convergence order two for solving transcendental as well as scalar equations. It is written for each $n = 0, 1, 2, \ldots$ as

$$x_0 \in \Omega,\ x_{n+1} = x_n - G'(x_n)^{-1}G(x_n),$$

where G' is the notation for the derivative of the operator G according to Fréchet [8,24,25]. The construction of Newton's algorithm is based on linearization. Let $x_0 \in \Omega$ be an initial point. If the operator G is Fréchet-differentiable, one can write the Ostrowski [21] representation

$$G'(x) = G(x_0) + G'(x_0)(x - x_0) + d(x, x_0),$$

where $\|d(x, x_0)\| = O(\|x - x_0\|)$ as $\|x - x_0\| \longrightarrow 0$. If $x^* \in \Omega$ is a solution of the Equation (1), it follows by the preceding representation that

$$G(x_0) + G'(x_0)(x^* - x_0) = -d(x^*, x_0).$$

If x^* is near to x_0, then one can neglect the supposedly small quantity $d(x^*, x_0)$, leading to a linear equation

$$G(x_0) + G'(x_0)(x - x_0) = 0.$$

It is said that this equation is obtained from Equation (1) by the technique of linearization or the tangent algorithm since this is the equation of the tangent line to the curve $y = G(x)$ at the point $(x_0, G(x_0))$ provided that G is a real function. It follows by this linearization that x_1 is a unique solution if $G'(x_0)^{-1}$ exists. In this case, we can write

$$x_1 = x_0 - G'(x_0)^{-1}G(x_0).$$

If x_0 is close to x^*, then maybe x_1 is even closer. Thus, this process can be repeated with x_1 replacing x_0 leading to Newton's algorithm. A main drawback with the implementation of Newton's algorithm is the inversion of the linear operator $G'(x_n)$ at each step of the iteration. We address this issue in this paper. Our methodology applies to all single-step, multistep methods, and quasi-Newton [5,10,20,26,27] using inverses along the same lines. We shall demonstrate our methodology to a large class of algorithms involving inverses including Newton's algorithm as a special case.

Let us consider the popular Newton-like algorithm defined for $y_0 \in \Omega$ and each $n = 0, 1, 2, ...$ by

$$y_0 \in \Omega,\ y_{n+1} = y_n - L_n^{-1}G(y_n) \tag{2}$$

Let $L : \Omega \longrightarrow \mathcal{L}(T_1, T_2)$, the space of bounded linear operators mapping the space T_1 into T_2. Notice that for $L_n = G'(y_0)$ or $L_n = G'(y_n)$, we obtain the modified and Newton's algorithm, respectively. There are two types of convergence usually studied for iterative algorithms: the semi-local and the local analysis. The former uses the condition on the initial guess y_0, and the operator F and the solution y^* are found in a neighborhood of x_0. The latter differs from the former since the convergence conditions depend on y^* and demonstrate how difficult is to choose the initial guess y_0. The main challenge of local analysis is that y^* is usually unknown. Numerous papers have been presented dealing with the semi-local as well as the local analysis of convergence for the Newton-like algorithm (2) [2,24–26,28]. The convergence conditions involve Lipchitz–Holder or generalized continuity conditions utilized to control the Fréchet derivative G' of the operator F. By L_n

in (2), we mean $L_n = L(y_n)$. The inversion of the linear operator at each step is computationally expensive or impossible in general. To essentially utilize the algorithm but without the inverse, we replace it with a finite sum of linear operators related to L as follows:

Suppose:

There exists $\Delta \in \mathcal{L}(T_1, T_2)$ such that $\Delta^{-1} \in \mathcal{L}(T_2, T_1)$ and $(I - \Delta^{-1}(\Delta - L(y))^{-1} \in \mathcal{L}(T_2, T_1)$. Then, the Newton-like algorithm (2) can be rewritten as

$$\begin{aligned} & y_0 \in \Omega, \\ y_{n+1} &= y_n - [I - \Delta^{-1}(\Delta - L_n)]^{-1}\Delta^{-1}G(y_n), \end{aligned} \tag{3}$$

where I denotes the identity operator. The iterates of algorithms (2) and (3) coincide, since

$$\begin{aligned} & [I - \Delta^{-1}(\Delta - L_n)]^{-1}\Delta^{-1} \\ = & [\Delta(I - \Delta^{-1}(\Delta - L_n))]^{-1} = L_n^{-1}. \end{aligned}$$

Even if we replace algorithm (2) with algorithm (3), we still need to invert the linear operator $(I - \Delta^{-1}(\Delta - L_n))$ at each step of the iteration. But we can avoid this inversion if we introduce for k a fixed natural number the operators $\Gamma = \Gamma(x) = \Delta^{-1}(\Delta - L(x))$ and

$$A_k(x) = A = I + \Gamma + \Gamma^2 + ... + \Gamma^k.$$

Then, consider the replacement of algorithm (2) defined for $x_0 = y_0 \in \Omega$ by

$$x_0 \in \Omega,\ x_{n+1} = x_n - A\Delta^{-1}G(x_n). \tag{4}$$

Algorithm (4) requires the inversion of the inversion of only the frozen linear operator Δ at each step. Notice also that A is a linear operator. By letting $k \longrightarrow +\infty$, and if $\lim_{k \to +\infty} A_k = L$ exists, then

$$L_n^{-1} = L\Delta^{-1}.$$

The condition

$$\|\Delta^{-1}(\Delta - L(x))\| < 1 \quad x \in \Omega$$

guarantees the existence of this limit [5,8,16]. A possible choice for $\Delta = I$. If $T_1 = T_2 = \mathbb{R}^i$, i a natural number, and H denotes the Hessian of the operator G, then we can choose $\Delta = H(x_0)$(semi-local case) or $\Delta = H(x^*)$(local case). The choice $\Delta = H(\bar{x})$ has been considered in [29], where $\bar{x} \in \Omega$ is an auxiliary point. In the more general setting of Banach space $\Delta \in L(T_1, T_2)$ which is the space of bounded linear operator form T_1 into T_2. As a further example, if $L_n = G'(x_n)$, then $\Delta = G'(x_n)$ (semi-local case) or $\Delta = G'(x^*)$ (local case) or $\Delta = G'(\bar{x})$. Other choices for Δ are possible as long as they satisfy the convergence conditions (H_4) and (H_5) (semi-local case) and (C_1) and (C_2) (local case) (see Section 2 and Section 3, respectively).

We also study the Kransnoselskij-like or the Picard-like algorithm [15,16]

$$x_0 \in \Omega,\ \ x_{n+1} = P_k(x_n), \tag{5}$$

where

$$P_k(x) = P(x) = (1 - \lambda)x - \lambda(A\Delta^{-1}G(x) - x),\ \lambda \in (0, 1]$$

for locating fixed points. If $\lambda = 1$, then algorithm (5) reduces to algorithm (4). The semi-local analysis of convergence for algorithm (4) relies on majorizing sequences [3,24,25]. But the analysis for algorithm (5) depends on the celebrated contraction mapping principle [3,17,21,22].

The preceding reasoning justifies the study of the semi-local and local analysis of convergence appearing in this paper. The rest of the paper is organized as follows: In Sections 2 and 3, we develop the semi-local and local analysis of convergence for algorithm (4). The convergence of the Krasnoselskij-like algorithm is presented in Section 4.

The numerical experimentations demonstrating the efficiency of the new hybrid algorithms are provided in Section 5. Concluding remarks and directions of future research complete this paper in Section 6.

2. Semi-Local Analysis

Throughout this paper, we use the symbol $U(x,\alpha)$ to denote the open ball centered at $x \in T_1$ with radius $\alpha > 0$, and $U[x,\alpha]$ is the closure of $U(x,\alpha)$. The following Banach Lemma is used to prove our results.

Theorem 1 (Banach Lemma on Invertible Operators [3,15]). *If M is a bounded linear operator in T_1, M^{-1} exists if and only if there is a bounded linear operator M_1 in T_1 such that M_1^{-1} exists and*

$$\|I - M_1 M\| < 1.$$

If M^{-1} exists, then

$$M^{-1} = \sum_{n=0}^{\infty} (I - M_1 M)^n M_1$$

and

$$\|M^{-1}\| \leq \frac{\|M_1\|}{1 - \|I - M_1 M\|}.$$

Further, we use majorizing sequences to prove the semi-local convergence. Recall the definition of a majorizing sequence.

Definition 1 ([3,5]). *Let $\{x_n\}$ be a sequence in a normed space X. Then a nonnegative scalar sequence $\{v_n\}$ for which*

$$\|x_{n+1} - x_n\| \leq v_{n+1} - v_n \;\; \forall n \geq 0 \tag{6}$$

holds is a majorizing sequence for $\{x_n\}$. Note that any majorizing sequence is necessarily nondecreasing. Moreover, if the sequence $\{v_n\}$ converges, then $\{x_n\}$ converges too, and for $v^* = \lim_{n\longrightarrow\infty} v_n$

$$\|x^* - x_n\| \leq v^* - v_n.$$

Hence, the study of the convergence of the sequence $\{x_n\}$ reduces to that of $\{v_n\}$.

Let $S = [0, +\infty)$.

Suppose:

(H_1) There exists parameters $\delta \geq 0, \gamma \in [0, \frac{1}{2})$, an element $x_0 \in \Omega$, and an invertible operator Δ such that

$$\|A\Delta^{-1}G(x_0)\| \leq \delta.$$

(H_2) There exists a function $w_0 : S \longrightarrow S$, which is nondecreasing as well as continuous (FNDC), such that the equation $w_0(t) - 1 = 0$ has the smallest positive solution. Denote such solution by ρ and set $S_0 = [0, \rho)$.

(H_3) There exist (FNDC) $w : S_0 \longrightarrow S$ and $w_1 : S_0 \longrightarrow S$. Define the scalar sequence $\{\alpha_n\}$ for $\alpha_0 = 0, \alpha_1 = \delta$ some $\gamma \geq 0, \bar{b} \geq 0$, and each $n = 0, 1, 2, \ldots$ by

$$\begin{aligned} q_{n+1} &= (1 + \gamma + \ldots + \gamma^k)(\int_0^1 w((1-\tau)(\alpha_{n+1} - \alpha_n)d\tau \\ &\quad + w_1(\alpha_n) + \bar{b}\gamma^{k+1}), \end{aligned}$$

and

$$\alpha_{n+2} = \alpha_{n+1} + q_{n+1}(\alpha_{n+1} - \alpha_n). \tag{7}$$

The sequence $\{\alpha_n\}$ is shown to be majorizing for algorithm (4) in Theorem 2. But first, the convergence conditions are given for the sequence $\{\alpha_n\}$.

(H_4) There exists $\rho_0 \in [0, \rho)$ such that for each $n = 0, 1, 2, ...$
$q_{n+1} < 1$ and $\alpha_n \leq \rho_0$.
This condition and the formula (7) imply that for each $n = 0, 1, 2, ...$

$$0 \leq \alpha_n \leq \alpha_{n+1} \leq \rho_0$$

and there exists $\alpha \in [0, \rho_0)$ such that $\lim_{n \to +\infty} = \alpha$. The parameter α is the unique least upper bound of the sequence $\{\alpha_n\}$.
It is worth noting that this sequence can be computed a priori and relates to the initial approximation. Such conditions are weaker than the usual convergence conditions given as functions of the starting point [3–5].
Next, we relate the scalar sequences and functions w_0, w to the operators on algorithm (4).

(H_5) $\|\Delta^{-1}(G'(v) - \Delta)\| \leq w_0(\|v - x_0\|)$ for each $v \in \Omega$.
Set $\Omega_0 = U(x_0, \rho) \cap \Omega$.

(H_6)

$$\|\Delta^{-1}(G'(v_2) - G'(v_1))\| \leq w(\|v_2 - v_1\|), \ \gamma = \|\Delta^{-1}(\Delta - L(v))\|,$$

and

$$\|\Delta^{-1}(G'(v) - L(v))\| \leq w_1(\|v - x_0\|)$$

for each $v, v_1, v_2 \in \Omega_0$.
Set $b = \gamma \frac{1-\gamma^k}{1-\gamma}$ and $\bar{b} = \frac{1}{1-b}$.
and

(H_7) $U[x_0, \alpha] \subset \Omega$.

The conditions (H_1)–(H_5) and the developed notations are utilized to show the main semi-local analysis of convergence for algorithm (4).

Theorem 2. *Suppose that the conditions (H_1)–(H_5) hold. Then, if the initial guess $x_0 \in \Omega$ the following assertions hold for the sequence $\{x_n\}$ generated by algorithm (4)*

$$\{x_n\} \subset B(x_0, \alpha) \tag{8}$$

and there exists a solution $x^* \in B[x_0, \alpha]$ of the equation $G(x) = 0$ such that

$$\|x^* - x_n\| \leq \alpha - \alpha_n, \tag{9}$$

where the sequence $\{\alpha_n\}$ is given by the formula (7) and α is its limit.

Proof. Notice that all the iterates $\{x_m\}$ (m a natural number) of algorithm (4) are well defined. We present a proof based on mathematical induction. In particular, we show that for each $n = 0, 1, 2, ...$

$$\|x_{n+1} - x_n\| \leq \alpha_{n+1} - \alpha_n. \tag{10}$$

Assertion (9) holds if $n = 0$, by (4), (7), and the condition (H_1), since

$$\|x_1 - x_0\| = \|A\Delta^{-1}G(x_0)\| \leq \alpha_1 - \alpha_0 < \alpha.$$

It also follows that the iterate $x_1 \in U(x_0, \alpha)$. Next, we show that the linear operator $A = A(x_m)$ is invertible by using the restrictions on $\gamma, b, \bar{b}$ given in the condition (H_5):

$$\begin{aligned} \|I - A(x_m)\| &= \|\Gamma + \Gamma^2 + ... + \Gamma^k\| \\ &\leq \|\Gamma\| + \|\Gamma\|^2 + ... + \|\Gamma\|^k = \|\Gamma\|\frac{1-\|\Gamma\|^k}{1-\|\Gamma\|} \\ &\leq \gamma\frac{1-\gamma^k}{1-\gamma} = b < 1 \end{aligned} \tag{11}$$

since $\gamma \in [0, \frac{1}{2})$ by (11) and the condition (H_1). Inequality (11) and Theorem 1 assure the invertibility of the operator A, and

$$\|A^{-1}\| \leq \frac{1}{1-b} = \bar{b}. \tag{12}$$

Then, we can write by algorithm (2) in turn that

$$\begin{aligned} G(x_{m+1}) &= G(x_{m+1}) - G(x_m) - \Delta A^{-1}(x_{m+1} - x_m) \\ &= G(x_{m+1}) - G(x_m) - G'(x_m)(x_{m+1} - x_m) \\ &\quad +(G'(x_m) - L_m)(x_{m+1} - x_m) \\ &\quad +(L_m - \Delta A^{-1})(x_{m+1} - x_m) \\ &= G(x_{m+1}) - G(x_m) - G'(x_m)(x_{m+1} - x_m) \\ &\quad +(G'(x_m) - L_m)(x_{m+1} - x_m) \\ &\quad +(L_m A - \Delta)A^{-1}(x_{m+1} - x_m). \end{aligned} \tag{13}$$

But we have

$$\begin{aligned} L_m A - \Delta &= L_m(I + \Gamma + \Gamma^2 + ... + \Gamma^k) - \Delta \\ &= L_m - \Delta + (-\Delta + \Delta + L_m)(\Gamma + \Gamma^2 + ... + \Gamma^k) \\ &= L_m - \Delta + \Delta(\Gamma + \Gamma^2 + ... + \Gamma^k) \\ &\quad -(\Delta - L_m)(\Gamma + \Gamma^2 + ... + \Gamma^k) \\ &= L_m - \Delta + \Delta\Gamma + \Delta(\Gamma^2 + ... + \Gamma^k) \\ &\quad -(\Delta - L_m)(\Gamma + \Gamma^2 + ... + \Gamma^k) \\ &= \Delta(\Gamma^2 + ... + \Gamma^k) - (\Delta - L_m)(\Gamma + \Gamma^2 + ... + \Gamma^k) \end{aligned} \tag{14}$$

since $L_m - \Delta + \Delta\Gamma = 0$ by the definition of Γ. Thus, we obtain from (14)

$$\begin{aligned} \Delta^{-1}(L_m A - \Delta) &= \Gamma^2 + ... + \Gamma^k - \Gamma(\Gamma + \Gamma^2 + ... + \Gamma^k) \\ &= -\Gamma^{k+1}. \end{aligned} \tag{15}$$

Then, it follows by (13), the conditions (H_4), (H_5), (13), the conditions (H_4), (H_5), (15), the inductions hypotheses, and the triangle inequality, in turn, that

$$\begin{aligned} \|\Delta^{-1}G(x_{m+1})\| &\leq \int_0^1 w((1-\tau)\|x_{m+1} - x_m\|)d\tau\|x_{m+1} - x_m\| \\ &\quad +w_1(\|x_m - x_0\|)\|x_{m+1} - x_m\| + \bar{b}\gamma^{k+1}\|x_{m+1} - x_m\| \\ &\leq \left[\int_0^1 w((1-\tau)(\alpha_{m+1} - \alpha_m)d\tau + w_1(\alpha_m) + \bar{b}\gamma^{k+1}\right] \\ &\quad \times(\alpha_{m+1} - \alpha_m). \end{aligned} \tag{16}$$

Consequently, by algorithms (4), (7), and (16), we obtain in turn that

$$\begin{aligned}\|x_{m+2}-x_{m+1}\| &\leq \|A(x_{m+1})\Delta^{-1}G(x_{m+1})\| \\ &\leq \|A(x_{m+1})\|\|\Delta^{-1}G(x_{m+1})\| \\ &\leq (1+\gamma+\ldots+\gamma^k)\left[\int_0^1 w((1-\tau)(\alpha_{m+1}-\alpha_m)d\tau\right. \\ &\left.+w_1(\alpha_m)+\bar{b}\gamma^{k+1}\right](\alpha_{m+1}-\alpha_m)=\alpha_{m+2}-\alpha_{m+1},\end{aligned}$$

and

$$\begin{aligned}\|x_{m+2}-x_0\| &\leq \|x_{m+2}-x_{m+1}\|+\|x_{m+1}-x_0\| \\ &\leq \alpha_{m+2}-\alpha_{m+1}+\alpha_{m+1}-\alpha_0=\alpha_{m+2}<\alpha.\end{aligned}$$

Hence, assertion (10) holds and the iterate $x_{m+2}\in B(x_0,\alpha)$. By the condition (H_4), the sequence $\{\alpha_m\}$ is complete as convergent to α. Therefore, by (10) the sequence $\{x_m\}$ is also complete in the Banach space T_1, and as such it converges to some $x^*\in B[x_0,\alpha]$(since $B[x_0,\alpha]$ is a closed set). By sending $m\longrightarrow+\infty$ in (16), and the continuity of the operator G, we deduce that $G(x^*)=0$. Finally, the estimation for i a natural number

$$\|x_{m+i}-x_m\|\leq\alpha_{m+i}-\alpha_m,$$

shows (9), if $i\longrightarrow+\infty$. □

Next, a set is specified that contains only one solution of the equation $G(x)=0$.

Proposition 1. *Suppose: There exists a solution $y\in B(x_0,\rho_1)$ of the equation $G(x)=0$ for some $\rho_4>0$; the condition (H_4) holds in the ball $B(x_0,\rho_1)$, and there exists $\rho_2\geq\rho_1$ such that*

$$\int_0^1 w_0(\tau\rho_1+(1-\tau)\rho_2)d\tau<1 \tag{17}$$

Set $\Omega_2=B[x_0,\rho_2]\cap\Omega$. Then, the element y is the only solution of the equation $G(x)=0$ in the set Ω_2.

Proof. Suppose that there exists a solution $z\in\Omega_2$ of the equation $G(x)=0$ with $z\neq y$. Define the linear operator $E=\int_0^1 G'(y+\tau(z-y))d\tau$. By using the condition (H_4) and (17), we obtain in turn

$$\begin{aligned}\|\Delta^{-1}(E-\Delta)\| &\leq \int_0^1 w_0(\tau\|y-x_0\|+(1-\tau)\|z-x_0\|)d\tau \\ &\leq \int_0^1 w_0(\tau\rho_2+(1-\tau)\rho_3)d\tau<1.\end{aligned}$$

Thus, the linear operator $E^{-1}\in\mathcal{L}(T_2,T_1)$, and from the identity we obtain

$$z-y=E^{-1}(G(z)-G(y))=E^{-1}(0)=0.$$

Therefore, we conclude that $z=y$. □

Remark 1.

(1) *The limit point α can be replaced by ρ in the condition (H_6).*
(2) *If all the conditions (H_1)-(H_6) hold in Proposition 1, take $y=x^*$ and $\rho_2=\alpha$.*
(3) *The second hypothesis in the condition (H_5) can be replaced as follows:*
Suppose that there exists (FNDC) $w_2:S\longrightarrow S$ such that equation $w_2(t)-1=0$ has an SPS. Denote such solution by ρ_3, and set $\gamma\in[0,w_2(\rho_3))$. Then, $\gamma\in[0,1)$. In this case, set $\bar{\alpha}=min\{\alpha,\rho_3\}$, and replace α by $\bar{\alpha}$ in the condition (H_6).

(4) *The results for algorithm (3) or (2) can be obtained if we let $k \longrightarrow +\infty$ in Theorem 1. A possible choice for $k = n$, although a smaller value $k = 1$ or $k = 2$ is preferred to reduce the computational cost (see also the numerical Section 6).*

3. Local Analysis

In Section 3, we exchange the role of x^* and the "w" functions with x_0, and the "ψ" function, respectively. But the computations are similar.

Suppose:

(C_1) There exists a solution $x^* \in \Omega$ of the equation $G(x) = 0$, and an invertible operator $\Delta \in \mathcal{L}(T_1, T_2)$ such that for each $\tau \in [0,1]$, $x \in \Omega$

$$\begin{aligned}&\|\int_0^1 \Delta^{-1}(G'(x^* + \tau(x - x^*)) - \Delta)d\tau\|\\ &\leq \int_0^1 \psi(\tau\|x - x^*\|)d\tau,\end{aligned}$$

for some (FNDC) $\psi : S \longrightarrow S$.

(C_2) There exists $\gamma \in [0, \frac{1}{2})$, such that

$$\|\Delta^{-1}(A(x) - \Delta)\| \leq \gamma.$$

Define the function $g : S \longrightarrow S$ by

$$g(t) = \frac{1-\gamma^{k+1}}{1-\gamma}\int_0^1 \psi(\tau t)d\tau + \gamma\frac{1-\gamma^k}{1-\gamma}.$$

(C_3) The equation $g(t) - 1 = 0$ has an SPS. Denote such a solution by ρ_4. and

(C_4) $U[x^*, \rho_4] \subset \Omega$.

Theorem 3. *Suppose that the conditions (C_1)–(C_4) hold. Then, the sequence $\{x_n\}$ with initial guess $x_0 \in U(x^*, \rho_4) - \{x^*\}$ exists in $U(x^*, \rho_4)$, stays in $U(x^*, \rho_4)$, and converges to x^* such that*

$$\|x_{n+1} - x^*\| \leq g(\|x_n - x^*\|)\|x_n - x^*\| \leq \|x_n - x^*\| \leq \rho_4. \quad (18)$$

Proof. Assertion (18) is shown by mathematical induction. Notice that all the iterates x_m exist by algorithm (4). We can also write, in turn, that

$$\begin{aligned}x_{m+1} - x^* &= x_m - x^* - A\Delta^{-1}G(x_m)\\ &= x_m - x^* - A\Delta^{-1}\int_0^1 G'(x^* + \tau(x_m - x^*))d\tau(x_m - x^*)\\ &= [I - A\Delta^{-1}(\int_0^1 G'(x^* + \tau(x_m - x^*)) - \Delta + \Delta)d\tau](x_m - x^*)\\ &= [(I - A - A\Delta^{-1}(\int_0^1 G'(x^* + \tau(x_m - x^*)) - \Delta)d\tau](x_m - x^*). \quad (19)\end{aligned}$$

But, we have as in the semi-local case

$$\|A\| \leq \frac{1-\gamma^{k+1}}{1-\gamma},$$

and

$$\|I - A\| \leq \gamma\frac{1-\gamma^k}{1-\gamma}. \quad (20)$$

Hence, (19) turns into

$$x_{m+1} - x^* = \left[(I - A) - A\left(\int_0^1 \Delta^{-1}(G'(x^* + \tau(x_m - x^*)) - \Delta)d\tau\right](x_m - x^*). \tag{21}$$

Using the conditions (C_1)–(C_3), we obtain

$$\begin{aligned} \|x_{m+1} - x^*\| &\leq \left[\gamma\frac{1-\gamma^k}{1-\gamma} + \frac{1-\gamma^{k+1}}{1-\gamma}\int_0^1 \psi(\tau\|x_m - x^*\|)d\tau\right]\|x_m - x^*\| \\ &\leq g(\|x_m - x^*\|)\|x_m - x^*\| \leq \|x_m - x^*\| < \rho_4. \end{aligned} \tag{22}$$

Thus, assertion (18) holds and the iterate $x_{m+1} \in U(x^*, \rho_4)$. Then, by (22), we obtain

$$\|x_{m+1} - x^*\| \leq d\|x_m - x^*\| \leq d^{m+1}\|x_0 - x^*\|, \tag{23}$$

where $d = g(\|x_0 - x^*\|) \in [0, 1)$. Therefore, we deduce from (23) that $\lim_{m\longrightarrow+\infty} = x^*$, and the iterate $x_{m+1} \in U(x^*, \rho_4)$. □

Next, the uniqueness of the x^* set is determined.

Proposition 2. *Suppose: There exists $\rho_5 > 0$ such that the condition (C_1) holds in the ball $U(x^*, \rho_5)$,*

$$\|\Delta^{-1}(G'(u) - M)\| \leq \psi_0(\|u - x^*\| \tag{24}$$

for each $u \in \Omega$ and some (FNDC) $\psi_0 : S \longrightarrow S$ and there exists $\rho_6 \geq \rho_5$ such that

$$\int_0^1 \psi_0(\tau\rho_6)d\tau < 1. \tag{25}$$

Define the set $\Omega_2 = U[x^*, \rho_6] \cap \Omega$. Then, the only solution of the equation $G(x) = 0$ in the set Ω_2 is x^*.

Proof. Suppose that there exists $y \in \Omega_3$ solving the equation $G(x) = 0$ and $y_0 \neq x^*$. Define the linear operator $E_1 = \int_0^1 G'(x^* + \tau(y_0 - x^*))d\tau$. By the condition ($C_1$), (24), and (25), we have in turn that

$$\begin{aligned} \|\Delta^{-1}(E_1 - \Delta)\| &\leq \int_0^1 \psi_0(\tau\|y_0 - x^*\|)d\tau \\ &\leq \int_0^1 \psi_0(\tau\rho_6)d\tau < 1, \end{aligned}$$

Hence, L_1 is invertible. Finally, from the identity $y_0 - x^* = E^{-1}(G(y_0) - G(x^*)) = E^{-1}(0) = 0$, we conclude that $y_0 = x^*$. □

Remark 2.

(1) *We can set in Proposition 2 $y_0 = x^*$ and $\rho_5 = \rho_4$.*
(2) *The parameter ρ_4 defined in the condition (C_3) is the radius of convergence for algorithm (4).*
(3) *As in the semi-local analysis if $k \longrightarrow +\infty$, we obtain the results for algorithm (3). Another choice for $k = n$. However, we shall choose a small value of k to save computational cost.*

4. Convergence of the Krasnoselskij-like Algorithm

The contraction mapping principle has been used extensively to find fixed points using iterative algorithms.

Theorem 4 ([15]). *Let $Q : T_1 \longrightarrow T_1$ be a contraction operator, with Lipchitz parameter $\zeta \in [0,1)$. Then, the operator Q has a fixed point $x^* \in T_1$ i.e., $Q(x^*) = x^*$, Moreover, for each initial guess $x_0 \in T_1$ the Picard algorithm or the algorithm of successive substitutions $x_{n+1} = Q(x_n)$ converges to $x^* \in T_1$.*

Theorem 4 cannot be used if the operator Q has more than one fixed point. The fixed points must be separated in this case. Let us consider Q_0 to be a closed substep of Q_0 with $Q_0 \neq \phi$. Then, the following result is available.

Theorem 5 ([15]). *Suppose that the operator $Q : Q_0 \longrightarrow Q_0$ is a contraction with constant $\zeta \in [0,1)$. Then, the operator Q has a unique fixed point $x^* \in Q_0$. Moreover, for each $x_0 \in Q_0$, the Picard algorithm converges to x^*. The convergence of algorithm (5) is based on Theorem 6.*

Theorem 6. *Let k be a fixed natural number. Suppose that the following conditions hold for $x, y \in \Omega$:*

$$\begin{aligned} \|\Delta^{-1}(L(x) - \Delta)\| &< 1, \\ \|P(x) - P(x_0)\| &\leq p_0\|x - x_0\|, \\ \|P(y) - P(x_0)\| &\leq p\|y - x\| \end{aligned}$$

and

$$\|P(x_0) - x_0\| \leq \frac{r}{1 - p_0},$$

for some invertible operator and $p_0, p \in [0,1)$, and $U(x_0, r) \subset \Omega$ provided that $r \in [0, \frac{1}{p_0})$. Then, the operator $P : \Omega \longrightarrow \Omega$ has a unique fixed point $x^* \in \Omega$, and the Krasnoselskij-like algorithm (5) converges to x^*.

Proof. Notice that

$$\begin{aligned} \|P(x) - x_0\| &\leq \|P(x) - P(x_0)\| + \|P(x_0) - x_0\| \\ &\leq p_0 r + \|P(x_0) - x_0\| \leq r. \end{aligned}$$

Thus, $P : U(x_0, r) \longrightarrow U(x_0, r)$ is a contraction operator with constant $p \in [0,1)$. Set $Q_0 = U[x_0, r]$. Then, the result follows from Theorem 5. □

5. Error Analysis

The sequences $\{y_n\}$ and $\{x_n\}$ are generated by formulas (2) and (4), respectively. We select a portion of the standard semi-local convergence result for the Newton-like algorithm (2) [28].

Theorem 7. *Let $G : \Omega \longrightarrow Y$ be Fréchet-differentiable and let $L(x) \in L(T_1, T_2)$ be an approximation to the linear operator $G'(x)$. Suppose that there exist an open convex subset $\tilde{\Omega}$ of Ω, $x_0 \in \tilde{\Omega}$, a bounded linear invertible operator $K_0(= L(x_0))$, and constants $\eta, K > 0, K_0, K_1, \mu, l \geq 0$ such that for all $x, y \in \tilde{\Omega}$ the following conditions hold:*

$$\begin{aligned} \|L_0^{-1}G(x_0)\| &\leq \eta, \\ \|K_0^{-1}(F'(y) - F'(x))\| &\leq K\|y - x\|, \\ \|K_0(F'(x) - K_1(x)) &\leq K_0\|x - x_0\| + \mu, \\ \|K_0(A(x) - K_0)\| &\leq K_1\|x - x_0\| + l, \\ l_1 &= \mu + l < 1. \end{aligned}$$

In addition, suppose that

$$h := \sigma\eta \leq \frac{1}{2}(1 - l_1)^2,$$

where $\sigma = max\{K, K_0 + K_1\}$, *and*

$$\bar{U} = U[x_0, t^*] \subset \tilde{\Omega}$$

where

$$t^* = (1 - l_1 - \sqrt{(1 - l_1)^2 - 2h)})/\sigma_0$$

Then, the following assertions hold

The sequence $\{y_n\}$ *generated by algorithm (2) remains in the ball* $U(x_0, t^*)$ *and converges to a solution* $x^* \in \bar{U}$ *of the equation* $G(x) = 0$ *and*

$$\|L_n^{-1}K_0\| \leq s_1 := \frac{1}{1 - (K_0 t^* + l)} \tag{26}$$

Next, the sequences $\{y_n\}$ *and* $\{x_n\}$ *are related to each other.*

Lemma 1. *Suppose that the conditions* $(H_1) - (H_5)$ *for* $(\Delta = K_0)$, *and those of Theorem 7 hold. Then, the following error assertion holds for each* $n = 0, 1, 2, ...$

$$\|x_{n+1} - y_{n+1}\| \leq e_n, \tag{27}$$

where

$$\begin{aligned} e_n \;=\; & s_1[(\int_0^1 w((1-\tau)\|x_n - y_n)d\tau + w(\|x_n - y_n\| + w_1(\|x_n - x_0\|)\|x_n - y_n\| \\ & + \bar{b}\gamma^{k+1}\|x_{n+1} - x_n\|]. \end{aligned} \tag{28}$$

Proof. Under the conditions of Theorems 2 and 7, the iterates $\{y_n\}$ and $\{x_n\}$ are well defined by formulas (2), and (4), respectively. By subtracting (2) from (4) and pulling out L_n^{-1}, we can write, in turn, that

$$\begin{aligned} x_{n+1} - y_{n+1} &= x_n - y_n + L_n^{-1}G(y_n) - A\Delta^{-1}G(x) \\ &= L_n^{-1}[L_n(x_n - y_n) - G(x_n) + G(x_n) + G(y_n) - L_n A\Delta^{-1}G(x_n)] \\ &= -L_n^{-1}(G(x_n) - G(y_n) - F'(y_n)(x_n - y_n) \\ &- L_n^{-1}(F'(y_n) - F'(x_n))(x_n - y_n) - L_n^{-1}(F'(x_n) - L_n)(x_n - y_n) \\ &+ L_n^{-1}(\Delta - L_n A)A^{-1}(x_{n+1} - x_n). \end{aligned} \tag{29}$$

We need, in turn, the following estimates obtained by the conditions of Theorems 2 and 7

$$\begin{aligned} & \|K_0^{-1}(G(x_n) - G(y_n) - F'(y_n)(x_n - y_n))\| \\ \leq\; & \|\int_0^1 sK_0^{-1}(F'(y_n + \tau(x_n - y_n)) - F'(y_n))d\tau(x_n - y_n)\| \\ \leq\; & \int_0^1 sw((1-\tau)\|x_n - y_n\|)d\tau\|x_n - y_n\| \end{aligned} \tag{30}$$

$$\|K_0^{-1}(F'(y_n) - F'(x_n))(x_n - y_n)\| \leq w(\|x_n - y_n\|)\|x_n - y_n\|, \tag{31}$$

$$\begin{aligned} \|K_0^{-1}(F'x_n) - L_n)(x_n - y_n)\| &\leq w_1(\|x_n - x_0\|)\|x_n - y_n\| \\ &\leq w_1(\alpha)\|x_n - y_n\| \end{aligned} \tag{32}$$

$$\|\Delta^{-1}(\Delta - L_n A)\|\Gamma^{k+1}\| \leq \gamma^{k+1}, \tag{33}$$

and

$$\|K_0 G(x_n)\| = \|B^{-1}(x_{n+1} - x_n)\| = \bar{b}\|x_{n+1} - x_n\| \tag{34}$$

By summing up (30)–(34) and using the triangle inequality in (34), we obtain (27). □

It is convenient for the next result to define the function $\varphi : [0,+\infty) \longrightarrow R$ by

$$\varphi(t) = s_1\left[\left(\int_0^1 w((1-\tau)t)d\tau + w(t) + w_1(\alpha)\right)t + s_2\right] - t, \text{where } s_2 = s_1\bar{b}\gamma^{k+1}\eta.$$

Proposition 3. *Let all the conditions of Lemma 1 hold. Suppose, in addition, that the equation $\psi(t) = 0$ has the smallest solution $s^* > s_2$. Then, the following assertion holds for each $n = 0,1,2,\ldots$*

$$\|x_n - y_n\| \le s^*. \tag{35}$$

Proof. The estimate (27) for $n = 0$ implies $\|x_1 - y_1\| \le s_2 \le s^*$, which is true by the choice of s^*. Suppose that (35) holds for all integer values m smaller or equal to m. Then, we have by the induction hypothesis, and (2) that

$$\|x_{m+1} - y_{m+1}\| \le e_m \le s_1(\int_0^1 w((1-\tau)s^*)d\tau + w(s^*) + w_1(\alpha))s^* + s_2 = s^*, \tag{36}$$

by the definition of s^*. □

6. Numerical Examples

The examples use $L_n = G'(x_n)$, $\Delta = I$, which are independent of x_0 and x^*.

Example 1. *The solution sought for the nonlinear system*

$$\begin{aligned} g_1 &= u - 0.1\sin u - 0.3\cos v + 0.4 \\ g_2 &= v - 0.2\cos u + 0.1\sin v + 0.3. \end{aligned}$$

Let $G = (g_1, g_1)$. Then, the system becomes

$$G(s) = 0 \ \ for \ \ s = (\theta_1, \theta_2)^T.$$

Then

$$G'((u,v)) = \begin{bmatrix} 1 - 0.1\cos(u) & 0.3\sin(v) \\ 0.2\sin(u) & 0.1\cos(v) + 1 \end{bmatrix}.$$

Algorithm (2)

$$x_{k+1} = x_k - G'(x_k)^{-1}G(x_k).$$

Algorithm (4), $k = 1, \ \Delta = I$

$$\begin{gathered} A_1(x) = I + (I - G'(x)), \\ p_1(x) = x - (I + (I - G'(x)))G(x), \\ x_{n+1} = p_1(x_n). \end{gathered} \tag{37}$$

Algorithm (4), $k = 2, \ \Delta = I$

$$\begin{gathered} A_2(x) = I + (I - G'(x)) + (I - G'(x))^2, \\ p_2(x) = x - A_2(x)G(x), \\ x_{n+1} = p_2(x_n). \end{gathered} \tag{38}$$

Algorithm (4), $k = 3, \ \Delta = I$

$$\begin{gathered} A_3(x) = I + (I - G'(x)) + (I - G'(x))^2 + (I - G'(x))^3, \\ p_3(x) = x - A_3(x)G(x), \\ x_{n+1} = p_3(x_n). \end{gathered} \tag{39}$$

Algorithm (4), $k = 4$, $\Delta = I$

$$A_4(x) = I + (I - G'(x)) + (I - G'(x))^2 + (I - G'(x))^3 + (I - G'(x))^4,$$
$$p_4(x) = x - A_4(x)G(x), \tag{40}$$
$$x_{n+1} = p_4(x_n).$$

Algorithm (4), $k = 5$, $\Delta = I$

$$A_5(x) = I + (I - G'(x)) + (I - G'(x))^2 + (I - G'(x))^3 + (I - G'(x))^4 + (I - G'(x))^5,$$
$$p_5(x) = x - A_5(x)G(x), \tag{41}$$
$$x_{n+1} = p_5(x_n).$$

Algorithm (4), $k = \overline{1,5}$, $\Delta = G'(x_0)$

$$x_{n+1} = x_n - A\Delta^{-1}G(x_n),$$
$$\Gamma = \Delta^{-1}(\Delta - G'(x)), \tag{42}$$
$$A = I + \sum_{i=1}^{k} \Gamma^i.$$

Thus, the comparison shows that the behavior of method (4) is essentially the same as Newton's method (2). However, the iterates of method (4) are cheaper to obtain than Newton's. As observed in Tables 1–4, the number of iterations required for the proposed methods with k ranging from 3 to 5 closely aligns with those of Newton's method.

Table 5 shows the results of calculations to determine the Computational Order of Convergence (COC) and the Approximated Computational Order of Convergence (ACOC) aiming to compare the convergence order of method (4) with the convergence order of Newton's method (2).

Table 1. Number of iterations to achieve tolerance $\varepsilon = 10^{-9}$ with initial guess $x_0 = (1,1)$ and $\|I - G'(x_0)\| = 0.3129 < 1$.

Algorithm	Iterations	Algorithm	Iterations
(2) Newton	4	(2) Newton	4
(37), $k = 1$	6	(42), $k = 1$	8
(38), $k = 2$	5	(42), $k = 2$	6
(39), $k = 3$	4	(42), $k = 3$	5
(40), $k = 4$	4	(42), $k = 4$	5
(41), $k = 5$	4	(42), $k = 5$	4

Table 2. Number of iterations to achieve tolerance $\varepsilon = 10^{-9}$ with initial guess $x_0 = (0,0)$ and $\|I - G'(x_0)\| = 0.1414 < 1$.

Algorithm	Iterations	Algorithm	Iterations
(2) Newton	3	(2) Newton	3
(37), $k = 1$	5	(42), $k = 1$	3
(38), $k = 2$	4	(42), $k = 2$	3
(39), $k = 3$	3	(42), $k = 3$	3
(40), $k = 4$	3	(42), $k = 4$	3
(41), $k = 5$	3	(42), $k = 5$	3

Table 3. Number of iterations to achieve tolerance $\varepsilon = 10^{-9}$, where $x_0 = (-15, -15)$ and $\|I - G'(x_0)\| = 0.257 < 1$.

Algorithm	Iterations	Algorithm	Iterations
(2) Newton	5	(2) Newton	5
(37), $k = 1$	7	(42), $k = 1$	9
(38), $k = 2$	5	(42), $k = 2$	7
(39), $k = 3$	5	(42), $k = 3$	6
(40), $k = 4$	5	(42), $k = 4$	6
(41), $k = 5$	5	(42), $k = 5$	5

Table 4. Number of iterations to achieve tolerance $\varepsilon = 10^{-12}$, where $x_0 = (-15, -15)$ and $\|I - G'(x_0)\| = 0.257 < 1$.

Algorithm	Iterations	Algorithm	Iterations
(2) Newton	7	(2) Newton	7
(37), $k = 1$	8	(42), $k = 1$	12
(38), $k = 2$	7	(42), $k = 2$	8
(39), $k = 3$	7	(42), $k = 3$	8
(40), $k = 4$	7	(42), $k = 4$	7
(41), $k = 5$	7	(42), $k = 5$	7

Definition 2. *The computational order of convergence of a sequence* $\{x_n\}_{n\geq 0}$ *is defined by*

$$\overline{\rho}_n = \frac{\ln |e_{n+1}/e_n|}{\ln |e_n/e_{n-1}|},$$

where x_{n-1}, x_n, x_{n+1} *are three consecutive iterations near the root* α *and* $e_n = x_n - \alpha$ *[6].*

Definition 3. *The approximated computational order of convergence of a sequence* $\{x_n\}_{n\geq 0}$ *is defined by*

$$\widehat{\rho}_n = \frac{\ln |\widehat{e}_{n+1}/\widehat{e}_n|}{\ln |\widehat{e}_n/\widehat{e}_{n-1}|},$$

where $\widehat{e}_n = x_n - x_{n-1}$. x_n, x_{n-1}, x_{n-2} *are three consecutive iterates [6].*

Table 5. Computational Order of Convergence and the Approximated Computational Order of Convergence, where $x_0 = (-15, -15)$, $\varepsilon = 10^{-12}$.

Algorithm	COC	ACOC
(2) Newton	1.8624	1.9697
(37), $k = 1$	0.863	1
(38), $k = 2$	0.2695	1.0438
(39), $k = 3$	1.9714	2.3569
(40), $k = 4$	1.8354	1.9453
(41), $k = 5$	1.8642	1.9661
(42), $k = 1$	0.9065	1.0118
(42), $k = 2$	0.5912	0.999
(42), $k = 3$	0.7321	0.9926
(42), $k = 4$	1.933	2.0151
(42), $k = 5$	1.8679	1.9578

Table 5 demonstrates that the convergence of the proposed methods closely corresponds with the convergence of Newton's method, particularly for values of k ranging from 4 to 5 with the convergence order closely approximating 2.

Example 2. *Let* $T_1 = T_2 = \mathbb{R}^3$ *and* $\Omega = U[x^*, 1]$. *The mapping G is defined on* Ω *for* $\Theta = (\Theta_1, \Theta_2, \Theta_3)^{tr} \in \mathbb{R}^3$ *as*

$$G(\Theta) = (\Theta_1, e^{\Theta_2} - 1, \frac{e-1}{2}\Theta_3^2 + \Theta_3)^{tr}.$$

Then, the definition of the derivative according to Fréchet [2,3,8,13,30] gives for the mapping G that

$$G'(\Theta) = \begin{bmatrix} 1 & 0 & 0 \\ 0 & e^{\Theta_2} & 0 \\ 0 & 0 & (e-1)\Theta_3 + 1 \end{bmatrix}.$$

The point $x^* = (0,0,0)^{tr}$ *solves the equation* $G(a) = 0$. *Moreover,* $G'(x^*) = I$. *The conditions of Theorem 3 hold, provided that* $\gamma = \gamma(t) = (e-1)t, \psi(t) = (e-1)t$, *and* $k = 1$. *Then, we can take* $\rho_4 \in (0, 0.2909883534)$.

Example 3. *Let* $U[0,1]$ *stand the space of continuous functions mapping the interval* $[0,1]$ *into the real number system. Let* $T_1 = T_2 = K[0,1]$ *and* $\Omega = U[x^*, 1]$ *with* $x^*(\kappa) = 0$. *The operator G is defined on* $U[0,1]$ *as*

$$G(z)(\kappa) = z(\kappa) - 4\int_0^1 \kappa z(\tau)^3 d\tau.$$

Then, the definition of the derivative according to Fréchet [2,3,8,13,30] gives for the operator G

$$G'(z(w))(\kappa) = w(\kappa) - 12\int_0^1 \kappa\tau z(\tau)^2 w(\tau) d\tau$$

for each $w \in U[0,1]$. *Therefore, the conditions of Theorem 3 hold for* $x^* = 0$, $k = 1$ $G'(x^*(\kappa)) = I$ *if we choose* $\gamma = \gamma(t) = 6t$ *and* $\psi(t) = 6t$. *Then, we obtain* $\rho_4 \in (0, 0.8\bar{3})$.

7. Concluding Remarks

The difficulty of implementing the Newton-like algorithms is addressed in this paper. In particular, the computation of L_n^{-1} required at each step of the Newton-like algorithms is avoided with the introduction of algorithm (4) (or algorithm (5)), where the inversion only once of a fixed linear operator is required to implement it. The inverse of the linear operator is exchanged with a finite sum of linear operators related to G'. Both the local and the semi-local convergence analysis of these algorithms is comparable to Newton's in the sense that the number of iteration steps to reach a predetermined tolerance of error is essentially the same. The numerical examples are used to demonstrate that algorithm (4) or algorithm (5) are reliable replacements of the Newton-like algorithms for all practical purposes. We plan to study extensions of the presented algorithms like

$$x_{n+1} = x_n - \tilde{G}(x_n)G(x_n),$$

where $\tilde{G}$ is a conscious approximation to the inverse of a linear operator (like L_n) which may be a divided difference or some other operator [18,30–35].

Author Contributions: Conceptualization, I.K.A., S.G., S.R. and C.I.A.; Algorithm, I.K.A., S.G., S.R. and C.I.A.; methodology, I.K.A., S.G., S.R. and C.I.A.; software, I.K.A., S.G., S.R. and C.I.A.; validation, I.K.A., S.G., S.R. and C.I.A.; formal analysis, I.K.A., S.G., S.R. and C.I.A.; investigation, I.K.A., S.G., S.R. and C.I.A.; resources, I.K.A., S.G., S.R. and C.I.A.; data curation, I.K.A., S.G., S.R. and C.I.A.; writing—original draft preparation, I.K.A., S.G., S.R. and C.I.A.; writing—review and editing, I.K.A., S.G., S.R. and C.I.A.; visualization, I.K.A., S.G., S.R. and C.I.A.; supervision, I.K.A., S.G., S.R. and C.I.A.; project administration, I.K.A., S.G., S.R. and C.I.A. All authors have read and agreed to the published version of the manuscript.

Funding: This research received no external funding.

Institutional Review Board Statement: Not applicable.

Data Availability Statement: Data are contained within the article.

Acknowledgments: We would like to thank graduate Mykhailo Havdiak from the Department of Optimal Processes, Ivan Franko National University of Lviv, Lviv, Ukraine, for providing Example 1 of this paper.

Conflicts of Interest: The authors declare that there are no conflicts of interest.

References

1. Driscoll, T.A.; Braun, R.J. *Fundamentals of Numerical Computation: Julia Edition*; SIAM: Philadelphia, PA, USA, 2022.
2. Ezquerro, J.A.; Gutierrez, J.M.; Hern, ez, M.A.; Romero, N.; Rubio, M.J. The Newton algorithm: From Newton to Kantorovich. *Gac. R. Soc. Mat. Esp.* **2010**, *13*, 53–76. (In Spanish)
3. Kantorovich, L.V.; Akilov, G. *Functional Analysis in Normed Spaces*; Fizmatgiz: Moscow, Russia, 1959; German translation, Akademie-Verlag: Berlin, Germany, 1964; English translation (2nd edition), Pergamon Press: London, UK, 1981, 1964.
4. Proinov, P.D. New general convergence theory for iterative processes and its applications to Newton- Kantarovich type theorems. *J. Complex.* **2010**, *25*, 3–42. [CrossRef]
5. Argyros, I.K. *The Theory and Applications of Iteration Methods with Applications*, 2nd ed.; Engineering Series; CRC Press: Boca Raton, FL, USA; Taylor and Francis Publ.: Abingdon, UK, 2022.
6. Candelario, G.; Cordero, A.; Torregrosa, J.R.; Vassileva, M.P. Generalized conformable fractional Newton-type method for solving nonlinear systems. *Numer. Algor.* **2023**, *93*, 1171–1208. [CrossRef]
7. Darve, E.; Wootters, M. *Numerical Linear Algebra with Julia*; SIAM: Philadelphia, PA, USA, 2021.
8. Deuflhard, P. *Newton Algorithms for Nonlinear Problems. Affine Invariance and Adaptive Algorithms*; Springer Series in Computational Mathematics; Springer: Berlin/Heidelberg, Germany, 2004; Volume 35.
9. Kelley, C.T. *Solving Nonlinear Equations with Iterative Methods*; Solvers and Examples in Julia, Fundamentals of Algorithms; SIAM: Philadelphia, PA, USA, 2023.
10. Pollock, S.; Rebholz, L.G.; Xiao, M. Anderson-accelerated convergence of Picard iterations for incompressible Navier–Stokes equations. *SIAM J. Numer. Anal.* **2019**, *57*, 615–637. [CrossRef]
11. Rackauckas, C.; Mishra, P.; Gowda, S.; Huang, L. *Sparse Diff Tools.jl*, Julia Package, 2020. Available online: https://github.com/JuliaDiff/SparseDiffTools.jl (accessed on 20 February 2024).
12. Rackauckas, C.; Nie, Q. Differentialequations. jl–a performant and feature-rich ecosystem for solving differential equations in Julia. *J. Open Res. Softw.* **2017**, *5*, 15. [CrossRef]
13. Revels, J. *Reverse Diff.jl, Julia Package*. 2020. Available online: https://github.com/JuliaDiff/ReverseDiff.jl (accessed on 20 February 2024).
14. Uecker, H. *Numerical Continuation and Bifurcation in Nonlinear PDEs*; SIAM: Philadelphia, PA, USA, 2021.
15. Berinde,V. *Iterative Approximation of Fixed Points*; Springer: New York, NY, USA, 2007.
16. Krasnoselskij, M.A. Two remarks on the algorithm of successive approximations. *Uspehi Mat. Nauk.* **1995**, *10*, 123–127. (In Russian)
17. Bian, W.; Chen, X.; Kelley, C.T. Anderson acceleration for a class of nonsmooth fixed-point problems. *SIAM J. Sci. Comp.* **2021**, *43*, S1–S20. [CrossRef]
18. De Dterck, H.; He, Y. Linear asymptotic convergence of Anderson acceleration: Fixed point analysis. *arXiv* **2021**, arXiv:14716v1.
19. Evans, C.; Pollock, S.; Rebholz, L.G.; Xiao, M. A proof that Anderson acceleration improves the convergence rate in linearly converging fixed-point methods (but not in those converging quadratically). *SIAM J. Numer. Anal.* **2020**, *58*, 788–810. [CrossRef]
20. Fang, H.R.; Saad, Y. Two classes of multisecant methods for nonlinear acceleration. *Numer. Linear Algebra Appl.* **2009**, *16*, 197–221. [CrossRef]
21. Pollock, S.; Rebholz, L.G. Anderson acceleration for contractive and noncontractive operators. *IMA Numer. Anal.* **2021**, *41*, 2841–2872. [CrossRef]
22. Padcharoen, A.; Kumam, P.; Chaipunya, P.; Shehu, Y. Convergence of inertial modified Krasnoselskii-Mann iteration with application to image recovery. *Thai J. Math.* **2020**, *18*, 126–142.
23. Zhang, J.; O'Donoghue, B.; Boyd, S. Globally convergent type-I Anderson acceleration for nonsmooth fixed-point iterations. *SIAM J. Optim.* **2020**, *30*, 3170–3197. [CrossRef]
24. Argyros, I.K.; George, S. On the complexity of extending the convergence region for Traub's algorithm. *J. Complex.* **2020**, *56*, 101423. [CrossRef]
25. Argyros, I.K.; George, S. On a unified convergence analysis for Newton-type algorithms solving generalized equations with the Aubin property. *J. Complex.* **2024**, *81*, 101817. [CrossRef]
26. Catinas, E. The inexact, inexact perturbed, and quasi-Newton algorithms are equivalent models. *Math. Comp.* **2005**, *74*, 291–301. [CrossRef]
27. Erfanifar, R.; Hajariah, M.A. A new multi-step method for solving nonlinear systems with high efficiency indices. *Numer. Algorithms* **2024**. [CrossRef]
28. Yamamoto, T. A convergence theorem for Newton-like algorithms in Banach spaces. *Numer. Math.* **1987**, *51*, 545–557. [CrossRef]
29. Ezquerro, J.A.; Hern, ez-Veron, M.A. Domains of global convergence for Newtons's algorithm from auxiliary points. *Appl. Math. Lett.* **2018**, *85*, 48–56. [CrossRef]

30. Ostrowski, A.M. *Solution of Equations in Euclidean and Banach Spaces*; Academic Press: New York, NY, USA, 1973.
31. Cances, E.; Kremlin, G.; Levitt, A. Convergence analysis of direct minimization and self-consistent iterations. *SIAM J. Sci. Comp.* **2021**, *42*, 243–274.
32. De Dterck, H.; He, Y. Anderson acceleration as a Krylov method with application to asymptotic convergence analysis. *arXiv* **2021**, arXiv:2109.14181v1.
33. Rheinboldt, W.C. A unified convergence theory for a class of iterative process. *SIAM J. Numer. Anal.* **1968**, *5*, 42–63. [CrossRef]
34. Singh, S. A third order iterative algorithm for inversion of cumulative beta distribution. *Numer. Algor.* **2023**, 1–23.
35. Traub, J.F.; Wozdiakowsi, H. Convegence and complexity of Newton iteration for operator equations. *J. Assoc. Comput. March.* **1979**, *26*, 250–258. [CrossRef]

algorithms

Article

The Algorithm of Gu and Eisenstat and D-Optimal Design of Experiments

Alistair Forbes

National Physical Laboratory, Teddington TW11 0LW, UK; alistair.forbes@npl.co.uk

Abstract: This paper addresses the following problem: given m potential observations to determine n parameters, $m > n$, what is the best choice of n observations. The problem can be formulated as finding the $n \times n$ submatrix of the complete $m \times n$ observation matrix that has maximum determinant. An algorithm by Gu and Eisenstat for a determining a strongly rank-revealing QR factorisation of a matrix can be adapted to address this latter formulation. The algorithm starts with an initial selection of n rows of the observation matrix and then performs a sequence of row interchanges, with the determinant of the current submatrix strictly increasing at each step until no further improvement can be made. The algorithm implements rank-one updating strategies, which leads to a compact and efficient algorithm. The algorithm does not necessarily determine the global optimum but provides a practical approach to designing an effective measurement strategy. In this paper, we describe how the Gu–Eisenstat algorithm can be adapted to address the problem of optimal experimental design and used with the QR algorithm with column pivoting to provide effective designs. We also describe implementations of sequential algorithms to add further measurements that optimise the information gain at each step. We illustrate performance on several metrology examples.

Keywords: design of experiment; D-optimality; metrology; QR factorisation; uncertainty evaluation

1. Introduction

While machine learning paradigms are based on access to large sets of data to extract the information of interest, many practical problems in science and technology arise in a context in which data are sparse and potentially expensive to generate. Given limited experimental resources, it is important that these resources are used effectively. The design of experiment (DoE) is an important area of study in applied mathematics and statistics. Many DoE studies relate to problems in which a response variable of interest, e.g., the yield of a crop, is possibly influenced by a number of control or stimulus variables, e.g., crop variety, soil type, irrigation regime, fertiliser treatment, etc. The form of the response function is usually unknown and the DoE problem is to design a number of experiments with the control variables set to different values, in order to estimate the response function from a set of candidate response functions and consequently choose control values that optimise the response; see, e.g., [1–8]. The DoE problem has been addressed in applications across many scientific disciplines, including the placement of sensors in a electrical power grid [9,10] and the measurement of thermal properties of materials [11,12], for example.

In this paper, we look at a particular aspect of the design problem: given a known response model, how do we choose a minimal number of measurements, taking into account the associated uncertainties, in order to estimate all the parameters of the system under study as accurately as possible. Thus, the problem can be posed as: given m potential observations to determine n parameters, $m > n$, what is the best choice of n observations. In the context of (linear) least squares estimation, a version of the problem can be formulated as finding the $n \times n$ submatrix of the complete $m \times n$ observation matrix that has maximum determinant. The problem can be posed as an example of the following general optimisation problem. Suppose $A = \{a_i, i = 1, \ldots, m\}$ is a set with m elements. Represent the set $P(A)$

of all subsets of A as the set of characteristic functions $\chi : A \longrightarrow \{0,1\}$: if $B \subset A$, then the corresponding function $\chi_B \in P(A)$ is such that $\chi_B(a_i) = 1$ if $a_i \in B$ and is 0 otherwise. Let $D : P(A) \longrightarrow \mathbb{R}$ be a real valued function defined on $P(A)$. The optimisation problem is

$$\min_{\chi \in P(A)} \quad D(\chi) \quad \text{subject to} \quad \sum_{i=1}^{m} \chi(a_i) = n. \tag{1}$$

In theory, the global solution to this DoE problem can be found by a brute force assessment of all choices of subsets, a total of $m!/(n!(m-n)!)$ possibilities. For all but small m and n, this approach will be computationally infeasible, even if the functional D is cheap to evaluate. Two other related approaches are commonly used to partially solve this type of combinatorial optimisation problem.

The first approach, with a long history, see e.g., [13–15], is to use a sequential approach that at each stage constructs a new candidate subset by removing one element from the subset, replacing it by a new element that leads to a strict improvement in D. The computational cost per step for such an algorithm will be given by the cost evaluating D and the cost of searching for a suitable exchange from the $n(m-n)$ possibilities. Ideally, the computational cost of an exchange should be no worse than $O(m)$ or $O(m \log m)$ as a function of m. The second approach is based on convex relaxation, replacing a combinatorial problem by a convex optimisation problem [16]. Let $0 \leq w_i = w(a_i) \leq 1$, where we think of w_i as specifying the degree to which the ith element is in the solution optimal subset. If $W(A)$ is the set of functions $A \longrightarrow [0,1]$, then the optimisation problem is defined by

$$\min_{w \in W(A)} \quad D(w) \quad \text{subject to} \quad \sum_{i=1}^{m} w(a_i) = n. \tag{2}$$

In this approach, the combinatorial constraint $\chi(a_i) \in \{0,1\}$ is relaxed to the convex constraint $0 \leq w(a_i) \leq 1$. Often, the optimal w is in a fact a characteristic function associated with an n-element subset. The optimisation algorithms for such problems are generally iterative and often also have a sequential element, such as in sequential quadratic programming [17] and semidefinite programming [18]. Convex relaxation approaches to DoE problems were considered in [4,9,10,16], for example.

In this paper, we show how two algorithms, both following a sequential approach and developed for numerical linear algebra applications, can be repurposed to address the design problem of choosing n observations from m possible observations.

The remainder of this paper is organised as follows. In Section 2, we give an overview of linear least squares estimation, while in Section 3, we describe commonly used aggregate measures of uncertainty and the corresponding criteria used in optimal experimental design. In Section 4, we describe two algorithms that address the DoE problem under consideration, as well as sequential algorithms for choosing additional observations in a step-wise optimal way. Example applications of the algorithms to design problems arising in metrology [19–22] are given in Section 5. Our concluding remarks are given in Section 6.

2. Least Squares Problems

For $m \times n$ observation matrix C, $m \times 1$ vector of measurements y $\boldsymbol{y}$, the least squares estimate $a\boldsymbol{a}$ is the solution of

$$\min_{\boldsymbol{\alpha}} \; (y - C\boldsymbol{\alpha})^\top (y - C\boldsymbol{\alpha}).$$

The QR factorisation of C expresses $C = QR$ as a product of an $m \times m$ orthogonal matrix Q and an upper triangular matrix R. If

$$C = [Q_1 \; Q_2] \begin{bmatrix} R_1 \\ \mathbf{0} \end{bmatrix} = Q_1 R_1,$$

where Q_1 is the submatrix constructed from the first n columns of Q and Q_2 that is constructed from the remaining $m-n$ columns, then

$$\begin{aligned} \boldsymbol{a} &= (C^\top C)^{-1} C^\top \boldsymbol{y} = R_1^{-1} Q_1^\top \boldsymbol{y}, \\ \hat{\boldsymbol{y}} = C\boldsymbol{a} &= C(C^\top C)^{-1} C^\top \boldsymbol{y} = Q_1 Q_1^\top \boldsymbol{y}, \\ \boldsymbol{r} = \boldsymbol{y} - C\boldsymbol{a} &= (I - Q_1 Q_1^\top)\boldsymbol{y} = Q_2 Q_2^\top \boldsymbol{y}, \end{aligned}$$

are the least squares solution, model approximant, and associated residual vector. The QR factorisation enables $\boldsymbol{a}$ to be given by the solution of $R_1\boldsymbol{a} = Q_1^\top \boldsymbol{y}$ and this solution can be constructed using back substitution, solving for a_n first, then a_{n-1}, etc., exploiting the upper-triangular form of R_1. The QR factorisation also avoids the numerical loss of accuracy that can occur in forming $C^\top C$ and its inverse in the case in which C is ill-conditioned [23] (Chapter 5). The computational cost of evaluation the least-squares solution in $O(mn^2)$.

The $n \times m$ matrix $S = R_1^{-1} Q_1^\top$ is the sensitivity matrix of the solution parameters $\boldsymbol{a}$ with respect to the data $\boldsymbol{y}$ with $s_{ji} = \partial a_j / \partial y_i$. If the variance matrix associated with $\boldsymbol{y}$ is $V_{\boldsymbol{y}}$, then the variance matrix associated with the solution $\boldsymbol{a}$ is

$$V_{\boldsymbol{a}} = S V_{\boldsymbol{y}} S^\top.$$

For the case $V_{\boldsymbol{y}} = I$,

$$V_{\boldsymbol{a}} = SS^\top = (R_1^\top R_1)^{-1} = (C^\top C)^{-1}.$$

For this case, the trace of the variance matrix is given by

$$\mathrm{Tr}(V_{\boldsymbol{a}}) = \sum_j \sum_i s_{ji}^2,$$

the variance associated with a_j is $u^2(a_j) = \sum_i s_{ji}^2$ and $\sum_j s_{ji}^2$ is the contribution of the uncertainty associated with the ith observation to the trace of $V_{\boldsymbol{a}}$.

The least-squares solution $\boldsymbol{a}$ is the maximum likelihood estimate of $\boldsymbol{\alpha}$ for the model $\boldsymbol{y} \in \mathcal{N}(C\boldsymbol{\alpha}, I)$. If $\boldsymbol{y}$ is associated with the variance matrix $V_{\boldsymbol{y}}$, i.e., $\boldsymbol{y} \in \mathcal{N}(C\boldsymbol{\alpha}, V_{\boldsymbol{y}})$, and $V_{\boldsymbol{y}}$ is full rank and has Cholesky decomposition $V_{\boldsymbol{y}} = LL^\top$, where L is lower triangular [23], then the maximum likelihood estimate $\boldsymbol{a}$ of $\boldsymbol{\alpha}$ solves the Gauss–Markov problem

$$\min_{\boldsymbol{\alpha}} (\boldsymbol{y} - C\boldsymbol{\alpha})^\top V_{\boldsymbol{y}}^{-1} (\boldsymbol{y} - C\boldsymbol{\alpha}). \tag{3}$$

If

$$\tilde{C} = L^{-1} C, \quad \tilde{\boldsymbol{y}} = L^{-1} \boldsymbol{y},$$

then the solution of (3) is the least-squares solution of $\tilde{C}\boldsymbol{\alpha} \approx \tilde{\boldsymbol{y}}$:

$$\boldsymbol{a} = \left(\tilde{C}^\top \tilde{C}\right)^{-1} \tilde{C}^\top \tilde{\boldsymbol{y}}.$$

In particular, if $V_{\boldsymbol{y}}$ is a diagonal matrix with $u_i^2 > 0$ in the ith diagonal element, then $\tilde{C}$ is constructed from the weighted rows of C, and similarly for $\tilde{\boldsymbol{y}}$:

$$\tilde{\boldsymbol{c}}_i = w_i \boldsymbol{c}_i, \quad \tilde{y}_i = w_i y_i, \quad w_i = 1/u_i.$$

In general, the cost of evaluating the Cholesky factorisation of $V_{\boldsymbol{y}}$ is $O(m^3)$, which could be problematic for large m. Often, $V_{\boldsymbol{y}}$ has a structure that enables the factorisation to be performed in $O(m)$, as in the simple but common case of a diagonal variance matrix.

3. Aggregate Measures of Uncertainty

A central issue in design of experiment analysis is to maximise some measure of the information gain subject to constraints on resources. In a metrology context, in particular, it is natural to express the information gain in terms of uncertainties associated with the fitted parameters; i.e., related to the $n \times n$ variance matrix $V_{\boldsymbol{a}}$. Given an aggregate measure of uncertainty derived from $V_{\boldsymbol{a}}$, one experiment will be judged to be more informative than another if it is associated with a lower aggregate uncertainty. There are a number of aggregate measures that are commonly used. In this paper, we are concerned with two: A-optimality and D-optimality.

A full rank variance matrix V is a symmetric positive definite matrix and has an eigenvalue decomposition with eigenvalues $\lambda_j, j = 1, \ldots n$. The A-measure is the trace of $\text{Tr}(V)$ of V, the sum of the diagonal elements of V, as well as the sum of the eigenvalues of V:

$$\text{Tr}(V) = \sum_{j=1}^{n} \lambda_j.$$

We note that $\text{Tr}(\sigma^2 V) = \sigma^2 \text{Tr}(V)$. The D-measure is the determinant $|V|$ of V, and also the product of the eigenvalues of V:

$$|V| = \prod_{j=1}^{n} \lambda_j.$$

We also evaluate

$$\bar{d}(V) = |V|^{1/n} = \left(\prod_{j=1}^{n} \lambda_j \right)^{1/n}, \tag{4}$$

the geometric mean of the eigenvalues, noting that $\bar{d}(\sigma^2 V) = \sigma^2 \bar{d}(V)$.

Arguments can be made for and against using each of these two measures. Given V, the A-measure can be evaluated in $O(n)$ steps, while the D-measure will usually involve $O(n^3)$ steps, using the Cholesky factorisation of the variance matrix to evaluate its determinant. The A-measure has the apparent advantage of being most easily interpreted as an aggregate uncertainty, being the sum of the variances $\sum_j u^2(a_j)$ associated with each of the parameter estimates a_j or, more precisely, the numerical values of these estimates. From a dimensional analysis point of view [24], the A-measure can only be meaningful if all the parameters in $\boldsymbol{a}$ have the same dimension, length , mass, etc. Otherwise, the trace involves adding quantities of different dimensions. The D-measure is consistent with a dimensional analysis.

The A-measure can also be regarded as the variance associated with the sum $\sum_j a_j$ of parameter estimates, *assuming* these estimates as statistically independent. In general, they will not be independent, and one disadvantage of the A-measure is that it takes no account of correlation, being defined solely in terms of the diagonal elements of V. The D-measure does take into account correlation. Consider the 2×2 variance matrix

$$V = \begin{bmatrix} 1 & \rho \\ \rho & 1 \end{bmatrix}, \quad \text{Tr}(V) = 2, \quad |V| = 1 - \rho^2,$$

associated with parameters a_1 and a_2. The A-measure is independent of the correlation coefficient ρ while the D-measure depends directly on ρ^2. If the result of an experiment is to change ρ from 0 to near 1, there is a gain of information, since $a_1 - a_2$ is now much more accurately estimated. The D-measure reflects this information gain, the A-measure does not.

The D-measure is invariant to re-parametrizations of the model, while the A-measure is not. Suppose $V_{\boldsymbol{a}} = (C^\top C)^{-1}$ and let $\boldsymbol{b} = B\boldsymbol{a}$, where B is $n \times n$ and full rank, so that $\boldsymbol{b}$ can be regarded as an alternative parametrization of the problem. For the same experimental design, the variance matrix associated with the estimates $\boldsymbol{b}$ is given by

$$V_{\boldsymbol{b}} = BV_{\boldsymbol{a}}B^{\top}.$$

We have

$$|V_{\boldsymbol{b}}| = |BV_{\boldsymbol{a}}B^{\top}| = |B|^2|V_{\boldsymbol{a}}|,$$

so that a strategy that is optimal for the D-measure for parametrization $\boldsymbol{a}$ will also be optimal for parametrization $\boldsymbol{b}$. For the A-measure,

$$\mathrm{Tr}(V_{\boldsymbol{b}}) = \mathrm{Tr}(BV_{\boldsymbol{a}}B^{\top}) = \mathrm{Tr}(B^{\top}BV_{\boldsymbol{a}}) = \mathrm{Tr}(V_{\boldsymbol{a}}B^{\top}B),$$

and, in general, there is no constant K such that $\mathrm{Tr}(BV_{\boldsymbol{a}}B^{\top}) = K\mathrm{Tr}(V_{\boldsymbol{a}})$, unless $B^{\top}B$ is a multiple of the identity matrix. In a similar vein, the D-measure is invariant with respect to changes in unit, the A-measure is not.

The eigenvalues λ_j of $V_{\boldsymbol{a}}$ are related to the size of the confidence ellipses centred about the parameter estimates $\boldsymbol{a}$. In fact, for a given confidence level, the lengths of the semi-axes are proportional to $\sqrt{\lambda_j}$. Hence, the D-measure is related to the square of the volume of the confidence ellipse, while the A-measure is related to the sum of the squares of the semi-axis lengths.

4. Choosing a *D*-Optimal Subset of *n* Measurements

Let C be an $m \times n$ observation matrix, $m > n$, representing m measurements that could potentially be made. If we are constrained only to use n measurements, which n should we choose, in order to minimise an aggregate measure of the uncertainties associated with the fitted parameters? It is assumed that there is at least one $n \times n$ submatrix constructed from the rows of C that is full rank, allowing all the parameters $\boldsymbol{a}$ to be estimated. In this section, we consider two algorithms described below in Sections 4.1 and 4.2, for addressing the issue of subset selection with a view to providing a D-optimal solution. Both algorithms were designed to address other problems but can be repurposed for design of experiment applications.

Let C_k be a full rank $n \times n$ submatrix of C defined by a subset of the rows of C indexed by row indices $I_k = \{i_1, \ldots, i_n\}$, and set $V_k = H_k^{-1}$, where $H = C_k^{\top}C_k$. Then,

$$|V_k| = |H_k|^{-1} = |C_k|^{-2},$$

so that choosing I_k to minimise $|V_k|$ amounts to choosing I_k to maximise $|C_k|$. For any square $n \times n$ matrix A with rows $\boldsymbol{a}_i$, $|A|$ is the volume of the parallelopiped spanned by the vectors $\boldsymbol{a}_i$ in $\mathbb{R}^n$. This volume will depend on the Euclidean lengths $\|\boldsymbol{a}_i\|_2$ of the vectors and also their mutual independence. For vectors of equal length, $|A|$ is maximised if the vectors $\boldsymbol{a}_i$ are orthogonal to each other (and can therefore be rotated or reflected to align with the coordinate axes). In terms of the observation matrix C, the lengths of the row vectors $\boldsymbol{c}_i$ reflect the accuracy associated with the measurement y_i—more accurate measurements will be accorded a greater weight and therefore larger row vector $\boldsymbol{c}_i$ in terms of Euclidean length. The orthogonality of the row vectors relates to how independent the sets of information represented by the measurements are from each other. Consider

$$\begin{bmatrix} y_1 \\ y_2 \end{bmatrix} = C\begin{bmatrix} a_1 \\ a_2 \end{bmatrix} + \begin{bmatrix} \epsilon_1 \\ \epsilon_2 \end{bmatrix}, \quad C = C(\theta) = \begin{bmatrix} 1 & 0 \\ \cos\theta & \sin\theta \end{bmatrix} \quad \epsilon_1, \epsilon_2 \in \mathcal{N}(0,1).$$

Both measurements y_1 and y_2 are associated with the equal information gain, since $\|\boldsymbol{c}_1\| = \|\boldsymbol{c}_2\| = 1$. The data point y_1 provides information about a_1 alone, while the data point y_2 potentially provides information about both a_1 and a_2. For $\theta \approx 0$, the observation equation involving y_2 only serves to provide more information about a_1, and a_2 remains poorly estimated with $|C(\theta)| = |\sin\theta| \approx 0$. For $\theta \approx \pi/2$, the information provided by y_2 provides significant new information about a_2 and $|C(\theta)| = |\sin\theta| \approx 1$. In this way, amongst a set of measurements of comparable accuracy, we look for subsets that are mutually maximally independent. Both of the algorithms described below involve trying to select a set of measurements that are sufficiently independent.

4.1. Subset Selection Using the QR Factorisation with Column Pivoting

The QR factorisation of an $m \times n$ matrix $C = QR$ has the following geometrical interpretation. The column vectors $\boldsymbol{c}_j$ of C define an n dimensional subspace $\mathcal{C}$ of $\mathbb{R}^m$. The orthogonal matrix Q defines an axis system for $\mathbb{R}^m$, such that the n columns of Q_1 define an axis system for $\mathcal{C}$ and the $m-n$ columns of Q_2 define an axis system for the space of vectors $\mathcal{C}^\perp$ orthogonal to $\mathcal{C}$. The columns for Q_1 are constructed so that $\boldsymbol{q}_1$ is aligned with $\boldsymbol{c}_1$, and so that there is an r_{11} such that $\boldsymbol{c}_1 = r_{11}\boldsymbol{q}_1$. The vector $\boldsymbol{q}_2$ is chosen to lie in the plane defined by $\boldsymbol{c}_1$ and $\boldsymbol{c}_2$, and so there are scalars r_{12} and r_{22} such that $\boldsymbol{c}_2 = r_{12}\boldsymbol{q}_1 + r_{22}\boldsymbol{q}_2$, etc. This gives the factorisation

$$[\boldsymbol{c}_1 \ \cdots \ \boldsymbol{c}_n] = [\boldsymbol{q}_1 \ \cdots \ \boldsymbol{q}_n]\begin{bmatrix} r_{11} & r_{12} & \cdots & r_{1n} \\ 0 & r_{22} & \cdots & r_{2n} \\ 0 & 0 & \ddots & \vdots \\ 0 & 0 & \cdots & r_{nn} \end{bmatrix},$$

i.e., in matrix notation $C = Q_1 R_1$.

An equivalent geometrical interpretation is that the orthogonal matrix $Q^\top$, a combination of rotations and reflections, transforms C so that its first column is aligned with the first axis in $\mathbb{R}^m$, the second column lies in the plane defined by the first and second axes in $\mathbb{R}^m$, the third column lies in the three-dimensional space defined by the first three axes in $\mathbb{R}^m$, and so on: $Q^\top C = R$.

For the standard QR algorithm, the columns of C are processed in the order they appear in the matrix C. For the QR algorithm with column pivoting [23], the order of the columns is changed to improve the numerical properties of the algorithm. The column of largest Euclidean length is aligned with the first axis. Of the remaining columns, the column that is furthest from the space defined by the first axis is then rotated to the plane defined by the first and second axes. At the $k+1$th stage, it is the column furthest from the subspace defined by the first k axes directions that is chosen to be transformed to lie in the space defined by the first $k+1$ axes directions. The output of the algorithm includes the specification of a permutation matrix P that permutes the columns of C, along with an orthogonal Q and upper triangular R such that $CP = QR$.

The first algorithm considered here is based on a subset selection algorithm described in [23] (Section 5.5). The issue addressed by the subset selection algorithm is that of trying to explain a measured response vector $\boldsymbol{y}$ in terms of basis vectors, usually representing covariates $\boldsymbol{x}$ and functions of covariates, stored in an $m \times p$ matrix C. It is assumed that $\boldsymbol{y}$ lies close to a n dimensional subspace defined by a subset of the columns of C. The problem is to estimate n and choose n columns of C that best explain the response, using the QR with column pivoting to select columns that are optimally mutually independent. This information can then be used to predict other, possibly future, responses based on knowledge of the variates represented by the selected columns. The algorithm uses the singular value decomposition (SVD) to estimate n and the QR algorithm with column pivoting to select an appropriate set of n columns.

Our design of experiment problem is different. The observation matrix C of all potential measurements is assumed to be full rank, so estimating the rank of C is not required. More importantly, our selection problem is to select from the rows of C, not the columns. However, the selection of a set of vectors that are maximally mutually independent is directly relevant. Suppose C has QR factorisation $C = Q_1 R_1$, where Q_1 is $m \times n$ and R_1 in $n \times n$ and upper-triangular and full rank. The matrix R_1 can be used to re-parametrize the problem in terms of $\boldsymbol{b} = R_1^{-1}\boldsymbol{a}$. The observation associated with parameters $\boldsymbol{b}$ is Q_1. Since D-optimal designs are independent of parametrization, it is sufficient to work with Q_1 as follows. We use column pivoting to find the QR factorisation of $Q_1^\top$ so that $Q_1^\top P = UT$, where U is an $n \times n$ orthogonal matrix and T is upper-triangular. The first n columns selected by the QR decomposition of $Q_1^\top$ specified by P also specify the n rows of C that are a good choice for the D-optimal design. This algorithm is summarised below (Algorithm 1). Its computational complexity is $O(mn^2)$.

Algorithm 1: SSQR: subset selection algorithm using QR with column pivoting to determine a favourable subset of an observation matrix in terms of the D-measure

Data: An $m \times n$ observation matrix C.

Result: Index set $I \subset \{i = 1, \ldots, m\}$ where the first n elements specify C_n that represent a good selection of the rows of C in terms of maximising the determinant.

Steps:

Determine the QR factorization of $C = Q_1 R_1$, where Q_1 is an $m \times n$ orthogonal matrix.

Use column pivoting to find the QR factorisation of $Q_1^\top$ so that $Q_1^\top P = UT$, where U is an $n \times n$ orthogonal matrix and T is upper-triangular.

Assign the index I by assigning $I(k) = i_k$ where i_k specifies the row index of the kth column of P such that $P(i_k, k) = 1$.

The SSQR algorithm is not targeted directly at determining a D-optimal design but is an approach that, at least heuristically, has elements that makes it likely to produce a design that is good in terms of the D-measure. The algorithm is an example of a sequential 'greedy algorithm', in that at each step, a choice is made that maximises some measure, irrespective of what previous steps have been made, and takes no account of what choices could arise in a future step. Once a row has been selected, there is no opportunity for the row to be deselected at a future iteration. The algorithm is largely independent of the ordering of the rows of the observation matrix C and, other than for reasons of rounding error effects and symmetries in the observation matrix, will arrive at the same solution for different row orderings. As has been mentioned, there is no guarantee that this solution is globally optimal. Example calculations using the algorithm are discussed in Section 5.

4.2. Exchange Approach Based on the Gu and Eisenstat Algorithm

The second algorithm described in this paper is directly targeted at determining a D-optimal design. It is based on the iterative algorithm of Gu and Eisenstat [25] for determining a strongly rank-revealing QR decomposition. Below, we show how it can be adapted to provide a reduction in the D-measure at each iteration through exchanging rows and to stop when no reduction is possible. It starts with a selection C_0 of n rows of the $m \times n$ observation matrix C, where C_0 is already full rank; e.g., that provided by the SSQR subset selection algorithm of Section 4.1.

We recall that minimising the D-measure is equivalent to maximising the absolute value of the determinant of the upper-triangular factor of the observation matrix. The idea of the algorithm is as follows. Partition $C^\top$ as $C^\top = \begin{bmatrix} A & B \end{bmatrix}$, where A is $n \times n$ and B is $n \times (m-n)$. We assume that $A = C_0^\top$ is full rank. The Gu–Eisenstat (GE) algorithm uses the following result:

Proposition 1. *For $n \times n$ full rank matrix A and $n \times p$ matrix B, let $\tilde{A}$ be the matrix constructed by replacing the ith column of A, $1 \le i \le n$, by the jth column of B, $1 \le j \le p$. Then,*

$$|\tilde{A}| = F_{i,j}|A|$$

where $F = A^{-1}B$.

Proof. For n-vectors $\boldsymbol{u}$ and $\boldsymbol{v}$,

$$|A + \boldsymbol{u}\boldsymbol{v}^\top| = |A|(1 + \boldsymbol{v}^\top A^{-1}\boldsymbol{u}); \tag{5}$$

see Appendix A.2. The result follows for $\boldsymbol{v} = \boldsymbol{e}_i$, the ith column of the identity matrix and $\boldsymbol{u} = \boldsymbol{b}_j - \boldsymbol{a}_i$, noting that $A^{-1}\boldsymbol{a}_i = \boldsymbol{e}_i$. □

This suggests the following approach, assuming that the first n rows of C are full rank (Algorithm 2).

Algorithm 2: Basic steps to determine a D-optimal subset of an observation matrix

---

Data: An $m \times n$ observation matrix C such that the first n rows are full rank, and tolerance factor $f > 1$.

Result: Index set $I \subset \{i = 1, \ldots, m\}$ where the first n elements specify C_n such that no row interchange will increase the determinant of C_n by more than f.

Steps:

Partition $C^\top = \begin{bmatrix} A & B \end{bmatrix}$ and form $F = A^{-1}B$.

Initialise $m \times 1$ index set I with $I(i) = i$.

while $\max_{i,j} |F_{i,j}| > f$ **do**

1. Choose (i, j) such that $|F_{i,j}|$ is maximised.
2. Row interchange: set $A := A + (\boldsymbol{b}_j - \boldsymbol{a}_i)\boldsymbol{e}_i^\top$, $B := B - (\boldsymbol{b}_j - \boldsymbol{a}_i)\boldsymbol{e}_j^\top$.
3. Update I, swapping the ith and $(n + j)$th elements.
4. Form $F = A^{-1}B$.

end

The main computational element of this algorithm is forming $F = A^{-1}B$ at each iteration. Forming F explicitly in this way involves $O(n^3)$ steps to evaluate A^{-1} and a further $O(mn)$ steps to evaluate the matrix product. Let Q be orthogonal such $\tilde{A} = Q^\top A$ is upper-triangular and set

$$Q^\top \begin{bmatrix} A & B \end{bmatrix} = \begin{bmatrix} \tilde{A} & \tilde{B} \end{bmatrix}. \tag{6}$$

Then, $F = A^{-1}B = \tilde{A}^{-1}\tilde{B}$, and the latter can be computed efficiently using back-substitution. Thus, step 4 in Algorithm 2 can be replaced by

4.1 Determine orthogonal Q such that $Q^\top A$ is upper-triangular.

4.2 Set

$$\begin{bmatrix} A & B \end{bmatrix} := Q^\top \begin{bmatrix} A & B \end{bmatrix}, \quad F = A^{-1}B.$$

Since $|Q^\top A| = |A|$, up to sign, the modified algorithm still determines the optimal index set. As described in the Gu and Eisenstat paper, these computations can be made much more efficient by exploiting the fact that at each iteration we are performing a rank-one modification to update an upper triangular matrix in step 2 in Algorithm 2 above. Suppose we have determined to replace ith column of the upper-triangular matrix A by the jth column of B. First, we move the ith column of A to the nth column, as shown schematically in (7).

$$\begin{matrix} * & * & i & * & * & * \\ & * & i & * & * & * \\ & & i & * & * & * \\ & & & * & * & * \\ & & & & * & * \\ & & & & & * \end{matrix} \quad \longrightarrow \quad \begin{matrix} * & * & * & * & * & i \\ & * & * & * & * & i \\ & & * & * & * & i \\ & & * & * & * & \\ & & & * & * & \\ & & & & * & \end{matrix} \tag{7}$$

This step can be written as

$$\begin{bmatrix} A_1 & \boldsymbol{a}_i & A_2 \end{bmatrix} \mapsto \begin{bmatrix} A_1 & A_2 & \boldsymbol{a}_i \end{bmatrix} = A\Pi_{i,n}$$

where $\Pi_{i,n}$ is the appropriate permutation matrix. The right-hand matrix is upper-triangular except for the sub-diagonal elements stored in A_2. Let Q be the orthogonal matrix where $\tilde{A} = Q^\top A\Pi_{i,n}$ is upper-triangular and set $\tilde{B} = Q^\top B$. Note that $Q^\top$ only acts on rows and columns i to n of $A\Pi_{i,n}$ and rows i to n of B. We have that

$$\tilde{F} = \tilde{A}^{-1}\tilde{B} = \Pi_{i,n}^\top F.$$

We can now replace the nth column $\tilde{\boldsymbol{a}}_n$ of $\tilde{A}$ by the jth column $\tilde{\boldsymbol{b}}_j$ of $\tilde{B}$, forming matrices

$$\bar{A} = \tilde{A} + \boldsymbol{u}\boldsymbol{e}_n^\top, \quad \bar{B} = \tilde{B} - \boldsymbol{u}\boldsymbol{e}_j^\top, \quad \boldsymbol{u} = \tilde{\boldsymbol{b}}_j - \tilde{\boldsymbol{a}}_n.$$

Note that the matrix $\bar{A}$ on the left remains upper-triangular. For the next iteration it remains to calculate $\bar{F} = \bar{A}^{-1}\bar{B}$. Using the Sherman–Morrison formula [26] (Appendix A.1), we have

$$\bar{A}^{-1} = \left(\tilde{A} + \boldsymbol{u}\boldsymbol{e}_n\right)^{-1} = \tilde{A}^{-1} - \frac{\tilde{\boldsymbol{u}}\tilde{\boldsymbol{v}}^\top}{1 + \boldsymbol{e}_n{}^\top\tilde{\boldsymbol{u}}}, \quad \tilde{\boldsymbol{u}} = \tilde{A}^{-1}\boldsymbol{u}, \quad \tilde{\boldsymbol{v}} = \tilde{A}^{-\top}\boldsymbol{e}_n,$$

so that

$$\begin{aligned} \bar{A}^{-1}\bar{B} &= \tilde{A}^{-1}\tilde{B} - \tilde{A}^{-1}\boldsymbol{u}\boldsymbol{e}_j^\top - \frac{\tilde{\boldsymbol{u}}(\tilde{\boldsymbol{v}}^\top\tilde{B})}{1 + \boldsymbol{e}_n{}^\top\tilde{\boldsymbol{u}}} + \frac{\tilde{\boldsymbol{u}}(\tilde{\boldsymbol{v}}^\top\boldsymbol{u})\boldsymbol{e}_j^\top}{1 + \boldsymbol{e}_n{}^\top\tilde{\boldsymbol{u}}}, \\ &= \tilde{F} + \tilde{\boldsymbol{u}}\boldsymbol{w}^\top, \end{aligned} \tag{8}$$

where

$$\boldsymbol{w} = \left(\frac{\tilde{\boldsymbol{v}}^\top\boldsymbol{u}}{1 + \boldsymbol{e}_n^\top\tilde{\boldsymbol{u}}} - 1\right)\boldsymbol{e}_j - \frac{\tilde{B}^\top\tilde{\boldsymbol{v}}}{1 + \boldsymbol{e}_n{}^\top\tilde{\boldsymbol{u}}}.$$

Thus, $\bar{F}$ is a rank-one update of $\tilde{F}$. The expressions above can be considerably simplified. We note that

$$\tilde{\boldsymbol{u}} = \tilde{F}(:,j) - \boldsymbol{e}_n, \quad \frac{\tilde{\boldsymbol{v}}^\top\boldsymbol{u}}{1 + \boldsymbol{e}_n^\top\tilde{\boldsymbol{u}}} - 1 = -1/\tilde{F}(n,j), \quad \frac{\tilde{B}^\top\tilde{\boldsymbol{v}}}{1 + \boldsymbol{e}_n{}^\top\tilde{\boldsymbol{u}}} = -\tilde{F}(n,:)^\top/\tilde{F}(n,j),$$

so that

$$\boldsymbol{w} = -\frac{1}{\tilde{F}_{n,j}}(\tilde{F}(n,:)^\top + \boldsymbol{e}_j).$$

Recalling that $\tilde{F} = \Pi_{i,n}^\top A^{-1}B$, we see that from (8), $\bar{A}^{-1}\bar{B}$ can be constructed as a rank-one update of the previously calculated $F = A^{-1}B$. We also note that $\tilde{F}(n,j) = F(i,j) > 1$ (since the update is only invoked to increase $|A|$) so that the rank-one update of F is numerically stable. The formation of $\tilde{A}$ and $\bar{A}$ requires $O(n^2)$ steps and the formation of $\bar{F}$ requires $O(mn)$ steps. These calculations lead to a computationally efficient and numerically stable version (Algorithm 3) of Algorithm 2:

Algorithm 3: GE: efficient algorithm to determine a D-optimal subset of an observation matrix

Data: An $m \times n$ observation matrix C such that the first n rows are full rank, and tolerance factor $f > 1$.

Result: Index set $I \subset \{i = 1, \ldots, m\}$ where the first n elements specify $C_n = C(I(1:n),:)$ such that no row interchange will increase the determinant of C_n by more that f; matrices $\begin{bmatrix} A & B \end{bmatrix}$ storing the upper-triangular factor of $C(I,:)^\top$.

Steps:

Partition $C^\top = \begin{bmatrix} A & B \end{bmatrix}$ and form $F = A^{-1}B$.

Initialise $m \times 1$ index set I with $I(i) = i$.

while $\max_{i,j} |F_{i,j}| > f$ **do**

1. Choose (i,j) such that $|F_{i,j}|$ is maximised. and store $\boldsymbol{a}_i$ and $\boldsymbol{b}_j$.
2. Apply permutation matrix $\Pi_{i,n}$ to move the ith column of A to the nth column $A := A\Pi_{i,n}$.
3. Apply $\Pi_{i,n}^\top$ to F: $F := \Pi_{i,n}F$.
4. Determine $(n-i+1) \times (n-i-1)$ orthogonal matrix Q such that $Q^\top A(i:n, i:n-1)$ is upper-triangular and update:
$$A(i:n, i:n) := Q^\top A(i:n, i:n), \quad B(i:n,:) := Q^\top B(i:n,:).$$
5. Set $\boldsymbol{a}_n = A(:,n)$, $\boldsymbol{b}_j = B(:,j)$, $\tilde{\boldsymbol{u}} = F(:,j) - \boldsymbol{e}_n$ and
$$\boldsymbol{w} = -\frac{1}{F_{n,j}}(F(n,:)^\top + \boldsymbol{e}_j).$$
6. Interchange columns and update I. $A(:,n) = \boldsymbol{b}_j$, $B(:,j) = \boldsymbol{a}_j$. Set $i_n = I(n)$, $i_j = I(n+j)$ and $I(n) = i_j$, $I(n+j) = i_n$.
7. Update F: $F := F + \tilde{\boldsymbol{u}}\boldsymbol{w}^\top$.

end

The algorithm is a sequential descent algorithm, with each exchange producing a decrease in the determinant of the variance matrix by at least a factor of $1/f^2 < 1$, ensuring that it stops in a finite number of steps. However, there is no guarantee that the algorithm will determine the globally D-optimal subset, only a locally D-optimal solution. A different ordering of the rows of the observation matrix is quite likely to produce a different solution. Consider the following matrix $C(a)$:

$$C(a) = \begin{bmatrix} 1.00 & 0.00 & 0.00 & 0.00 \\ 0.00 & 1.00 & 0.00 & 0.00 \\ 0.00 & 0.00 & 1.00 & 0.00 \\ 0.00 & 0.00 & 0.00 & a \\ 0.50 & 0.50 & 0.50 & 0.50 \\ 0.17 & -0.83 & 0.17 & 0.50 \\ 0.17 & 0.17 & -0.83 & 0.50 \\ -0.83 & 0.17 & 0.17 & 0.50 \end{bmatrix}. \tag{9}$$

The bottom four rows form an orthogonal matrix Q with $|Q| = 1$. However, for $a > 0.5$, there is no row interchange that will allow the determinant of the first four rows to be increased, and the GE algorithms stops with the first four rows with determinant a, sub-optimal if $a < 1.0$. For $0.5 < a < 1.0$, the SSQR algorithm, by contrast, selects the bottom four rows, in this case, the globally optimal solution.

4.3. Combined SSQR-GE Algorithm

An effective approach in practice is to combine the SSQR and GE algorithms, using the SSQR algorithm to find an initial selection of columns, and to use the GE interchange algorithm to improve on, if possible, this selection. This combined algorithm has the advantages of being largely independent of the row ordering and will deliver a local minimum based on the SSQR algorithm starting point. Again, there is no guarantee that the solution will be a global optimum, but the examples discussed in Section 5 show that the combined algorithm works well.

4.4. Sequential Approach for Choosing a Set of Additional Observations

The algorithms described in Sections 4.1 and 4.2 are designed to determine a minimal set of observations that produce D-optimal estimates of the model parameters $\boldsymbol{a}$. We now consider the case in which estimates of $\boldsymbol{a}$ have already been established. Suppose the (full rank) variance matrix $V = LL^\top = (R^\top R)^{-1}$ represents the current knowledge about parameters $\boldsymbol{a}$ and up to m (new) measurements are potentially available, modelled according to

$$y_i \sim \mathcal{N}(\boldsymbol{c}_i^\top \boldsymbol{a}, 1), \quad i = 1, \ldots, m.$$

(As before, we assume that, for each i, y_i and $\boldsymbol{c}_i$ have been weighted so that the uncertainty associated with y_i is 1.) Which measurements should we make in order to best improve our knowledge about $\boldsymbol{a}$? We suppose that resources exist to make p more observations. An exhaustive search of all combinations is only feasible for small p and modest m. The sequential algorithms below add the observations one at a time, with the choice at each stage designed to maximise the information. These are further examples of so-called 'greedy algorithms'.

Let C be the $m \times n$ observation matrix whose ith row is $\boldsymbol{c}_i^\top$ and V_i the variance matrix associated with $\boldsymbol{a}$ given that the ith measurement (and only the ith measurement) is made. Then,

$$V_i = (R^\top R + \boldsymbol{c}_i \boldsymbol{c}_i^\top)^{-1} = V - \frac{V\boldsymbol{c}_i (V\boldsymbol{c}_i)^\top}{1 + \boldsymbol{c}_i^\top V \boldsymbol{c}_i},$$

using the Sherman–Morrison formula [26] (Appendix A.1). Let

$$\boldsymbol{f}_i = V\boldsymbol{c}_i, \quad g_i^2 = \boldsymbol{c}_i^\top \boldsymbol{f}_i = \boldsymbol{c}_i^\top V \boldsymbol{c}_i. \tag{10}$$

Then,

$$V_i = V - \frac{\boldsymbol{f}_i \boldsymbol{f}_i^\top}{1 + g_i^2} = V - \boldsymbol{u}_i \boldsymbol{u}_i^\top, \quad \boldsymbol{u}_i = \frac{1}{\sqrt{1 + g_i^2}} \boldsymbol{f}_i. \tag{11}$$

so that V_i is a symmetric rank-one update of V.

The eigenvalues $\lambda_{j,i}$ of V_i can be related to the eigenvalues λ_j of V through the following result [23] (Section 8.1), [27] (p. 94):

Proposition 2. *Suppose $W = V + \rho \boldsymbol{u}\boldsymbol{u}^\top$, where V is an $n \times n$ symmetric matrix, $\boldsymbol{u}$ is an n-vector with $\boldsymbol{u}^\top \boldsymbol{u} = 1$ and ρ is a constant. Suppose V and W have eigenvalues*

$$\lambda_1(V) \geq \lambda_1(V) \geq \lambda_2(V) \geq \cdots \geq \lambda_n(V),$$

and

$$\lambda_1(W) \geq \lambda_1(W) \geq \lambda_2(W) \geq \cdots \geq \lambda_n(W).$$

If $\rho \geq 0$, then

$$\lambda_i(W) \in [\lambda_i(V), \lambda_{i-1}(V)], \quad i = 2, \ldots, n,$$

while if $\rho \leq 0$, then

$$\lambda_i(W) \in [\lambda_{i+1}(V), \lambda_i(V)], \quad i = 1, \ldots, n-1.$$

In either case, there exist n constants γ_i, $\gamma_i \geq 0$, *with* $\gamma_1 + \cdots \gamma_n = 1$, *such that*

$$\lambda_i(W) = \lambda_i(V) + \gamma_i \rho.$$

4.4.1. Sequential D-Optimality Algorithm

We use the relationship (5) to write

$$\begin{aligned} |V_i| &= |(R^\top R + \boldsymbol{c}_i \boldsymbol{c}_i^\top|^{-1}, \\ &= |R^\top R|^{-1} \left(1 + \boldsymbol{c}_i^\top (R^\top R)^{-1} \boldsymbol{c}_i\right)^{-1}, \\ &= |V| \left(1 + \boldsymbol{c}_i^\top V \boldsymbol{c}_i\right)^{-1}, \\ &= \frac{1}{1+g_i^2}|V|, \quad g_i^2 = \boldsymbol{c}_i^\top V \boldsymbol{c}_i. \end{aligned}$$

A sequential algorithm can therefore be designed which at each stage chooses the i that maximises g_i^2. The algorithm can be run in two modes: the first in which the algorithm can choose any row at most one time, the second in which there is no limit in the number of times any row can be selected. Having a row repeated k times is equivalent to having a single more accurate observation with its uncertainty reduced by a factor of $1/\sqrt{k}$. In the second mode, it is often the case that additional measurements correspond to repeat measurements chosen from the D-optimal set of n points. In the algorithm below (Algorithm 4), the mode is controlled by the flag i_R:

Algorithm 4: Sequential approach to optimising the D-measure in choosing p observations from a possible $m > p$ observations

---

Data:
- C, an $m \times n$ observation matrix C, $m > n$,
- V, an $n \times n$ full rank variance matrix V encoding the current state of knowledge of the parameters,
- p, an integer $1 \leq p < m$ specifying the maximum number of observations to be made, and
- flag $i_R = 0$ or 1. If $i_R = 1$, rows of C can be selected more than once.

Result: A $p \times 1$ index set $I \subset \{i = 1, \ldots, m\}$, updated variance matrix V representing the variance in the parameters as a result of using observations $C_p = C(I,:)$, the p rows of C specified by I, and $p \times 1$ $\boldsymbol{t}$ where $0 < t_q < 1$, the reduction in the determinant of V at each iteration: $|V_{q+1}| := t_q|V_q|$. The rows are chosen to minimise t_q at each iteration.

Set $F = VC^\top$ and $n \times m$ matrix G with $G_{j,i} = F_{j,i}C_{j,i}$ with $g_i = \|G(:,i)\|$, $i = 1, \ldots, m$.

for $q = 1$ **to** p **do**

1. Evaluate $\tau_i = 1/(1+g_i^2)$, $i = 1, \ldots, m$.
2. If $q > 1$ and $i_R = 0$, set $\tau_i = 2$, $i = I(1), \ldots, I(q-1)$.
3. Find k such that $\tau_k \leq \tau_i$, $i = 1, \ldots, m$.
4. Set $I(q) = k$, $t_q = \tau_k$, $\boldsymbol{u} = F(:,k)/\sqrt{1+g_k^2}$, and $\boldsymbol{w} = C\boldsymbol{u}_k$.
5. Update $V := V - \boldsymbol{u}\boldsymbol{u}^\top$, $F = F - \boldsymbol{u}\boldsymbol{w}^\top$, and $g_i = [(g_i - w_i)(g_i + w_i)]^{1/2}$, $i = 1, \ldots, m$.

end

For each q, the only operations required are matrix–vector multiplications and similar, requiring at most $O(mn)$ steps. This compares with $O(mn^3)$ steps for an algorithm not exploiting rank-one updates.

4.4.2. Expected Information Gain Calculations

The quantities t_q calculated at each stage give the reduction in determinant of the variance matrix at the qth step. The following argument can be used to provide an expected value for t_q, under the assumption that the rows of C each provide approximately the same amount of information. Suppose C consists of rows with a 1 in one column and zeros elsewhere. A D-optimal design would select rows so that the ones are distributed equally amongst the columns. If a total of q rows are selected, the variance matrix V_q is such that

$$V_q \approx \frac{n}{q} I, \quad |V_q| \approx \left(\frac{n}{q}\right)^n.$$

From this argument, the expected value of t_q, $q > n$, is such that

$$t_q \approx \tilde{t}(q) = \left(\frac{q-1}{q}\right)^n. \tag{12}$$

In terms of the geometric mean $\bar{d}$ of the eigenvalues of the variance matrices, we have

$$\bar{d}(V_q) = \frac{q-1}{q} \bar{d}(V_{q-1}).$$

For $q = n+1$, we have

$$\frac{1}{2} \geq \left(\frac{n}{n+1}\right)^n \geq \frac{1}{e} \approx 0.37,$$

where the left hand value corresponds to the case $n = 1$, and $1/e$ is the limiting value as $n \to \infty$.

4.4.3. Sequential A-Optimality Algorithm

We can also construct a counterpart of Algorithm 4 for A-optimality. For any $\boldsymbol{v}$, $\mathrm{Tr}(\boldsymbol{v}\boldsymbol{v}^\top) = \boldsymbol{v}^\top \boldsymbol{v} = \|\boldsymbol{v}\|^2$. With $\boldsymbol{f}_i$ and g_i as in (10), if

$$\tau_i^2 = f_i^2/(1+g_i^2), \quad f_i^2 = \boldsymbol{f}_i^\top \boldsymbol{f}_i,$$

then

$$\mathrm{Tr}(V_i) = \mathrm{Tr}(V) - \tau_i^2.$$

This last expression calculates the change in the A-measure for a given choice of one additional measurement. A sequential algorithm can therefore be designed which at each stage chooses the i that maximises τ_i^2. Each stage requires the calculation of the quantities f_i^2 and g_i^2, $i = 1, \ldots, m$. If V is the current variance matrix, then the f_i^2 are given by the sum of the squares of row elements of $F = VC^\top$. Suppose measurement k maximises τ_i^2. Then, the update of F is given by

$$F = (V - \boldsymbol{u}_k \boldsymbol{u}_k^\top) C^\top = F - \boldsymbol{u}_k \boldsymbol{w}^\top, \quad \boldsymbol{w} = C\boldsymbol{u}_k,$$

and the updated of g_i^2 is given by $g_i^2 - w_i^2$, with the latter evaluated as $(g_i - w_i)(g_i + w_i)$ for better numerical accuracy. These steps are summarised in (Algorithm 5):

Algorithm 5: Sequential approach to optimising the A-measure in choosing p observations from a possible $m > p$ observations

Data:
- C, an $m \times n$ observation matrix C, $m > n$,
- V, an $\times n$ full rank variance matrix V encoding the current state of knowledge of the parameters,
- p, an integer $1 \leq p < m$ specifying the maximum number of observations to be made, and
- flag $i_R = 0$ or 1. If $i_R = 1$, rows of C can be selected more than once.

Result: A $p \times 1$ index set $I \subset \{i = 1, \ldots, m\}$, updated variance matrix V representing the variance in the parameters as a result of using observations $C_p = C(I,:)$, the p rows of C specified by I, and $p \times 1$ $\boldsymbol{t}$ where $t_q = \tau_q^2$, the reduction in the trace of V at the qth iteration. The rows are chosen to maximise τ_i^2 at each iteration.

Set $F = VC^\top$, $f_i = \|F(:,i)\|$, $n \times m$ matrix G with $G_{j,i} = F_{j,i}C_{j,i}$ with $g_i = \|G(:,i)\|$, $i = 1, \ldots, m$.

for $q = 1$ **to** p **do**

1. Evaluate $\tau_i^2 = f_i^2/(1 + g_i^2)$, $i = 1, \ldots, m$.
2. If $q > 1$ and $i_R = 0$, set $\tau_i^2 = 0$, $i = I(1), \ldots, I(q-1)$.
3. Find k such that $\tau_k^2 \geq \tau_i^2$, $i = 1, \ldots, m$.
4. Set $I(q) = k$, $t_q = \tau_k^2$, $\boldsymbol{u} = F(:,k)/\sqrt{1 + g_k^2}$, and $\boldsymbol{w} = C\boldsymbol{u}$.
5. Update $V := V - \boldsymbol{u}\boldsymbol{u}^\top$, $F = F - \boldsymbol{u}\boldsymbol{w}^\top$, $f_i = \|F(:,i)\|$ and $g_i = [(g_i - w_i)(g_i + w_i)]^{1/2}$, $i = 1, \ldots, m$.

end

As for Algorithm 4, the computational requirement for each step is $O(mn)$.

5. Numerical Examples

5.1. Polynomial Calibration Curves

Many instrument response functions are modelled as polynomial curves, most commonly, of order 2 (degree 1), i.e., modelling a linear response, but higher orders arise, e.g., in the calibration of platinum resistance thermometers [28], or in the calibration of a stage motion [29] where polynomials of order 10 are involved. (Design problems for problems for polynomial models augmented by Gaussian process models are considered in [4].) Suppose a polynomial response function of order n is defined on the interval [−1,1], and that it is desired to calibrate the response function using n calibration points $x_i \in [-1,1]$. Which choice of n points is D-optimal? For a linear (quadratic) response $n = 2$ ($n = 3$), it is intuitively clear that the end points $x = \pm 1$ (and $x = 0$) are optimal. For higher orders, the optimal choice is not obvious. In fact, the D-optimal set [30] for order n is given by solutions of

$$(1 - x^2)\dot{L}_n(x) = 0, \tag{13}$$

where $\dot{L}_n(x)$ is the derivative of Legendre polynomial of order n (degree $n-1$). These solutions can be approximated by the so-called 'arcsine' points

$$x_i = \cos\left(\pi \frac{n-1-i}{n-1}\right), \quad i = 0, 1, \ldots, n-1. \tag{14}$$

Table 1 gives the results of applying the SSQR algorithm and the combined SSQR-GE algorithm to estimate the D-optimal calibration points for the cases $n = 4, 5, \ldots, 11$ choosing from $m = 2001$ possible points equally spaced in the interval $[-1,1]$ in steps of 0.001. For these calculations, we have used Chebyshev polynomial basis functions [31] for numerical stability (but the optimal choice of points is independent of the choice of basis functions.) The arcsine estimates are also given in the table, along with the D-optimal points derived from (13); see Appendix B. The results show that, in terms of closeness to the D-optimal points, the SSQR algorithm improves on the arcsine estimates, while the GE algorithm improves the SSQR solution and is accurate to three decimals in almost all cases (there is one exception for the case $n = 11$). Table 2 gives the computed geometric mean measure $\bar{d}(V_{\boldsymbol{a}})$ given in (4) for evenly spaced calibration points, the arcsine solutions, the SSQR solutions, and the SSQR-GE solutions given in Table 1. Also shown is the number of GE exchanges undertaken to improve the SSQR solutions. The table reflects the improvements provided by the SSQR and SSQR-GE algorithms. The number of GE exchanges is small compared to $m = 2001$. The evenly spaced calibration points are markedly worse than the other solutions for larger n.

Table 1. D-optimal calibration points for polynomial calibration curves of order $n = 4, \ldots, 11$ defined in the interval $[-1, 1]$. For each set of four columns, the first column, labelled x_0, gives the arcsine estimates of the calibration points in (14); the second column, labelled $x_{\rm QR}$, gives the optimal the calibration points determined by the SSQR algorithm; the third column, labelled $x_{\rm GE}$, gives the estimates determined by the combined SSQR-GE algorithm; while the fourth column, labelled x^*, gives the optimal the calibration points derived from (13).

x_0	$x_{\rm QR}$	$x_{\rm GE}$	x^*	x_0	$x_{\rm QR}$	$x_{\rm GE}$	x^*
	$n = 4$				$n = 5$		
−1.000	−1.000	−1.000	−1.000	−1.000	−1.000	−1.000	−1.000
−0.500	−0.488	−0.447	−0.447	−0.707	−0.669	−0.655	−0.655
0.500	0.437	0.447	0.447	0.000	0.006	0.000	0.000
1.000	1.000	1.000	1.000	0.707	0.686	0.655	0.655
				1.000	1.000	1.000	0.000
	$n = 6$				$n = 7$		
−1.000	−1.000	−1.000	−1.000	−1.000	−1.000	−1.000	−1.000
−0.809	−0.786	−0.765	−0.765	−0.866	−0.845	−0.830	−0.830
−0.309	−0.286	−0.285	−0.285	−0.500	−0.484	−0.469	−0.469
0.309	0.308	0.285	0.285	0.000	0.002	0.000	0.000
0.809	0.779	0.765	0.765	0.500	0.493	0.469	0.469
1.000	1.000	1.000	1.000	0.866	0.841	0.830	0.830
				1.000	1.000	1.000	0.000
	$n = 8$				$n = 9$		
−1.000	−1.000	−1.000	−1.000	−1.000	−1.000	−1.000	−1.000
−0.901	−0.880	−0.872	−0.872	−0.924	−0.908	−0.900	−0.900
−0.623	−0.613	−0.592	−0.592	−0.707	−0.692	−0.677	−0.677
−0.223	−0.211	−0.209	−0.209	−0.383	−0.383	−0.363	−0.363
0.223	0.225	0.210	0.209	0.000	−0.002	0.000	0.000
0.623	0.608	0.592	0.592	0.383	0.376	0.363	0.363
0.901	0.882	0.872	0.872	0.707	0.695	0.677	0.677
1.000	1.000	1.000	1.000	0.924	0.906	0.900	0.900
				1.000	1.000	1.000	0.000
	$n = 10$				$n = 11$		
−1.000	−1.000	−1.000	−1.000	−1.000	−1.000	−1.000	−1.000
−0.940	−0.925	−0.920	−0.920	−0.951	−0.938	−0.934	−0.934
−0.766	−0.753	−0.739	−0.739	−0.809	−0.796	−0.784	−0.784
−0.500	−0.493	−0.478	−0.478	−0.588	−0.580	−0.565	−0.565
−0.174	−0.177	−0.165	−0.165	−0.309	−0.311	−0.296	−0.296
0.174	0.168	0.165	0.165	0.000	−0.001	0.000	0.000
0.500	0.497	0.478	0.478	0.309	0.307	0.296	0.296
0.766	0.751	0.739	0.739	0.588	0.582	0.566	0.565
0.940	0.925	0.920	0.920	0.809	0.795	0.785	0.784
1.000	1.000	1.000	1.000	0.951	0.939	0.934	0.934
				1.000	1.000	1.000	0.000

Table 2. For $n = 4, 5, \ldots, 11$, rows two to five give the computed geometric mean measure $\bar{d}(V_a)$ given in (4) for evenly spaced calibration points, the arcsine solutions, the SSQR solutions, and the SSQR-GE solutions given in Table 1. The final row gives the number of GE exchanges undertaken to improve the SSQR solutions.

n	4	5	6	7	8	9	10	11
$\bar{d}_0$	0.4871	0.4152	0.3748	0.3511	0.3379	0.3316	0.3304	0.3332
$\bar{d}_{AS}$	0.4714	0.3789	0.3175	0.2734	0.2403	0.2143	0.1935	0.1763
$\bar{d}_{QR}$	0.4682	0.3746	0.3130	0.2691	0.2362	0.2107	0.1901	0.1733
$\bar{d}_{GE}$	0.4673	0.3735	0.3119	0.2682	0.2354	0.2099	0.1894	0.1726
n_{GE}	3	6	7	9	12	17	19	21

Figure 1 graphs the values of t_q, $q = 1, \ldots, 100$, and its analytical approximation $\tilde{t}(q)$ given by (12), determined by the sequential D-optimal Algorithm 4, without repeat measurements, for the cases $n = 4$, upper curve, and $n = 11$, lower curve, starting with the SSQR-GE solution. The values of t_q are batched with a significant jump after every n additional measurement. Without the repeat measurement, the algorithm tends to select available points closest to the D-optimal n points. If the algorithm is run with repeat measurements allowed, the algorithm selects the same set of D-optimal n points again and again. This behaviour is reflected in the graphs of t_q for this case; see Figure 2.

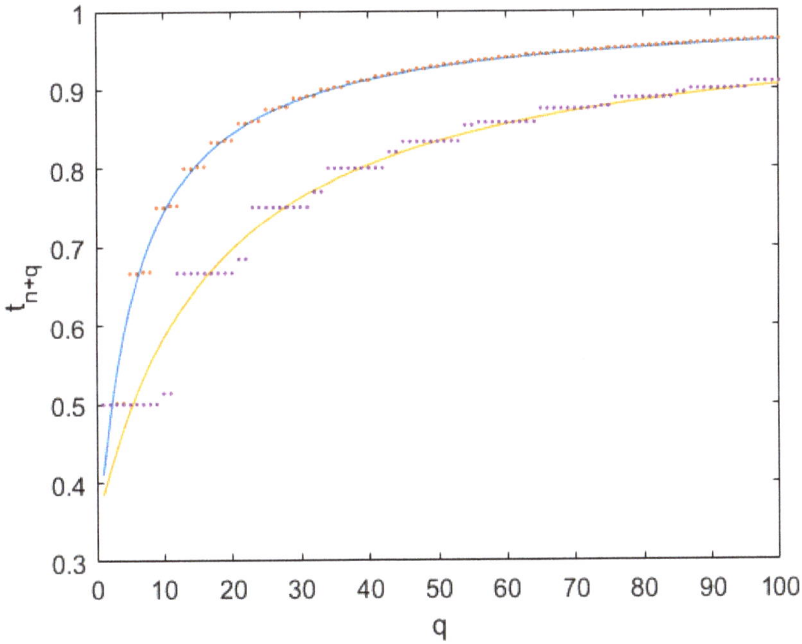

Figure 1. Polynomial calibration points: values of t_q, $q = 1, \ldots, 100$, and its analytical approximation $\tilde{t}(q)$ given by (12), determined by the sequential D-optimal Algorithm 4, for the cases $n = 4$, upper curve, and $n = 11$, lower curve.

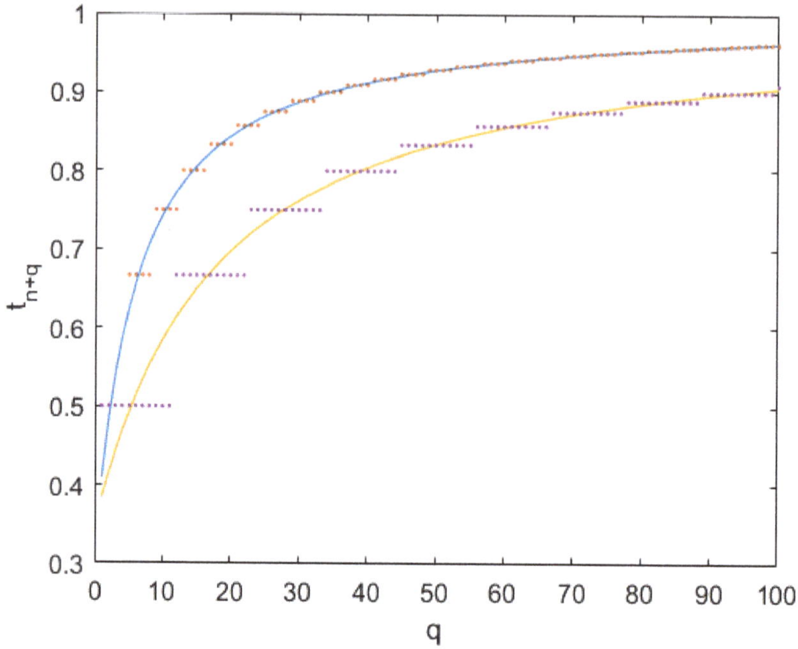

Figure 2. As Figure 1, but for the case where repeated measurements are permitted.

5.2. Tensor Product of Polynomials

Our second example arose from a problem relating to the estimation of heat distribution on an plate [32]. The application involved modelling the heat distribution as a Gaussian process model [33] involving tensor product polynomials, along with spatially correlated effects. A sub-problem is related to choosing a set of calibration points from a 10×14 grid of possible sensor locations in order to determine the tensor product mean function. The tensor product model is of the form

$$f(x, y, \boldsymbol{a}) = \sum_{i=1}^{n_x} \sum_{j=1}^{n_y} a_{ij} f_i(x) g_j(y),$$

where $f_i(x)$ and $g_j(y)$ are Chebyshev polynomial basis functions. Figure 3 shows the solution calibration points for $n_x = n_y = 5$ constructed from polynomial basis functions up to degree 4. Two solutions are shown: the first in which the calibration points are selected from a 131×91 finer grid of 11,921 points, the second in which the calibration points are selected from a 14×10 grid of 140 points. The solutions found for the fine grid coincide with a grid 5×5 grid of points (x_k^*, y_l^*), where the x_k^*, $k = 1, \ldots, 5$ are the D-optimal solutions derived from the solution of (13) on the interval $[0, 20]$ and the y_l^* are the D-optimal solutions on the interval $[0, 10]$. For this arrangement, the observation matrix C associated with the polynomial fit is the tensor product $C = C_x \otimes C_y$. For square, full rank matrices A and B of size $n_A \times n_A$ and $n_B \times n_B$

$$|A \otimes B| = |A|^{n_A} |B|^{n_B}.$$

Hence, it is consistent that the D-optimal points for the tensor product are constructed from the D-optimal points for the x- and y-basis functions. The solution points for the coarse grid are seen to approximate the solution for the fine grid, given the constraints imposed by the coarse grid. The coarse grid solution points do not represent a regular grid in this case.

For the case of the fine grid, the GE algorithm took 16 iterations to reach a local optimum, starting with the SSQR solution. For the coarse grid, the GE algorithm confirmed that the SSQR solution is locally optimal.

Figures 4 and 5 plot the values of t_q, $q = 1, \ldots, 200$, and its analytical approximation $\tilde{t}(q)$ given by (12), determined by the sequential D-optimal Algorithm 4, for the fine and coarse grids, respectively, with and without repeat measurements. For a coarse grid and no repeat measurements, the additional calibration points that are permitted consistently provide less information compared to the expected information gain given by $\tilde{t}$ in (12). For a coarse grid of just 140 points, later additional measurements have to be selected from locations that are less informative in terms of improving the D-measure. With repeat measurements on the coarse grid, the actual information gain is in line with the expected information gain.

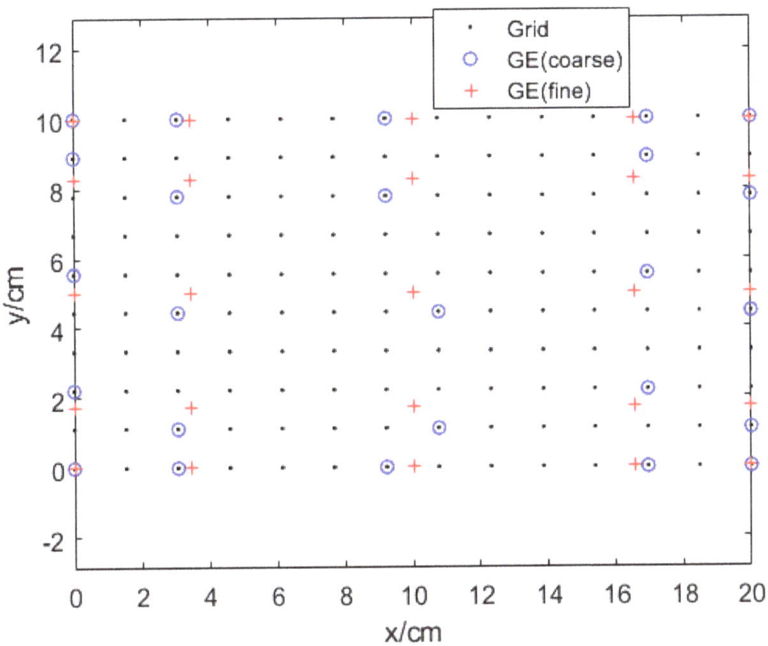

Figure 3. Tensor product polynomial calibration points found by the SSQR-GE algorithm. The points marked with a '+' are those selected from a fine grid, those marked 'o' from a coarse grid.

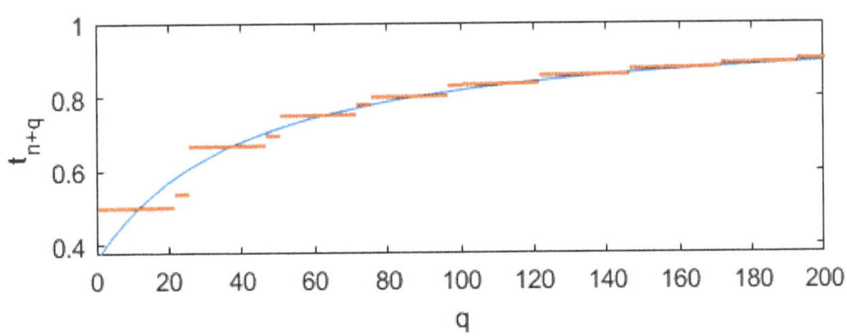

Figure 4. *Cont.*

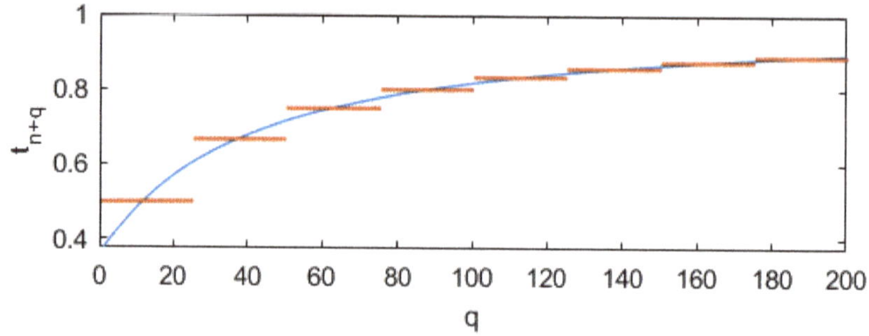

Figure 4. Tensor product polynomial calibration points: values of t_q, $q = 1, \ldots, 100$, and its analytical approximation $\tilde{t}(q)$ given by (12), determined by the sequential D-optimal Algorithm 4, for the fine grid, with no repeat measurements, upper curve, and with repeat measurements, lower curve.

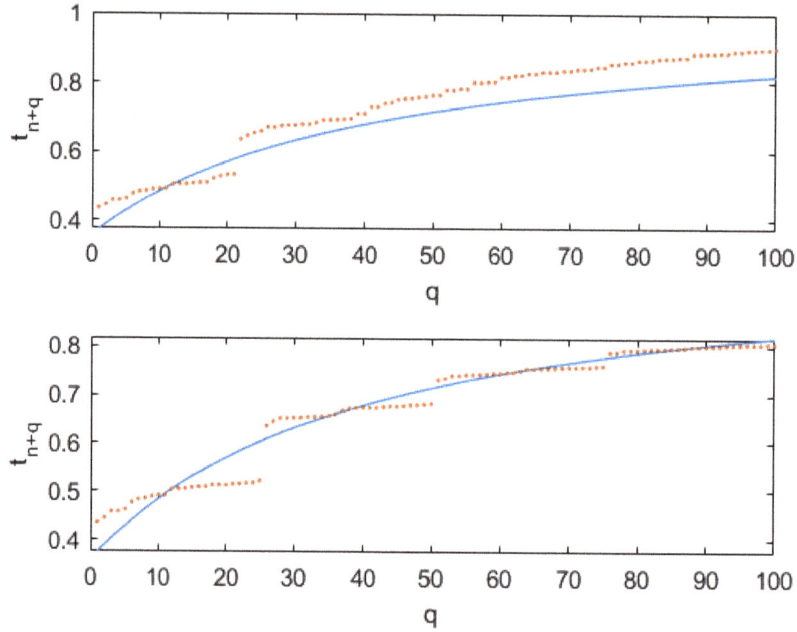

Figure 5. Tensor product polynomial calibration points: values of t_q, $q = 1, \ldots, 100$, and its analytical approximation $\tilde{t}(q)$ given by (12), determined by the sequential D-optimal Algorithm 4, for the coarse grid, with no repeat measurements, upper curve, and with repeat measurements, lower curve.

5.3. Coordinate Metrology

In coordinate metrology, the size and shape of a geometric surface $\mathcal{S}$ is determined from the measurement of the coordinates $\boldsymbol{x}_i$ of points lying on the surface. If the geometric surface $\mathcal{S} = \mathcal{S}(\boldsymbol{a})$ is parametrized by $\boldsymbol{a} = (a_1, \ldots, a_n)^\top$ and $d(\boldsymbol{x}, \boldsymbol{a})$ is a measure of the distance from a point $\boldsymbol{x}$ to $\mathcal{S}(\boldsymbol{a})$, then for measurement points $\boldsymbol{x}_{1:m} = \{\boldsymbol{x}_i, i = 1, \ldots, m\}$, $m \geq n$, estimates of $\boldsymbol{a}$ are determined by solving the least squares optimisation problem

$$\min_{\boldsymbol{a}} = \sum_{i=1}^{m} d^2(\boldsymbol{x}_i, \boldsymbol{a}).$$

If the variance matrix associated with the measured points coordinates is $\sigma^2 I$, then the variance matrix $V_{\boldsymbol{a}}$ associated with the fitted parameters is approximated by

$$V_{\boldsymbol{a}} = \sigma^2 \left(J^\top J \right)^{-1}, \quad J_{ij} = \frac{\partial d}{\partial a_j}(\boldsymbol{x}_i, \boldsymbol{a}),$$

involving the Jacobian matrix J. Thus, it is possible to determine optimal measurement strategies, i.e., where to measure on the surface $\mathcal{S}$, in terms of the determinant of the Jacobian matrix J. Often, the case of choosing a minimal set of points with $m = n$ is of interest. For simple geometries, an optimal set of points is straightforward to arrive at, e.g., three points equally spaced around a circle, but for more complex geometries, the choice of an optimal set may not be obvious, as considered below.

Figure 6 shows a reference artefact designed and calibrated by the Czech Metrology Institute (CMI) [34,35]. The form of the geometric surface is a hyperbolic paraboloid given by the equation

$$z = 24 + (x/8 - 1)(y/8 - 1). \tag{15}$$

A paraboloid with its axis approximately parallel to the z-axis is parametrized in terms of two rotation angles α and β and six further parameters $\boldsymbol{b}$ such that

$$\hat{z} = b_1 \hat{x}^2 + b_2 \hat{y}^2 + b_3 \hat{x}\hat{y} + b_4 \hat{x} + b_5 \hat{y} + b_6, \quad \hat{\boldsymbol{x}} = R(\alpha, \beta)\boldsymbol{x},$$

where $R(\alpha, \beta) = R_x(\alpha) R_y(\beta)$ is the rotation matrix constructed from the plane rotation matrices

$$R_x(\alpha) = \begin{bmatrix} 1 & 0 & 0 \\ 0 & \cos\alpha & -\sin\alpha \\ 0 & \sin\alpha & \cos\alpha \end{bmatrix}, \quad R_y(\beta) = \begin{bmatrix} \cos\beta & 0 & \sin\beta \\ 0 & 1 & 0 \\ -\sin\beta & 0 & \cos\beta \end{bmatrix}.$$

For the surface (15), nominally $\boldsymbol{b}^\top = (0, 0, 1/64, -1/8, -1/8, 25)$. Figures 7–9 show the distribution of the x- and y-coordinates for three sets X_k, $k = 1, 2, 3$, of eight calibration points. The first two are given by 'expert judgement'; that is, possible choices for a measurement strategy based on engineering practice. The first expert choice, X_1, Figure 7, represents an approximately uniform distribution, a first guess at a good distribution. This distribution of points is sufficient to determine all the parameters, and the associated 8×8 Jacobian matrix is full rank. (If the x- and y- coordinates of the data points are rotated through 45 degrees about the z-axis, the associated Jacobian matrix is rank deficient by three.) The second set X_2 came about through experimenting with moving the inner four points in dataset X_1 closer or further from the origin of the xy-plane. It turned out that moving the four points further from the origin was better, resulting in dataset X_2, Figure 8. At first glance, the distribution of points in X_2 does not look a good choice. The third dataset is that obtained using the SSQR-GE algorithm. The GE algorithm performed 11 interchanges starting from the SSQR solution. The optimal choice has seven points along the boundary in the xy-plane and one point in the interior. Any interior point not too far from the origin would also result in a good selection. In terms of the aggregate measure $\bar{d}_k$ defined in (4) associated with the three datasets, we have $\bar{d}_1 = 6.0\bar{d}_3$ and $\bar{d}_2 = 2.2\bar{d}_3$.

Table 3 gives the square roots v_i of the diagonal elements of $(J_k^\top J_k)^{-1}$, where J_k is the Jacobian matrix associated with the kth dataset. If the variance matrix associated with the measurements is $\sigma^2 I$, then $v_i = u(a_j)/\sigma$, where $u(a_j)$ is the estimate of the standard uncertainty associated with the jth parameter [36,37]. In terms of standard uncertainties, the D-optimal dataset X_3 is far better for estimating the surface parameters than the other two datasets.

Figure 10 plots the values of t_q, $q = 1, \ldots, 100$, and its analytical approximation $\tilde{t}(q)$ given by (12), determined by the sequential D-optimal Algorithm 4, with and without repeat measurements. For higher values of q, there is good agreement between t_q and $\tilde{t}_q$.

Table 3. Standard uncertainty factors $u(a_j)/\sigma$ associated with the hyperbolic paraboloid parameters $\boldsymbol{a} = (\alpha, \beta, \boldsymbol{b}^\top)^\top$ for three eight-point datasets X_k, $k = 1,2,3$.

$u(a_j)/\sigma$	α	β	b_1	b_2	b_3	b_4	b_5	b_6
X_1	2.3867	2.3867	0.0049	0.0049	0.0139	2.6219	2.6219	10.5328
X_2	0.7259	0.7259	0.0042	0.0042	0.0035	0.4999	0.4999	2.8188
X_3	0.2049	0.2413	0.0012	0.0014	0.0014	0.2883	0.2520	1.4690

Figure 6. The CMI hyperbolic paraboloid reference artefact.

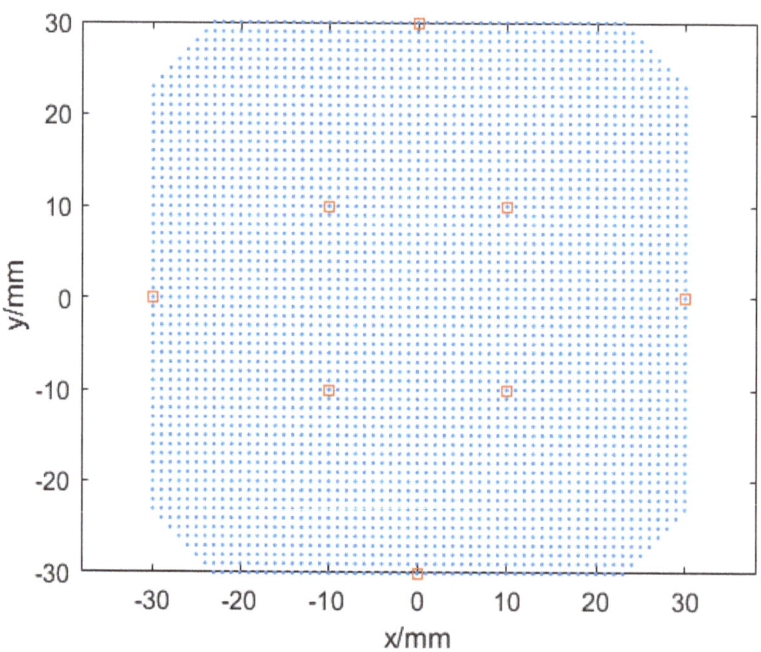

Figure 7. Hyperbolic paraboloid calibration points: 'expert' guess at a good set of calibration points.

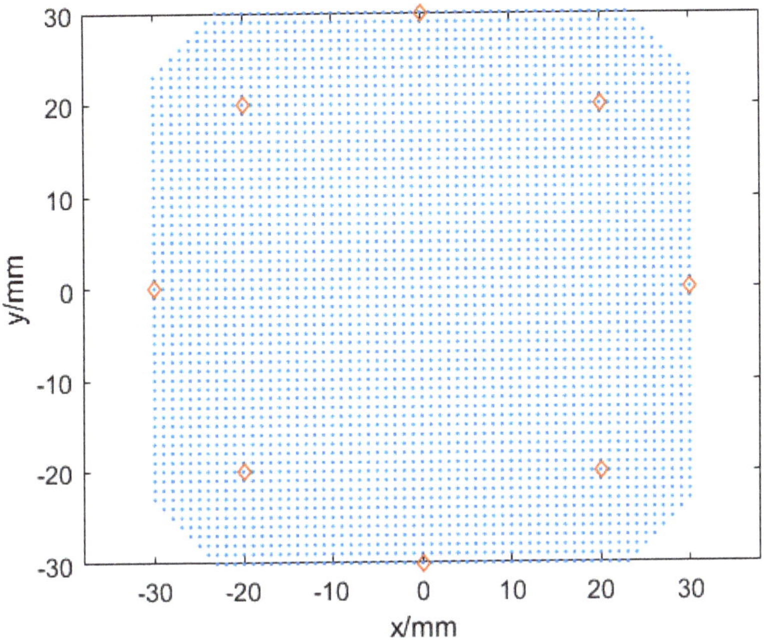

Figure 8. Hyperbolic paraboloid calibration points: second 'expert' guess at a good set of calibration points, a modification of X_1, Figure 7.

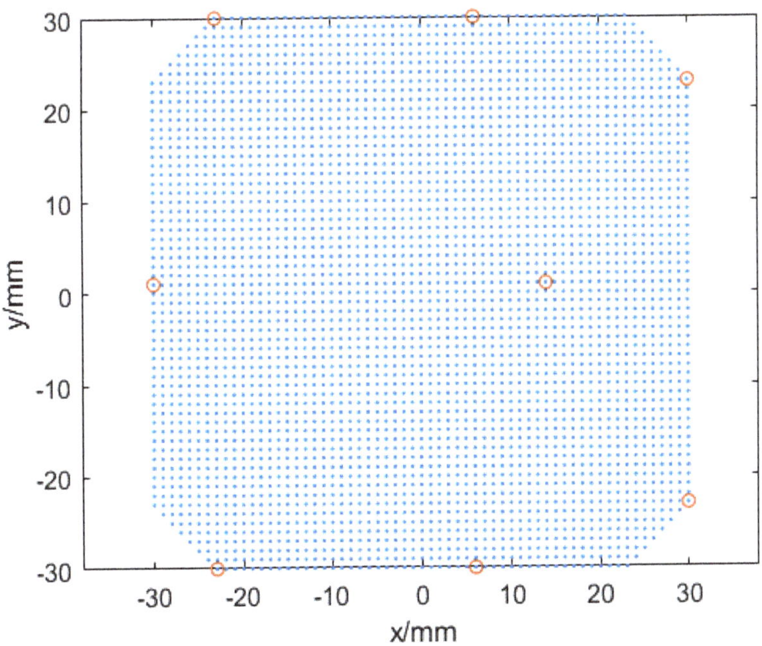

Figure 9. Hyperbolic paraboloid calibration points: D-optimal points determined by the SSQR-GE algorithm.

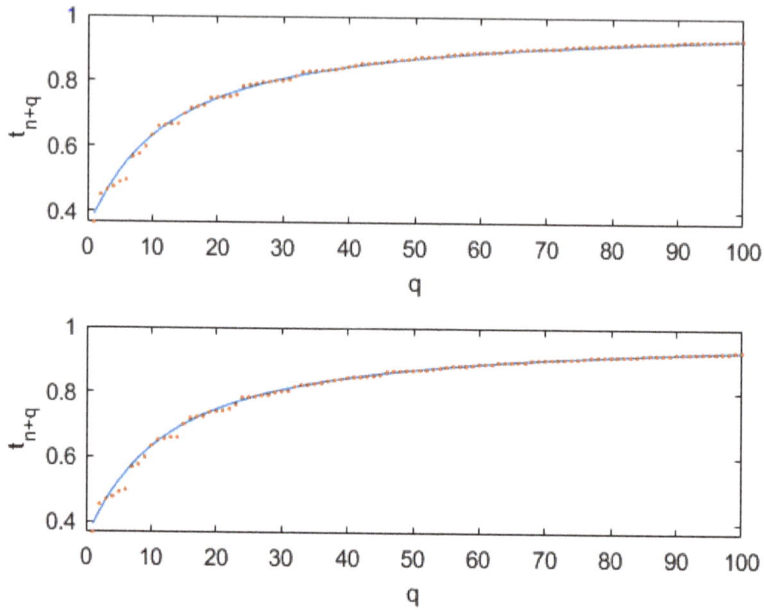

Figure 10. Hyperbolic paraboloid calibration points: values of t_q, $q = 1, \ldots, 100$, and its analytical approximation $\tilde{t}(q)$ given by (12), determined by the sequential D-optimal Algorithm 4, with no repeat measurements, upper curve, and with repeat measurements, lower curve.

5.4. Calibration of a Network of Standards Using a Comparator

In this section, we consider the calibration of a number of standards, starting with one calibrated standard and using a comparator to calibrate the other standards relative to the calibrated standard. An example application is in mass calibration using a mass balance [38–40]. Given a calibrated standard A_1, nominally of 1 kg, and two uncalibrated masses A_2 and A_3 each of nominal mass 0.5 kg, A_2 and A_3 can be calibrated using two measurements, the first comparing A_1 with the combined mass of A_2 and A_3, and the second comparing A_2 and A_3. A second example from length metrology relates to the calibration of gauge blocks [41], where it is possible to align two gauge blocks end to end, a process called 'wringing', to define a combined length. This discussion is relevant to the calibration of any set of artefacts defining an extensive quantity where the artefacts can be grouped to define a quantity that is (nominally) the sum of the individual quantities. We use the term 'network' to reflect the fact that the estimates of all the quantities are statistically correlated and an experiment that updates the information about one artefact will also update information about the other artefacts.

Let $\boldsymbol{a} = (a_1, \ldots, a_n)^\top$ represent the values of the quantities associated with the artefacts where the nominal values $\boldsymbol{a}^\sharp = (a_1^\sharp, \ldots, a_n^\sharp)^\top$ of the quantities are such that the artefacts can be grouped into sets of two groups, with each group being associated with the same nominal value. Thus, there are disjoint subsets L_i, $R_i \subset I = \{1, 2, \ldots, n\}$, $L_i \cap R_i = \emptyset$, $i = 1, \ldots, m$, such that

$$\sum_{q \in L_i} a_q \approx \sum_{q \in R_i} a_q.$$

An example considered further below involves a set of artefacts having nominal values 1.00, 0.50, 0.50, 0.20, 0.20, 0.10, 0.10, 0.05, and 0.05. The fact that the two subsets are associated with the same nominal value allows a comparator measurement to be made, with the associated observation equation

$$0 \approx y_i = \sum_{q \in L_i} a_q - \sum_{q \in R_i} a_q + \epsilon_i, \quad \epsilon_i \in \mathcal{N}(0, \sigma_i^2),$$

where ϵ_i represents a random effect associated with the comparator measurement. The design problem is to define measurement strategies encoded by L_i and R_i, $i = 1, \ldots, n$, that provide the most accurate calibration of the standards. For the example considered here, there are almost 400 viable experiments.

Tables 4 and 5 show five designs for the calibration of the $n = 9$ standards. A '1' in column j indicates that $j \in L_i$, while a '-1' indicates that $j \in R_i$. Each array of 9×9 elements is in fact the unweighted observation matrix C associated with the particular design. The first row in each design represents the observation associated with the calibration of the first standard using an absolute measurement system. This first observation is the only source of information about the first standard and must be included in the optimal solution. The uncertainties associated with the remaining standards scale with the uncertainty associated with the first observation.

The first design, given in Table 4, is one defined by 'expert judgement', and represents a typical set of experiments that would be performed in practice. (In fact, for this set of standards, it is not entirely obvious how to choose a design that will calibrate all of the standards.) The next design, indicated by the first nine rows in Table 5, is that determined by the SSQR-GE algorithm with $\sigma_1 = \sigma_C = 1.0$ and $\sigma_i = \sigma_R = 0.5$, $i = 2, \ldots, n$. The most striking and, at first sight, surprising feature of this second design is that almost all the artefacts are involved in experiments 2 to n. The reason for this is that, in some sense, the comparator uncertainties represented by $\sigma_i = \sigma_R$ are partitioned amongst the artefacts involved, and more artefacts being included in an experiment means that each artefact is assigned a smaller uncertainty.

The two designs considered so far are based on the assumption that all the comparator measurements are associated with the same uncertainty σ_R, irrespective of the number of artefacts and the nominal values of the artefacts. A more plausible assignment of uncertainties would take into account both the number and nominal values. For a mass balance, it would be usual for the associated uncertainty to have a component that varied in proportion to the mass. Similarly, a length measuring device will usually have one or more influencing factors, e.g., refractive index effects for the case of a laser interferometric comparator [41], that vary in proportion to length. The use of multiple artefacts in each experiment will also likely introduce effects that will add to the uncertainties.

For each experiment, let n_i be the total number of artefacts involved and let

$$v_i = \sum_{q \in L_i \cup R_i} a_q^{\sharp},$$

be a measure of the total value of the quantities involved in the ith experiment. Given σ_R, σ_V and σ_N, we can assign σ_i according to

$$\sigma_i^2 = \sigma_R^2 + n_{i,2}\sigma_N^2 + v_i^2\sigma_V^2, \quad n_{i,2} = \max(n_i - 2, 0), \quad i = 2, \ldots, n. \tag{16}$$

Table 5 presents the optimal designs calculated using the SSQR-GE algorithm with σ_i calculated as in (16) for the four different values of σ_R, σ_N and σ_V shown in Table 6. The third design in Table 5 is the result penalising experiments with a large number of artefacts, while the fourth design in the table is the result of penalising experiments with a larger value of v_i.

Table 6 gives the uncertainties $u(a_i)$ associated with a_i for four different assignments of σ_i, $i = 2, \ldots, n = 9$ for the designs in Tables 4 and 5. Also shown in the table is the aggregate measure $\bar{d} = |V_{\boldsymbol{a}}|^{1/n}$ of uncertainty, the total number $\sum_i n_i$ of artefacts and a measure $\sum_i v_i$ of the total nominal values involved in each set of experiments. The last three rows give the values of σ_R, σ_N and σ_V used to calculate σ_i, $i = 2, \ldots, n$. In all cases, the SSQR-GE algorithm leads to significant improvements over the expert judgement design.

Table 4. Design of a calibration experiment given by 'expert judgement'.

1.00	0.50	0.50	0.20	0.20	1.0	1.0	0.05	0.05
1	0	0	0	0	0	0	0	0
1	−1	−1	0	0	0	0	0	0
0	1	−1	0	0	0	0	0	0
0	1	0	−1	−1	−1	0	0	0
0	0	1	−1	−1	0	−1	0	0
0	0	0	1	−1	0	0	0	0
0	0	0	1	0	0	0	−1	−1
0	0	0	0	0	1	0	−1	−1
0	0	0	0	0	0	0	1	−1

Table 5. Designs of four calibration experiments calculated using the SSQR-GE algorithm with uncertainties σ_i calculated according to (16) for the different values of σ_R, σ_N, and σ_V given in Table 6.

1.00	0.50	0.50	0.20	0.20	1.0	1.0	0.05	0.05
1	0	0	0	0	0	0	0	0
1	0	−1	−1	−1	−1	−1	1	1
1	−1	−1	1	−1	1	−1	1	−1
1	−1	−1	1	−1	−1	1	−1	1
1	−1	−1	−1	1	1	−1	−1	1
1	−1	−1	−1	1	−1	1	1	−1
0	1	0	−1	−1	0	0	−1	−1
0	1	−1	0	1	−1	0	−1	−1
0	0	1	−1	0	−1	−1	−1	−1
1	0	0	0	0	0	0	0	0
1	−1	−1	0	0	0	0	0	0
0	1	0	−1	−1	−1	−1	1	1
0	1	−1	1	−1	1	−1	−1	1
0	1	−1	0	1	−1	0	−1	−1
0	1	−1	0	−1	1	1	1	−1
0	0	1	−1	−1	0	0	−1	−1
0	0	0	1	0	−1	−1	1	−1
0	0	0	1	−1	−1	1	−1	1
1	0	0	0	0	0	0	0	0
1	−1	−1	0	0	0	0	0	0
0	1	0	−1	−1	0	−1	0	0
0	1	−1	0	0	0	0	0	0
0	0	0	1	0	−1	−1	0	0
0	0	0	1	−1	0	0	0	0
0	0	0	0	0	1	0	−1	−1
0	0	0	0	0	1	−1	0	0
0	0	0	0	0	0	0	1	−1
1	0	0	0	0	0	0	0	0
1	−1	−1	0	0	0	0	0	0
0	1	0	−1	−1	0	−1	0	0
0	1	−1	0	0	0	0	0	0
0	0	0	1	0	−1	−1	0	0
0	0	0	1	−1	0	0	0	0
0	0	0	0	0	1	0	−1	−1
0	0	0	0	0	1	−1	0	0
0	0	0	0	0	0	0	1	−1

Table 6. Uncertainties associated with a_i for four different assignments of σ_i, $i = 2, \ldots, n = 9$. The columns labelled E give the uncertainties associated with 'expert judgement' design given in Table 4, while the columns labelled 'O' give those associated with the optimal designs, given, in Table 5, calculated by the SSQR-GE algorithm. Also shown in the table are the aggregate measure $\bar{d} = |V_{\boldsymbol{a}}|^{1/n}$ of uncertainty, the total number $\sum_i n_i$ of artefacts, and a measure $\sum_i v_i$ of the total nominal values involved in each set of experiments. The last three rows give the values of σ_R, σ_N, and σ_V used to calculate σ_i, $i = 2, \ldots, n$.

$a_i^\sharp$	E	O1	E	02	E	O3	E	04
1.00	1.00	1.00	1.00	1.00	1.00	1.00	1.00	1.00
0.50	0.61	0.56	0.66	0.64	0.69	0.69	1.04	1.04
0.50	0.61	0.55	0.66	0.64	0.69	0.69	1.04	1.04
0.20	0.39	0.31	0.43	0.40	0.60	0.58	0.50	0.57
0.20	0.49	0.29	0.52	0.39	0.61	0.58	0.54	0.60
0.10	0.57	0.26	0.61	0.32	0.90	0.44	0.57	0.36
0.10	0.91	0.27	1.03	0.36	1.64	0.45	1.34	0.34
0.05	0.35	0.20	0.36	0.28	0.40	0.48	0.29	0.27
0.05	0.35	0.20	0.36	0.28	0.40	0.48	0.29	0.27
$\bar{d}$	0.17	0.06	0.21	0.12	0.21	0.13	0.21	0.15
$\sum v_i$	7.0	17.4	7.0	11.0	7.0	6.3	7.0	6.3
$\sum n_i$	24	62	24	48	24	22	24	22
σ_R	0.50	0.50	0.50	0.50	0.20	0.20	0.20	0.20
σ_N	0.00	0.00	0.20	0.20	0.80	0.80	0.20	0.20
σ_V	0.00	0.00	0.20	0.20	0.20	0.20	0.80	0.80

6. Concluding Remarks

In this paper, we have looked at the experimental design problem: given m potential observations to determine n parameters, $m > n$, what is the best choice of n observations. In the context of least squares estimation, the problem was formulated as finding the $n \times n$ submatrix of the complete $m \times n$ observation matrix that has maximum determinant, corresponding to the D-optimality criterion. We described two algorithms, the SSQR and GE algorithms, to address this problem. Both were adapted from numerical linear algebra algorithms associated with the QR factorisation of a matrix. We also described algorithms for updating estimates of the problem parameters that are sequentially optimal with the respect to the D-optimality and A-optimality criteria. We illustrated the behaviour of the algorithms on a number of applications drawn from the field of metrology. These algorithms are straightforward to implement. The SSQR algorithm, in particular, is a minor adaption of the standard QR factorisation algorithm with pivoting. The algorithms enable computationally efficient implementations, exploiting rank-one matrix updating techniques. The algorithms are descent-type algorithms that will converge to a local minimum that is not necessarily a global minimum. In the examples considered, the algorithms determined effective experimental designs, often with much better performance than 'hand-crafted' designs based on expert judgement.

Funding: This work was supported by the UK's National Measurement System programme for Data Science.

Data Availability Statement: No new data were created or analyzed in this study. Data sharing is not applicable to this article

Acknowledgments: The author is grateful to the anonymous referees for their helpful comments, and to the Czech Metrology Institute for permission to use their photograph of the hyperbolic paraboloid, Figure 6.

Conflicts of Interest: The author declares no conflict of interest.

Appendix A. Numerical Linear Algebra

Appendix A.1. Sherman-Morrison Formula

See [23]. For $n \times n$ full rank matrix A and n-vectors $\boldsymbol{u}$ and $\boldsymbol{v}$

$$\left(A + \boldsymbol{u}\boldsymbol{v}^\top\right)^{-1} = A^{-1} - \frac{A^{-1}\boldsymbol{u}\boldsymbol{v}^\top A^{-1}}{1 + \boldsymbol{v}^\top A^{-1}\boldsymbol{u}}.$$

If $\tilde{\boldsymbol{u}} = A^{-1}\boldsymbol{u}$ and $\tilde{\boldsymbol{v}} = A^{-\top}\boldsymbol{v}$, then

$$\left(A + \boldsymbol{u}\boldsymbol{v}^\top\right)^{-1} = A^{-1} - \frac{\tilde{\boldsymbol{u}}\tilde{\boldsymbol{v}}^\top}{1 + \boldsymbol{v}^\top \tilde{\boldsymbol{u}}} = A^{-1} - \frac{\tilde{\boldsymbol{u}}\tilde{\boldsymbol{v}}^\top}{1 + \tilde{\boldsymbol{v}}^\top \boldsymbol{u}}.$$

Appendix A.2. Determinant of a Rank One Update of a Matrix

See, e.g., [42,43]. For A and square matrix and n-vectors $\boldsymbol{u}$ and $\boldsymbol{v}$

$$|A + \boldsymbol{u}\boldsymbol{v}^\top| = |A|(1 + \boldsymbol{v}^\top A^{-1}\boldsymbol{u}).$$

Writing

$$A + \boldsymbol{u}\boldsymbol{v}^\top = A(I + (A^{-1}\boldsymbol{u})\boldsymbol{v}^\top),$$

it is sufficient to consider the case $A = I$: $|I + \boldsymbol{u}\boldsymbol{v}^\top| = 1 + \boldsymbol{v}^\top\boldsymbol{u}$. From the matrix equation

$$\begin{bmatrix} I & 0 \\ \boldsymbol{v}^\top & 1 \end{bmatrix}\begin{bmatrix} I + \boldsymbol{u}\boldsymbol{v}^\top & \boldsymbol{u} \\ 0 & 1 \end{bmatrix}\begin{bmatrix} I & 0 \\ -\boldsymbol{v}^\top & 1 \end{bmatrix} = \begin{bmatrix} I & \boldsymbol{u} \\ 0 & 1 + \boldsymbol{v}^\top\boldsymbol{u} \end{bmatrix}$$

it is seen that the determinant of the matrix on the right-hand side is equal to the determinant of the matrix in the middle of the left-hand side, establishing the result for the case $A = I$.

Appendix B. Legendre Polynomials

See, e.g., [44]. *Note*: in this section, the parameter vector $\boldsymbol{a} = (a_0, \ldots, a_n)^\top$ is an $n+1$ vector indices starting from 0, not 1.

The D-optimal points for calibration experiments involving a polynomial response function of degree n are defined in terms of solutions of the polynomial equation

$$F(x) = 0, \quad F(x) = (1 - x^2)\dot{L}_{n-1}(x) = 0,$$

where $\dot{L}_{n-1}(x)$ is the derivative of Legendre polynomial of degree $n-1$. Given an estimate x of a solution, an updated solution is given by Newton's method [17,45] according to

$$x := x - F(x)/\dot{F}(x), \quad \dot{F}(x) = -2x\dot{L}_{n-1} + (1 - x^2)\ddot{L}_{n-1}(x)$$

where $\ddot{L}$ denotes the second derivative of $L(x)$ with respect to x. Starting estimates for the $n+1$ solutions are given by the arcsine points:

$$x_i = \cos\left(\pi\frac{n-i}{n}\right), \quad i = 0, 1, \ldots, n.$$

Legendre polynomials $L_j(x)$ are orthogonal polynomial basis functions defined on the interval $[-1, 1]$ and can be evaluated using the three-term recurrence relationship, starting with $L_0(x) = 1$, $L_1(x) = x$, and for $j \geq 2$,

$$jL_j(x) = (2j-1)xL_{j-1}(x) - (j-1)L_{j-1}(x).$$

If $f(x)$ is a degree n polynomial given by a sum of Legendre polynomial basis functions

$$f(x, \boldsymbol{a}) = \sum_{j=0}^{n} a_j L_j(x),$$

then $\dot{f}(x)$, the derivative of f with respect to x, is a degree $n-1$ polynomial and can be written as

$$\dot{f}(x, \dot{\boldsymbol{a}}) = \sum_{j=0}^{n-1} \dot{a}_j L_j(x),$$

where $\dot{\boldsymbol{a}} = \dot{A}(a_1, \ldots, a_n)^\top$ is defined in terms of the latter n elements of the $n+1$ vector $\boldsymbol{a} = (a_0, a_1, \ldots, a_n)^\top$, and $\dot{A}$ is the $n \times n$ matrix constructed from diagonals formed from the sequence $1, 3, 5, \ldots, 2n-1$. The matrix $\dot{A}$ for the case $n = 10$ is

$$\dot{A} = \begin{bmatrix} 1 & 0 & 1 & 0 & 1 & 0 & 1 & 0 & 1 & 0 \\ 0 & 3 & 0 & 3 & 0 & 3 & 0 & 3 & 0 & 3 \\ 0 & 0 & 5 & 0 & 5 & 0 & 5 & 0 & 5 & 0 \\ 0 & 0 & 0 & 7 & 0 & 7 & 0 & 7 & 0 & 7 \\ 0 & 0 & 0 & 0 & 9 & 0 & 9 & 0 & 9 & 0 \\ 0 & 0 & 0 & 0 & 0 & 11 & 0 & 11 & 0 & 11 \\ 0 & 0 & 0 & 0 & 0 & 0 & 13 & 0 & 13 & 0 \\ 0 & 0 & 0 & 0 & 0 & 0 & 0 & 15 & 0 & 15 \\ 0 & 0 & 0 & 0 & 0 & 0 & 0 & 0 & 17 & 0 \\ 0 & 0 & 0 & 0 & 0 & 0 & 0 & 0 & 0 & 19 \end{bmatrix}$$

Similarly, the second derivative $\ddot{f}(x)$ can be written as

$$\ddot{f}(x, \ddot{\boldsymbol{a}}) = \sum_{j=0}^{n-2} \ddot{a}_j L_j(x),$$

where $\ddot{\boldsymbol{a}} = \ddot{A}(a_2, \ldots, a_n)^\top$ is defined in terms of the latter $n-1$ elements of $\boldsymbol{a} = (a_0, a_1, \ldots, a_n)^\top$, and $\ddot{A}$ is the $(n-1) \times (n-1)$ matrix with

$$\ddot{A} = \dot{A}(1:n-1, 1:n-1)\dot{A}(2:n, 2:n).$$

References

1. Atkinson, A.C.; Donev, A.N.; Tobias, R.D. *Optimum Experimental Designs, with SAS*; Oxford University Press: Oxford, UK, 2007.
2. Box, G.E.P.; Hunter, W.G.; Hunter, J.S. *Statistics for Experimenters : Design, Innovation and Discovery*, 2nd ed.; Wiley: Hoboken, NJ, USA, 2005.
3. Chaloner, K. Optimal Bayesian experimental design for linear models. *Ann. Stat.* **1984**, *12*, 283–300. [CrossRef]
4. Forbes, A.B.; Minh, H.D. Design of linear calibration experiments. *Measurement* **2013**, *46*, 3730–3736. [CrossRef]
5. Goos, P. *The Optimal Design of Blocked and Split-Plot Experiments*; Springer: New York, NY, USA, 2002.
6. Goos, P.; Jones, B. *Optimal Design of Experiments: A Case Study Approach*; John Wiley & Sons: New York, NY, USA, 2011.
7. Jones, B.; Nachtsheim, C.J. Effective Model Selection for Definitive Screening Designs. *Technometrics* **2017**, *59*, 319–329. [CrossRef]
8. Montgomery, D.C. *Design and Analysis of Experiments*, 8th ed.; John Wiley & Sons: New York, NY, USA, 2013.
9. Chretien, S.; Clarkson, P. A fast algorithm for the semi-definite relaxation of the state estimation problem in power grids. *J. Ind. Manag. Optim.* **2020**, *16*, 431–443. [CrossRef]
10. Kekatos, V.; Giannakis, G.B. A convex relaxation approach to optimal placement of phasor measurement units. In Proceedings of the 2011 4th IEEE International Workshop on Computational Advances in Multi-Sensor Adaptive Processing (CAMSAP), San Juan, PR, USA, 13–16 December 2011; pp. 145–148.
11. Berger, J.; Busser, T.; Dutykh, D.; Nathan Mendes, N. An efficient method to estimate sorption isotherm curve coefficients. *Inverse Probl. Sci. Eng.* **2019**, *27*, 735–772. [CrossRef]
12. Berger, J.; Dutykh, D.; Mendes, N. On the optimal experiment design for heat and moisture parameter estimation. *Exp. Therm. Fluid Sci.* **2017**, *81*, 109–122. .: 10.1016/j.expthermflusci.2016.10.008. [CrossRef]
13. Chernoff, H. *Sequential Analysis and Optimal Design*; Society for Industrial and Applied Mathematics: Philadelphia, PA, USA, 1972. [CrossRef]

14. Meyer, R.K.; Nachtsheim, C.J. The Coordinate Exchange Algorithm for Constructing Exact Optimal Designs. *Technometrics* **1995**, *37*, 60–69. [CrossRef]
15. Wald, A. Sequential Tests of Statistical Hypotheses. *Ann. Math. Stat.* **1945**, *16*, 117–186. [CrossRef]
16. Boyd, S.; Vandenberghe, L. *Convex Optimization*; Cambridge University Press: Cambridge, UK, 2004.
17. Gill, P.E.; Murray, W.; Wright, M.H. *Practical Optimization*; Academic Press: London, UK, 1981.
18. Vandenberghe, L.; Boyd, S. Semidefinite Programming. *SIAM Rev.* **1996**, *38*, 49–95. [CrossRef]
19. Barker, R.M.; Cox, M.G.; Forbes, A.B.; Harris, P.M. *Software Support for Metrology Best Practice Guide No. 4: Modelling Discrete Data and Experimental Data Analysis*. Technical Report DEM-ES 018; National Physical Laboratory: Teddington, UK, 2007.
20. BIPM. *The International System of Units (SI Brochure (EN))*, 9th ed.; BIPM: Sèvres, France, 2019.
21. BIPM; IEC; IFCC; ILAC; ISO; IUPAC; IUPAP; OIML. *Evaluation of Measurement Data—Guide to the Expression of Uncertainty in Measurement*; JCGM 100:2008; Joint Committee for Guides in Metrology: Sèvres, France, 2008.
22. BIPM; IEC; IFCC; ILAC; ISO; IUPAC; IUPAP; OIML. *Evaluation of Measurement Data—Supplement 2 to the "Guide to the Expression of Uncertainty in Measurement"—Extension to Any Number of Output Quantities*; JCGM 102:2011; Joint Committee for Guides in Metrology: Sèvres, France, 2011.
23. Golub, G.; Van Loan, C. *Matrix Computations*, 4th ed.; Johns Hopkins University Press: Baltimore, MD, USA, 2013.
24. Hart, G.W. *Multidimensional Analysis: Algebras and Systems for Science and Engineering*; Springer: Berlin, Germany, 1995.
25. Gu, M.; Eisenstat, S.C. Efficient Algorithms for Computing a Strong Rank-Revealing QR Factorization. *SIAM J. Sci. Comput.* **1996**, *17*, 848–869. [CrossRef]
26. Sherman, J.; Morrison, W.J. Adjustment of an Inverse Matrix Corresponding to a Change in One Element of a Given Matrix. *Ann. Math. Stat.* **1950**, *21*, 124–127. [CrossRef]
27. Wilkinson, J.H. *The Algebraic Eigenvalue Problem*; Oxford University Press, Inc.: New York, NY, USA, 1988.
28. Preston-Thomas, H. The International Temperature Scale of 1990 (ITS-90). *Metrologia* **1990**, *27*, 3–10. [CrossRef]
29. Bartlett, G.; Forbes, A.; Heaps, E.; Raby, A.C.; Yacoot, A. Spatial positioning correction for multi-axis nanopositioning stages. In Proceedings of the ASPE Convention and Expo, Indianapolis, IN, USA, 16–21 September 2022.
30. Pukelsheim, F. *Optimal Design of Experiments*; SIAM: Philadelphia, PA, USA, 2006; Reproduction of 1993 book published by John Wiley and Sons, New York.
31. Handscombe, D.C.; Mason, J.C. *Chebyshev Polynomials*; Chapman & Hall/CRC Press: London, UK, 2003.
32. Forbes, A.B.; Jagan, K.; Dunlevy, J.; Sousa, J.A. Optimization of sensor distribution using Gaussian processes. *Meas. Sens.* **2021**, *18*, 100128. [CrossRef]
33. Rasmussen, C.E.; Williams, C.K.I. *Gaussian Processes for Machine Learning*; MIT Press: Cambridge, MA, USA, 2006.
34. Linkeová, I.; Zelený, V. Application of ruled surfaces in freeform and gear metrology. *Acta Polytech.* **2021**, *61*, 99–109. [CrossRef]
35. Zelený, V.; Linkeová, I.; Skalnik, P. Calibration of freeform standard. In Proceedings of the 15th International Conference of the European Society for Precision Engineering and Nanotechnology, EUSPEN 2015, Leuven, Belgium, 1–5 June 2015; pp. 147–148.
36. Forbes, A.B. Parameter estimation based on least squares methods. In *Data Modeling for Metrology and Testing in Measurement Science*; Pavese, F., Forbes, A.B., Eds.; Birkhäuser-Boston: New York, NY, USA, 2009; pp. 147–176.
37. Forbes, A.B. Sensitivity analysis for Gaussian associated features. *Appl. Sci.* **2022**, *12*, 2808. [CrossRef]
38. Grabe, M. Note on the Application of the Method of Least Squares. *Metrologia* **1978**, *14*, 143. [CrossRef]
39. Hotelling, H. Some Improvements in Weighing and Other Experimental Techniques. *Ann. Math. Stat.* **1944**, *15*, 297–306. [CrossRef]
40. Nielsen, L. Evaluation of mass measurements in accordance with the GUM. *Metrologia* **2014**, *51*, S183. [CrossRef]
41. Lewis, A.J.; Hughes, E.B.; Aldred, P.J.E. Long term study of gauge block interferometer performance and gauge blocks stability. *Metrologia* **2010**, *47*, 473–486. [CrossRef]
42. Ding, J.; Zhou, A. Eigenvalues of rank-one updated matrices with some applications. *Appl. Math. Lett.* **2007**, *20*, 1223–1226. [CrossRef]
43. Hogben, L. (Ed.) *Handbook of Linear Algebra*; Chapman & Hall/CRC: Boca Raton, FL, USA, 2007.
44. Abramowitz, M.; Stegun, I.A. *Handbook of Mathematical Functions*; Dover: New York, NY, USA, 1964.
45. Fletcher, R. *Practical Methods of Optimization*, 2nd ed.; John Wiley and Sons: Chichester, UK, 1987.

algorithms MDPI

Article

Solving Least-Squares Problems via a Double-Optimal Algorithm and a Variant of the Karush–Kuhn–Tucker Equation for Over-Determined Systems

Chein-Shan Liu [1], Chung-Lun Kuo [1] and Chih-Wen Chang [2,*]

[1] Center of Excellence for Ocean Engineering, National Taiwan Ocean University, Keelung 20224, Taiwan; csliu@mail.ntou.edu.tw (C.-S.L.); eji1215@gmail.com (C.-L.K.)
[2] Department of Mechanical Engineering, National United University, Miaoli 36063, Taiwan
* Correspondence: cwchang@nuu.edu.tw

Abstract: A double optimal solution (DOS) of a least-squares problem $\mathbf{Ax} = \mathbf{b}$, $\mathbf{A} \in \mathbb{R}^{q\times n}$ with $q \neq n$ is derived in an $m + 1$-dimensional varying affine Krylov subspace (VAKS); two minimization techniques exactly determine the $m + 1$ expansion coefficients of the solution $\mathbf{x}$ in the VAKS. The minimal-norm solution can be obtained automatically regardless of whether the linear system is consistent or inconsistent. A new double optimal algorithm (DOA) is created; it is sufficiently time saving by inverting an $m \times m$ positive definite matrix at each iteration step, where $m \ll \min(n, q)$. The properties of the DOA are investigated and the estimation of residual error is provided. The residual norms are proven to be strictly decreasing in the iterations; hence, the DOA is absolutely convergent. Numerical tests reveal the efficiency of the DOA for solving least-squares problems. The DOA is applicable to least-squares problems regardless of whether $q < n$ or $q > n$. The Moore–Penrose inverse matrix is also addressed by adopting the DOA; the accuracy and efficiency of the proposed method are proven. The $m + 1$-dimensional VAKS is different from the traditional m-dimensional affine Krylov subspace used in the conjugate gradient (CG)-type iterative algorithms CGNR (or CGLS) and CGRE (or Craig method) for solving least-squares problems with $q > n$. We propose a variant of the Karush–Kuhn–Tucker equation, and then we apply the partial pivoting Gaussian elimination method to solve the variant, which is better than the original Karush–Kuhn–Tucker equation, the CGNR and the CGNE for solving over-determined linear systems. Our main contribution is developing a double-optimization-based iterative algorithm in a varying affine Krylov subspace for effectively and accurately solving least-squares problems, even for a dense and ill-conditioned matrix $\mathbf{A}$ with $q \ll n$ or $q \gg n$.

Keywords: linear least-squares problems; varying affine Krylov subspace; double optimal algorithm; minimal-norm solution; residual orthogonality; absolute convergence; Moore–Penrose inverse; Karush–Kuhn–Tucker equation

MSC: 15A06; 65F10

1. Introduction

One of the most important numerical methods to solve linear systems is the Krylov subspace method [1–7]; since the appearance of pioneering works in [8,9], it has been studied extensively. A lot of Krylov subspace methods were reported in [10–16].

Many algorithms have been used to solve the least-squares problem with a minimum norm [17,18]:

$$\min \|\mathbf{x}\| \;\; \text{s.t.} \;\; \mathbf{x} \in \arg\min_{\mathbf{x}} \|\mathbf{b} - \mathbf{Ax}\|, \tag{1}$$

where $\|\cdot\|$ denotes the Euclidean norm, $\mathbf{x} \in \mathbb{R}^n$ is unknown, and $\mathbf{A} \in \mathbb{R}^{q\times n}$ and $\mathbf{b} \in \mathbb{R}^q$ are given.

We are concerned with a varying affine Krylov subspace (VAKS) solution and its corresponding iterative algorithm for solving

$$\mathbf{A}\mathbf{x} = \mathbf{b}, \tag{2}$$

which is an over-determined system when $q > n$ and an under-determined system when $q < n$.

Most results for Equation (2) are directly extendable to Equation (1), whose solution is called the minimum-norm or pseudo-inverse solution and is given by $\mathbf{x} = \mathbf{A}^{\dagger}\mathbf{b}$, where $\mathbf{A}^{\dagger}$ denotes the pseudo-inverse of $\mathbf{A}$ satisfying the following Penrose equations:

$$\mathbf{A}\mathbf{A}^{\dagger}\mathbf{A} = \mathbf{A}, \tag{3}$$
$$\mathbf{A}^{\dagger}\mathbf{A}\mathbf{A}^{\dagger} = \mathbf{A}^{\dagger}, \tag{4}$$
$$(\mathbf{A}\mathbf{A}^{\dagger})^{\mathrm{T}} = \mathbf{A}\mathbf{A}^{\dagger}, \tag{5}$$
$$(\mathbf{A}^{\dagger}\mathbf{A})^{\mathrm{T}} = \mathbf{A}^{\dagger}\mathbf{A}. \tag{6}$$

The pseudo-inverse has been investigated and applied to solve linear least-squares problems, and some computational methods have been developed to compute the Moore–Penrose inverse matrix [19–24]. In terms of the Moore–Penrose inverse matrix $\mathbf{A}^{\dagger}$, the general solution of the least-squares problem is given by

$$\mathbf{x} = \mathbf{A}^{\dagger}\mathbf{b} + (\mathbf{I} - \mathbf{R}_A)\mathbf{z}, \tag{7}$$
$$\mathbf{R}_A = \mathbf{A}^{\dagger}\mathbf{A}, \tag{8}$$

where $\mathbf{z}$ is arbitrary. Because $\mathbf{A}^{\dagger}\mathbf{b}$ and $(\mathbf{I} - \mathbf{R}_A)\mathbf{z}$ are orthogonal, it follows that $\mathbf{x} = \mathbf{A}^{\dagger}\mathbf{b}$ is the unique solution of Equation (1).

QR factorization is a good method to solve Equation (2). The normalized Gram–Schmidt process is a modification of the original Gram–Schmidt orthogonalization by normalizing each vector after orthogonalization. Let $\mathbf{Q}$ denote a $q \times n$ matrix obtained by applying the normalized Gram–Schmidt process on the column vectors of $\mathbf{A}$:

$$\mathbf{Q} := [\mathbf{q}_1, \ldots, \mathbf{q}_n], \tag{9}$$

which possesses the following orthonormal property:

$$\mathbf{Q}^{\mathrm{T}}\mathbf{Q} = \mathbf{I}_n. \tag{10}$$

Hence, by $\mathbf{Q}\mathbf{R}\mathbf{x} = \mathbf{b}$, it is easy to solve $\mathbf{x}$ by applying the backward substitution method to

$$\mathbf{R}\mathbf{x} = \mathbf{Q}^{\mathrm{T}}\mathbf{b}. \tag{11}$$

However, the QR method is only applicable to an over-determined system with $q > n$. For an under-determined system with $q < n$, the QR method is not applicable because $\mathbf{R}$ is no longer an upper triangular matrix.

Recently, many scholars have proposed many algorithms to tackle least-squares problems, like relaxed greedy randomized coordinate descent methods [25], the parallel approximate pseudo-inverse matrix technique in conjunction with the explicit preconditioned conjugate gradient least squares scheme [26], the new randomized Gauss–Seidel method [27], the QR–Cholesky method, the implicit QR method, the bidiagonal implicit QR method [28], the randomized block coordinate descent algorithm [29], an iterative pre-conditioning technique [30], splitting-based randomized iterative methods [31], the projection method [32], the index-search-method-based inner–outer iterative algorithm [33], the greedy double subspace coordinate descent method [34], a distributed algorithm [35], and multivalued projections [36].

As just mentioned, there are many algorithms for solving linear least-squares problems. For small- and medium-sized problems, one can use QR-based methods, and, if the problem is rank-deficient or ill conditioned, singular value decomposition (SVD) can be adopted [37,38]. For large-sized least-squares problems, Krylov subspace methods are more efficient than direct methods if the matrix is sparse or otherwise structured. In the present paper, we plan to develop an efficient and accurate Krylov subspace method for large-sized least-squares problems with dense and maybe ill-conditioned matrices, not limited to sparse matrices. It is also interesting that the proposed iterative algorithm can solve least-squares problems with extremal cases of $q \ll n$ or $q \gg n$. In fact, for under-determined systems, many methods are not applicable, for instance, QR and conjugate gradient (CG)-type methods, for which the basis vectors of the Krylov subspace are orthogonalized implicitly.

In addition to Equation (1), Liu [39] considered the following minimization

$$\min_{\mathbf{x}} \left\{ f = \frac{\mathbf{x}^{\mathrm{T}}\mathbf{A}^{\mathrm{T}}\mathbf{A}\mathbf{x}}{(\mathbf{b}^{\mathrm{T}}\mathbf{A}\mathbf{x})^2} \right\} \tag{12}$$

for solving ill-posed linear systems of Equation (2) with $q = n$. Liu [40,41] employed the double optimization technique to iteratively solve inverse problems. Liu [42] replaced the Krylov subspace by the column space of $\mathbf{A}$ and derived an effective iterative algorithm based on the maximal projection in Equation (12). These results are quite promising. However, for a non-square system (2), there exists no similar method of double optimization in the literature.

It is well known that the least-squares solution of the minimization in Equation (1) is obtained by

$$\mathbf{A}^{\mathrm{T}}\mathbf{A}\mathbf{x} = \mathbf{A}^{\mathrm{T}}\mathbf{b}. \tag{13}$$

When the derivative of f in Equation (12) vs. $\mathbf{x}$ is equal to zero, we can obtain

$$\mathbf{A}^{\mathrm{T}}\mathbf{A}\mathbf{x} = \frac{\mathbf{x}^{\mathrm{T}}\mathbf{A}^{\mathrm{T}}\mathbf{A}\mathbf{x}}{\mathbf{b}^{\mathrm{T}}\mathbf{A}\mathbf{x}}\mathbf{A}^{\mathrm{T}}\mathbf{b}. \tag{14}$$

Taking the transpose of Equation (2) to $\mathbf{x}^{\mathrm{T}}\mathbf{A}^{\mathrm{T}} = \mathbf{b}^{\mathrm{T}}$ and inserting it into Equation (14), we can immediately obtain

$$\mathbf{A}^{\mathrm{T}}\mathbf{A}\mathbf{x} = \frac{\mathbf{b}^{\mathrm{T}}\mathbf{A}\mathbf{x}}{\mathbf{b}^{\mathrm{T}}\mathbf{A}\mathbf{x}}\mathbf{A}^{\mathrm{T}}\mathbf{b}, \tag{15}$$

which implies Equation (13) by $\mathbf{b}^{\mathrm{T}}\mathbf{A}\mathbf{x}/\mathbf{b}^{\mathrm{T}}\mathbf{A}\mathbf{x} = 1$. Many numerical methods for least-squares problems are based on Equation (1); no numerical methods for least-squares problems are based on Equation (12).

To solve the least-squares problem (2), the usual approach is the Karush–Kuhn–Tucker equation method, which sets Equation (2) to a square linear system but with a larger dimension $q + n$. Let

$$\mathbf{r} = \mathbf{b} - \mathbf{A}\mathbf{x} \tag{16}$$

be the residual vector. By considering the enforcing constraint:

$$\mathbf{A}^{\mathrm{T}}\mathbf{r} = \mathbf{0}, \tag{17}$$

which is equivalent to the normal form of Equation (2), Equations (16) and (17) constitute a $q + n$-dimensional linear system:

$$\begin{bmatrix} \mathbf{I}_q & \mathbf{A} \\ \mathbf{A}^{\mathrm{T}} & \mathbf{0} \end{bmatrix} \begin{bmatrix} \mathbf{r} \\ \mathbf{x} \end{bmatrix} = \begin{bmatrix} \mathbf{b} \\ \mathbf{0} \end{bmatrix}. \tag{18}$$

The advantage of the so-called Karush–Kuhn–Tucker equation [15] is that one can employ a linear solver to solve Equation (18) to obtain the least-squares solution of Equation (2); however, a disadvantage is that the dimension is enlarged to $q + n$.

A major novelty of the present paper is that we employ both Equations (1) and (12) to develop a novel numerical method in a varying affine Krylov subspace. The innovative contributions are as follows: this is the first time a new concept of varying affine Krylov subspaces has been introduced, and a double-optimal iterative algorithm has been derived based on the double optimization method to solve least-squares problems. A new variant of the Karush–Kuhn–Tucker equation can significantly improve the accuracy of solving least-squares problems by adopting the partial pivoting Gaussian elimination method.

In Section 2, the solution of Equation (2) is expanded in an $m+1$-dimensional varying affine Krylov subspace (VAKS), whose $m+1$ unknown coefficients are optimized in Section 3 by two merit functions. The double optimal algorithm (DOA) is developed in Section 4; we prove a key property of absolute convergence, and the decreasing quantity of residual squares is derived explicitly. Examples of linear least-squares problems are given and compared in Section 5; in addition, Moore–Penrose inverses of non-square matrices are discussed. In Section 6, we discuss a variant of the Karush–Kuhn–Tucker equation for solving over-determined linear systems, which is compared to conjugate-gradient-type iterative methods. Finally, Section 7 describes the conclusions.

2. A Varying Affine Krylov Subspace Method

We consider an $m+1$-dimensional varying affine Krylov subspace (VAKS) by

$$\begin{aligned}\mathcal{K}_m &:= \text{span}\{\mathbf{A}^{\mathrm{T}}\mathbf{A}\mathbf{u}_0,\ldots,(\mathbf{A}^{\mathrm{T}}\mathbf{A})^m\mathbf{u}_0\},\\ \mathcal{K}'_m &:= \text{span}\{\mathbf{u}_0,\mathcal{K}_m\},\end{aligned} \tag{19}$$

where

$$\mathbf{u}_0 := \mathbf{A}^{\mathrm{T}}\mathbf{b} \in \mathbb{R}^n, \tag{20}$$

and $\mathcal{K}'_m$ is equivalent to $\mathcal{K}_{m+1}(\mathbf{A}^{\mathrm{T}}\mathbf{A},\mathbf{u}_0)$.

Then, we expand the solution $\mathbf{x}$ of Equation (2) in $\mathcal{K}'_m$ by

$$\mathbf{x} = \alpha_0\mathbf{u}_0 + \sum_{k=1}^{m}\alpha_k\mathbf{u}_k \in \mathcal{K}'_m, \tag{21}$$

where $\mathbf{u}_k := (\mathbf{A}^{\mathrm{T}}\mathbf{A})^k\mathbf{u}_0$, $k=1,\ldots,m$, and α_0 and α_k are constants to be determined. We take $m \ll \min(n,q)$.

Let

$$\mathbf{U} := [\mathbf{u}_1,\ldots,\mathbf{u}_m] = [\mathbf{A}^{\mathrm{T}}\mathbf{A}\mathbf{u}_0,\ldots,(\mathbf{A}^{\mathrm{T}}\mathbf{A})^m\mathbf{u}_0] \tag{22}$$

be an $n\times m$ matrix. The Arnoldi process [15] as a normalized Gram–Schmidt process is used to normalize and orthogonalize the Krylov vectors $(\mathbf{A}^{\mathrm{T}}\mathbf{A})^j\mathbf{u}_0$, $j=1,\ldots,m$, such that the resultant vectors $\mathbf{u}_i$, $i=1,\ldots,m$ satisfy $\mathbf{u}_i^{\mathrm{T}}\mathbf{u}_j=\delta_{ij}$, $i,j=1,\ldots,m$, where δ_{ij} is the Kronecker delta symbol. Select m, give an initial $\mathbf{v}_0=\mathbf{A}^{\mathrm{T}}\mathbf{A}\mathbf{u}_0$, and compute $\mathbf{u}_1=\frac{\mathbf{v}_0}{\|\mathbf{v}_0\|}$:

$$\begin{aligned}&\text{Do } j=1:m-1,\\ &\mathbf{w}_j=\mathbf{A}^{\mathrm{T}}\mathbf{A}\mathbf{u}_j,\\ &\text{Do } i=1:j,\\ &h_{ij}=\mathbf{u}_i^{\mathrm{T}}\mathbf{w}_j,\\ &\mathbf{w}_j=\mathbf{w}_j-h_{ij}\mathbf{u}_i,\\ &\text{End do of } i,\\ &h_{j+1,j}=\|\mathbf{w}_j\|;\ \text{if } h_{j+1,j}=0\ \text{stop},\\ &\mathbf{u}_{j+1}=\frac{\mathbf{w}_j}{h_{j+1,j}},\\ &\text{End do of } j.\end{aligned} \tag{23}$$

Hence, Equation (21) can be written as

$$\mathbf{x} = \alpha_0 \mathbf{u}_0 + \mathbf{U}\boldsymbol{\alpha}, \tag{24}$$

where $\boldsymbol{\alpha} := (\alpha_1, \ldots, \alpha_m)^{\mathrm{T}}$ and $\mathbf{U}$ satisfies the orthonormal property:

$$\mathbf{U}^{\mathrm{T}}\mathbf{U} = \mathbf{I}_m. \tag{25}$$

For a consistent system, the exact solution $\mathbf{x}^*$ satisfies $\mathbf{A}\mathbf{x}^* = \mathbf{b}$ exactly. For an inconsistent system, the exact solution $\mathbf{x}^*$ exactly satisfies Equation (1) given by $\mathbf{x}^* = \mathbf{A}^{\dagger}\mathbf{b}$, because the equality in $\mathbf{A}\mathbf{x}^* = \mathbf{b}$ no longer holds.

On this occasion, we briefly recall the m-dimensional conjugate gradient (CG)-type iterative method with an approximation to Equation (2) with $q > n$. The CGNR (or CGLS) minimizes the energy function of the error [15,43]:

$$f(\mathbf{x}) = (\mathbf{A}^{\mathrm{T}}\mathbf{A}(\mathbf{x}^* - \mathbf{x}), (\mathbf{x}^* - \mathbf{x})), \tag{26}$$

where $(\mathbf{a}, \mathbf{b})$ denotes the inner product of vectors $\mathbf{a}$ and $\mathbf{b}$, and $\mathbf{x}^*$ is the exact solution of Equation (2). Overall, $\mathbf{x}$ in the affine Krylov subspace (AKS) is:

$$\mathbf{x}_0 + \mathcal{K}_m(\mathbf{A}^{\mathrm{T}}\mathbf{A}, \mathbf{A}^{\mathrm{T}}\mathbf{r}_0) := \mathbf{x}_0 + \mathrm{span}\{\mathbf{A}^{\mathrm{T}}\mathbf{r}_0, \mathbf{A}^{\mathrm{T}}\mathbf{A}\mathbf{A}^{\mathrm{T}}\mathbf{r}_0, \ldots, (\mathbf{A}^{\mathrm{T}}\mathbf{A})^{m-1}\mathbf{A}^{\mathrm{T}}\mathbf{r}_0\}, \tag{27}$$

where $\mathbf{r}_0 = \mathbf{b} - \mathbf{A}\mathbf{x}_0$. There is some extension of the CGNR (or CGLS) to bounded perturbation resilient iterative methods [44].

Craig's idea was to seek the solution of Equation (2) with $q > n$ by

$$\mathbf{x} = \arg\min_{\mathbf{z}\in\mathbb{R}^n} \|\mathbf{z}\| \quad \text{subject to } \mathbf{A}\mathbf{z} = \mathbf{b}. \tag{28}$$

By introducing $\mathbf{x} = \mathbf{A}^{\mathrm{T}}\mathbf{y}$, we seek $\mathbf{y}$ in the AKS by

$$\mathbf{y}_0 + \mathcal{K}_m(\mathbf{A}\mathbf{A}^{\mathrm{T}}, \mathbf{r}_0), \tag{29}$$

to minimize

$$f(\mathbf{y}) = (\mathbf{A}\mathbf{A}^{\mathrm{T}}(\mathbf{y}^* - \mathbf{y}), (\mathbf{y}^* - \mathbf{y})), \tag{30}$$

where $\mathbf{y}^*$ is the exact solution of $\mathbf{A}\mathbf{A}^{\mathrm{T}}\mathbf{y} = \mathbf{b}$. The above statement for $\mathbf{x}^*$ is also valid for $\mathbf{y}^*$ with $\mathbf{A}\mathbf{x}^* = \mathbf{b}$ replaced by $\mathbf{A}\mathbf{A}^{\mathrm{T}}\mathbf{y}^* = \mathbf{b}$. The resulting CG iterative algorithm is known as the CGNE (or the Craig method), which minimizes the error function [15,43].

Notice that the affine Krylov subspaces in Equations (27) and (29) are fixed AKSs upon setting $\mathbf{x}_0$ and $\mathbf{y}_0$, while in Equation (21), the AKS is varying, denoted as a VAKS, owing to the translation vector $\alpha_0\mathbf{u}_0$ not being fixed to $\mathbf{u}_0$:

$$\mathcal{K}_m' = \mathcal{K}_{m+1}(\mathbf{A}^{\mathrm{T}}\mathbf{A}, \mathbf{A}^{\mathrm{T}}\mathbf{b}) = \mathrm{span}\{\mathbf{A}^{\mathrm{T}}\mathbf{b}, \mathbf{A}^{\mathrm{T}}\mathbf{A}\mathbf{A}^{\mathrm{T}}\mathbf{b}, \ldots, (\mathbf{A}^{\mathrm{T}}\mathbf{A})^m\mathbf{A}^{\mathrm{T}}\mathbf{b}\}. \tag{31}$$

Upon comparing Equation (21), whose coefficient α_0 before $\mathbf{u}_0$ is an unknown constant, to Equations (27) and (29), whose coefficients before $\mathbf{x}_0$ and $\mathbf{y}_0$ are fixed to one, the affine Krylov subspace in Equation (21) is a varying affine Krylov subspace (VAKS). The dimension of the VAKS is one more than that of the AKS.

3. A Double Optimal Solution

In this section, we adopt the new concept of the VKAS to derive a double optimal solution of Equation (2). The starting point is Equation (21), rather than $\mathbf{x} = \mathbf{u}_0 + \sum_{k=1}^{m} \alpha_k \mathbf{u}_k \in \mathbf{u}_0 + \mathcal{K}_m$.

3.1. Two Minimized Functions

Let

$$\mathbf{y} := \mathbf{A}\mathbf{x}, \tag{32}$$

such that the approximation to $\mathbf{b}$ in Equation (2) is accounted by the error vector:

$$\mathbf{e} := \mathbf{b} - \mathbf{b}^{\mathrm{T}}\left(\frac{\mathbf{y}}{\|\mathbf{y}\|}\right)\frac{\mathbf{y}}{\|\mathbf{y}\|}; \tag{33}$$

the squared-norm of $\mathbf{e}$ reads as

$$\|\mathbf{e}\|^2 = \|\mathbf{b}\|^2 - \frac{(\mathbf{b}^{\mathrm{T}}\mathbf{y})^2}{\|\mathbf{y}\|^2} \geq 0. \tag{34}$$

We maximize

$$\max_{\mathbf{y}}\left\{\frac{(\mathbf{b}^{\mathrm{T}}\mathbf{y})^2}{\|\mathbf{y}\|^2}\right\} \tag{35}$$

as the maximal projection of $\mathbf{b}$ on the direction $\mathbf{y}/\|\mathbf{y}\|$, such that $\|\mathbf{e}\|^2$ is minimized. Equation (35) is equivalent to the minimization of

$$\min_{\mathbf{y}}\left\{f := \frac{\|\mathbf{y}\|^2}{(\mathbf{b}^{\mathrm{T}}\mathbf{y})^2}\right\} \geq \frac{1}{\|\mathbf{b}\|^2}. \tag{36}$$

Let $\mathbf{J}$ be

$$\mathbf{J} := \mathbf{A}\mathbf{U} \in \mathbb{R}^{q\times m}. \tag{37}$$

Then, with the aid of Equation (24), Equation (32) can be written as

$$\mathbf{y} = \mathbf{y}_0 + \mathbf{J}\boldsymbol{\alpha}, \tag{38}$$

where

$$\mathbf{y}_0 := \alpha_0\mathbf{A}\mathbf{u}_0 \in \mathbb{R}^q. \tag{39}$$

Inserting Equation (38) for $\mathbf{y}$ into Equation (36) yields

$$\min_{\boldsymbol{\alpha}}\left\{f = \frac{\|\mathbf{y}\|^2}{(\mathbf{b}^{\mathrm{T}}\mathbf{y})^2} = \frac{\|\mathbf{y}_0 + \mathbf{J}\boldsymbol{\alpha}\|^2}{(\mathbf{b}^{\mathrm{T}}\mathbf{y}_0 + \mathbf{b}^{\mathrm{T}}\mathbf{J}\boldsymbol{\alpha})^2}\right\}, \tag{40}$$

which is used to find $\boldsymbol{\alpha}$. However, α_0 in $\mathbf{y}_0 = \alpha_0\mathbf{A}\mathbf{u}_0$ is still an unknown value to be determined by imposing the minimization of squared-norm of residuals, shortened to the residual square:

$$\min_{\alpha_0}\{\|\mathbf{r}\|^2 = \|\mathbf{b} - \mathbf{A}\mathbf{x}\|^2\}. \tag{41}$$

3.2. Mathematical Preliminaries

To determine the unknowns α_j, $j = 0, 1, \ldots, m$, the following main theorem is an extension of [39] to a non-square system (2).

Theorem 1. *Assume that $m_0 = rank(\mathbf{A}) \leq \min(n, q)$. $\mathbf{U}$ satisfies Equation (25) with $m < m_0$. $\mathbf{x} \in \mathcal{K}'_m$, approximately satisfying Equation (2) derived from two minimizations (40) and (41), is a double optimal solution (DOS), given by*

$$\mathbf{x} = \alpha_0(\mathbf{u}_0 - \mathbf{V}\mathbf{A}\mathbf{u}_0 + \lambda_0\mathbf{V}\mathbf{b}), \tag{42}$$

where

$$\mathbf{u}_0 = \mathbf{A}^T\mathbf{b},\ \mathbf{J} = \mathbf{A}\mathbf{U},\ \mathbf{C} = \mathbf{J}^T\mathbf{J},\ \mathbf{D} = \mathbf{C}^{-1},\ \mathbf{V} = \mathbf{U}\mathbf{D}\mathbf{J}^T,\ \mathbf{P} = \mathbf{A}\mathbf{V} = \mathbf{J}\mathbf{D}\mathbf{J}^T,$$

$$\lambda_0 = \frac{\|\mathbf{A}\mathbf{u}_0\|^2 - \mathbf{u}_0^T\mathbf{A}^T\mathbf{P}\mathbf{A}\mathbf{u}_0}{\mathbf{b}^T\mathbf{A}\mathbf{u}_0 - \mathbf{b}^T\mathbf{P}\mathbf{A}\mathbf{u}_0}, \tag{43}$$

$$\mathbf{w} = \lambda_0\mathbf{P}\mathbf{b} + \mathbf{A}\mathbf{u}_0 - \mathbf{P}\mathbf{A}\mathbf{u}_0, \tag{44}$$

$$\alpha_0 = \frac{\mathbf{w}^T\mathbf{b}}{\|\mathbf{w}\|^2}. \tag{45}$$

To guarantee the existence of $\mathbf{D} = \mathbf{C}^{-1}$*, the dimension* m *of* $\mathcal{K}_m$ *follows* $m < m_0$.

Proof. In view of Equation (38), we can write

$$\mathbf{b}^{\mathrm{T}}\mathbf{y} = \mathbf{b}^{\mathrm{T}}\mathbf{y}_0 + \mathbf{b}^{\mathrm{T}}\mathbf{J}\boldsymbol{\alpha}, \tag{46}$$

$$\|\mathbf{y}\|^2 = \|\mathbf{y}_0\|^2 + 2\mathbf{y}_0^{\mathrm{T}}\mathbf{J}\boldsymbol{\alpha} + \boldsymbol{\alpha}^{\mathrm{T}}\mathbf{C}\boldsymbol{\alpha} \tag{47}$$

for f in Equation (40), where

$$\mathbf{C} := \mathbf{J}^{\mathrm{T}}\mathbf{J} \tag{48}$$

is an $m \times m$ positive definite matrix; hence, $\mathbf{C}^{-1}$ exists. For the rank deficiency matrix $\mathbf{A}$, we suppose that its rank is $m_0 = \mathrm{rank}(\mathbf{A})$, where $m_0 < \min(n,q)$; to guarantee that $\mathbf{C}$ is an $m \times m$ positive definite matrix, m must be smaller than m_0.

The minimality condition of Equation (40) is

$$(\mathbf{b}^{\mathrm{T}}\mathbf{y})^2\nabla_{\boldsymbol{\alpha}}\|\mathbf{y}\|^2 - 2\mathbf{b}^{\mathrm{T}}\mathbf{y}\|\mathbf{y}\|^2\nabla_{\boldsymbol{\alpha}}(\mathbf{b}^{\mathrm{T}}\mathbf{y}) = \mathbf{0}; \tag{49}$$

$\nabla_{\boldsymbol{\alpha}}$ is the gradient with respect to $\boldsymbol{\alpha}$. Cancelling $\mathbf{b}^{\mathrm{T}}\mathbf{y}$ in Equation (49) yields

$$\mathbf{b}^{\mathrm{T}}\mathbf{y}\mathbf{y}_2 - 2\|\mathbf{y}\|^2\mathbf{y}_1 = \mathbf{0}, \tag{50}$$

where

$$\mathbf{y}_1 := \nabla_{\boldsymbol{\alpha}}(\mathbf{b}^{\mathrm{T}}\mathbf{y}) = \mathbf{J}^{\mathrm{T}}\mathbf{b}, \tag{51}$$

$$\mathbf{y}_2 := \nabla_{\boldsymbol{\alpha}}\|\mathbf{y}\|^2 = 2\mathbf{J}^{\mathrm{T}}\mathbf{y}_0 + 2\mathbf{C}\boldsymbol{\alpha}. \tag{52}$$

From Equation (50), $\mathbf{y}_2$ is observed to be proportional to $\mathbf{y}_1$ in the m-dimensional space $\mathbb{R}^m$, which is supposed to be

$$\mathbf{y}_2 = \frac{2\|\mathbf{y}\|^2}{\mathbf{b}^{\mathrm{T}}\mathbf{y}}\mathbf{y}_1 = 2\lambda\mathbf{y}_1, \tag{53}$$

where 2λ is a proportional factor determined below. The second equality leads to

$$\|\mathbf{y}\|^2 = \lambda\mathbf{b}^{\mathrm{T}}\mathbf{y}. \tag{54}$$

Then, by Equations (51)–(53), we have

$$\boldsymbol{\alpha} = \lambda\mathbf{D}\mathbf{J}^{\mathrm{T}}\mathbf{b} - \mathbf{D}\mathbf{J}^{\mathrm{T}}\mathbf{y}_0, \tag{55}$$

where

$$\mathbf{D} := \mathbf{C}^{-1} = (\mathbf{J}^{\mathrm{T}}\mathbf{J})^{-1} > 0 \tag{56}$$

is an $m \times m$ matrix. It follows from Equations (46), (47) and (55) that

$$\mathbf{b}^{\mathrm{T}}\mathbf{y} = \mathbf{b}^{\mathrm{T}}\mathbf{y}_0 + \lambda\mathbf{b}^{\mathrm{T}}\mathbf{P}\mathbf{b} - \mathbf{b}^{\mathrm{T}}\mathbf{P}\mathbf{y}_0, \tag{57}$$

$$\|\mathbf{y}\|^2 = \lambda^2\mathbf{b}^{\mathrm{T}}\mathbf{P}\mathbf{b} + \|\mathbf{y}_0\|^2 - \mathbf{y}_0^{\mathrm{T}}\mathbf{P}\mathbf{y}_0, \tag{58}$$

where

$$\mathbf{P} := \mathbf{J}\mathbf{D}\mathbf{J}^{\mathrm{T}} \tag{59}$$

is a $q \times q$ positive semi-definite matrix. $\mathbf{P}$ is indeed a projection operator.

From Equations (54), (57) and (58), λ is solved from

$$\|\mathbf{y}_0\|^2 - \mathbf{y}_0^{\mathrm{T}}\mathbf{P}\mathbf{y}_0 = \lambda(\mathbf{b}^{\mathrm{T}}\mathbf{y}_0 - \mathbf{b}^{\mathrm{T}}\mathbf{P}\mathbf{y}_0) \tag{60}$$

as follows:

$$\lambda = \frac{\|\mathbf{y}_0\|^2 - \mathbf{y}_0^{\mathrm{T}}\mathbf{P}\mathbf{y}_0}{\mathbf{b}^{\mathrm{T}}\mathbf{y}_0 - \mathbf{b}^{\mathrm{T}}\mathbf{P}\mathbf{y}_0}. \tag{61}$$

Inserting this into Equation (55) generates

$$\boldsymbol{\alpha} = \frac{\|\mathbf{y}_0\|^2 - \mathbf{y}_0^{\mathrm{T}}\mathbf{P}\mathbf{y}_0}{\mathbf{b}^{\mathrm{T}}\mathbf{b}\mathbf{y}_0 - \mathbf{b}^{\mathrm{T}}\mathbf{P}\mathbf{y}_0}\mathbf{D}\mathbf{J}^{\mathrm{T}}\mathbf{b} - \mathbf{D}\mathbf{J}^{\mathrm{T}}\mathbf{y}_0. \tag{62}$$

By inserting the above $\boldsymbol{\alpha}$ into Equation (24) and using Equation (39) for $\mathbf{y}_0$, we can obtain

$$\mathbf{x} = \alpha_0(\mathbf{u}_0 - \mathbf{V}\mathbf{A}\mathbf{u}_0 + \lambda_0\mathbf{V}\mathbf{b}), \tag{63}$$

where

$$\mathbf{V} := \mathbf{U}\mathbf{D}\mathbf{J}^{\mathrm{T}}, \tag{64}$$

$$\lambda_0 := \frac{\|\mathbf{A}\mathbf{u}_0\|^2 - \mathbf{u}_0^{\mathrm{T}}\mathbf{A}^{\mathrm{T}}\mathbf{P}\mathbf{A}\mathbf{u}_0}{\mathbf{b}^{\mathrm{T}}\mathbf{A}\mathbf{u}_0 - \mathbf{b}^{\mathrm{T}}\mathbf{P}\mathbf{A}\mathbf{u}_0}. \tag{65}$$

Upon letting

$$\mathbf{v} := \mathbf{u}_0 - \mathbf{V}\mathbf{A}\mathbf{u}_0 + \lambda_0\mathbf{V}\mathbf{b}, \tag{66}$$

$\mathbf{x}$ in Equation (63) can be expressed as

$$\mathbf{x} = \alpha_0\mathbf{v}, \tag{67}$$

where α_0 is determined by inserting Equation (67) into Equation (41):

$$\|\mathbf{b} - \mathbf{A}\mathbf{x}\|^2 = \|\alpha_0\mathbf{A}\mathbf{v} - \mathbf{b}\|^2 = \alpha_0^2\|\mathbf{w}\|^2 - 2\alpha_0\mathbf{w}^{\mathrm{T}}\mathbf{b} + \|\mathbf{b}\|^2, \tag{68}$$

where

$$\mathbf{w} := \mathbf{A}\mathbf{v} = \mathbf{A}\mathbf{u}_0 - \mathbf{A}\mathbf{V}\mathbf{A}\mathbf{u}_0 + \lambda_0\mathbf{A}\mathbf{V}\mathbf{b}. \tag{69}$$

In view of Equations (37) and (64), $\mathbf{P}$ in Equation (59) can be written as

$$\mathbf{P} = \mathbf{A}\mathbf{V}, \tag{70}$$

such that $\mathbf{w}$ can be written as

$$\mathbf{w} = \lambda_0\mathbf{P}\mathbf{b} + \mathbf{A}\mathbf{u}_0 - \mathbf{P}\mathbf{A}\mathbf{u}_0.$$

This is just $\mathbf{w}$ in Equation (44).

Set the differential of Equation (68) vs. α_0 to zero, obtaining

$$\alpha_0 = \frac{\mathbf{w}^{\mathrm{T}}\mathbf{b}}{\|\mathbf{w}\|^2}; \tag{71}$$

hence, the proof of Equation (42) in Theorem 1 is complete. □

In the proof of Theorem 1, we have used the minimality condition of Equation (40) to determine $\boldsymbol{\alpha}$ in Equation (62), and then the minimality condition of Equation (41) was used to determine α_0 in Equation (71). Two optimizations were used to derive $\mathbf{x}$ in Equation (42), which is thus named a double optimal solution (DOS).

Theorem 2. *The two factors λ_0 in Equation (43) and α_0 in Equation (45) satisfy*

$$\alpha_0\lambda_0 = 1. \tag{72}$$

Proof. It follows from Equations (59) and (56) that

$$\mathbf{P}^2 = \mathbf{P}; \tag{73}$$

hence, $\mathbf{P}$ is a projection operator.

In Equation (44), $\mathbf{w}$ can be written as

$$\mathbf{w} = \lambda_0 \mathbf{P}\mathbf{b} + (\mathbf{I}_q - \mathbf{P})\mathbf{A}\mathbf{u}_0. \tag{74}$$

Utilizing Equation (73), $\mathbf{P}^{\mathrm{T}} = \mathbf{P}$, and $(\mathbf{I}_q - \mathbf{P})^2 = \mathbf{I}_q - \mathbf{P}$, it follows that

$$\|\mathbf{w}\|^2 = \lambda_0^2 \mathbf{b}^{\mathrm{T}}\mathbf{P}\mathbf{b} + \mathbf{u}_0^{\mathrm{T}}\mathbf{A}^{\mathrm{T}}\mathbf{A}\mathbf{u}_0 - \mathbf{u}_0^{\mathrm{T}}\mathbf{A}^{\mathrm{T}}\mathbf{P}\mathbf{A}\mathbf{u}_0, \tag{75}$$

$$\mathbf{b}^{\mathrm{T}}\mathbf{w} = \lambda_0 \mathbf{b}^{\mathrm{T}}\mathbf{P}\mathbf{b} + \mathbf{b}^{\mathrm{T}}\mathbf{A}\mathbf{u}_0 - \mathbf{b}^{\mathrm{T}}\mathbf{P}\mathbf{A}\mathbf{u}_0. \tag{76}$$

With the aid of Equation (65), Equation (76) is further reduced to

$$\mathbf{b}^{\mathrm{T}}\mathbf{w} = \lambda_0 \mathbf{b}^{\mathrm{T}}\mathbf{P}\mathbf{b} + \frac{1}{\lambda_0}(\mathbf{u}_0^{\mathrm{T}}\mathbf{A}^{\mathrm{T}}\mathbf{A}\mathbf{u}_0 - \mathbf{u}_0^{\mathrm{T}}\mathbf{A}^{\mathrm{T}}\mathbf{P}\mathbf{A}\mathbf{u}_0). \tag{77}$$

Now, after inserting Equation (77) into Equation (71), we can obtain

$$\alpha_0\lambda_0\|\mathbf{w}\|^2 = \lambda_0^2 \mathbf{b}^{\mathrm{T}}\mathbf{P}\mathbf{b} + \mathbf{u}_0^{\mathrm{T}}\mathbf{A}^{\mathrm{T}}\mathbf{A}\mathbf{u}_0 - \mathbf{u}_0^{\mathrm{T}}\mathbf{A}^{\mathrm{T}}\mathbf{P}\mathbf{A}\mathbf{u}_0, \tag{78}$$

which is just equal to $\|\mathbf{w}\|^2$ by Equation (75). Hence, we obtain Equation (72) readily. The proof of $\alpha_0\lambda_0 = 1$ is complete. □

Remark 1. *The DOS is derived in the VAKS $\mathcal{K}'_m$ in a way to stick onto Equation (21). If one takes $\alpha_0 = 1$ in Equation (21), $\mathcal{K}'_m$ reduces to the usual affine Krylov subspace $\mathbf{u}_0 + \mathcal{K}_m$. However, in this situation, Theorem 1 no longer holds. If one takes $\mathbf{u}_0 = \mathbf{0}$ and works directly in the $(m+1)$-dimensional Krylov $\mathcal{K}_{m+1}$ by expanding $\boldsymbol{\alpha}$ in an $(m+1)$-dimensional subspace, it would result in failure, because after inserting $\mathbf{y}_0 = \mathbf{A}\mathbf{u}_0 = \mathbf{0}$ into Equation (62), $\boldsymbol{\alpha}$ cannot be defined. We can conclude that the DOS directly benefits from the VAKS.*

The following corollaries are proven to help us understand the properties of DOS whose bases appeared in Theorems 1 and 2 and equations appeared in the proof of Theorem 1.

Corollary 1. *The multiplier λ defined by Equation (61) satisfies*

$$\lambda = 1, \tag{79}$$

and the double optimal solution (42) can be written as

$$\mathbf{x} = \alpha_0\mathbf{u}_0 - \alpha_0\mathbf{V}\mathbf{A}\mathbf{u}_0 + \mathbf{V}\mathbf{b}. \tag{80}$$

Proof. Inserting Equation (39) for $\mathbf{y}_0$ into Equation (61), we have

$$\lambda = \alpha_0 \frac{\|\mathbf{A}\mathbf{u}_0\|^2 - \mathbf{u}_0^{\mathrm{T}}\mathbf{A}^{\mathrm{T}}\mathbf{P}\mathbf{A}\mathbf{u}_0}{\mathbf{b}^{\mathrm{T}}\mathbf{A}\mathbf{u}_0 - \mathbf{b}^{\mathrm{T}}\mathbf{P}\mathbf{A}\mathbf{u}_0}, \tag{81}$$

in view of Equation (65), which can be written as

$$\lambda = \alpha_0\lambda_0. \tag{82}$$

Hence, $\lambda = 1$ follows immediately from Equation (72). Then, using Equations (72) and (42) proves Equation (80). □

Remark 2. *Inserting Equation (20) for* $\mathbf{u}_0$ *into Equation (80) yields*

$$\mathbf{x} = [\mathbf{V} + \alpha_0\mathbf{A}^T - \alpha_0\mathbf{V}\mathbf{A}\mathbf{A}^T]\mathbf{b},$$

which is a pseudo-inverse solution obtained from the DOS.

Corollary 2. *If* $m = n < q$ *and* $\mathbf{A}^T\mathbf{A}$ *is non-singular, then the double optimal solution* $\mathbf{x}$ *of Equation (2) is the Penrose solution, given by*

$$\mathbf{x} = \mathbf{A}^\dagger\mathbf{b}, \tag{83}$$

where $\mathbf{A}^\dagger$ *is the Moore–Penrose pseudo-inverse of* $\mathbf{A}$.

Proof. If $m = n$, then $\mathbf{J} = \mathbf{A}\mathbf{U}$ defined by Equation (37) is a $q \times n$ matrix. Meanwhile, $\mathbf{U}$ defined by Equation (22) is a rank n orthogonal matrix due to Equation (25):

$$\mathbf{U}^\mathrm{T}\mathbf{U} = \mathbf{U}\mathbf{U}^\mathrm{T} = \mathbf{I}_n. \tag{84}$$

From Equations (56) and (84), it follows that

$$\mathbf{D} = \mathbf{U}^\mathrm{T}(\mathbf{A}^\mathrm{T}\mathbf{A})^{-1}\mathbf{U}. \tag{85}$$

Inserting $\mathbf{J} = \mathbf{A}\mathbf{U}$, $\mathbf{V} = \mathbf{U}\mathbf{D}\mathbf{J}^\mathrm{T}$ and Equation (85) into Equation (80), we have

$$\mathbf{x} = \alpha_0\mathbf{u}_0 + \mathbf{U}\mathbf{U}^\mathrm{T}(\mathbf{A}^\mathrm{T}\mathbf{A})^{-1}\mathbf{U}\mathbf{U}^\mathrm{T}\mathbf{A}^\mathrm{T}\mathbf{b} - \alpha_0\mathbf{U}\mathbf{U}^\mathrm{T}(\mathbf{A}^\mathrm{T}\mathbf{A})^{-1}\mathbf{U}\mathbf{U}^\mathrm{T}\mathbf{A}^\mathrm{T}\mathbf{A}\mathbf{u}_0. \tag{86}$$

By using Equation (84), we can derive

$$\mathbf{x} = (\mathbf{A}^\mathrm{T}\mathbf{A})^{-1}\mathbf{A}^\mathrm{T}\mathbf{b}, \tag{87}$$

which is a Penrose solution of Equation (2). □

Corollary 2 reveals that the DOS can give the pseudo-inverse solution. SVD can offer a pseudo-inverse solution as follows [37]. In SVD, the matrix $\mathbf{A}$ is decomposed by

$$\mathbf{A} = \mathbf{U}\mathbf{D}\mathbf{V}^\mathrm{T}, \tag{88}$$

where $\mathbf{U}$ is a $q \times q$ orthonormal matrix, $\mathbf{V}$ is an $n \times n$ orthonormal matrix, and $\mathbf{D}$ is a $q \times n$ pseudo-diagonal matrix with non-zero singular values on the diagonal and all other elements are zero. Suppose that $\mathbf{A}^\mathrm{T}\mathbf{A}$ is non-singular. Then, Equation (87) is a pseudo-inverse solution, which in terms of SVD can be derived as follows:

$$\mathbf{x} = \mathbf{V}(\mathbf{D}^\mathrm{T}\mathbf{D})^{-1}\mathbf{D}^\mathrm{T}\mathbf{U}^\mathrm{T}\mathbf{b} = \mathbf{V}\mathbf{D}^\dagger\mathbf{U}^\mathrm{T}\mathbf{b}; \tag{89}$$

letting

$$\mathbf{A}^\dagger = \mathbf{V}\mathbf{D}^\dagger\mathbf{U}^\mathrm{T}, \tag{90}$$

we can derive Equation (83) in Corollary 2. Mathematically, the DOS is equivalent to the SVD solution in Equation (89), even though they are different approaches to the least-squares problem.

3.3. Estimation of Residual Error

Equation (80) is written as

$$\mathbf{x} = \alpha_0(\mathbf{u}_0 - \mathbf{V}\mathbf{A}\mathbf{u}_0) + \mathbf{V}\mathbf{b}, \tag{91}$$

where

$$\alpha_0 = \frac{1}{\lambda_0} = \frac{\mathbf{b}^\mathrm{T}\mathbf{A}\mathbf{u}_0 - \mathbf{b}^\mathrm{T}\mathbf{P}\mathbf{A}\mathbf{u}_0}{\|\mathbf{A}\mathbf{u}_0\|^2 - \mathbf{u}_0^\mathrm{T}\mathbf{A}^\mathrm{T}\mathbf{P}\mathbf{A}\mathbf{u}_0}, \tag{92}$$

by using Equations (65) and (72).

Let us investigate the residual square:

$$\|\mathbf{b}-\mathbf{A}\mathbf{x}\|^2 = \|\mathbf{y}-\mathbf{b}\|^2 = \|\mathbf{y}\|^2 - 2\mathbf{b}^{\mathrm{T}}\mathbf{y} + \|\mathbf{b}\|^2, \tag{93}$$

where

$$\mathbf{y} = \mathbf{A}\mathbf{x} = \alpha_0\mathbf{A}\mathbf{u}_0 - \alpha_0\mathbf{P}\mathbf{A}\mathbf{u}_0 + \mathbf{P}\mathbf{b}. \tag{94}$$

is obtained from Equation (91) by using Equation (70).

From Equation (94), it follows that

$$\|\mathbf{y}\|^2 = \alpha_0^2(\mathbf{u}_0^{\mathrm{T}}\mathbf{A}^{\mathrm{T}}\mathbf{A}\mathbf{u}_0 - \mathbf{u}_0^{\mathrm{T}}\mathbf{A}^{\mathrm{T}}\mathbf{P}\mathbf{A}\mathbf{u}_0) + \mathbf{b}^{\mathrm{T}}\mathbf{P}\mathbf{b}, \tag{95}$$

$$\mathbf{b}^{\mathrm{T}}\mathbf{y} = \alpha_0\mathbf{b}^{\mathrm{T}}\mathbf{A}\mathbf{u}_0 - \alpha_0\mathbf{b}^{\mathrm{T}}\mathbf{P}\mathbf{A}\mathbf{u}_0 + \mathbf{b}^{\mathrm{T}}\mathbf{P}\mathbf{b}. \tag{96}$$

In the first equation, we take Equation (73) into account. Inserting them into Equation (93) leads to

$$\|\mathbf{b}-\mathbf{A}\mathbf{x}\|^2 = \alpha_0^2(\mathbf{u}_0^{\mathrm{T}}\mathbf{A}^{\mathrm{T}}\mathbf{A}\mathbf{u}_0 - \mathbf{u}_0^{\mathrm{T}}\mathbf{A}^{\mathrm{T}}\mathbf{P}\mathbf{A}\mathbf{u}_0) + 2\alpha_0(\mathbf{b}^{\mathrm{T}}\mathbf{P}\mathbf{A}\mathbf{u}_0 - \mathbf{b}^{\mathrm{T}}\mathbf{A}\mathbf{u}_0) + \|\mathbf{b}\|^2 - \mathbf{b}^{\mathrm{T}}\mathbf{P}\mathbf{b}. \tag{97}$$

Finally, inserting Equation (92) for α_0 into the above equation yields

$$\|\mathbf{b}-\mathbf{A}\mathbf{x}\|^2 = \|\mathbf{b}\|^2 - \mathbf{b}^{\mathrm{T}}\mathbf{P}\mathbf{b} - \frac{(\mathbf{b}^{\mathrm{T}}\mathbf{A}\mathbf{u}_0 - \mathbf{b}^{\mathrm{T}}\mathbf{P}\mathbf{A}\mathbf{u}_0)^2}{\mathbf{u}_0^{\mathrm{T}}\mathbf{A}^{\mathrm{T}}\mathbf{A}\mathbf{u}_0 - \mathbf{u}_0^{\mathrm{T}}\mathbf{A}^{\mathrm{T}}\mathbf{P}\mathbf{A}\mathbf{u}_0}. \tag{98}$$

If $\mathbf{P} \approx \mathbf{I}_q$, the value of $\|\mathbf{b}-\mathbf{A}\mathbf{x}\|^2$ in Equation (98) is made as small as possible:

$$\|\mathbf{b}-\mathbf{A}\mathbf{x}\|^2 \approx 0. \tag{99}$$

Theorem 3. *For the double optimal solution $\mathbf{x} \in \mathcal{K}'_m$ derived in Equation (42) of the least-squares problem (2), the squared-norm of the error vector $\mathbf{e}$:*

$$\|\mathbf{e}\|^2 = \|\mathbf{b}\|^2 - \frac{(\mathbf{b}^{\mathrm{T}}\mathbf{A}\mathbf{x})^2}{\|\mathbf{A}\mathbf{x}\|^2}, \tag{100}$$

and the squared-norm of the residual vector $\mathbf{r} = \mathbf{b} - \mathbf{A}\mathbf{x}$ are equal, i.e.,

$$\|\mathbf{e}\|^2 = \|\mathbf{b}-\mathbf{A}\mathbf{x}\|^2. \tag{101}$$

Moreover,

$$\|\mathbf{r}\| < \|\mathbf{b}\|. \tag{102}$$

Proof. Inserting Equation (92) for α_0 into Equations (95) and (96), we can obtain

$$\|\mathbf{y}\|^2 = \frac{(\mathbf{b}^{\mathrm{T}}\mathbf{A}\mathbf{u}_0 - \mathbf{b}^{\mathrm{T}}\mathbf{P}\mathbf{A}\mathbf{u}_0)^2}{\mathbf{u}_0^{\mathrm{T}}\mathbf{A}^{\mathrm{T}}\mathbf{A}\mathbf{u}_0 - \mathbf{u}_0^{\mathrm{T}}\mathbf{A}^{\mathrm{T}}\mathbf{P}\mathbf{A}\mathbf{u}_0} + \mathbf{b}^{\mathrm{T}}\mathbf{P}\mathbf{b}, \tag{103}$$

$$\mathbf{b}^{\mathrm{T}}\mathbf{y} = \frac{(\mathbf{b}^{\mathrm{T}}\mathbf{A}\mathbf{u}_0 - \mathbf{b}^{\mathrm{T}}\mathbf{P}\mathbf{A}\mathbf{u}_0)^2}{\mathbf{u}_0^{\mathrm{T}}\mathbf{A}^{\mathrm{T}}\mathbf{A}\mathbf{u}_0 - \mathbf{u}_0^{\mathrm{T}}\mathbf{A}^{\mathrm{T}}\mathbf{P}\mathbf{A}\mathbf{u}_0} + \mathbf{b}^{\mathrm{T}}\mathbf{P}\mathbf{b}. \tag{104}$$

Consequently, we have

$$\|\mathbf{y}\|^2 = \mathbf{b}^{\mathrm{T}}\mathbf{y}. \tag{105}$$

It follows from Equations (98) and (103) that

$$\|\mathbf{b}-\mathbf{A}\mathbf{x}\|^2 = \|\mathbf{b}\|^2 - \|\mathbf{y}\|^2. \tag{106}$$

By Equations (105) and (34),

$$\|\mathbf{e}\|^2 = \|\mathbf{b}\|^2 - \|\mathbf{y}\|^2. \tag{107}$$

Equation (101) was proven upon comparing Equations (106) and (107).

Using $\mathbf{r} = \mathbf{b} - \mathbf{A}\mathbf{x}$ and Equations (101) and (107) leads to

$$\|\mathbf{r}\|^2 = \|\mathbf{e}\|^2 = \|\mathbf{b}\|^2 - \|\mathbf{y}\|^2. \tag{108}$$

Due to $\|\mathbf{y}\|^2 > 0$, Equation (102) follows readily. □

4. A Numerical Algorithm

In this section, we develop an iterative algorithm to solve the least-squares problem, starting from an initial guess $\mathbf{x}_0$. We assume that the value of $\mathbf{x}_k$ at the kth step is already known, and $\mathbf{x}_{k+1}$ is computed at the next step via the iterative algorithm. According to the value $\mathbf{x}_k$, the kth step residual $\mathbf{r}_k = \mathbf{b} - \mathbf{A}\mathbf{x}_k$ can be computed.

When the initial guess $\mathbf{x}_0$ is given, an initial residual is written as follows:

$$\mathbf{r}_0 = \mathbf{b} - \mathbf{A}\mathbf{x}_0. \tag{109}$$

Upon letting

$$\mathbf{z} = \mathbf{x} - \mathbf{x}_0, \tag{110}$$

Equation (2) is equivalent to solving $\mathbf{z}$ from

$$\mathbf{A}\mathbf{z} = \mathbf{r}_0, \tag{111}$$

which is deduced by

$$\mathbf{A}\mathbf{z} = \mathbf{A}\mathbf{x} - \mathbf{A}\mathbf{x}_0 = \mathbf{b} - \mathbf{A}\mathbf{x}_0 = \mathbf{r}_0.$$

For system (111), we seek $\mathbf{z}$ in the VAKS by

$$\mathbf{z} = \alpha_0\mathbf{u}_0 + \sum_{k=1}^{m} \alpha_k\mathbf{u}_k = \alpha_0\mathbf{u}_0 + \mathbf{U}\boldsymbol{\alpha} \in \mathcal{K}'_m, \tag{112}$$

where

$$\mathbf{u}_0 = \mathbf{A}^{\mathrm{T}}\mathbf{r}_0, \tag{113}$$

and $\mathbf{u}_k := (\mathbf{A}^{\mathrm{T}}\mathbf{A})^k\mathbf{u}_0$, $k = 1, \ldots, m$. The constants α_k, $k = 1, \ldots, m$ and α_0 are determined by the following two minimizations:

$$\min_{\mathbf{z}}\left\{ f = \frac{\|\mathbf{A}\mathbf{z}\|^2}{(\mathbf{r}_0^{\mathrm{T}}\mathbf{A}\mathbf{z})^2} \right\}, \tag{114}$$

$$\min_{\mathbf{z}} r = \|\mathbf{r}_0 - \mathbf{A}\mathbf{z}\|^2. \tag{115}$$

After inserting Equation (112) for $\mathbf{z}$, the first minimization can derive $\boldsymbol{\alpha}$, while the second minimization can derive α_0.

Since two minimizations in Equations (114) and (115) are adopted to determine the descent vector $\mathbf{z}$ in Equation (112), the resulting iterative algorithm to solve the least-square problem might be labeled as a double optimal algorithm (DOA).

To treat the rank deficient least-squares problem, the dimension m of the VAKS must be $m < m_0 = \text{rank}(\mathbf{A})$, such that $\text{rank}(\mathbf{J}_k) = m$ and $\mathbf{C}_k = \mathbf{J}_k^{\mathrm{T}}\mathbf{J}_k > \mathbf{0}$ is an $m \times m$ invertible matrix. Consequently, the DOA is depicted by (i), giving $\mathbf{A}$, $\mathbf{b}$, $m < m_0$, an initial value $\mathbf{x}_0$ and the convergence criterion ε and (ii) doing $k = 0, 1, \ldots$,

$$
\begin{aligned}
&\mathbf{r}_k = \mathbf{b} - \mathbf{A}\mathbf{x}_k, \\
&\mathbf{u}_0^k = \mathbf{A}^{\mathrm{T}}\mathbf{r}_k, \\
&\mathbf{u}_j^k = (\mathbf{A}^{\mathrm{T}}\mathbf{A})^j\mathbf{u}_0^k,\ j = 1, \ldots, m, \\
&\mathbf{U}_k = [\mathbf{u}_1^k, \ldots, \mathbf{u}_m^k] \text{ (By Arnoldi procedure)}, \\
&\mathbf{J}_k = \mathbf{A}\mathbf{U}_k, \\
&\mathbf{C}_k = \mathbf{J}_k^{\mathrm{T}}\mathbf{J}_k, \\
&\mathbf{D}_k = \mathbf{C}_k^{-1}, \\
&\mathbf{V}_k = \mathbf{U}_k\mathbf{D}_k\mathbf{J}_k^{\mathrm{T}}, \\
&\mathbf{P}_k = \mathbf{A}\mathbf{V}_k, \\
&\lambda_k = \frac{\|\mathbf{A}\mathbf{u}_0^k\|^2 - (\mathbf{u}_0^k)^{\mathrm{T}}\mathbf{A}^{\mathrm{T}}\mathbf{P}_k\mathbf{A}\mathbf{u}_0^k}{\mathbf{r}_k^{\mathrm{T}}\mathbf{A}\mathbf{u}_0^k - \mathbf{r}_k^{\mathrm{T}}\mathbf{P}_k\mathbf{A}\mathbf{u}_0^k}, \\
&\mathbf{z}_k = \mathbf{V}_k\mathbf{r}_k + \frac{1}{\lambda_k}(\mathbf{u}_0^k - \mathbf{V}_k\mathbf{A}\mathbf{u}_0^k), \\
&\mathbf{x}_{k+1} = \mathbf{x}_k + \mathbf{z}_k,
\end{aligned} \tag{116}
$$

until the convergence with $\|\mathbf{x}_{k+1} - \mathbf{x}_k\| < \varepsilon$ or $\|\mathbf{r}_{k+1}\| < \varepsilon$. We call $\|\mathbf{x}_{k+1} - \mathbf{x}_k\|$ the relative residual.

In the above, $\mathbf{U}_k$ is $n \times m$ matrix, $\mathbf{J}_k$ is a $q \times m$ matrix, $\mathbf{C}_k$ and $\mathbf{D}_k$ are $m \times m$ matrices, $\mathbf{V}_k$ is an $n \times q$ matrix and $\mathbf{P}_k$ is a $q \times q$ matrix. The computational cost is expanded to compute $\mathbf{D}_k$, which is however quite time-saving because m is a small number. $\mathbf{A}^{\mathrm{T}}\mathbf{A}$ is an $n \times n$ fixed matrix computed once and used in the construction of $\mathbf{U}_k$, which requires mn operations. $\mathbf{J}_k$ requires mnq operations; $\mathbf{C}_k$ requires m^2q operations; $\mathbf{D}_k$ requires m^2q operations; $\mathbf{V}_k$ requires nm^2q operations; and $\mathbf{P}_k$ requires nq^2 operations. In each iteration, there are $mn + mnq + 2m^2q + nm^2q + nq^2$ operations. Denote the number of iterations by NI. The computational complexity is in total NI$\times[mn + mnq + 2m^2q + nm^2q + nq^2]$.

It is known that m is a key parameter in the Krylov subspace. A proper choice of m can significantly enhance the convergence speed and accuracy. For ill-posed linear least-squares problems, there exists the best value of m, but for well-posed linear least-squares problems, small values of m may slow down convergence. When m is increased, both the convergence speed and accuracy are increased. However, when m is increased, more computational power is required in the construction of the projection operator $\mathbf{P}$ and the inversion matrix $\mathbf{C}^{-1}$.

We make some comments about the initial value of $\mathbf{x}_0$. For under-determined systems, there are many solutions; hence, different choices of $\mathbf{x}_0$ would generate different solutions. In general, as required in Equation (1), the minimal-norm solution can be obtained by setting $\mathbf{x}_0 = \mathbf{0}$, whose norm is zero $\|\mathbf{x}_0\| = 0$. For over-determined systems, we take $\mathbf{x}_0 = \mathbf{0}$ such that the initial residual is a non zero vector $\mathbf{r}_0 = \mathbf{b}$ and the DOA based on the residual $\mathbf{r}_k$ is solvable. Indeed, the DOA is not sensitive to the initial value of $\mathbf{x}_0$; we will take $\mathbf{x}_0 = \mathbf{0}$ for most problems unless otherwise specified.

The following corollaries help the clarification of the DOA and are based on the DOS used in the residual Equation (111).

Corollary 3. $\mathbf{r}_{k+1}$ *is orthogonal to* $\mathbf{A}\mathbf{z}_k$ *in Equation (111) and* $\mathbf{r}_{k+1} - \mathbf{r}_k$, *i.e.,*

$$\mathbf{r}_{k+1}^T(\mathbf{A}\mathbf{z}_k) = 0, \tag{117}$$

$$\mathbf{r}_{k+1}^T(\mathbf{r}_{k+1} - \mathbf{r}_k) = 0. \tag{118}$$

Proof. By Equation (111), the next $\mathbf{r}_{k+1}$ is given by

$$\mathbf{r}_{k+1} = \mathbf{r}_k - \mathbf{y}_k, \tag{119}$$

where $\mathbf{y}_k = \mathbf{A}\mathbf{z}_k$. By Equation (116), we can also derive Equation (119) as follows:

$$\mathbf{b} - \mathbf{A}\mathbf{x}_{k+1} = \mathbf{b} - \mathbf{A}\mathbf{x}_k - \mathbf{A}\mathbf{z}_k.$$

For Equation (111), Equation (105) is written as

$$\mathbf{r}_k^{\mathrm{T}}\mathbf{y}_k - \|\mathbf{y}_k\|^2 = 0; \tag{120}$$

and taking the $\mathbf{y}_k$ inner product to Equation (119), we have

$$\mathbf{r}_{k+1}^{\mathrm{T}}\mathbf{y}_k = \mathbf{r}_{k+1}^{\mathrm{T}}(\mathbf{A}\mathbf{z}_k) = \mathbf{r}_k^{\mathrm{T}}\mathbf{y}_k - \|\mathbf{y}_k\|^2 = 0. \tag{121}$$

Inserting $\mathbf{A}\mathbf{z}_k = \mathbf{r}_k - \mathbf{r}_{k+1}$ into Equation (117), we can prove Equation (118). The DOA is a good approximation of Equation (2) with a better descent direction $\mathbf{z}_k$ in the VAKS. □

Corollary 4. *For the DOA, the convergence rate is*

$$\frac{\|\mathbf{r}_k\|}{\|\mathbf{r}_{k+1}\|} = \frac{1}{\sin\theta} > 1, \;\; \theta \neq \frac{\pi}{2}. \tag{122}$$

Proof. It follows from Equations (119) and (120) that

$$\|\mathbf{r}_{k+1}\|^2 = \|\mathbf{r}_k\|^2 - \|\mathbf{r}_k\|\|\mathbf{y}_k\|\cos\theta, \tag{123}$$

where θ is the intersection angle between $\mathbf{r}_k$ and $\mathbf{y}_k = \mathbf{A}\mathbf{z}_k$. With help from Equation (120), we have

$$\cos\theta = \frac{\|\mathbf{y}_k\|}{\|\mathbf{r}_k\|}. \tag{124}$$

Obviously, $\theta \neq \pi/2$ because $\|\mathbf{y}_k\| > 0$. Then, Equation (123) can be further reduced to

$$\|\mathbf{r}_{k+1}\|^2 = \|\mathbf{r}_k\|^2[1 - \cos^2\theta] = \|\mathbf{r}_k\|^2\sin^2\theta. \tag{125}$$

Dividing both sides by $\|\mathbf{r}_{k+1}\|^2$ and taking the square-roots of both sides, we can obtain Equation (122). □

Corollary 5. *The residual is decreased step by step,*

$$\|\mathbf{r}_{k+1}\|^2 = \|\mathbf{r}_k\|^2 - \|\mathbf{A}\mathbf{z}_k\|^2 \Rightarrow \|\mathbf{r}_{k+1}\| < \|\mathbf{r}_k\|, \tag{126}$$

Proof. Taking the squared norms of Equation (119) and using Equation (120) generates

$$\|\mathbf{r}_{k+1}\|^2 = \|\mathbf{r}_k\|^2 - \|\mathbf{y}_k\|^2; \tag{127}$$

and after inserting $\mathbf{y}_k = \mathbf{A}\mathbf{z}_k$, Equation (126) is proven. Corollary 5 is easily deduced by noting $\|\mathbf{A}\mathbf{z}_k\|^2 > 0$. □

The property in Equation (126) is very important, which guarantees that the DOA is absolutely convergent at each iteration.

Remark 3. *At the very beginning, the DOS and DOA are developed in the VAKS $\mathcal{K}'_m$ similar to those in Equation (21). If one takes $\alpha_0 = 1$ in Equation (21), $\mathcal{K}'_m$ reduces to the usual affine Krylov subspace $\mathbf{u}_0 + \mathcal{K}_m$. However, in this situation, several good properties similar to those in Theorem 3 and Corollaries 3–5 will be lost; convergence cannot be guaranteed in the resulting iterative algorithm. On the other hand, if we take $\mathbf{u}_0 = \mathbf{0}$ and directly work in the $(m+1)$-dimensional Krylov subspace $\mathcal{K}_{m+1}$, by extending $\boldsymbol{\alpha}$ to an $(m+1)$-dimensional vector, this would result in failure because after inserting $\mathbf{y}_0 = \mathbf{A}\mathbf{u}_0 = \mathbf{0}$ into Equation (62), $\boldsymbol{\alpha}$ cannot be defined.*

Corollary 6. *In the iterative algorithm (116), we may encounter a slowly varying point:*

$$\mathbf{x}_{k+1} \approx \mathbf{x}_k, \tag{128}$$

when $\mathbf{x}_0$ *is given such that* $\mathbf{r}_0 = \mathbf{b} - \mathbf{A}\mathbf{x}_0$ *satisfies* $\mathbf{A}^T\mathbf{A}\mathbf{r}_0 = c_0\mathbf{r}_0$ *for some* $c_0 \gg 1$.

Proof. Inserting $\mathbf{b} = \mathbf{r}_0$ and

$$\mathbf{u}_0 = \mathbf{A}^T\mathbf{r}_0$$

into Equation (92) yields

$$\alpha_0 = \frac{1}{c_0}.$$

when $c_0 \gg 1$, $\alpha_0 \approx 0$, and $\mathbf{x}_{k+1} \approx \mathbf{x}_k$ by Equation (116). □

5. Numerical Examples

To demonstrate the efficiency and accuracy of the present iterative DOA, several examples are presented. All the numerical computations were carried out in Microsoft Developer Studio with an Intel Core I7-3770, CPU 2.80 GHz and 8 GB memory. The precision is 10^{-16}.

5.1. Example 1

In this example, we find a minimal-norm and least-squares solution of the following consistent equation:

$$\begin{bmatrix} 1 & 2 & 3 & -1 \\ 3 & 2 & 1 & -1 \\ 2 & 3 & 1 & 1 \end{bmatrix} \begin{bmatrix} x_1 \\ x_2 \\ x_3 \\ x_4 \end{bmatrix} = \begin{bmatrix} 1 \\ 1 \\ 1 \end{bmatrix}. \tag{129}$$

Although Example 1 is simple, we use it to test the DOS and DOA for finding the minimal-norm solution for a consistent system.

First we apply the DOS with $m = 2$ to solve this problem, where we can find a quite accurate solution of $x_1 = 1.48176 \times 10^{-1}$, $x_2 = 1.92567 \times 10^{-1}$, $x_3 = 1.48141 \times 10^{-1}$, and $x_4 = -2.22689 \times 10^{-2}$, for which the three residual errors are $r_1 = 2.012 \times 10^{-6}$, $r_2 = 7.125 \times 10^{-5}$ and $r_3 = -7.51756 \times 10^{-5}$, respectively.

Although the DOS is an accurate solution with an error on the order of 10^{-5}, the accuracy can be improved by the DOA. In the application of the iterative DOA, initial guesses are given by $x_1^0 = x_2^0 = x_3^0 = x_4^0 = c_0$ for some constant. Then, the DOA with $m = 1$ solves this problem with $c_0 = 0$ and $\varepsilon = 10^{-12}$. As shown in Figure 1a, the DOA converges very fast in only seven steps. Furthermore, we can find very accurate solutions of $x_1 = 1.48148 \times 10^{-1}$, $x_2 = 1.925926 \times 10^{-1}$, $x_3 = 1.48148 \times 10^{-1}$, and $x_4 = -2.22222 \times 10^{-2}$; the norm of $\mathbf{x}$ is $\|\mathbf{x}\| = 0.28545$. The three residual errors are $r_1 = -5.63993 \times 10^{-14}$, $r_2 = -5.61773 \times 10^{-14}$ and $r_3 = -5.63993 \times 10^{-14}$, respectively.

On the other hand, if we take $c_0 = 1$, $r_1 = 5.6 \times 10^{-13}$, $r_2 = 5.59 \times 10^{-13}$ and $r_3 = 8.47 \times 10^{-13}$ are obtained, which show that the solution $(x_1, x_2, x_3, x_4) = (0.4815, -0.2741, 0.4815, 0.3778)$ is accurate. However, the norm $\|\mathbf{x}\| = 0.82552$ is not the minimal one. Therefore, the correct one for the minimal-norm solution is $c_0 = 0$. Unless otherwise specified, we will take the initial guess by $\mathbf{x}_0 = \mathbf{0}$ for most problems computed below.

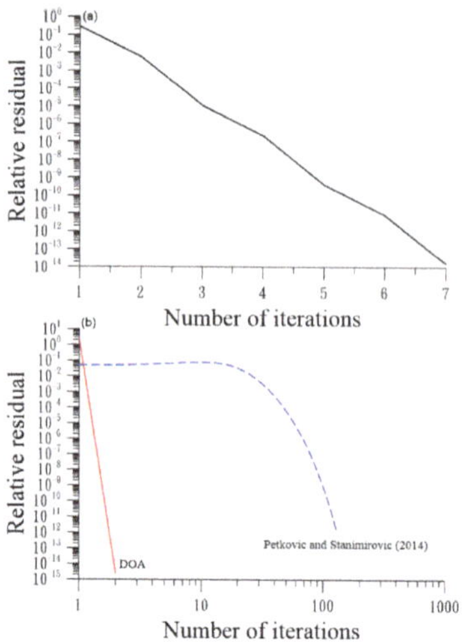

Figure 1. Examples 1 and 2, showing the relative residuals with respect to the number of iterations; the latter is compared with [20].

5.2. Example 2

A minimal norm and least-squares solution is sought for of the following inconsistent equation:

$$\begin{bmatrix} 1 & 1 & 0 \\ 1 & 0 & 1 \\ -1 & 0 & 0 \\ 1 & 1 & 1 \end{bmatrix} \begin{bmatrix} x_1 \\ x_2 \\ x_3 \end{bmatrix} = \begin{bmatrix} 0 \\ 0 \\ 1 \\ 2 \end{bmatrix}. \tag{130}$$

Although Example 2 is simple, we use it to test the DOS and DOA for finding the minimal-norm solution for an inconsistent system.

We first apply the DOS with $m = 1$, where we obtain a very accurate solution with $x_1 = -1.25$, $x_2 = 1.5$, and $x_3 = 1.5$. The residuals are $(r_1, r_2, r_3, r_4) = (0.25, 0.25, 0.25, -0.25)$ with the residual norm being 0.5.

Then, the DOA with $m = 1$ and $\varepsilon = 10^{-12}$ solves this problem. Starting from the initial guesses $x_1^0 = x_2^0 = x_3^0 = 0$, as shown in Figure 1b, the DOA converges very fast in only two steps, where we obtain a very accurate solution with $x_1 = -1.25$, $x_2 = 1.5$, and $x_3 = 1.5$. The residuals are $(r_1, r_2, r_3, r_4) = (0.25, 0.25, 0.25, -0.25)$, with the residual norm being 0.5. For the inconsistent system, the equality in Equation (130) no longer holds, such that the residual is not a zero vector.

The Moore–Penrose inverse of the coefficient matrix $\mathbf{A}$ in Equation (130) has an exact solution:

$$\mathbf{A}^{\dagger} = \begin{bmatrix} \frac{1}{4} & \frac{1}{4} & -\frac{3}{4} & -\frac{1}{4} \\ \frac{1}{2} & -\frac{1}{2} & \frac{1}{2} & \frac{1}{2} \\ -\frac{1}{2} & \frac{1}{2} & \frac{1}{2} & \frac{1}{2} \end{bmatrix}. \tag{131}$$

Then, we find that the exact solution of Equation (130) is given by $\mathbf{x} = \mathbf{A}^{\dagger}\mathbf{b}$, as just $(x_1, x_2, x_3) = (-1.25, 1.5, 1.5)$ mentioned above.

We apply the iterative method developed by Petkovic and Stanimirovic [20] to find the Moore–Penrose inverse, which has a simpler form:

$$\mathbf{X}_{k+1} = (1 + \beta)\mathbf{X}_k - \beta \mathbf{X}_k \mathbf{A} \mathbf{X}_k, \;\; \mathbf{X}_0 = \beta \mathbf{A}^{\mathrm{T}}, \tag{132}$$

based on the Penrose Equations (4) and (6). Under the same convergence criterion, this iteration process is convergent in 130 steps, as shown in Figure 1b by a dashed line.

Note that the DOA can be also used to find the Moore–Penrose inverse $\mathbf{A}^\dagger$ by taking the right-hand side $\mathbf{b}$ to be a zero vector and the k-th element to be one, of which the corresponding solution of $\mathbf{x}$ denoted by $\mathbf{x}^k$ constitutes the kth column of $\mathbf{A}^\dagger$. Then, we can run the DOA q times from $k = 1, \ldots, q$ to obtain

$$\mathbf{A}^\dagger = \begin{bmatrix} x_1^1 & \ldots & x_1^q \\ x_2^1 & \ldots & x_2^q \\ \vdots & \ldots & \ldots \\ x_n^1 & \ldots & x_n^q \end{bmatrix}. \tag{133}$$

The dimension of $\mathbf{A}^\dagger$ is $n \times q$. By applying the DOA with $m = 1$ under the same convergence criterion $\varepsilon = 10^{-12}$, we can find a very accurate solution of $\mathbf{A}^\dagger$, as shown in Equation (131) with the maximum error of all elements being 10^{-14}, which is achieved in a total of 26 steps and converges faster than the method developed by Petkovic and Stanimirovic [20].

5.3. Example 3

We consider an over-determined Hilbert linear problem:

$$\begin{aligned} A_{ij} &= \frac{1}{i+j-1}, \ 1 \le i \le q, \ 1 \le j \le n, \\ b_i &= \sum_{j=1}^{n} \frac{1}{i+j-1} x_j, \ 1 \le i \le q, \end{aligned} \tag{134}$$

where we fix $q = 6$ and $n = 5$, and the exact solutions $x_j = 1/j, \ 1 \le j \le n$ are given.

The DOA with $m = 4$ and $\varepsilon = 10^{-13}$ can solve this problem very fast, starting from the initial guesses $x_1^0 = x_2^0 = x_3^0 = x_4^0 = x_5^0 = 0$, in only four steps, as shown in Figure 2a. Furthermore, we can find a very accurate solution, as shown in Figure 2b with the maximum error (ME) being 8.91×10^{-12}. If we consider an under-determined system with $q = 5$ and $n = 6$, the DOA with $m = 3$ is still workable with ME = 2.22×10^{-8}.

For least-squares problems, QR is a famous method to directly solve Equation (2). This process is shown in Equations (9)–(11). However, we find that ME = 1.05×10^{-6} obtained by QR is less accurate than that obtained via the DOA in the Hilbert least-squares problem with $q = 6$ and $n = 5$. We have checked the accuracy of the orthogonality in Equation (10) by

$$\|\mathbf{Q}^\mathrm{T}\mathbf{Q} - \mathbf{I}_n\|,$$

which is on the order of 8.65×10^{-12}.

The Hilbert matrix is known as a notorious example of a highly ill-conditioned matrix. For this problem, the condition number is 2.5×10^5. Then, in the floating point, $\mathbf{b}$ has a rounding error on the the order of 10^{-16}. The algorithm introduces further floating point errors which are of the same order of magnitude. Perturbation theory for the least-squares problem [38] shows that one can only expect to find a solution that is approximately within $2 \times 10^{(5-16)} = 2 \times 10^{-11}$ of the exact solution. The error $\|\mathbf{Q}^\mathrm{T}\mathbf{Q} - \mathbf{I}_n\| = 8.65 \times 10^{-12}$ is within this range; however, the ME = 1.05×10^{-6} obtained by QR is not within this range. The reason for this is that when we use Equation (11) to find the solution in the QR method, very small singular values appearing in the denominator will enlarge the rounding error and the error from $\mathbf{Q}$; for example, the last two singular values 4.803×10^{-4} and 1.734×10^{-5}, upon dividing by those two small values sequentially, lead to ME = 1.05×10^{-6}, which is not within the range of 2×10^{-11}. In contrast, the proposed DOA method with an error of 8.91×10^{-12} is within this range. For highly ill-conditioned least-squares problems, the QR method loses some accuracy, and even $\mathbf{Q}$ well satisfies Equation (10) within the range of 2×10^{-11}.

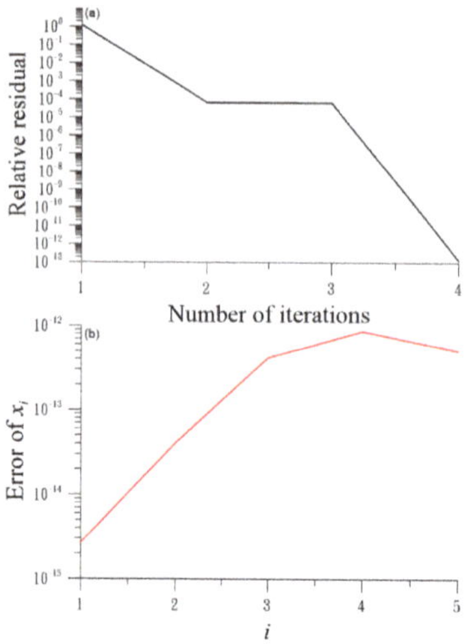

Figure 2. An over-determined Hilbert linear problem solved by the DOA: (**a**) relative residual, and (**b**) numerical error.

We raise q to 10, and Table 1 lists the computed ME for different values of n. We also list the error $\|\mathbf{Q}^{\mathrm{T}}\mathbf{Q} - \mathbf{I}_n\|$ for the QR method, which is worse when n is increased. Based on Equations (88)–(90), this problem is solved by SVD. The DOA is competitive with SVD, and for all cases, it is slightly more accurate than SVD.

Table 1. Example 3: ME obtained by the DOA and QR for different n and a fixed $q = 10$.

n	2	3	4	5
$\|\mathbf{Q}^{\mathrm{T}}\mathbf{Q} - \mathbf{I}_n\|$	1.63×10^{-15}	2.48×10^{-14}	3.11×10^{-13}	5.24×10^{-12}
QR	8.33×10^{-15}	3.55×10^{-12}	6.77×10^{-10}	2.27×10^{-7}
SVD	5.55×10^{-16}	1.60×10^{-14}	1.89×10^{-13}	9.09×10^{-13}
DOA	1.11×10^{-16}	3.55×10^{-15}	9.09×10^{-14}	7.81×10^{-13}

To investigate the influence of m on the DOA, Table 2 compares the ME and the iteration number (IN). For ill-posed Hilbert problems, there exists the best value $m = 3$.

Table 2. Example 3: the ME and NI obtained by the DOA for different m, where $(q, n) = (20, 8)$ and $\varepsilon = 10^{-8}$.

m	2	3	4	5	6
ME	4.94×10^{-6}	5.30×10^{-8}	1.34×10^{-6}	2.90×10^{-6}	6.28×10^{-6}
IN	14	4	4	4	5

5.4. Example 4

We find the Moore–Penrose inverse of the following rank deficient matrix [45]:

$$\mathbf{A} = \begin{bmatrix} -1 & 0 & 1 & 2 \\ -1 & 1 & 0 & -1 \\ 0 & -1 & 1 & 3 \\ 0 & 1 & -1 & -3 \\ 1 & -1 & 0 & 1 \\ 1 & 0 & -1 & -2 \end{bmatrix}. \tag{135}$$

We apply the iterative method of the DOA as demonstrated in Section 5.2 to find the Moore–Penrose inverse, which has an exact solution:

$$\mathbf{A}^{\dagger} = \frac{1}{102} \begin{bmatrix} -15 & -18 & 3 & -3 & 18 & 15 \\ 8 & 13 & -5 & 5 & -13 & -8 \\ 7 & 5 & 2 & -2 & -5 & -7 \\ 6 & -3 & 9 & -9 & 3 & -6 \end{bmatrix}. \tag{136}$$

Under the same convergence criterion $\varepsilon = 10^{-9}$ as used by Xia et al. [45], the iteration process of the DOA with $m = 1$ converges very fast in a total of 12 steps, as shown in Figure 3. In Table 3, we compare the DOA with other numerical methods specified by Xia et al. [45] and Petkovic and Stanimirovic [20] to asses the performance of the DOA measured by four numerical errors of the Penrose Equations (3)–(6) and the iteration number (IN).

Table 3. Computed results of a rank-deficient matrix in Example 4.

Alg.	IP	$\|\mathbf{A}\mathbf{A}^{\dagger}\mathbf{A} - \mathbf{A}\|^2$	$\|\mathbf{A}^{\dagger}\mathbf{A}\mathbf{A}^{\dagger} - \mathbf{A}^{\dagger}\|^2$	$\|(\mathbf{A}\mathbf{A}^{\dagger})^{\mathrm{T}} - \mathbf{A}\mathbf{A}^{\dagger}\|^2$	$\|(\mathbf{A}^{\dagger}\mathbf{A})^{\mathrm{T}} - \mathbf{A}^{\dagger}\mathbf{A}\|^2$	IN
[45]	$10^{-5}\mathbf{A}^{\mathrm{T}}$	8.92×10^{-16}	3.36×10^{-16}	7.85×10^{-17}	1.97×10^{-15}	50
[45]	$0.1\mathbf{A}^{\mathrm{T}}$	9.17×10^{-20}	6.99×10^{-16}	7.85×10^{-17}	2.86×10^{-15}	60
[20]	$0.1\mathbf{A}^{\mathrm{T}}$	1.53×10^{-14}	9.21×10^{-17}	4.13×10^{-31}	1.17×10^{-30}	165
DOA	$\mathbf{x}_0 = 10^{-16}\mathbf{1}$	5.21×10^{-27}	2.57×10^{-29}	3.82×10^{-27}	1.61×10^{-27}	12

In the above, Alg. is the shorthand of algorithm, IP is the shorthand of initial point and IN is the shorthand of "iteration number". $\mathbf{1} = (1, \ldots, 1)^{\mathrm{T}}$. The first two algorithms were reported by Xia et al. [45]. It can be seen that the DOA converges much faster and is more accurate than the other algorithms.

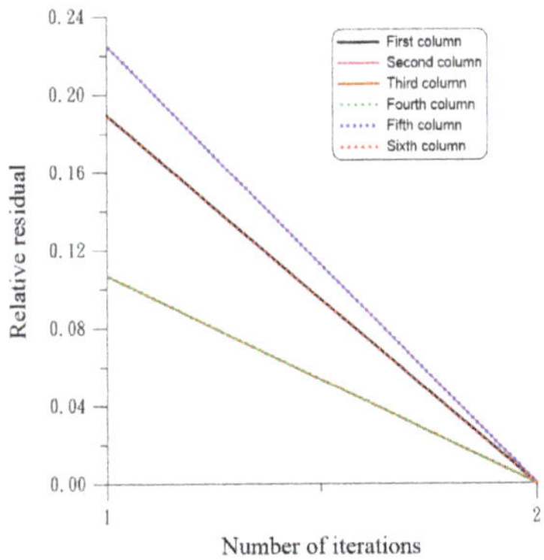

Figure 3. A computation of the Moore–Penrose inverse by the DOA, showing the relative residuals in the computation of each column of the Moore–Penrose inverse matrix.

In the computation of the pseudo-inverse, the last two singular values close to zero can be avoided to appear in the matrix $\mathbf{C} \in \mathbb{R}^{m \times m}$ if we take m small enough. For example, we take $m = 1$ in Table 3, such that $\mathbf{D} = \mathbf{C}^{-1}$ can be computed accurately without inducing a zero singular value to appear in the denominator to increase the rounding error.

5.5. Example 5

In this example, we find the Moore–Penrose inverse of the Hilbert matrix in Equation (134), which is more difficult than the previous example. Here, we fix $q = 3$, $n = 50$, and $q = 50$, $n = 3$. The numerical errors of the Penrose Equations (3)–(6) are compared in Tables 4 and 5, respectively.

Table 4. Computed results of the Hilbert matrix $q = 3, n = 50$ in Example 5.

Alg.	IP	$\|\mathbf{AA}^\dagger\mathbf{A} - \mathbf{A}\|^2$	$\|\mathbf{A}^\dagger\mathbf{AA}^\dagger - \mathbf{A}^\dagger\|^2$	$\|(\mathbf{AA}^\dagger)^{\mathrm{T}} - \mathbf{AA}^\dagger\|^2$	$\|(\mathbf{A}^\dagger\mathbf{A})^{\mathrm{T}} - \mathbf{A}^\dagger\mathbf{A}\|^2$	IN
[20]	$0.8\mathbf{A}^{\mathrm{T}}$	1.4×10^{-28}	3.7×10^{-21}	4.1×10^{-30}	3.5×10^{-27}	34
DOA	$\mathbf{x}_0 = \mathbf{0}$	9.8×10^{-28}	2×10^{-20}	4.1×10^{-24}	9.3×10^{-28}	3

Table 5. Computed results of the Hilbert matrix $q = 50, n = 3$ in Example 5.

Alg.	IP	$\|\mathbf{AA}^\dagger\mathbf{A} - \mathbf{A}\|^2$	$\|\mathbf{A}^\dagger\mathbf{AA}^\dagger - \mathbf{A}^\dagger\|^2$	$\|(\mathbf{AA}^\dagger)^{\mathrm{T}} - \mathbf{AA}^\dagger\|^2$	$\|(\mathbf{A}^\dagger\mathbf{A})^{\mathrm{T}} - \mathbf{A}^\dagger\mathbf{A}\|^2$	IN
[20]	$0.8\mathbf{A}^{\mathrm{T}}$	1.2×10^{-28}	3.7×10^{-21}	3.3×10^{-27}	6.3×10^{-28}	34
DOA	$\mathbf{x}_0 = \mathbf{0}$	9×10^{-29}	2.6×10^{-22}	3.2×10^{-23}	5.5×10^{-29}	3

5.6. Example 6

In this example, we consider a non-harmonic boundary value problem on an amoeba-like boundary:

$$\Delta u = u_{xx} + u_{yy} = 0, \ \ (x,y) \in \Omega, \tag{137}$$

$$u(x,y) = f(x,y) = x^2 y^3, \ \ (x,y) \in \Gamma, \tag{138}$$

$$\Gamma := \{\rho(\theta)\cos\theta, \rho(\theta)\sin\theta, \ \ 0 \le \theta \le 2\pi\},$$

$$\rho = \exp(\sin\theta)\sin^2(2\theta) + \exp(\cos\theta)\cos^2(2\theta), \tag{139}$$

where $\Delta f(x,y) = 2y^3 + 6x^2 y \neq 0$, and the contour is displayed in Figure 4a by a solid black line. Feng et al. [46] have pointed out that "non-harmonic boundary data" mean that the solution does not have a harmonic extension to the whole plane and a solution is very difficult to achieve.

Let

$$v(x,y) = u(x,y) - f(x,y) \tag{140}$$

be a new variable, and then we come to a Poisson equation under a homogeneous boundary condition:

$$\text{Poisson equation}: \begin{cases} \Delta v = F(x,y) = -\Delta f(x,y), \\ v(x,y)|_{(x,y)\in\Gamma} = 0, \end{cases} \tag{141}$$

where $F(x,y) \neq 0$, because $f(x,y)$ is a non-harmonic boundary function. When $v(x,y)$ is solved, we can find $u(x,y) = v(x,y) + f(x,y)$.

In order to obtain an accurate solution of $v(x,y)$, we use the multiple-scale Pascal triangle polynomial expansion method developed by Liu and Kuo [47]:

$$v(x,y) = \sum_{i=1}^{m_0} \sum_{j=1}^{i} c_{ij} s_{ij} \left(\frac{x}{R_0}\right)^{i-j} \left(\frac{y}{R_0}\right)^{j-1}. \tag{142}$$

After collocating q points to satisfy the governing equation and boundary condition (141), we have a non-square linear system (2), where the scales s_{ij} are determined such that each column of the coefficient matrix **A** has the same norm.

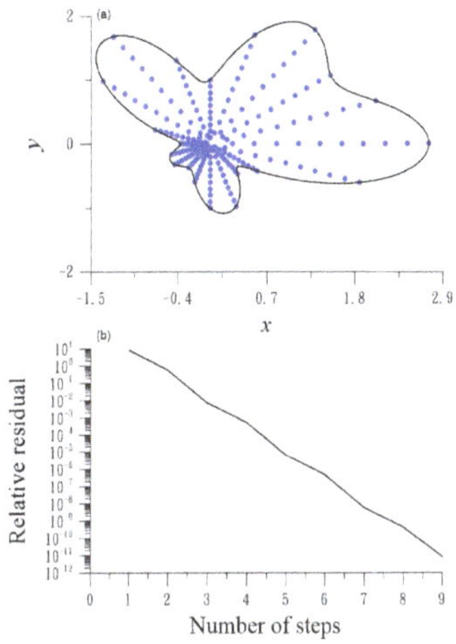

Figure 4. An amoeba-like boundary under a non-harmonic boundary function solved by the DOA: (**a**) the contour and distribution of collocation points, and (**b**) the convergence rate.

By using the DOA, we take $m = 4$, $q = 200$, $m_0 = 5$, $R_0 = 1000$ and $c_{ij}^0 = 0$. Hence, the dimension of **A** is 200×15, for which system (2) is an over-determined system. The distribution of collocation points is shown in Figure 4a, while the convergence rate is shown in Figure 4b. It is amazing that the DOA converges with nine steps even under a stringent convergence criterion $\varepsilon = 10^{-10}$. In Figure 5a, we compare the recovered boundary function with the exact function at 1000 points along the boundary. They are almost coincident, and thus we plot the absolute error in Figure 5b, whose ME = 2.95×10^{-12} and the root mean square error (RMSE) = 1.2×10^{-12}. This result is much better than that computed by Feng et al. [46].

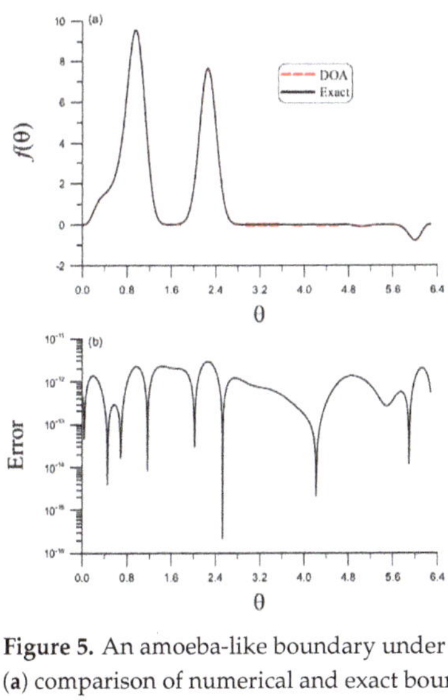

Figure 5. An amoeba-like boundary under a non-harmonic boundary function solved by the DOA: (**a**) comparison of numerical and exact boundary conditions, and (**b**) the numerical error.

5.7. Example 7

The method of fundamental solutions (MFS) is adopted to solve Equation (137) by using

$$u(\mathbf{z}) = \sum_{j=1}^{n} c_j \ln \|\mathbf{z} - \mathbf{s}_j\|, \;\; \mathbf{z} \in \Omega, \;\; \mathbf{s}_j \in \bar{\Omega}^c, \tag{143}$$

where $\bar{\Omega}^c$ is the complementary set of $\bar{\Omega}$, and

$$\mathbf{s}_j = (R\cos\theta_j, R\sin\theta_j)^{\mathrm{T}}, \; \theta_j = \frac{2j\pi}{n} \tag{144}$$

are source points, in which $R > 0$ is a parameter of the circle on which are the source points placed on.

We consider an exact solution $u(x, y) = x^2 - y^2$ and specify the Dirichlet boundary condition on the boundary given by Equation (139). In Equation (144), we fix $R = 5$. By using the DOA, we take $m = 11$ and $c_j^0 = 0$.

Table 6 lists the computed ME for different values of (n, q) obtained by the DOA. It can be seen that the DOA is applicable to least-squares problems regardless of whether $q < n$ or $q > n$.

Table 6. For the MFS on the 2D Laplace equation, ME with different (n, q) obtained by the DOA.

(n, q)	(100,80)	(100,90)	(100,110)	(100,120)	(100,150)
ME	1.88×10^{-8}	7.25×10^{-9}	7.67×10^{-8}	2.13×10^{-8}	4.45×10^{-8}

When we apply the QR to solve this problem with $n = 100$ and $q = 150$, ME = 42.81 is obtained. When we take $n = 150$ and $q = 400$, ME = 2.11 is obtained. Obviously, the QR is not applicable to this problem.

5.8. Example 8

In this example, we consider Equation (2) with a cyclic matrix, whose first row is given by $(1, \ldots, n)$. The coefficient matrix can be generated by the following procedure:

$$\begin{aligned}
&\text{Do } i = 1 : n, \\
&\text{Do } j = 1 : n, \\
&\text{If } i = 1, \text{ then } B_{i,j} = j; \text{ otherwise}, \\
&B_{i,j} = B_{i-1,j} + 1, \\
&\text{If } B_{i,j} > n, \text{ then } B_{i,j} = B_{i,j} - n, \\
&\text{Enddo.}
\end{aligned} \tag{145}$$

The exact solution is $x_i = 1, \; i = 1, \ldots, n$. The non-square matrix $\mathbf{A}$ is obtained by taking the first q rows from $\mathbf{B}$.

Since the right-hand side vector $\mathbf{b}$ is obtained from Equation (2) after inserting $\mathbf{A}$ and $\mathbf{x} = \mathbf{1} = (1, \ldots, 1)^{\mathrm{T}}$, the non-square system is consistent. Therefore, we can compute the maximal error (ME) by

$$\text{ME} := \max_{i=1,\ldots,n} |x_i^e - x_i^n|,$$

where $x_i^e = 1, \; i = 1, \ldots, n$ are the exact solutions and x_i^n are numerical solutions.

For a large matrix of $\mathbf{A}$ with $n = 2000$ and $q = 100$, Table 7 lists the computed ME for different values of m and the INs under $\varepsilon = 10^{-5}$ with the initial guesses $x_i^0 = 1 + 0.1i, \; i = 1, \ldots, n$. If we take the initial guesses to be $x_i^0 = 0, \; i = 1, \ldots, n$, the accuracy is poor with ME = 0.961 and the convergence is very slow. As shown in Corollary 6, for the zero initial value, we have $\mathbf{r}_0 = \mathbf{b}$ and for the cyclic matrix, we have $\mathbf{r}_0 = \mathbf{b} = c\mathbf{1}$, where $c_1 = \sum_{j=1}^{n} j$. Thus, $\mathbf{A}^{\mathrm{T}}\mathbf{A}\mathbf{r}_0 = c_0^2 c_2 \mathbf{r}_0$, where $c_2 = \sum_{j=1}^{q} j$. Hence, $c_0 = 2001000^2 \times 5050$ is a huge value, which causes the failure of the DOA with a zero initial value for this problem. The same situation happens for the constant initial guesses $x_i^0 = c$, $i = 1, \dots, n$. It is apparent that the DOA is quite accurate and converges faster. It is interesting that a small m is enough to achieve accurate solutions; when m is smaller, the accuracy is better, but it converges slower. In contrast, when m is larger, the accuracy is worse but it converges faster.

Table 7. Example 8 with $n = 2000$ and $q = 100$: ME obtained by the DOA and INs for different m.

m	**5**	**8**	**10**	**12**	**15**
ME	7.24×10^{-5}	1.73×10^{-4}	2.95×10^{-4}	1.02×10^{-3}	4.99×10^{-3}
IN	174	42	23	16	10

Next, we construct a large-sized matrix $\mathbf{B}$ from Equation (145) with n replaced by n_q. The non-square matrix $\mathbf{A}$ is obtained by taking the first n columns from $\mathbf{B}$. With $n = 500$ and $q = 2000$, Table 8 lists the computed ME for different values of m, and IN under $\varepsilon = 10^{-5}$ with the initial guesses $x_i^0 = 1 + 0.1i$, $i = 1, \dots, n$. It is apparent that the DOA is quite accurate and converges faster. When m is increased, the accuracy is better and convergence is faster.

Table 8. Example 8 with $n = 500$ and $q = 2000$: ME obtained by the DOA and INs for different m.

m	**10**	**12**	**15**	**18**	**20**
ME	1.39×10^{-4}	6.82×10^{-5}	3.20×10^{-5}	1.70×10^{-5}	1.26×10^{-5}
IN	123	72	41	25	19

As mentioned, the QR is not applicable to the least-squares problem with $q < n$; however, for $q > n$, the QR is applicable. Table 9 lists the computed ME for different values of (q, n). In the DOA, we take $m = 30$, $\varepsilon = 10^{-12}$, and $x_i^0 = 1 + 0.1i$, $i = 1, \dots, n$. The DOA converges within 25 iterations for the first four cases, and 79 iterations for the last case. The DOA can improve the accuracy by about three and four orders compared to the QR. We also list the error $\|\mathbf{Q}^{\mathrm{T}}\mathbf{Q} - \mathbf{I}_n\|$ for the QR method, which is very accurate.

Table 9. Example 8 with different (q, n), $q > n$: ME obtained by QR and the DOA.

(q, n)	**(1000,500)**	**(1500,500)**	**(1500,1000)**	**(2000,500)**	**(2500,1000)**
$\|\mathbf{Q}^{\mathrm{T}}\mathbf{Q} - \mathbf{I}_n\|$	1.81×10^{-12}	2.55×10^{-12}	8.99×10^{-12}	2.20×10^{-12}	1.14×10^{-11}
QR	3.18×10^{-10}	6.91×10^{-10}	1.73×10^{-9}	1.61×10^{-9}	4.56×10^{-9}
DOA	2.49×10^{-13}	2.66×10^{-13}	2.46×10^{-13}	1.77×10^{-13}	1.24×10^{-13}

If we take $\alpha_0 = 1$ in Equation (21), $\mathcal{K}'_m$ reduces to the usual affine Krylov subspace $\mathbf{u}_0 + \mathcal{K}_m$. Equation (80) becomes

$$\mathbf{x} = \mathbf{u}_0 + \lambda_0 \mathbf{V}\mathbf{b} - \mathbf{V}\mathbf{A}\mathbf{u}_0.$$

As shown in Table 10, the DOA is more accurate than the DOA with $\alpha_0 = 1$.

Table 10. Example 8 with $n = 500$, $n_q = 2000$, $\varepsilon = 10^{-12}$, ME1 and IN1 obtained by taking $\alpha_0 = 1$ in the DOA and ME2 and IN2 obtained by the DOA for different m.

m	20	22	25	28	30	35
ME1	1.13×10^{-12}	8.50×10^{-13}	3.33×10^{-13}	2.0×10^{-3}	1.88×10^{-13}	2.15×10^{-13}
IN1	62	48	34	25	21	15
ME2	9.15×10^{-13}	8.59×10^{-13}	3.94×10^{-13}	2.71×10^{-13}	1.77×10^{-13}	1.41×10^{-13}
IN2	64	48	33	25	21	16

6. A Variant of the Karush–Kuhn–Tucker Equation

In many meshless collocation methods to solve linear partial differential equations, we may collocate more q points than the number of coefficients n in the expansion of the solution, which results in an over-determined system in Equation (2) with $q > n$. The residual vector $\mathbf{r} = \mathbf{b} - \mathbf{A}\mathbf{x}$ together with Equation (2) can be written as

$$\begin{bmatrix} \mathbf{I}_q & \mathbf{A} \\ \mathbf{A}^{\mathrm{T}} & \mathbf{0} \end{bmatrix} \begin{bmatrix} \mathbf{r} \\ \mathbf{x} \end{bmatrix} = \begin{bmatrix} \mathbf{b} \\ \mathbf{0} \end{bmatrix}. \tag{146}$$

This is called the Karush–Kuhn–Tucker equation [15].

Instead of Equation (146), we may consider a permutation of the Karush–Kuhn–Tucker equation:

$$\begin{bmatrix} \mathbf{A} & \mathbf{I}_q \\ \mathbf{0} & \mathbf{A}^{\mathrm{T}} \end{bmatrix} \begin{bmatrix} \mathbf{x} \\ \mathbf{r} \end{bmatrix} = \begin{bmatrix} \mathbf{b} \\ \mathbf{0} \end{bmatrix}, \tag{147}$$

which is a $(q+n) \times (q+n)$ system. Equation (147) is a variant of the Karush–Kuhn–Tucker Equation (146) by re-ordering the unknown vector from $(\mathbf{r}, \mathbf{x})$ to $(\mathbf{x}, \mathbf{r})$. When the dimension $q+n$ is a moderate value, we can apply the partial pivoting Gaussian elimination method (GEM) to solve Equations (146) and (147). We name the first partial pivoting Gaussian elimination method on Equation (146) as GEM1, and on Equation (147) as GEM2. It would be convincing if medium-size least-squares problems shown below the partial pivoting Gaussian elimination method can be an effective and accurate method. Moreover, the variant of the Karush–Kuhn–Tucker equation is more accurate than the original Karush–Kuhn–Tucker equation, even they have the same condition numbers.

6.1. Example 9

We consider Example 7 again, which is now solved by the above four methods. Table 11 lists the computed ME for different values of (n, q) obtained by GEM1 and GEM2. Here, CGNR is the conjugate gradient method applied to the normal equation with minimal residuals, while CGNE is the conjugate gradient method applied to the normal equation with minimal errors [15]. CGNR is also known as the CGLS (least-squares), and CGNE is Craig's method. In this test, we find that when the partial pivoting Gaussian elimination method is applied to Equation (147) as GEM2, the accuracy is the best.

Table 11. The MFS on a 2D Laplace equation: ME with different (n, q) obtained by different methods.

(n, q)	(100,120)	(100,150)	(100,170)	(100,200)	(150,180)
GEM1	7.76×10^{-6}	1.81×10^{-6}	6.12×10^{-7}	2.99×10^{-6}	5.73×10^{-6}
GEM2	2.17×10^{-11}	3.82×10^{-14}	2.24×10^{-14}	1.99×10^{-14}	3.52×10^{-14}
CGNR	8.33×10^{-11}	8.23×10^{-11}	7.18×10^{-10}	6.62×10^{-9}	7.98×10^{-10}
CGNE	1.84×10^{-11}	4.78×10^{-7}	5.44×10^{-6}	6.46×10^{-10}	2.58×10^{-6}

6.2. Example 10

We consider Example 3 again, which is now solved by the above four methods, the QR method and the DOA with $\varepsilon = 10^{-14}$. Based on Equations (88)–(90), this problem is solved via SVD. As shown in Table 12, the GEM2, DOA and SVD are competitive and are more accurate than other methods.

We also list the errors $\|\mathbf{Q}^{\mathrm{T}}\mathbf{Q} - \mathbf{I}_n\|$ for the QR method in Table 12, which is very accurate. However, since quite small singular values are present for the highly ill-conditioned Hilbert matrix, the QR method gradually deteriorates when the condition number is increased from 7.04×10^4 to 2.1×10^7.

Table 12. Example 10: ME with different (n, q) obtained by GEM1, GEM2, CGNR, CGNE, the DOA and QR.

(n, q)	(5,6)	(5,8)	(5,10)	(7,25)	(8,25)
GEM1	1.80×10^{-6}	5.18×10^{-7}	4.57×10^{-7}	2.20×10^{-3}	1.55×10^{-3}
GEM2	3.15×10^{-12}	2.27×10^{-13}	1.02×10^{-12}	1.29×10^{-10}	4.70×10^{-9}
CGNR	4.98×10^{-8}	1.62×10^{-7}	3.05×10^{-7}	4.83×10^{-8}	7.64×10^{-8}
CGNE	3.52×10^{-12}	1.64×10^{-12}	1.88×10^{-12}	4.83×10^{-8}	7.64×10^{-8}
DOA	1.88×10^{-13}	2.55×10^{-12}	7.81×10^{-13}	2.72×10^{-9}	1.48×10^{-9}
SVD	4.85×10^{-12}	1.52×10^{-12}	9.09×10^{-13}	3.08×10^{-10}	1.52×10^{-8}
$\|\mathbf{Q}^{\mathrm{T}}\mathbf{Q} - \mathbf{I}_n\|$	8.65×10^{-12}	4.57×10^{-12}	5.24×10^{-12}	5.66×10^{-10}	1.89×10^{-8}
QR	1.05×10^{-6}	3.50×10^{-7}	2.27×10^{-7}	2.46×10^{-3}	5.68×10^{-1}

7. Conclusions

Least-squares problems arise in a lot of applications, and many iterative algorithms are already available to seek their solutions. In this paper, a new concept of a varying affine Krylov subspace was introduced, which is different to the fixed-type affine Krylov subspace used in the CG-type numerical solution of over-determined least-squares problems. In the $m+1$-dimensional varying affine Krylov subspace, a closed-form double optimal solution in a simple form in Equation (42) was derived and further reduced to Equation (80), which was obtained by two minimizations in Equations (40) and (41). We analyzed a key equation (72) to link these two optimizations together. The double optimal solution is an excellent approximation, as verified for least-squares problems. The iterative DOA was developed, which leads to step-by-step absolute convergence. In each iterative step, by merely inverting an $m \times m$ matrix with a small value of m, the computational cost of the DOA is very low, lower than other methods. The Moore–Penrose inverses of non-square matrices were also derived by using the varying affine Krylov subspace method. For over-determined least-squares problems, we proposed a variant of the Karush–Kuhn–Tucker equation, which uses the partial pivoting Gaussian elimination method and is better than the original Karush–Kuhn–Tucker equation and CG-type numerical methods of CGNR (CGLS) and CGNE (Craig's method). It is very important that the DOA can be applied to both rectangular systems with $q > n$ or $q < n$.

The novelties involved in this paper are as follows:

- A double-optimization-based iterative algorithm for least-squares problems was developed in a varying affine Krylov subspace.
- For dense large-size least-squares problems, the proposed iterative DOA is efficient and accurate.
- The DOA can be applied to both rectangular systems with extremal cases of $q \ll n$ or $q \gg n$.
- A variant of the Karush–Kuhn–Tucker equation was presented, which is an improved version of the original Karush–Kuhn–Tucker equation.

The idea of varying the affine Krylov subspace is useful to find a better solution of linear algebraic equations based on the mechanism of double optimization. In the future, we may extend the new methods of DOSs and the DOA to other linear matrix equations. The limitations are that there are several matrix multiplications needed to construct the projection operator **P** in the Krylov subspace and the high computational cost for inversion of the matrix $\mathbf{C}^{-1}$ with dimension m.

Author Contributions: Conceptualization, C.-S.L. and C.-W.C.; Methodology, C.-S.L. and C.-W.C.; Software, C.-S.L., C.-W.C. and C.-L.K.; Validation, C.-S.L., C.-W.C. and C.-L.K.; Formal analysis, C.-S.L. and C.-W.C.; Investigation, C.-S.L., C.-W.C. and C.-L.K.; Resources, C.-S.L. and C.-W.C.; Data curation, C.-S.L., C.-W.C. and C.-L.K.; Writing—original draft, C.-S.L.; Writing—review & editing, C.-W.C.; Visualization, C.-S.L., C.-W.C. and C.-L.K.; Supervision, C.-S.L. and C.-W.C.; Project administration, C.-W.C.; Funding acquisition, C.-W.C. All authors have read and agreed to the published version of the manuscript.

Funding: This research received no external funding.

Data Availability Statement: The data presented in this study are available on request from the corresponding authors.

Acknowledgments: The authors would like to express their thanks to the reviewers, who supplied valuable opinions to improve the quality of this paper.

Conflicts of Interest: The authors declare no conflicts of interest.

References

1. Liu, C.S. An optimal multi-vector iterative algorithm in a Krylov subspace for solving the ill-posed linear inverse problems. *Comput. Mater. Contin.* **2013**, *33*, 175–198.
2. Dongarra, J.; Sullivan, F. Guest editors' introduction to the top 10 algorithms. *Comput. Sci. Eng.* **2000**, *2*, 22–23. [CrossRef]
3. Simoncini, V.; Szyld, D.B. Recent computational developments in Krylov subspace methods for linear systems. *Numer. Linear Alg. Appl.* **2007**, *14*, 1–59. [CrossRef]
4. Saad, Y.; Schultz, M.H. GMRES: A generalized minimal residual algorithm for solving nonsymmetric linear systems. *SIAM J. Sci. Stat. Comput.* **1986**, *7*, 856–869. [CrossRef]
5. Saad, Y. Krylov subspace methods for solving large unsymmetric linear systems. *Math. Comput.* **1981**, *37*, 105–126. [CrossRef]
6. Freund, R.W.; Nachtigal, N.M. QMR: A quasi-minimal residual method for non-Hermitian linear systems. *Numer. Math.* **1991**, *60*, 315–339. [CrossRef]
7. van Den Eshof, J.; Sleijpen, G.L.G. Inexact Krylov subspace methods for linear systems. *SIAM J. Matrix Ana. Appl.* **2004**, *26*, 125–153. [CrossRef]
8. Hestenes, M.R.; Stiefel, E.L. Methods of conjugate gradients for solving linear systems. *J. Res. Nat. Bur. Stand.* **1952**, *49*, 409–436. [CrossRef]
9. Lanczos, C. Solution of systems of linear equations by minimized iterations. *J. Res. Nat. Bur. Stand.* **1952**, *49*, 33–53. [CrossRef]
10. Paige, C.C.; Saunders, M.A. Solution of sparse indefinite systems of linear equations. *SIAM J. Numer. Anal.* **1975**, *12*, 617–629. [CrossRef]
11. Fletcher, R. *Conjugate Gradient Methods for Indefinite Systems*; Lecture Notes in Math; Springer: Berlin/Heidelberg, Germany, 1976; Volume 506, pp. 73–89.
12. Sonneveld, P. CGS: A fast Lanczos-type solver for nonsymmetric linear systems. *SIAM J. Sci. Stat. Comput.* **1989**, *10*, 36–52. [CrossRef]
13. van der Vorst, H.A. Bi-CGSTAB: A fast and smoothly converging variant of Bi-CG for the solution of nonsymmetric linear systems. *SIAM J. Sci. Stat. Comput.* **1992**, *13*, 631–644. [CrossRef]
14. Saad, Y.; van der Vorst, H.A. Iterative solution of linear systems in the 20th century. *J. Comput. Appl. Math.* **2000**, *123*, 1–33. [CrossRef]
15. Saad, Y. *Iterative Methods for Sparse Linear Systems*, 2nd ed.; SIAM: Philadelphia, PA, USA, 2003.
16. van der Vorst, H.A. *Iterative Krylov Methods for Large Linear Systems*; Cambridge University Press: New York, NY, USA, 2003.
17. Golub, G. Numerical methods for solving linear least squares problems. *Numer. Math.* **1965**, *7*, 206–216. [CrossRef]
18. Choi, S.C.; Paige, C.C.; Saunders, M.A. MINRES-QLP: A Krylov subspace method for indefinite or singular symmetric systems. *SIAM J. Sci. Comput.* **2011**, *33*, 1810–1836. [CrossRef]
19. Petkovic, M.D.; Stanimirovic, P.S. Iterative method for computing Moore-Penrose inverse based on Penrose equations. *J. Comput. Appl. Math.* **2011**, *235*, 1604–1613. [CrossRef]
20. Petkovic, M.D.; Stanimirovic, P.S. Two improvements of the iterative method for computing Moore-Penrose inverse based on Penrose equations. *J. Comput. Appl. Math.* **2014**, *267*, 61–71. [CrossRef]

21. Katsikis, V.N.; Pappas, D.; Petralias, A. An improved method for the computation of the Moore-Penrose inverse matrix. *Appl. Math. Comput.* **2011**, *217*, 9828–9834. [CrossRef]
22. Stanimirovic, I.; Tasic, M. Computation of generalized inverse by using the LDL* decomposition. *Appl. Math. Lett.* **2012**, *25*, 526–531. [CrossRef]
23. Sheng, X.; Wang, T. An iterative method to compute Moore-Penrose inverse based on gradient maximal convergence rate. *Filomat* **2013**, *27*, 1269–1276. [CrossRef]
24. Toutounian, F.; Ataei, A. A new method for computing Moore-Penrose inverse matrices. *J. Comput. Appl. Math.* **2009**, *228*, 412–417. [CrossRef]
25. Zhang, J.; Guo, J. On relaxed greedy randomized coordinate descent methods for solving large linear least-squares problems. *Appl. Numer. Math.* **2020**, *157*, 372–384. [CrossRef]
26. Lipitakis, A.-D.; Filelis-Papadopoulos, C.K.; Gravvanis, G.A.; Anagnostopoulos, D. A note on parallel approximate pseudoinverse matrix techniques for solving linear least squares problems. *J. Comput. Sci.* **2020**, *41*, 101092. [CrossRef]
27. Niu, Y.-Q.; Zhang, B. A new randomized Gauss–Seidel method for solving linear least-squares problems. *Appl. Math. Lett.* **2021**, *116*, 107057. [CrossRef]
28. Bojanczyk, A.W. Algorithms for indefinite linear least squares problems. *Linear Algebra Appli.* **2021**, *623*, 104–127. [CrossRef]
29. Du, K.; Ruan, C.-C.; Sun, X.-H. On the convergence of a randomized block coordinate descent algorithm for a matrix least squares problem. *Appl. Math. Lett.* **2022**, *124*, 107689. [CrossRef]
30. Chakrabarti, K.; Gupta, N.; Chopra, N. Iterative pre-conditioning for expediting the distributed gradient-descent method: The case of linear least-squares problem. *Automatica* **2022**, *137*, 110095. [CrossRef]
31. Zhang, Y.; Li, H. Splitting-based randomized iterative methods for solving indefinite least squares problem. *Appl. Math. Comput.* **2023**, *446*, 127892. [CrossRef]
32. Pes, F.; Rodriguez, G. A projection method for general form linear least-squares problems. *Appl. Math. Lett.* **2023**, *145*, 108780. [CrossRef]
33. Kuo, Y.-C.; Liu, C.-S. An index search method based inner-outer iterative algorithm for solving nonnegative least squares problems. *J. Comput. Appl. Math.* **2023**, *424*, 114954. [CrossRef]
34. Jin, L.-L.; Li, H.B. Greedy double subspaces coordinate descent method for solving linear least-squares problems. *J. Comput. Sci.* **2023**, *70*, 102029. [CrossRef]
35. Jahvani, M.; Guay, M. Solving least-squares problems in directed networks: A distributed approach. *Comput. Chem. Eng.* **2024**, *185*, 108654. [CrossRef]
36. Laura Arias, M.; Contino, M.; Maestripieri, A.; Marcantognini, S. Matrix representations of multivalued projections and least squares problems. *J. Math. Anal. Appl.* **2024**, *530*, 127631. [CrossRef]
37. Golub, G.H.; Reinsch, C. Singular value decomposition and least squares solutions. *Numer. Math.* **1970**, *14*, 403–420. [CrossRef]
38. Björck, A. *Numerical Methods for Least Squares Problems*; SIAM Publisher: Philadelphia, PA, USA, 1996.
39. Liu, C.S. A doubly optimized solution of linear equations system expressed in an affine Krylov subspace. *J. Comput. Appl. Math.* **2014**, *260*, 375–394. [CrossRef]
40. Liu, C.S. Optimal algorithms in a Krylov subspace for solving linear inverse problems by MFS. *Eng. Anal. Bound. Elem.* **2014**, *44*, 64–75. [CrossRef]
41. Liu, C.S. A double optimal descent algorithm for iteratively solving ill-posed linear inverse problems. *Inv. Prob. Sci. Eng.* **2015**, *23*, 38–66. [CrossRef]
42. Liu, C.S. A maximal projection solution of ill-posed linear system in a column subspace, better than the least squares solution. *Comput. Math. Appl.* **2014**, *67*, 1998–2014. [CrossRef]
43. Papez, J.; Tichy, P. Estimating error norms in CG-like algorithms for least-squares and least-norm problems. *Numer. Algorithms* **2024** , 1–28. [CrossRef]
44. Abbasi, M.; Nikazad, T. Bounded perturbations resilient iterative methods for linear systems and least squares problems: Operator-based approaches, analysis, and performance evaluation. *BIT Numer. Math.* **2024**, *64*, 15. [CrossRef]
45. Xia, Y.; Chen, T.; Shan, J. A novel iterative method for computing generalized inverse. *Neural Comput.* **2014**, *26*, 449–465. [CrossRef] [PubMed]
46. Feng, G.; Li, M.; Chen, C.S. On the ill-conditioning of the MFS for irregular boundary data with sufficient regularity. *Eng. Anal. Bound. Elem.* **2014**, *41*, 98–102. [CrossRef]
47. Liu, C.S.; Kuo, C.L. A multiple-scale Pascal polynomial triangle solving elliptic equations and inverse Cauchy problems. *Eng. Anal. Bound. Elem.* **2016**, *62*, 35–43. [CrossRef]

algorithms

MDPI

Article

Re-Orthogonalized/Affine GMRES and Orthogonalized Maximal Projection Algorithm for Solving Linear Systems

Chein-Shan Liu [1], Chih-Wen Chang [2,*] and Chung-Lun Kuo [1]

[1] Center of Excellence for Ocean Engineering, National Taiwan Ocean University, Keelung 202301, Taiwan; csliu@ntou.edu.tw (C.-S.L.); ejil1215@gmail.com (C.-L.K.)
[2] Department of Mechanical Engineering, National United University, Miaoli 36063, Taiwan
* Correspondence: cwchang@nuu.edu.tw

Abstract: GMRES is one of the most powerful and popular methods to solve linear systems in the Krylov subspace; we examine it from two viewpoints: to maximize the decreasing length of the residual vector, and to maintain the orthogonality of the consecutive residual vector. A stabilization factor, η, to measure the deviation from the orthogonality of the residual vector is inserted into GMRES to preserve the orthogonality automatically. The re-orthogonalized GMRES (ROGMRES) method guarantees the absolute convergence; even the orthogonality is lost gradually in the GMRES iteration. When $\eta < 1/2$, the residuals' lengths of GMRES and GMRES(m) no longer decrease; hence, $\eta < 1/2$ can be adopted as a stopping criterion to terminate the iterations. We prove $\eta = 1$ for the ROGMRES method; it automatically keeps the orthogonality, and maintains the maximality for reducing the length of the residual vector. We improve GMRES by seeking the descent vector to minimize the residual in a larger space of the affine Krylov subspace. The resulting orthogonalized maximal projection algorithm (OMPA) is identified as having good performance. We further derive the iterative formulas by extending the GMRES method to the affine Krylov subspace; these equations are slightly different from the equations derived by Saad and Schultz (1986). The affine GMRES method is combined with the orthogonalization technique to generate a powerful affine GMRES (A-GMRES) method with high performance.

Keywords: GMRES; absolute convergence; orthogonality of consecutive residual vector; maximal decreasing length of residual vector; restarted and re-orthogonalized GMRES; affine GMRES; orthogonalized maximal projection algorithm

MSC: 65F10; 15A06

1. Introduction

In the paper, we revisit GMRES and GMRES(m) to solve

$$\mathbf{A}\mathbf{x} = \mathbf{b},\ \mathbf{x}, \mathbf{b} \in \mathbb{R}^n,\ \mathbf{A} \in \mathbb{R}^{n\times n}. \tag{1}$$

In terms of an initial guess, $\mathbf{x}_0$, and the initial residual $\mathbf{r}_0 = \mathbf{b} - \mathbf{A}\mathbf{x}_0$, Equation (1) is equivalent to

$$\mathbf{A}\mathbf{u} = \mathbf{r}_0, \tag{2}$$

where

$$\mathbf{u} = \mathbf{x} - \mathbf{x}_0 \tag{3}$$

is the descent vector or called a corrector.

Equation (2) can be used to search a proper descent vector, $\mathbf{u}$, which is inserted into Equation (3) to produce a solution, $\mathbf{x}$, closer to the exact solution. We suppose an m-dimensional Krylov subspace:

$$\mathcal{K}_m(\mathbf{A}, \mathbf{r}_0) := \text{Span}\{\mathbf{r}_0, \mathbf{A}\mathbf{r}_0, \ldots, \mathbf{A}^{m-1}\mathbf{r}_0\}. \tag{4}$$

Krylov subspaces are named after A.N. Krylov in a paper [1], who described a method for solving a secular equation to determine the frequencies of small vibrations of material systems.

Many well-developed Krylov subspace methods for solving Equation (1) have been discussed in the text books [2–5]. Godunov and Gordienko [6] used an optimal representation of vectors in the Krylov space with the help of a variational problem; the extremum of the variational problem is the solution of the Kalman matrix equation, in which the 2-norm minimal solution is used as a characteristic of the Krylov space.

One of the most important classes of numerical methods for iteratively solving Equation (1) is the Krylov subspace methods [7,8], and the preconditioned Krylov subspace methods [9]. Recently, Bouyghf et al. [10] presented a comprehensive framework for unifying the Krylov subspace methods for Equation (1).

The GMRES method in [11] used the Petrov–Galerkin method to search $\mathbf{u} \in \mathcal{K}_m$ via a perpendicular property:

$$\mathbf{r}_0 - \mathbf{A}\mathbf{u} \perp \mathcal{L}_m = \mathbf{A}\mathcal{K}_m. \tag{5}$$

The descent vector $\mathbf{u} \in \mathcal{K}_m$ is achieved by minimizing the length of the residual vector [2]:

$$\min \|\mathbf{r}\|, \tag{6}$$

where

$$\mathbf{r} = \mathbf{b} - \mathbf{A}\mathbf{x} = \mathbf{r}_0 - \mathbf{A}\mathbf{u}, \tag{7}$$

and $\|\mathbf{r}\|$ is the Euclidean norm of the residual vector, $\mathbf{r}$.

In many Krylov subspace methods, the central idea is the projection. The method of GMRES is a special case of the Krylov subspace methods with an oblique projection. Let $\mathbf{V}$ and $\mathbf{W} = \mathbf{A}\mathbf{V}$ be the matrix representations of $\mathcal{K}_m$ and $\mathcal{L}_m$, respectively. We project the residual Equation (2) from an n-dimensional space to an m-dimensional subspace:

$$\mathbf{W}^{\mathrm{T}}\mathbf{A}\mathbf{u} = \mathbf{W}^{\mathrm{T}}\mathbf{r}_0. \tag{8}$$

We seek $\mathbf{u}$ in the subspace $\mathcal{K}_m$ by

$$\mathbf{u} = \mathbf{V}\mathbf{y}; \tag{9}$$

hence, Equation (8) becomes

$$\mathbf{W}^{\mathrm{T}}\mathbf{W}\mathbf{y} = \mathbf{W}^{\mathrm{T}}\mathbf{r}_0. \tag{10}$$

Solving $\mathbf{y}$ from Equation (10) and inserting it into Equation (9), we can obtain $\mathbf{u}$, and, hence, the next $\mathbf{x} = \mathbf{x}_0 + \mathbf{u}$; $\mathbf{x}$ is a correction of $\mathbf{x}_0$, and we hope $\|\mathbf{r}\| < \|\mathbf{r}_0\|$.

Meanwhile, Equation (7) becomes

$$\mathbf{r} = \mathbf{r}_0 - \mathbf{W}\mathbf{y}, \tag{11}$$

where $\mathbf{W}\mathbf{y} = \mathbf{A}\mathbf{u}$. By means of Equations (5), (7), and (11), we enforce the perpendicular property by

$$\mathbf{r}_0 - \mathbf{W}\mathbf{y} \perp \mathcal{L}_m; \tag{12}$$

it is equivalent to the projection of $\mathbf{r}_0$ on $\mathcal{L}_m$. $\mathbf{W}\mathbf{y} = \mathbf{A}\mathbf{u}$ is the projection of $\mathbf{r}_0$ on $\mathcal{L}_m$, and $\mathbf{r}$ is perpendicular to $\mathcal{L}_m$, which ensures that the length of the residual vector is decreasing. For seeking a fast convergent iterative algorithm, we must keep the orthogonality, and simultaneously maximize the projection quantity $\|\mathbf{Wy}\| = \|\mathbf{Au}\|$. Later on, we will modify the GMRES method from these two aspects.

Liu [12] derived an effective iterative algorithm based on the maximal projection method to solve Equation (1) in the column space of $\mathbf{A}$. Recently, Liu et al. [13] applied the maximal projection method and the minimal residual norm method to solve the least-squares problems.

There are two reasons to cause the slowdown and even the stagnation of GMRES: one is the loss of orthogonality, and other is that its descent vector, $\mathbf{u} = \mathbf{Vy}$, is not the best one. There are different methods to accelerate and improve GMRES [14–17]. The restart technique is often applied to the Krylov subspace methods, which, however, generally slows their convergence. Another way to accelerate the convergence is updating the initial value [18,19]. The preconditioning technology is often used to speed up GMRES and other Krylov subspace methods [20–24]. The multipreconditioned GMRES [25] is also proposed by using two or more preconditioners simultaneously. The polynomial preconditioned GMRES and GMRES-DR were also developed in [26]. As an augment of the Krylov subspace, some authors sought better descent vectors in a larger space to accelerate the convergence of GMRES [27–29]. Recently, Carson et al. [30] presented mathematically important and practically relevant phenomena that uncover the behavior of GMRES through a discussion of computed examples; they considered the conjugate gradient (CG) and GMRES methods crucially important for further development and practical applications, as well as other Krylov subspace methods.

Contribution and Novelty:

The major contribution and novelty of the present paper are that we propose a quite simple strategy to overcome the slowdown and stagnation behavior of GMRES by inserting a stabilization factor, $\eta = \mathbf{r}^{\mathrm{T}}\mathbf{Au}/\|\mathbf{Au}\|^2$, into the iterative equation. In doing so, the orthogonality property of the residual vector and the stepwise decreasing property of the residual length are automatically preserved. We extend the space of GMRES to an affine Krylov subspace for seeking a better descent vector. When the maximal projection method and the affine technique are combined into the orthogonalization technique, two very powerful iterative algorithms for solving linear systems with high performance are developed.

Highlight:

- Examine the iterative algorithm for linear systems from the viewpoint of maximizing the decreasing quantity of the residual and maintaining the orthogonality of the consecutive residual vector.
- A stabilization factor to measure the deviation from the orthogonality is inserted into GMRES to preserve the orthogonality automatically.
- The re-orthogonalized GMRES guarantees the preservation of orthogonality, even if the orthogonality is gradually lost in the iterations by means of GMRES.
- Improve GMRES by seeking the descent vector to minimize the length of the residual vector in a larger space of the affine Krylov subspace.
- The new algorithms are all absolute convergence.

Outline:

The Arnoldi process and the conventional GMRES and GMRES(m) are described in Section 2. In Section 3, we propose a new algorithm modified from GMRES by inserting a stabilization factor, η, which is named a re-orthogonalized GMRES (ROGMRES). We discuss ROGMRES from two aspects: preserving the orthogonality and maximally decreasing the length of the residual vector; we also prove that ROGMRES can automatically maintain $\eta = 1$, and preserve the good property of orthogonality and the maximality of reducing the residual. In Section 4, a better descent vector is sought in a larger affine Krylov subspace by using the maximal projection method; upon combining it with the orthogonalization technique, we propose a new algorithm, the orthogonalized maximal projection algorithm (OMPA). The iterative equations in GMRES are extended to that in the affine Krylov subspace by using the minimum residual method in Section 5, which results in an affine

GMRES (A-GMRES) method. The numerical tests of some examples are given in Section 6. Section 7 concludes the achievements, novelties, and contributions of this paper.

2. GMRES and GMRES(m)

The Arnoldi process [2] is often used to orthonormalize the Krylov vectors $\mathbf{A}^j\mathbf{r}_0$, $j = 0, \ldots, m-1$ in Equation (4), such that the resultant vectors $\mathbf{v}_i$, $i = 1, \ldots, m$ satisfy $\mathbf{v}_i \cdot \mathbf{v}_j = \delta_{ij}$, $i, j = 1, \ldots, m$, where δ_{ij} is the Kronecker delta symbol. The full orthonormalization procedure, known as the Algorithm 1, is set up as follows.

Algorithm 1: Arnoldi process

Select m and give an initial $\mathbf{r}_0$
$\mathbf{v}_1 = \frac{\mathbf{r}_0}{\|\mathbf{r}_0\|}$
Do $j = 1 : m$
$\mathbf{w}_j = \mathbf{A}\mathbf{v}_j$
    Do $i = 1 : j$
    $h_{ij} = \mathbf{v}_i \cdot \mathbf{w}_j$
    $\mathbf{w}_j = \mathbf{w}_j - h_{ij}\mathbf{v}_i$
    Enddo of i
$h_{j+1,j} = \|\mathbf{w}_j\|$
$\mathbf{v}_{j+1} = \frac{\mathbf{w}_j}{h_{j+1,j}}$
Enddo of j

A dot between two vectors represents the inner product, like as $\mathbf{a} \cdot \mathbf{b} = \mathbf{a}^{\mathrm{T}}\mathbf{b}$. $\mathbf{V}$ denotes the Arnoldi matrix whose jth column is $\mathbf{v}_j$:

$$\mathbf{V} := [\mathbf{v}_1, \ldots, \mathbf{v}_m] \in \mathbb{R}^{n \times m}. \tag{13}$$

After m steps of the Arnoldi process, we can construct an upper Hessenberg matrix, $\mathbf{H}_m$, as follows:

$$\mathbf{H}_m = \begin{bmatrix} h_{11} & h_{12} & \cdots & h_{1m} \\ h_{21} & h_{22} & \cdots & h_{2m} \\ & h_{32} & \cdots & h_{3m} \\ & \ddots & & \vdots \\ & & h_{m,m-1} & h_{mm} \end{bmatrix}. \tag{14}$$

Utilizing $\mathbf{H}_m$, the Arnoldi process can be expressed as a matrix product form:

$$\mathbf{A}\mathbf{V} = \mathbf{V}\mathbf{H}_m + h_{m+1,m}\mathbf{v}_{m+1}\mathbf{e}_m^{\mathrm{T}}, \tag{15}$$

where $\mathbf{H}_m \in \mathbb{R}^{m \times m}$, and $\mathbf{e}_m = [0, \ldots, 0, 1]^{\mathrm{T}} \in \mathbb{R}^m$. Now, the augmented Hessenberg upper matrix, $\bar{\mathbf{H}}_m \in \mathbb{R}^{(m+1) \times m}$, can be formulated as

$$\bar{\mathbf{H}}_m = \begin{bmatrix} \mathbf{H}_m \\ h_{m+1,m}\mathbf{e}_m^{\mathrm{T}} \end{bmatrix} = \begin{bmatrix} h_{11} & h_{12} & \cdots & h_{1m} \\ h_{21} & h_{22} & \cdots & h_{2m} \\ & h_{32} & \cdots & h_{3m} \\ & \ddots & & \vdots \\ & & h_{m,m-1} & h_{mm} \\ 0 & \cdots & 0 & h_{m+1,m} \end{bmatrix}. \tag{16}$$

According to [2], the Algorithm 2 method is given as follows.

Algorithm 2: GMRES

1: Give m_0, $\mathbf{x}_0$, and ε
2: Do $m = m_0, \ldots$ $(k = m - m_0)$, until $\|\mathbf{r}_k\| < \varepsilon$
3: $\mathbf{r}_k = \mathbf{b} - \mathbf{A}\mathbf{x}_k$
4: $\mathbf{v}_1^k = \frac{\mathbf{r}_k}{\|\mathbf{r}_k\|}$
5: $\mathbf{V}_k = [\mathbf{v}_1^k, \ldots, \mathbf{v}_m^k]$ (by Arnoldi process)
6: Solve $(\bar{\mathbf{H}}_k^{\mathrm{T}} \bar{\mathbf{H}}_k)\mathbf{y}_k = \|\mathbf{r}_k\| \bar{\mathbf{H}}_k^{\mathrm{T}} \mathbf{e}_1$
7: $\mathbf{x}_{k+1} = \mathbf{x}_k + \mathbf{V}_k \mathbf{y}_k$

In the above, $\mathbf{x}_k$ is the kth step value of $\mathbf{x}$; $\mathbf{r}_k$ is the kth residual vector; $\mathbf{y}_k$ is the vector of m expansion coefficients used in $\mathbf{V}_k\mathbf{y}_k$; $\mathbf{e}_1$ is the first column of $\mathbf{I}_{m+1}$; $\mathbf{V}_k$ denotes the kth step $\mathbf{V}$ in Equation (13); $\bar{\mathbf{H}}_k \in \mathbb{R}^{(m+1)\times m}$ is the kth step augmented Hessenberg upper matrix in Equation (16); and $\mathbf{V}_k\mathbf{y}_k$ is the kth step descent vector. Notice that $\mathbf{V}_k\mathbf{y}_k \in \mathcal{K}_m(\mathbf{A}, \mathbf{r}_k)$.

In addition to Algorithm 3, there is a restarted version (GMRES(m)) [11,31] described as follows.

Algorithm 3: GMRES(m)

1: Give m_0, m_1, $\mathbf{x}_0$, and ε
2: Do $j = 1, \ldots$
3: Do $m = m_0, \ldots, m_1$ $(k = m - m_0)$
4: $\mathbf{r}_k = \mathbf{b} - \mathbf{A}\mathbf{x}_k$
5: $\mathbf{v}_1^k = \frac{\mathbf{r}_k}{\|\mathbf{r}_k\|}$
6: $\mathbf{V}_k = [\mathbf{v}_1^k, \ldots, \mathbf{v}_m^k]$ (by Arnoldi process)
7: Solve $(\bar{\mathbf{H}}_k^{\mathrm{T}} \bar{\mathbf{H}}_k)\mathbf{y}_k = \|\mathbf{r}_k\| \bar{\mathbf{H}}_k^{\mathrm{T}} \mathbf{e}_1$
8: $\mathbf{x}_{k+1} = \mathbf{x}_k + \mathbf{V}_k \mathbf{y}_k$
9: If $\|\mathbf{r}_{k+1}\| < \varepsilon$, stop
10: Otherwise, $\mathbf{x}_0 = \mathbf{x}_{k+1}$; go to 2

Instead of $m_0 = 1$ used in the original GMRES and GMRES(m), a suitable value of $m_0 \neq 1$ can speed up the convergence; hence, $m = m_1 - m_0 + 1$ is the frequency for restart. Rather than the original GMRES and GMRES(m) [2,11,31], we take $\mathbf{x}_k$, not $\mathbf{x}_0$, in (7) of Algorithm GMRES and in (8) of Algorithm GMRES(m). Because $\mathbf{x}_0$ is not updated in the original GMRES and GMRES(m), they converge slowly. In Algorithm GMRES(m), if we take $m_1 = m_0$, then we return to the usual iterative algorithm for GMRES, with a fixed-dimension, $m = m_0$, of the Krylov subspace.

The restart remedies the long-term recurrence and accumulations of the round-off errors of GMRES. In order to improve the restart process, several improvement techniques have been proposed, as mentioned above. Some techniques based on adaptively determining the restart frequency can be found in [32–35]. For a quicker convergence, we can update the current value of $\mathbf{x}$ at each iterative step. Because $\mathbf{x}_k$ is new information for determining the next value of $\mathbf{x}_{k+1}$, the result of $\mathbf{x}_k$ cannot be wasted for saving computational cost.

3. Maximally Decreasing Residual and Automatically Preserving Orthogonality

In this section, we develop a new iterative algorithm based on GMRES to solve Equation (1). The iterative form is

$$\mathbf{x}_{k+1} = \mathbf{x}_k + \mathbf{u}_k, \tag{17}$$

where $\mathbf{u}_k$ is the kth step descent vector, which is determined by the iterative algorithm. For instance, $\mathbf{u}_k$ in GMRES is $\mathbf{u}_k = \mathbf{V}_k\mathbf{y}_k$.

Lemma 1. *A better iterative scheme than that in Equation (17) is*

$$\mathbf{x}_{k+1} = \mathbf{x}_k + \frac{\mathbf{r}_k^T \mathbf{A}\mathbf{u}_k}{\|\mathbf{A}\mathbf{u}_k\|^2}\mathbf{u}_k. \tag{18}$$

Proof. Let

$$\mathbf{x}_{k+1} = \mathbf{x}_k + \eta \mathbf{u}_k; \tag{19}$$

it follows that

$$\mathbf{r}_{k+1} = \mathbf{r}_k - \eta \mathbf{A}\mathbf{u}_k. \tag{20}$$

Taking the squared norms of both sides yields

$$\|\mathbf{r}_{k+1}\|^2 = \|\mathbf{r}_k\|^2 - (2\eta \mathbf{r}_k^{\mathrm{T}} \mathbf{A}\mathbf{u}_k - \eta^2 \|\mathbf{A}\mathbf{u}_k\|^2) =: \|\mathbf{r}_k\|^2 - h. \tag{21}$$

To maximally reduce the residual we consider

$$\max_{\eta}\left\{h = 2\eta \mathbf{r}_k^{\mathrm{T}} \mathbf{A}\mathbf{u}_k - \eta^2 \|\mathbf{A}\mathbf{u}_k\|^2\right\}, \tag{22}$$

which leads to

$$\eta_k = \frac{\mathbf{r}_k^{\mathrm{T}} \mathbf{A}\mathbf{u}_k}{\|\mathbf{A}\mathbf{u}_k\|^2}. \tag{23}$$

Inserting η_k into Equation (19), the proof of Equation (18) is finished. □

Lemma 2. *In Equation (18) the following orthogonal property holds:*

$$\mathbf{r}_{k+1} \cdot (\mathbf{A}\mathbf{u}_k) = 0. \tag{24}$$

Proof. Apply $\mathbf{A}$ to Equation (18):

$$\mathbf{A}\mathbf{x}_{k+1} = \mathbf{A}\mathbf{x}_k + \frac{\mathbf{r}_k^{\mathrm{T}} \mathbf{A}\mathbf{u}_k}{\|\mathbf{A}\mathbf{u}_k\|^2}\mathbf{A}\mathbf{u}_k. \tag{25}$$

Using $\mathbf{A}\mathbf{x}_{k+1} = \mathbf{b} - \mathbf{r}_{k+1}$ and $\mathbf{A}\mathbf{x}_k = \mathbf{b} - \mathbf{r}_k$, Equation (25) changes to

$$\mathbf{r}_{k+1} = \mathbf{r}_k - \frac{\mathbf{r}_k^{\mathrm{T}} \mathbf{A}\mathbf{u}_k}{\|\mathbf{A}\mathbf{u}_k\|^2}\mathbf{A}\mathbf{u}_k. \tag{26}$$

Taking the inner product with $\mathbf{A}\mathbf{u}_k$ yields

$$\mathbf{r}_{k+1} \cdot (\mathbf{A}\mathbf{u}_k) = \mathbf{r}_k^{\mathrm{T}} \mathbf{A}\mathbf{u}_k - \frac{\mathbf{r}_k^{\mathrm{T}} \mathbf{A}\mathbf{u}_k}{\|\mathbf{A}\mathbf{u}_k\|^2}\|\mathbf{A}\mathbf{u}_k\|^2 = 0. \tag{27}$$

Equation (24) is proven. □

It follows from Equation (5) that $\eta = 1$ is implied by GMRES. However, during the GMRES iteration the numerical values of η may deviate from 1 to a great extent, even being zero or a negative value. Therefore, we propose a re-orthogonalized version of GMRES, namely the ROGMRES method (Algorithm 4), as follows.

Algorithm 4: ROGMRES

1: Give m_0, $\mathbf{x}_0$, and ε
2: Do $m = m_0, \ldots$ $(k = m - m_0)$, until $\|\mathbf{r}_k\| < \varepsilon$
3: $\mathbf{r}_k = \mathbf{b} - \mathbf{A}\mathbf{x}_k$
4: $\mathbf{v}_1^k = \frac{\mathbf{r}_k}{\|\mathbf{r}_k\|}$
5: $\mathbf{V}_k = [\mathbf{v}_1^k, \ldots, \mathbf{v}_m^k]$ (by Arnoldi process)
6: Solve $(\bar{\mathbf{H}}_k^{\mathrm{T}} \bar{\mathbf{H}}_k)\mathbf{y}_k = \|\mathbf{r}_k\| \bar{\mathbf{H}}_k^{\mathrm{T}} \mathbf{e}_1$
7: $\mathbf{u}_k = \mathbf{V}_k \mathbf{y}_k$
8: $\mathbf{x}_{k+1} = \mathbf{x}_k + \frac{\mathbf{r}_k^{\mathrm{T}} \mathbf{A}\mathbf{u}_k}{\|\mathbf{A}\mathbf{u}_k\|^2} \mathbf{u}_k$

Correspondingly, the ROGMRES(m) Algorithm 5 is given as follows.

Algorithm 5: ROGMRES(m)

1: Give m_0, m_1, $\mathbf{x}_0$, and ε
2: Do $j = 1, \ldots$
3: Do $m = m_0, \ldots, m_1$ $(k = m - m_0)$
4: $\mathbf{r}_k = \mathbf{b} - \mathbf{A}\mathbf{x}_k$
5: $\mathbf{v}_1^k = \frac{\mathbf{r}_k}{\|\mathbf{r}_k\|}$
6: $\mathbf{V}_k = [\mathbf{v}_1^k, \ldots, \mathbf{v}_m^k]$ (by Arnoldi process)
7: Solve $(\bar{\mathbf{H}}_k^{\mathrm{T}} \bar{\mathbf{H}}_k)\mathbf{y}_k = \|\mathbf{r}_k\| \bar{\mathbf{H}}_k^{\mathrm{T}} \mathbf{e}_1$
8: $\mathbf{u}_k = \mathbf{V}_k \mathbf{y}_k$
9: $\mathbf{x}_{k+1} = \mathbf{x}_k + \frac{\mathbf{r}_k^{\mathrm{T}} \mathbf{A}\mathbf{u}_k}{\|\mathbf{A}\mathbf{u}_k\|^2} \mathbf{u}_k$
10: If $\|\mathbf{r}_{k+1}\| < \varepsilon$, stop
11: Otherwise, $\mathbf{x}_0 = \mathbf{x}_{k+1}$; go to 2

Upon comparing with GMRES(m) in Section 2, the computational cost of ROGMRES(m) is slightly increased by computing an extra term, $\mathbf{r}_k^{\mathrm{T}} \mathbf{A}\mathbf{u}_k / \|\mathbf{A}\mathbf{u}_k\|^2$, at each iteration; it needs one matrix-vector product and two inner products of vectors.

Theorem 1. *For GMRES, if $\eta_k \neq 1$ is happened at the kth step, the orthogonality in Equation (5) cannot be continued after the kth step, i.e.,*

$$\mathbf{r}_{k+1} \cdot (\mathbf{A}\mathbf{u}_k) \neq 0; \tag{28}$$

moreover, if $\eta_k < 1/2$, the residual does not decrease, i.e.,

$$\|\mathbf{r}_{k+1}\| > \|\mathbf{r}_k\|. \tag{29}$$

For ROGMRES in Equation (18), the residual is absolutely decreased:

$$\|\mathbf{r}_{k+1}\| < \|\mathbf{r}_k\|, \tag{30}$$

and the orthogonality $\mathbf{r}_{k+1} \cdot (\mathbf{A}\mathbf{u}_k) = 0$ holds.

Proof. Applying $\mathbf{A}$ to $\mathbf{x}_{k+1} = \mathbf{x}_k + \mathbf{V}_k \mathbf{y}_k$ in Algorithm GMRES yields

$$\mathbf{r}_{k+1} = \mathbf{r}_k - \mathbf{A}\mathbf{u}_k, \tag{31}$$

where $\mathbf{u}_k = \mathbf{V}_k \mathbf{y}_k$.

It follows from Equation (23) that

$$(\mathbf{r}_k - \eta_k \mathbf{A}\mathbf{u}_k) \cdot (\mathbf{A}\mathbf{u}_k) = 0, \tag{32}$$

which can be rearranged to

$$(\mathbf{r}_k - \mathbf{A}\mathbf{u}_k) \cdot (\mathbf{A}\mathbf{u}_k) = (\eta_k - 1)\|\mathbf{A}\mathbf{u}_k\|^2, \tag{33}$$

and then by Equation (31),

$$\mathbf{r}_{k+1} \cdot (\mathbf{A}\mathbf{u}_k) = (\eta_k - 1)\|\mathbf{A}\mathbf{u}_k\|^2. \tag{34}$$

If $\eta_k \neq 1$, we can prove Equation (28). Taking the squared norm of Equation (31) and using Equation (23) generates

$$\|\mathbf{r}_{k+1}\|^2 = \|\mathbf{r}_k\|^2 - 2\mathbf{r}_k \cdot (\mathbf{A}\mathbf{u}_k) + \|\mathbf{A}\mathbf{u}_k\|^2 = \|\mathbf{r}_k\|^2 + (1 - 2\eta_k)\|\mathbf{A}\mathbf{u}_k\|^2. \tag{35}$$

If $\eta_k < 1/2$, $(1 - 2\eta_k)\|\mathbf{A}\mathbf{u}_k\|^2 > 0$ and Equation (29) is proven.

For ROGMRES, taking the squared norm of Equation (26), we have

$$\|\mathbf{r}_{k+1}\|^2 = \|\mathbf{r}_k\|^2 - \frac{(\mathbf{r}_k^{\mathrm{T}}\mathbf{A}\mathbf{u}_k)^2}{\|\mathbf{A}\mathbf{u}_k\|^2}. \tag{36}$$

Because $(\mathbf{r}_k^{\mathrm{T}}\mathbf{A}\mathbf{u}_k)^2/\|\mathbf{A}\mathbf{u}_k\|^2 > 0$, Equation (30) is proven. The orthogonality $\mathbf{r}_{k+1} \cdot (\mathbf{A}\mathbf{u}_k) = 0$ was proven in Equation (24) of Lemma 2. □

Corollary 1. *For GMRES, the orthogonality of the consecutive residual vector is equivalent to the maximality of h in Equation (22), where*

$$\eta_k = 1 \Leftrightarrow \mathbf{r}_k^T\mathbf{A}\mathbf{u}_k = \|\mathbf{A}\mathbf{u}_k\|^2. \tag{37}$$

Proof. By means of Equation (34), to satisfy the orthogonality condition of the residual vector we require $\eta_k = 1$; it implies Equation (37) by the definition in Equation (23). At the same time, Equation (36) changes to

$$\|\mathbf{r}_{k+1}\|^2 = \|\mathbf{r}_k\|^2 - \|\mathbf{A}\mathbf{u}_k\|^2. \tag{38}$$

Inserting $\eta_k = 1$ into Equation (22), we have

$$\max h = 2\mathbf{r}_k^{\mathrm{T}}\mathbf{A}\mathbf{u}_k - \|\mathbf{A}\mathbf{u}_k\|^2 = \|\mathbf{A}\mathbf{u}_k\|^2, \tag{39}$$

of which $\mathbf{r}_k^{\mathrm{T}}\mathbf{A}\mathbf{u}_k = \|\mathbf{A}\mathbf{u}_k\|^2$ in Equation (37) was used. For GMRES, obtaining the maximal value of h and preserving the orthogonality are the same. □

Remark 1. *To maintain $\eta_k = 1$ in GMRES is a key issue for preserving the orthogonality and for the maximality of reducing the residual. However, for the traditional GMRES, it is not always true that it can maintain $\eta_k = 1$ during the iteration process.*

Theorem 2. *Let*

$$\mathbf{w}_k = \frac{\mathbf{r}_k^T\mathbf{A}\mathbf{u}_k}{\|\mathbf{A}\mathbf{u}_k\|^2}\mathbf{u}_k \tag{40}$$

be the descent vector of ROGMRES in

$$\mathbf{x}_{k+1} = \mathbf{x}_k + \mathbf{w}_k. \tag{41}$$

The following identity holds:

$$\eta_k = \frac{\mathbf{r}_k^T\mathbf{A}\mathbf{w}_k}{\|\mathbf{A}\mathbf{w}_k\|^2} = 1; \tag{42}$$

such that ROGMRES can automatically maintain the orthogonality of the residual vector, and achieve the maximality of the decreasing length of the residual vector.

Proof. Let

$$\gamma = \frac{\mathbf{r}_k^{\mathrm{T}} \mathbf{A} \mathbf{u}_k}{\|\mathbf{A}\mathbf{u}_k\|^2}; \tag{43}$$

Equation (40) is written as

$$\mathbf{w}_k = \gamma \mathbf{u}_k. \tag{44}$$

Inserting it into Equation (42) produces

$$\eta_k = \frac{1}{\gamma} \frac{\mathbf{r}_k^{\mathrm{T}} \mathbf{A} \mathbf{u}_k}{\|\mathbf{A}\mathbf{u}_k\|^2}; \tag{45}$$

by means of Equation (43), it follows that

$$\eta_k = \frac{\|\mathbf{A}\mathbf{u}_k\|^2}{\mathbf{r}_k^{\mathrm{T}} \mathbf{A} \mathbf{u}_k} \frac{\mathbf{r}_k^{\mathrm{T}} \mathbf{A} \mathbf{u}_k}{\|\mathbf{A}\mathbf{u}_k\|^2} = 1. \tag{46}$$

Thus, Equation (42) is proven. The proof to satisfy the orthogonality condition of the consecutive residual vector and the maximality of h is the same as that given in Corollary 1. □

Remark 2. *The properties in ROGMRES are crucial so that it can automatically maintain $\eta_k = 1$, for preserving its good properties of the orthogonality, and the maximality for reducing the residual. The method in Equation (18) can be viewed as an automatically orthogonality preserving method. Numerical experiments will verify these points.*

Theorem 3. *For ROGMRES in Equation (18), the iterative point, x_k, is located on an invariant manifold:*

$$\|\mathbf{r}_k\|^2 = \frac{\|\mathbf{r}_0\|^2}{Q_k}, \tag{47}$$

where $Q_0 = 1$; $Q_k > 0$ with $Q_{k+1} > Q_k$ is an increasing sequence.

Proof. We begin with the following time-invariant manifold:

$$\|\mathbf{r}\|^2 = \frac{\|\mathbf{r}_0\|^2}{Q(t)}, \tag{48}$$

where $Q(0) = 1$, and $Q(t)$ is an increasing function of t.

We suppose that the evolution of $\mathbf{x}$ in time is driven by a descent vector, $\mathbf{u}$:

$$\dot{\mathbf{x}} = \gamma \mathbf{u}. \tag{49}$$

Equation (48) is a time-invariant manifold, such that its time differential is zero, i.e.,

$$\frac{1}{2Q(t)} \dot{Q}(t) \|\mathbf{r}\|^2 - \mathbf{r} \cdot (\mathbf{A}\dot{\mathbf{x}}) = 0, \tag{50}$$

where $\dot{\mathbf{r}} = -\mathbf{A}\dot{\mathbf{x}}$ was used.

Inserting Equation (49) into Equation (50) yields

$$\gamma = \frac{\dot{Q}(t)}{2Q(t)} \frac{\|\mathbf{r}\|^2}{\mathbf{r}^{\mathrm{T}} \mathbf{A} \mathbf{u}}. \tag{51}$$

Inserting γ into Equation (49) and using $\mathbf{v} = \mathbf{A}\mathbf{u}$, we can derive

$$\dot{\mathbf{x}} = q(t)\frac{\|\mathbf{r}\|^2}{\mathbf{r}^{\mathrm{T}}\mathbf{v}}\mathbf{u}, \tag{52}$$

where

$$q(t) := \frac{\dot{Q}(t)}{2Q(t)} > 0. \tag{53}$$

By applying the forward Euler scheme to Equation (52), we have

$$\mathbf{x}(t+\Delta t) = \mathbf{x}(t) + \beta\frac{\|\mathbf{r}\|^2}{\mathbf{r}^{\mathrm{T}}\mathbf{v}}\mathbf{u}, \tag{54}$$

where

$$\beta = q(t)\Delta t > 0. \tag{55}$$

An iterative form of Equation (54) is

$$\mathbf{x}_{k+1} = \mathbf{x}_k + \beta_k\frac{\|\mathbf{r}_k\|^2}{\mathbf{r}_k^{\mathrm{T}}\mathbf{v}_k}\mathbf{u}_k. \tag{56}$$

It generates the residual form:

$$\mathbf{r}_{k+1} = \mathbf{r}_k - \beta_k\frac{\|\mathbf{r}_k\|^2}{\mathbf{r}_k^{\mathrm{T}}\mathbf{v}_k}\mathbf{v}_k. \tag{57}$$

Taking the squared norms of both sides yields

$$\|\mathbf{r}_{k+1}\|^2 = \|\mathbf{r}_k\|^2 - 2\beta_k\|\mathbf{r}_k\|^2 + \beta_k^2\frac{\|\mathbf{r}_k\|^4}{(\mathbf{r}_k^{\mathrm{T}}\mathbf{v}_k)^2}\|\mathbf{v}_k\|^2. \tag{58}$$

Dividing both sides by $\|\mathbf{r}_k\|^2$ renders

$$a_k\beta_k^2 - 2\beta_k + 1 - \frac{\|\mathbf{r}_{k+1}\|^2}{\|\mathbf{r}_k\|^2} = 0, \tag{59}$$

where

$$a_k := \frac{\|\mathbf{r}_k\|^2\|\mathbf{v}_k\|^2}{(\mathbf{r}_k^{\mathrm{T}}\mathbf{v}_k)^2} \geq 1. \tag{60}$$

By means of Equation (47), Equation (59) changes to

$$a_k\beta_k^2 - 2\beta_k + 1 - \frac{Q_k}{Q_{k+1}} = 0. \tag{61}$$

If we take

$$\beta_k = \frac{1}{a_k} = \frac{(\mathbf{r}_k^{\mathrm{T}}\mathbf{v}_k)^2}{\|\mathbf{r}_k\|^2\|\mathbf{v}_k\|^2}, \quad \frac{Q_k}{Q_{k+1}} = 1 - \frac{1}{a_k}, \tag{62}$$

Equation (61) is satisfied. In terms of β_k, Equation (56) is recast to

$$\mathbf{x}_{k+1} = \mathbf{x}_k + \frac{(\mathbf{r}_k^{\mathrm{T}}\mathbf{v}_k)^2}{\|\mathbf{r}_k\|^2\|\mathbf{v}_k\|^2}\frac{\|\mathbf{r}_k\|^2}{\mathbf{r}_k^{\mathrm{T}}\mathbf{v}_k}\mathbf{u}_k = \mathbf{x}_k + \frac{\mathbf{r}_k^{\mathrm{T}}\mathbf{v}_k}{\|\mathbf{v}_k\|^2}\mathbf{u}_k. \tag{63}$$

Noticing $\mathbf{v}_k = \mathbf{A}\mathbf{u}_k$, Equation (63) is just the iterative Equation (18) for ROGMRES. □

Remark 3. *For the factor defined in Equation (23), η_k plays a vital role in Equation (34) to dominate the orthogonal property of GMRES, and also in Equation (35) to control the residual decreasing property of GMRES. When GMRES blows up for some problems, by introducing η_k in GMRES,* *we can stabilize the iterations to avoid blowing up. In this viewpoint, η_k is a stabilization factor for GMRES.*

Inspired by Remark 3, two simple methods can be developed for GMRES and GMRES(m): we can compute η_k at each step, and consider $|\eta_k - 1| > 10^{-10}$ or $\eta_k < 1/2$ to be the stopping criteria of the iterations; they are labeled as GMRES(η) and GMRES(m,η), respectively.

Remark 4. *h given in Equation (21) is a decreasing quantity of the residual, whose maximal value by means of Equation (36) is*

$$\max h = \frac{(\mathbf{r}_k^T \mathbf{A}\mathbf{u}_k)^2}{\|\mathbf{A}\mathbf{u}_k\|^2}. \tag{64}$$

Upon comparing with a_k in Equation (60), we have

$$\max h = \frac{\|\mathbf{r}_k\|^2}{a_k}. \tag{65}$$

Because $a_k \geq 1$, the best value of $\max h$ that can be obtained is $\|\mathbf{r}_k\|^2$ when $a_k = 1$; in this situation, it will directly lead to the exact solution at the $k+1$ step, owing to $\|\mathbf{r}_{k+1}\| = 0$, which is obtained by inserting $\mathbf{A}\mathbf{u}_k = \mathbf{r}_k$ into Equation (38). In this sense, a_k can also be an objective function. a_0 correlates to $f = (\mathbf{r}_0^T \mathbf{A}\mathbf{u})^2/\|\mathbf{A}\mathbf{u}\|^2$, which is the quantity of the projection from $\mathbf{r}_0$ on $\mathcal{L}_m = \mathbf{A}\mathcal{K}_m$, by

$$a_0 = \frac{\|\mathbf{r}_0\|^2}{f}. \tag{66}$$

In the next section, we will maximize f, i.e., minimize a_0, to find the best descent vector for $\mathbf{u}$.

4. Orthogonalized Maximal Projection Algorithm

As seen from Equation (38), we must make $\|\mathbf{A}\mathbf{u}\|^2$ as large as possible. $\mathbf{A}\mathbf{u} = \mathbf{A}\mathbf{V}\mathbf{y}$, mentioned in Section 1, is the projection of $\mathbf{r}_0$ on $\mathcal{L}_m = \mathbf{A}\mathcal{K}_m$. Therefore we consider

$$\max_{\mathbf{u}} \left\{ f = \frac{(\mathbf{r}_0 \cdot \mathbf{v})^2}{\|\mathbf{v}\|^2} \right\} \tag{67}$$

to be the maximal projection of $\mathbf{r}_0$ on the direction $\mathbf{v}/\|\mathbf{v}\|$, where $\mathbf{v} = \mathbf{A}\mathbf{u}$. In view of Equation (66), the maximization of f is equivalent to the minimization of a_0.

We can expand $\mathbf{u} \in \mathbf{r}_0 + \mathcal{K}_m$ via

$$\mathbf{u} = \mathbf{r}_0 + \mathbf{V}\mathbf{y}, \tag{68}$$

which is slightly different from the descent vector $\mathbf{u} = \mathbf{V}\mathbf{y} \in \mathcal{K}_m$ used in GMRES.

Theorem 4. *For $\mathbf{u} \in \mathbf{r}_0 + \mathcal{K}_m(\mathbf{A}, \mathbf{r}_0)$, $\mathbf{r}_0 \neq \mathbf{0}$, the optimal $\mathbf{u}$ approximately satisfying Equation (2) and subjecting to Equation (67) is given in Equation (68), where*

$$\mathbf{W} = \mathbf{A}\mathbf{V},\ \mathbf{C} = \mathbf{W}^T\mathbf{W}, \tag{69}$$

$$\mathbf{C}\mathbf{y} = \mathbf{W}^T(\mathbf{r}_0 - \mathbf{A}\mathbf{r}_0). \tag{70}$$

Proof. Due to $\mathbf{v} = \mathbf{A}\mathbf{u}$ and Equation (68),

$$\mathbf{v} = \mathbf{v}_0 + \mathbf{W}\mathbf{y}, \tag{71}$$

where $\mathbf{W} = \mathbf{A}\mathbf{V}$ and

$$\mathbf{v}_0 = \mathbf{A}\mathbf{r}_0. \tag{72}$$

By Equation (71), one has

$$\mathbf{r}_0 \cdot \mathbf{v} = \mathbf{r}_0 \cdot \mathbf{v}_0 + \mathbf{r}_0^{\mathrm{T}} \mathbf{W} \mathbf{y}, \tag{73}$$

$$\|\mathbf{v}\|^2 = \|\mathbf{v}_0\|^2 + 2\mathbf{v}_0^{\mathrm{T}} \mathbf{W} \mathbf{y} + \mathbf{y}^{\mathrm{T}} \mathbf{C} \mathbf{y}, \tag{74}$$

where

$$\mathbf{C} := \mathbf{W}^{\mathrm{T}} \mathbf{W} > \mathbf{0} \in \mathbb{R}^{m \times m}. \tag{75}$$

Resorting to the maximality condition for f in Equation (67), we can obtain

$$\mathbf{r}_0 \cdot \mathbf{v} \mathbf{y}_2 - \|\mathbf{v}\|^2 \mathbf{y}_1 = \mathbf{0}, \tag{76}$$

where

$$\mathbf{y}_1 := \nabla_{\mathbf{y}} (\mathbf{r}_0 \cdot \mathbf{v}) = \mathbf{W}^{\mathrm{T}} \mathbf{r}_0, \tag{77}$$

$$\mathbf{y}_2 := \frac{1}{2} \nabla_{\mathbf{y}} \|\mathbf{v}\|^2 = \mathbf{W}^{\mathrm{T}} \mathbf{v}_0 + \mathbf{C} \mathbf{y}. \tag{78}$$

From Equation (76), $\mathbf{y}_2$ is equal to $\mathbf{y}_1$:

$$\mathbf{y}_2 = \eta^{-1} \mathbf{y}_1 \Rightarrow \mathbf{y}_2 = \mathbf{y}_1, \tag{79}$$

because of

$$\eta^{-1} = \frac{\|\mathbf{v}\|^2}{\mathbf{r}_0 \cdot \mathbf{v}} = \frac{\|\mathbf{A}\mathbf{u}\|^2}{\mathbf{r}_0^{\mathrm{T}} \mathbf{A} \mathbf{u}} = 1, \tag{80}$$

according to Corollary 1.

Then, it follows from Equations (77)–(79) that

$$\mathbf{C} \mathbf{y} + \mathbf{W}^{\mathrm{T}} \mathbf{v}_0 = \mathbf{W}^{\mathrm{T}} \mathbf{r}_0, \tag{81}$$

which can be arranged to Equation (70). □

We pull back the minimization problem used in the GMRES method to the following one:

$$\min\{\|\mathbf{r}_0 - \mathbf{v}\|^2\} = \min\{\|\mathbf{r}_0 - \mathbf{A}\mathbf{u}\|^2\}, \tag{82}$$

which is derived from Equations (6) and (7).

Theorem 5. *For* $\mathbf{u} \in \mathbf{r}_0 + \mathcal{K}_m(\mathbf{A}, \mathbf{r}_0)$, *derived from the minimization problem in Equation (82),* $\mathbf{u}$ *in Equation (2) is optimized to be that in Theorem 4.*

Proof. From

$$\nabla_{\mathbf{y}} \|\mathbf{r}_0 - \mathbf{v}\|^2 = \nabla_{\mathbf{y}} \|\mathbf{v}\|^2 - 2\nabla_{\mathbf{y}} (\mathbf{r}_0 \cdot \mathbf{v}) = \mathbf{0}, \tag{83}$$

and Equations (78) and (77) it follows that

$$\mathbf{W}^{\mathrm{T}} \mathbf{v}_0 + \mathbf{C} \mathbf{y} = \mathbf{W}^{\mathrm{T}} \mathbf{r}_0, \tag{84}$$

which is just Equation (70). □

According to Theorems 4 and 5, the maximal projection is equivalent to the minimization of residual. Therefore, the maximal projection algorithm (MPA) is an extension of GMRES to a larger space of the affine Krylov subspace, not merely in the Krylov subspace.

By considering Lemma 1 and Theorem 5, we come to an improvement of ROGMRES as well as the double-improvement of GMRES to the following iterative formulas for the orthogonalized maximal projection algorithm (OMPA):

$$(\mathbf{W}^{\mathrm{T}}\mathbf{W})\mathbf{y} = \mathbf{W}^{\mathrm{T}}(\mathbf{r}_0 - \mathbf{A}\mathbf{r}_0), \tag{85}$$

$$\mathbf{u} = \mathbf{r}_0 + \mathbf{V}\mathbf{y}, \tag{86}$$

$$\mathbf{x} = \mathbf{x}_0 + \frac{\mathbf{r}_0^{\mathrm{T}}\mathbf{A}\mathbf{u}}{\|\mathbf{A}\mathbf{u}\|^2}\mathbf{u}. \tag{87}$$

Equation (85) can be further simplified in the next section. The algorithms of OMPA (Algorithm 6) and OMPA(m) (Algorithm 7) are given as follows.

Algorithm 6: OMPA

Give m_0, $\mathbf{x}_0$, and ε
Do $m = m_0, \ldots$ $(k = m - m_0)$, until $\|\mathbf{r}_k\| < \varepsilon$
$\mathbf{r}_k = \mathbf{b} - \mathbf{A}\mathbf{x}_k$
$\mathbf{v}_1^k = \frac{\mathbf{r}_k}{\|\mathbf{r}_k\|}$
$\mathbf{V}_k = [\mathbf{v}_1^k, \ldots, \mathbf{v}_m^k]$ (by Arnoldi process)
$\mathbf{W}_k = \mathbf{A}\mathbf{V}_k$
Solve $(\mathbf{W}_k^{\mathrm{T}}\mathbf{W}_k)\mathbf{y}_k = \mathbf{W}_k^{\mathrm{T}}(\mathbf{r}_k - \mathbf{A}\mathbf{r}_k)$
$\mathbf{u}_k = \mathbf{r}_k + \mathbf{V}_k\mathbf{y}_k$
$\mathbf{x}_{k+1} = \mathbf{x}_k + \frac{\mathbf{r}_k^{\mathrm{T}}\mathbf{A}\mathbf{u}_k}{\|\mathbf{A}\mathbf{u}_k\|^2}\mathbf{u}_k$

Algorithm 7: OMPA(m)

Give m_0, m_1, $\mathbf{x}_0$, and ε
Do $j = 1, \ldots$
Do $m = m_0, \ldots, m_1$ $(k = m - m_0)$
$\mathbf{r}_k = \mathbf{b} - \mathbf{A}\mathbf{x}_k$
$\mathbf{v}_1^k = \frac{\mathbf{r}_k}{\|\mathbf{r}_k\|}$
$\mathbf{V}_k = [\mathbf{v}_1^k, \ldots, \mathbf{v}_m^k]$ (by Arnoldi process)
$\mathbf{W}_k = \mathbf{A}\mathbf{V}_k$
Solve $(\mathbf{W}_k^{\mathrm{T}}\mathbf{W}_k)\mathbf{y}_k = \mathbf{W}_k^{\mathrm{T}}(\mathbf{r}_k - \mathbf{A}\mathbf{r}_k)$
$\mathbf{u}_k = \mathbf{r}_k + \mathbf{V}_k\mathbf{y}_k$
$\mathbf{x}_{k+1} = \mathbf{x}_k + \frac{\mathbf{r}_k^{\mathrm{T}}\mathbf{A}\mathbf{u}_k}{\|\mathbf{A}\mathbf{u}_k\|^2}\mathbf{u}_k$
If $\|\mathbf{r}_{k+1}\| < \varepsilon$, stop
Otherwise, $\mathbf{x}_0 = \mathbf{x}_{k+1}$; go to 2

In Algorithm OMPA(m), if we take $m_1 = m_0$, then we return to the usual iterative algorithm with a fixed-dimension $m = m_0$ of the Krylov subspace. In Algorithm OMPA(m), $m = m_1 - m_0 + 1$ is the frequency for restart.

Remark 5. *In view of Equations (67) and (80), f and $\eta = 1$ have the following relation:*

$$f = \frac{\mathbf{r}_0 \cdot \mathbf{v}}{\eta} = \mathbf{r}_0 \cdot \mathbf{v} = \|\mathbf{A}\mathbf{u}\|^2. \tag{88}$$

Therefore, the maximization of f is equivalent to the maximization of the length of the projection vector $\mathbf{A}\mathbf{u}$. By means of Equation (64),

$$f = \max h; \tag{89}$$

hence,

$$\max_{\mathbf{u}} f = \max_{\mathbf{u}} \max_{\eta} \left\{ 2\eta \mathbf{r}_0^T \mathbf{A}\mathbf{u} - \eta^2 \|\mathbf{A}\mathbf{u}\|^2 \right\} \tag{90}$$

follows from Equation (22). Through a double-maximization, we have derived the maximal projection algorithm.

5. An Affine GMRES Method

In the derivation of GMRES, the following technique was employed [11]:

$$\min_{\mathbf{y}} \|r_0 \mathbf{e}_1 - \bar{\mathbf{H}}_m \mathbf{y}\|, \tag{91}$$

where $r_0 = \|\mathbf{r}_0\|$.

Instead of Equation (91) used in the GMRES method to develop the iterative algorithm from $\mathbf{u} \in \mathcal{K}_m(\mathbf{A}, \mathbf{r}_0)$, we seek the descent vector, $\mathbf{u}$, in a larger subspace of $\mathbf{u} \in \mathbf{r}_0 + \mathcal{K}_m(\mathbf{A}, \mathbf{r}_0)$ by

$$\min_{\mathbf{y}} \|\mathbf{r}_0 - \mathbf{A}\mathbf{r}_0 - \mathbf{W}\mathbf{y}\|, \tag{92}$$

which is obtained from Equation (82) by inserting Equation (86) for $\mathbf{u}$, and using $\mathbf{W} = \mathbf{A}\mathbf{V}$.

Now, Equation (92) is re-written as

$$\min_{\mathbf{y}} \|\mathbf{r}_0 - \mathbf{A}\mathbf{r}_0 - \mathbf{A}\mathbf{V}_m \mathbf{y}\|. \tag{93}$$

We denote $\mathbf{V}_m$ the $m \times m$-dimensional Arnoldi matrix, and $\mathbf{V}_{m+1}$ the $(m+1) \times (m+1)$-dimensional Arnoldi matrix. It is known that

$$\mathbf{A}\mathbf{V}_m = \mathbf{V}_{m+1} \bar{\mathbf{H}}_m, \ \mathbf{V}_m^{\mathrm{T}} \mathbf{V}_m = \mathbf{I}_m, \ \mathbf{V}_{m+1}^{\mathrm{T}} \mathbf{V}_{m+1} = \mathbf{I}_{m+1}. \tag{94}$$

Theorem 6. *For $\mathbf{u} \in \mathbf{r}_0 + \mathcal{K}_m(\mathbf{A}, \mathbf{r}_0)$, derived from Equation (93), the orthogonalized affine GMRES possesses the following iterative formulas:*

$$(\bar{\mathbf{H}}_m^T \bar{\mathbf{H}}_m)\mathbf{y} = r_0 \bar{\mathbf{H}}_m^T \mathbf{e}_1 - r_0 [\bar{\mathbf{H}}_m^T \bar{\mathbf{H}}_m]_1, \tag{95}$$

$$\mathbf{u} = \mathbf{r}_0 + \mathbf{V}_m \mathbf{y}, \tag{96}$$

$$\mathbf{x} = \mathbf{x}_0 + \frac{\mathbf{r}_0^T \mathbf{A}\mathbf{u}}{\|\mathbf{A}\mathbf{u}\|^2} \mathbf{u}, \tag{97}$$

where $r_0 = \|\mathbf{r}_0\|$, and $[\bar{\mathbf{H}}_m^T \bar{\mathbf{H}}_m]_1$ is the first column of $\bar{\mathbf{H}}_m^T \bar{\mathbf{H}}_m$.

Proof. The objective $\mathbf{r}_0 - \mathbf{A}\mathbf{r}_0 - \mathbf{A}\mathbf{V}_m \mathbf{y}$ in Equation (93) can be simplified as follows. In terms of the first Krylov vector $\mathbf{v}_1 = \mathbf{r}_0 / r_0$, where $r_0 = \|\mathbf{r}_0\|$, and using the first one in Equation (94), we have

$$\mathbf{r}_0 - \mathbf{A}\mathbf{r}_0 - \mathbf{A}\mathbf{V}_m \mathbf{y} = r_0 \mathbf{v}_1 - r_0 \mathbf{A}\mathbf{v}_1 - \mathbf{V}_{m+1} \bar{\mathbf{H}}_m \mathbf{y}. \tag{98}$$

We write

$$\mathbf{v}_1 = \mathbf{V}_{m+1} \mathbf{e}_1, \ \mathbf{A}\mathbf{v}_1 = \mathbf{A}\mathbf{V}_m \mathbf{e}_2 = \mathbf{V}_{m+1} \bar{\mathbf{H}}_m \mathbf{e}_2, \tag{99}$$

where $\mathbf{e}_2$ is the first column of $\mathbf{I}_m$. Then, it follows that

$$\mathbf{r}_0 - \mathbf{A}\mathbf{r}_0 - \mathbf{A}\mathbf{V}_m \mathbf{y} = \mathbf{V}_{m+1} [r_0 \mathbf{e}_1 - r_0 \bar{\mathbf{H}}_m \mathbf{e}_2 - \bar{\mathbf{H}}_m \mathbf{y}]. \tag{100}$$

Taking the squared norms of both sides yields

$$
\begin{aligned}
&\|\mathbf{r}_0 - \mathbf{A}\mathbf{r}_0 - \mathbf{A}\mathbf{V}_m\mathbf{y}\|^2 \\
&= [r_0\mathbf{e}_1 - r_0\bar{\mathbf{H}}_m\mathbf{e}_2 - \bar{\mathbf{H}}_m\mathbf{y}]^{\mathrm{T}}\mathbf{V}_{m+1}^{\mathrm{T}}\mathbf{V}_{m+1}[r_0\mathbf{e}_1 - r_0\bar{\mathbf{H}}_m\mathbf{e}_2 - \bar{\mathbf{H}}_m\mathbf{y}] \\
&= [r_0\mathbf{e}_1 - r_0\bar{\mathbf{H}}_m\mathbf{e}_2 - \bar{\mathbf{H}}_m\mathbf{y}]^{\mathrm{T}}[r_0\mathbf{e}_1 - r_0\bar{\mathbf{H}}_m\mathbf{e}_2 - \bar{\mathbf{H}}_m\mathbf{y}] \\
&= \|r_0\mathbf{e}_1 - r_0\bar{\mathbf{H}}_m\mathbf{e}_2 - \bar{\mathbf{H}}_m\mathbf{y}\|^2,
\end{aligned}
\tag{101}
$$

where the last one in Equation (94) was used.

The minimization problem in Equation (93) is now simplified to

$$\min_{\mathbf{y}} \|r_0\mathbf{e}_1 - r_0\bar{\mathbf{H}}_m\mathbf{e}_2 - \bar{\mathbf{H}}_m\mathbf{y}\|. \tag{102}$$

To satisfy the minimality condition, we can derive Equation (95); using the technique in Lemma 1, Equation (97) can be derived. □

Remark 6. *Notice that the minimization problem in Equation (102) is slightly different from that in Equation (91), where an extra term, $r_0\bar{\mathbf{H}}_m\mathbf{e}_2$, appeared in Equation (102). However, the resulting affine GMRES (A-GMRES) (Algorithm 8) in the affine Krylov subspace can significantly improve the performance rather than GMRES, when the orthogonality is taken into account. We remind that (6)–(8) in the following Algorithm A-GMRES are slightly more computational expensive than (6) and (7) in Algorithm GMRES. Also, (7)–(9) in the following Algorithm A-GMRES(m) (Algorithm 9) are slightly more computational expensive than (7) and (8) in Algorithm GMRES(m).*

Algorithm 8: A-GMRES

1: Give m_0, $\mathbf{x}_0$, and ε
2: Do $m = m_0, \ldots$ $(k = m - m_0)$
3: $\mathbf{r}_k = \mathbf{b} - \mathbf{A}\mathbf{x}_k$, $r_k = \|\mathbf{r}_k\|$
4: $\mathbf{v}_1^k = \frac{\mathbf{r}_k}{\|\mathbf{r}_k\|}$
5: $\mathbf{V}_k = [\mathbf{v}_1^k, \ldots, \mathbf{v}_m^k]$ (by Arnoldi process)
6: $(\bar{\mathbf{H}}_k^{\mathrm{T}}\bar{\mathbf{H}}_k)\mathbf{y}_k = r_k\bar{\mathbf{H}}_k^{\mathrm{T}}\mathbf{e}_1 - r_k[\bar{\mathbf{H}}_k^{\mathrm{T}}\bar{\mathbf{H}}_k]_1$
7: $\mathbf{u}_k = \mathbf{r}_k + \mathbf{V}_k\mathbf{y}_k$
8: $\mathbf{x}_{k+1} = \mathbf{x}_k + \frac{\mathbf{r}_k^{\mathrm{T}}\mathbf{A}\mathbf{u}_k}{\|\mathbf{A}\mathbf{u}_k\|^2}\mathbf{u}_k$
9: If $\|\mathbf{r}_{k+1}\| < \varepsilon$, stop

Algorithm 9: A-GMRES(m)

1: Give m_0, m_1, $\mathbf{x}_0$, and ε
2: Do $j = 1, \ldots$
3: Do $m = m_0, \ldots, m_1$ $(k = m - m_0)$
4: $\mathbf{r}_k = \mathbf{b} - \mathbf{A}\mathbf{x}_k$, $r_k = \|\mathbf{r}_k\|$
5: $\mathbf{v}_1^k = \frac{\mathbf{r}_k}{\|\mathbf{r}_k\|}$
6: $\mathbf{V}_k = [\mathbf{v}_1^k, \ldots, \mathbf{v}_m^k]$ (by Arnoldi process)
7: $(\bar{\mathbf{H}}_k^{\mathrm{T}}\bar{\mathbf{H}}_k)\mathbf{y}_k = r_k\bar{\mathbf{H}}_k^{\mathrm{T}}\mathbf{e}_1 - r_k[\bar{\mathbf{H}}_k^{\mathrm{T}}\bar{\mathbf{H}}_k]_1$
8: $\mathbf{u}_k = \mathbf{r}_k + \mathbf{V}_k\mathbf{y}_k$
9: $\mathbf{x}_{k+1} = \mathbf{x}_k + \frac{\mathbf{r}_k^{\mathrm{T}}\mathbf{A}\mathbf{u}_k}{\|\mathbf{A}\mathbf{u}_k\|^2}\mathbf{u}_k$
10: If $\|\mathbf{r}_{k+1}\| < \varepsilon$, stop
11: Otherwise, $\mathbf{x}_0 = \mathbf{x}_{k+1}$; go to 2

6. Examples

In this section, we apply the Algorithms GMRES, GMRES(m), A-GMRES, A-GMRES(m), ROGMRES, ROGMRES(m), OMPA, and MPA(m) to solve some examples endowed with a symmetric cyclic matrix, randomized matrix, ill-posed Cauchy problem, diagonal dominant matrix, and highly ill-conditioned matrix. They are subjected to the convergence criterion with $\|\mathbf{r}_k\| < \varepsilon$, where ε is the error tolerance.

6.1. Example 1: Cyclic Matrix Linear System

A cyclic matrix, **A**, is generated from the first row $(1, \ldots, n)$. The exact solution is assumed to be $x_i = 1$, $i = 1, \ldots, n$, and **b** can be obtained from Equation (1). First we apply GMRES to this problem with $n = 300$ and $\varepsilon = 10^{-10}$. The initial values are $x_i^0 = 1 + 1/i$, $i = 1, \ldots, n$. For this problem, we encounter the situation that $\eta_k < 1$ after 150 steps, not $\eta_k = 1$ as shown in Figure 1a by a solid line. As shown in Figure 1b, the residuals blow up for GMRES.

This problem shows that the original GMRES method cannot be applied to solve this linear system, since $\eta_k < 1$ happened after 150, and the iteration of GMRES blows up after around 157 steps. In this situation, by Equation (35),

$$\|\mathbf{r}_{k+1}\| > \|\mathbf{r}_k\|,$$

the residual obtained by GMRES grows step-by step; hence, the GMRES method is no longer stable after 157 steps.

To overcome this drawback of GMRES, we can either relax the convergence criterion, or employ the version of GMRES(η) before it diverges.

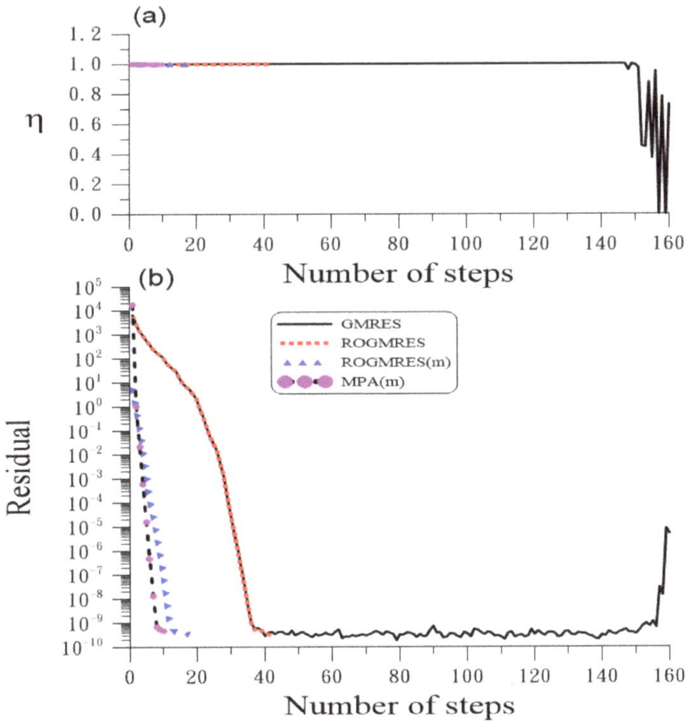

Figure 1. For example 1: (**a**) η of GMRES, ROGMRES, ROGMRES(m), and MPA(m), and (**b**) the residuals blowing up obtained by GMRES, ROGMRES, ROGMRES(m), and MPA(m) do not blow up.

Table 1 compares the maximum error (ME) and number of steps (NS) for different methods under $\varepsilon = 3 \times 10^{-10}$. $m_0 = 10$ is used in GMRES and ROGMRES; $m_0 = 1$ is used in GMRES(η); $m_0 = 30$ and $m_1 = 40$ are used in GMRES(m); $m_0 = 35$ and $m_1 = 40$ are used in ROGMRES(m) and GMRES(m,η); $m_0 = 50$ and $m_1 = 50$ are used in MPA(m) and A-GMRES(m). As shown in Figure 1a by a dashed line and a dashed–dotted line, both ROGMRES and ROGMRES(m) can keep the value of $\eta_k = 1$ up to the termination; as shown in Figure 1b, they converge faster than GMRES. ROGMRES(m) converges faster than ROGMRES. This example shows that ROGMRES can stabilize GMRES when it is unstable, as shown in Figure 1b.

It is interesting that the MPA(m), with a fixed value $m = m_0 = m_1 = 50$, can attain a highly accurate solution with ten steps, as shown in Figure 1b; A-GMRES(m) obtains ME = 5.88×10^{-14} with 15 steps by using $m = m_0 = m_1 = 50$. If the orthogonality is not adopted in A-GMRES(m), ME = 7.51×10^{-14} is obtained with 14 steps; it means that $\eta_k = 1$ can be kept, even though the orthogonalization technique was not employed. Within ten steps, A-GMRES with $m_0 = 50$ can obtain ME = 7.06×10^{-14}. Within nine steps, A-GMRES(m) with $m_0 = 50$ and $m_0 = 51$ can obtain ME = 8.02×10^{-14}.

Table 1. For example 1, ME and NS obtained by different methods.

	GMRES(m)	ROGMRES	ROGMRES(m)	GMRES(η)	GMRES(m,η)	MPA(m)	A-GMRES(m)
ME	7.39×10^{-14}	7.22×10^{-14}	6.50×10^{-14}	8.22×10^{-14}	9.04×10^{-14}	5.66×10^{-14}	5.88×10^{-14}
NS	28	44	18	16	16	10	15

6.2. Example 2: Random Matrix Linear System

The coefficient matrix $\mathbf{A}$ of size 50×50 is randomly generated with $1 < a_{ij} < 2$. The exact solution is $x_i = 1$, $i = 1, \ldots, 50$, and $\mathbf{b}$ can be obtained by Equation (1).

We consider the initial values $x_i^0 = 0.5$ and take $\varepsilon = 10^{-10}$. Table 2 compares ME and NS for different methods under $\varepsilon = 10^{-10}$. $m_0 = 10$ is used in ROGMRES, OMPA, A-GMRES, and GMRES(η); $m_0 = 30$ and $m_1 = 40$ are used in GMRES(m), ROGMRES(m) and GMRES(m,η). ROGMRES is better than OMPA and A-GMRES; other methods are not accurate. For this example, the restart ROGMRES(m) is not good.

Table 2. For example 2, ME and NS obtained by different methods.

	GMRES(m)	ROGMRES	OMPA	A-GMRES	ROGMRES(m)	GMRES(η)	GMRES(m,η)
ME	3.51×10^{-2}	9.77×10^{-14}	1.88×10^{-13}	3.36×10^{-9}	3.51×10^{-2}	3.65×10^{-2}	3.25×10^{-2}
NS	>550	51	43	52	>550	45	163

6.3. Example 3: Inverse Cauchy Problem and Boundary Value Problem

Consider the inverse Cauchy problem:

$$\Delta u = u_{rr} + \frac{1}{r}u_r + \frac{1}{r^2}u_{\theta\theta} = 0, \tag{103}$$

$$u(\rho, \theta) = h(\theta), \ 0 \leq \theta \leq \pi, \tag{104}$$

$$u_n(\rho, \theta) = g(\theta), \ 0 \leq \theta \leq \pi. \tag{105}$$

The method of fundamental solutions is taken as

$$u(\mathbf{x}) = \sum_{j=1}^{n} c_j U(\mathbf{x}, \mathbf{s}_j), \ \ \mathbf{s}_j \in \Omega^c. \tag{106}$$

where

$$U(\mathbf{x}, \mathbf{s}_j) = \ln r_j, \ \ r_j = \|\mathbf{x} - \mathbf{s}_j\|. \tag{107}$$

We consider

$$u(x, y) = \cos x \cosh y + \sin x \sinh y, \tag{108}$$
$$\rho(\theta) = \exp(\sin\theta)\sin^2(2\theta) + \exp(\cos\theta)\cos^2(2\theta), \tag{109}$$

where $u(x, y)$ is an exact solution of the Laplace equation, and $\rho(\theta)$, $0 \le \theta \le 2\pi$ describes the boundary contour in the polar coordinates.

By GMRES with $n = 60$, through 60 iterations the residual is shown in Figure 2a, which does not satisfy the convergence criterion with $\varepsilon = 10^{-2}$. If we run it to 70 steps, GMRES will blow up. In contrast, ROGMRES converges with 72 steps. As compared in Figure 2b, ME obtained by ROGMRES is smaller than 0.27, as shown in Figure 2c. However, GMRES obtains ME = 0.5 with a worse result.

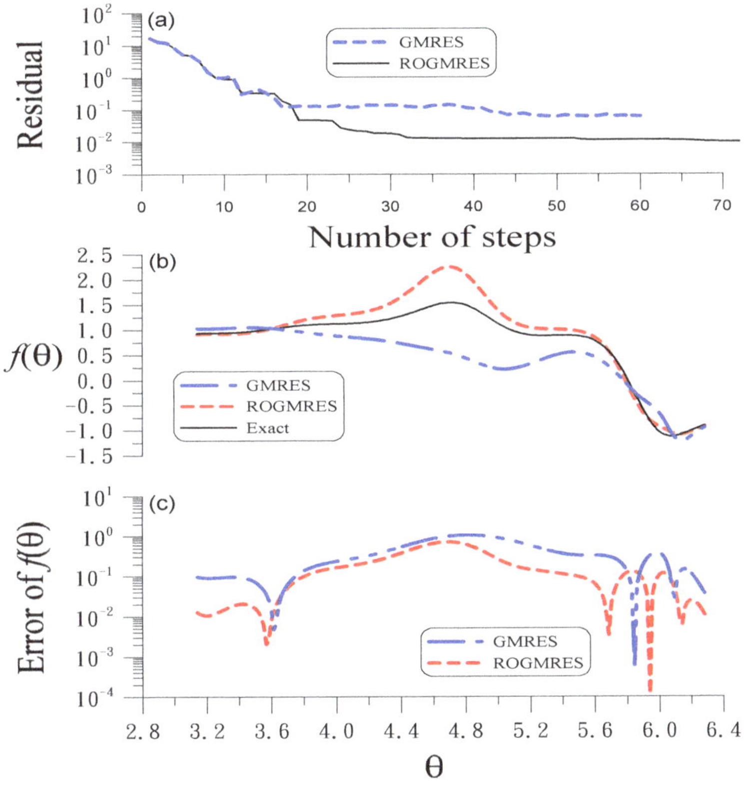

Figure 2. For example 3 of the inverse Cauchy problem: (**a**) the residuals, (**b**) comparing solutions, and (**c**) the errors obtained by GMRES and ROGMRES.

Because the inverse Cauchy problem is a highly ill-posed problem, GMRES does not converge within a loose convergence criterion $\varepsilon = 10^{-2}$, and after 70 steps it will blow up. It is apparent that GMRES cannot solve the Cauchy problem adequately. The ROGMRES method is stable for the Cauchy problem, and can improve the accuracy to 0.27; however, it leaves room to improve the methods based on the Krylov subspace for solving the Cauchy problem.

Next, we consider the mixed boundary value problem. Under $\varepsilon = 10^{-3}$, for GMRES(m) with $m_0 = 3$ and $m_1 = 10$, as shown in Figure 3a, η_k after 12 steps is irregular; it causes the residuals to decrease non-monotonically, as shown in Figure 3b. The GMRES(m) does not converge within 200 steps, and obtains an incorrect solution with ME = 0.92. If $\eta_k < 1/2$ is used as a stopping criterion, through 14 steps, ME = 6.64×10^{-3} is obtained. When the ROGMRES method is applied, we obtain ME = 4.16×10^{-4} with 187 steps. As shown in Figure 3a by a dashed line, $\eta_k = 1$ is kept well, and the residuals monotonically decrease, as shown in Figure 3b by a dashed line for ROGMRES. ME = 4.87×10^{-4} is obtained through 167 steps by OMPA with $m_0 = 4$, which is convergent faster than ROGMRES. ME = 4.09×10^{-4} is obtained through 109 steps by A-GMRES with $m_0 = 4$, which is convergent faster than ROGMRES and OMPA.

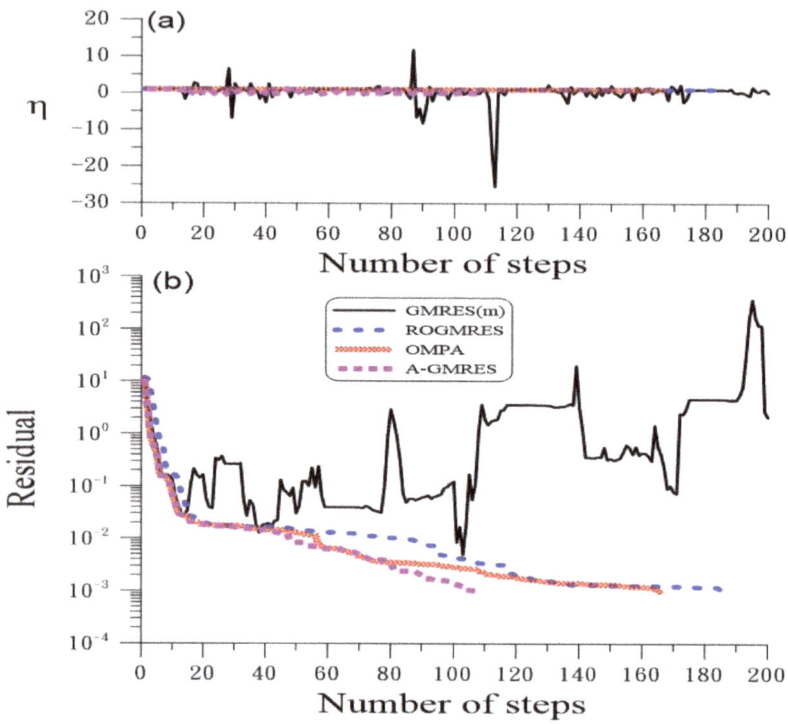

Figure 3. For example 3 with mixed boundary conditions: (**a**) η and (**b**) the residuals obtained by GMRES(m), ROGMRES, OMPA, and A-GMRES; the residuals obtained by GMRES(m) fail.

6.4. Example 4: Diagonal Dominant Linear System

Consider an example borrowed from Morgan [27]. The matrix **A** consists of main diagonal elements from 1 to 1000; the super diagonal elements are 0.1, and other elements are zero. We take $\mathbf{b} = \mathbf{1}$.

Table 3 compares the number of steps (NS) for different methods under $\varepsilon = 10^{-10}$. $m_0 = 1$ is used in OMPA and A-GMRES; $m_0 = 1, m_1 = 25$ are used in original GMRES(m), ROGMRES(m) and current GMRES(m). The NS of ROGMRES(m), and current GMRES(m) are the same, because of $\eta_k = 1$ for the current GMRES(m) before convergence. Figure 4 compares the residuals.

Table 3. For example 4, NS obtained by different methods.

	Original GMRES(m)	ROGMRES(m)	OMPA	A-GMRES	[27]	Current GMRES(m)
NS	>300	44	32	31	300	44

The method in Equation (18) is an automatically orthogonality preserving method. When the OMPA and A-GMRES methods do not consider the automatically orthogonality preserving method, the results are drastically different, as shown in Figure 5. Because $\eta = 1$ is not preserved, both OMPA and A-GMRES diverge very quickly after 40 steps.

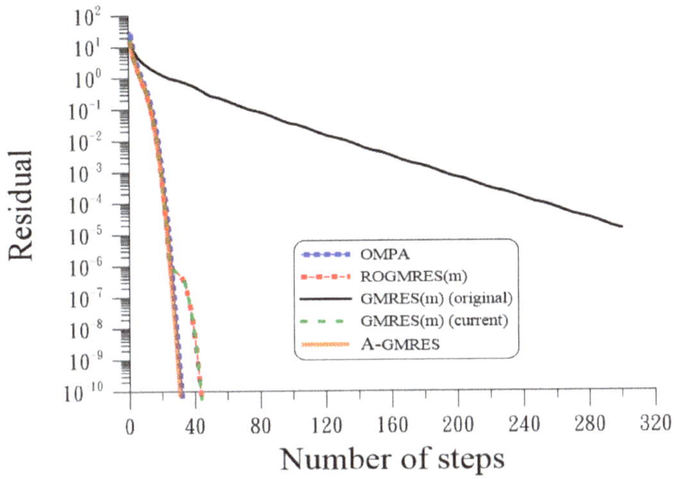

Figure 4. For example 4, the residuals obtained by OMPA, A-GMRES, ROGMRES(m), original GMRES(m), and current GMRES(m), which is coincident with that obtained by ROGMRES(m).

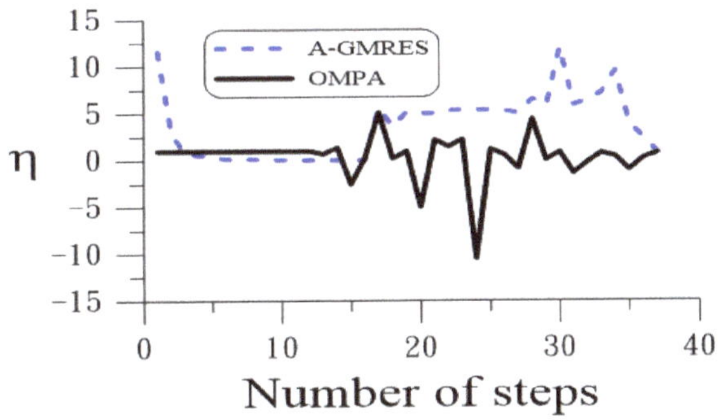

Figure 5. For example 4, η obtained by OMPA and A-GMRES. If the automatically orthogonality preserving method is not taken into account, they fail.

6.5. Example 5: Densely Clustered Non-Symmetric Linear System

Consider an example with $a_{ij} = (i + j/2)/n$, $i, j = 1, \ldots, n$, where $n = 2000$. Suppose that $x_i = 1$, $i = 1, \ldots, 2000$ are exact solutions, and the initial values are $x_i^0 = 0$, $i = 1, \ldots, 2000$. The values of matrix elements are densely clustered in a narrow range with $a_{ij} \in [0.00075, 1.5]$. This problem is a highly ill-conditioned non-symmetric linear system with the condition number over the order 10^{15}. Table 4 compares the maximal errors (MEs) and number of steps (NS) for different methods under $\varepsilon = 10^{-10}$. $m_0 = 1$ is used in OMPA and A-GMRES; $m_0 = 1, m_1 = 3$ are used in current GMRES(m), and $m_0 = 2, m_1 = 3$ are used in ROGMRES(m). ROGMRES(m) is convergent faster than other algorithms. Even for this highly ill-conditioned problem, the proposed novel methods are highly efficient in finding accurate solutions.

Table 4. For example 5, ME and NS obtained by different methods.

Method	OMPA	A-GMRES	Current GMRES(m)	ROGMRES(m)
NS	4	3	9	2
ME	2.10×10^{-9}	1.46×10^{-11}	1.18×10^{-13}	5.98×10^{-13}

7. Conclusions

In this paper the GMRES method was re-examined from the two aspects of preserving the orthogonality and maximizing the decreasing length of the residual vector, both of which can significantly enhance the stability and accelerate the convergence speed of GMRES. If $\eta_k = \mathbf{r}_k^{\mathrm{T}} \mathbf{A} \mathbf{u}_k / \|\mathbf{A} \mathbf{u}_k\|^2$ in GMRES is kept to $\eta_k = 1$ during the iterations, it is stable; otherwise, the GMRES is unstable. For any iterative form of $\mathbf{x}_{k+1} = \mathbf{x}_k + \mathbf{u}_k$, we can improve it to $\mathbf{x}_{k+1} = \mathbf{x}_k + \mathbf{u}_k (\mathbf{r}_k^{\mathrm{T}} \mathbf{A} \mathbf{u}_k) / \|\mathbf{A} \mathbf{u}_k\|^2$, such that the orthogonality is preserved automatically, and simultaneously the length of the residual vector is reduced maximally. It brings the GMRES method to a re-orthogonalized GMRES (ROGMRES) method, preserving the orthogonality automatically and having the property of absolute convergence. We derived the new iterative form $\mathbf{x}_{k+1} = \mathbf{x}_k + \mathbf{u}_k (\mathbf{r}_k^{\mathrm{T}} \mathbf{A} \mathbf{u}_k) / \|\mathbf{A} \mathbf{u}_k\|^2$ also from the invariant-manifold theory.

In order to find a better descent vector, we solved a maximal projection problem in a larger affine Krylov subspace to derive the descent vector. We proved that the maximal projection is equivalent to the minimal residual length in the m-dimensional affine Krylov subspace. After taking the automatically orthogonality preserving method into account, a new orthogonalized maximal projection algorithm (OMPA) method was developed. We derived formulas similar to GMRES in the affine Krylov subspace, namely the affine GMRES (A-GMRES) method, which had superior performance for the problems tested in the paper; it is even better than OMPA for some problems. The algorithm A-GMRES possessed two advantages: solving a simpler least squares problem in the affine Krylov subspace, and preserving the orthogonality automatically.

Through the studies conducted in the paper, the novelty points and main contributions are summarized as follows.

- The GMRES was examined from the viewpoints of maximizing the decreasing quantity of the residual and to maintain the orthogonality of the consecutive residual vector.
- A stabilization factor, η, to maintain orthogonality was inserted into GMRES to preserve the orthogonality automatically.
- GMRES was improved in a larger space of the affine Krylov subspace.
- A new orthogonalized maximal projection algorithm, OMPA, was derived; a new affine A-GMRES was derived.
- The new algorithms ROGMRES, OMPA, and A-GMRES guarantee the absolute convergence.
- Through examples, we showed that the orthogonalization techniques are useful to stabilize the methods of GMRES, MPA, and A-GMRES.
- Numerical testings for five different problems confirmed that the methods of ROGMRES, ROGMRES(m), OMPA, OMPA(m), A-GMRES, and A-GMRES(m) can significantly accelerate the convergence speed compared with the original methods of GMRES and GMRES(m).

Author Contributions: Conceptualization, C.-S.L. and C.-W.C.; Methodology, C.-S.L. and C.-W.C.; Software, C.-S.L., C.-W.C. and C.-L.K.; Validation, C.-S.L., C.-W.C. and C.-L.K.; Formal analysis, C.-S.L. and C.-W.C.; Investigation, C.-S.L., C.-W.C. and C.-L.K.; Resources, C.-S.L. and C.-W.C.; Data curation, C.-S.L., C.-W.C. and C.-L.K.; Writing—original draft, C.-S.L. and C.-W.C.; Writing—review & editing, C.-W.C.; Visualization, C.-S.L., C.-W.C. and C.-L.K.; Supervision, C.-S.L. and C.-W.C.; Project administration, C.-W.C.; Funding acquisition, C.-W.C. All authors have read and agreed to the published version of the manuscript.

Funding: This research received no external funding.

Data Availability Statement: The data presented in this study are available on request from the corresponding authors.

Conflicts of Interest: The authors declare no conflicts of interest.

References

1. Krylov, A.N. On the numerical solution of equation by which are determined in technical problems the frequencies of small vibrations of material systems. *Izv. Akad. Nauk. SSSR* **1931**, *7*, 491–539.
2. Saad, Y. *Iterative Methods for Sparse Linear Systems*, 2nd ed.; SIAM: Pittsburgh, PA, USA, 2003.
3. van der Vorst, H.A. *Iterative Krylov Methods for Large Linear Systems*; Cambridge University Press: New York, NY, USA, 2003.
4. Meurant, G.; Tabbens, J.D. *Krylov Methods for Non-Symmetric Linear Systems: From Theory to Computations*; Springer Series in Computational Mathematics; Springer: Berlin/Heidelberg, Germany, 2020; Volume 57.
5. Godunov, S.K.; Rozhkovskaya, T. *Modern Aspects of Linear Algebra*; Translations of Mathematical Monographs; AMS: Little Rock, AR, USA, 1998.
6. Godunov, S.K.; Gordienko, V.M. The Krylov space and the Kalman equation. *Sib. Zhurnal Vychislitel'Noi Mat.* **1998**, *1*, 5–10.
7. van Den Eshof, J.; Sleijpen, G.L.G. Inexact Krylov subspace methods for linear systems. *SIAM J. Matrix Anal. Appl.* **2004**, *26*, 125–153. [CrossRef]
8. Bai, Z.Z. Motivations and realizations of Krylov subspace methods for large sparse linear systems. *J. Comput. Appl. Math.* **2015**, *283*, 71–78. [CrossRef]
9. Simoncini, V.; Szyld, D.B. Recent computational developments in Krylov subspace methods for linear systems. *Numer. Linear Algebra Appl.* **2007**, *14*, 1–59. [CrossRef]
10. Bouyghf, F.; Messaoudi, A.; Sadok, H. A unified approach to Krylov subspace methods for solving linear systems. *Numer. Algor.* **2024**, *96*, 305–332. [CrossRef]
11. Saad, Y.; Schultz, M.H. GMRES: A generalized minimal residual algorithm for solving nonsymmetric linear systems. *SIAM J. Sci. Stat. Comput.* **1986**, *7*, 856–869. [CrossRef]
12. Liu, C.S. A maximal projection solution of ill-posed linear system in a column subspace, better than the least squares solution. *Comput. Math. Appl.* **2014**, *67*, 1998–2014. [CrossRef]
13. Liu, C.S.; Kuo, C.L.; Chang, C.W. Solving least-squares problems via a double-optimal algorithm and a variant of Karush-Kuhn-Tucker equation for over-determined system. *Algorithms* **2024**, *17*, 211. [CrossRef]
14. Morgan, R.B. Implicitly restarted GMRES and Arnoldi methods for nonsymmetric systems of equations. *SIAM J. Matrix Anal. Appl.* **2000**, *21*, 1112–1135. [CrossRef]
15. Baker, A.H.; Jessup, E.R.; Manteuffel, T. A technique for accelerating the convergence of restarted GMRES. *SIAM J. Matrix Anal. Appl.* **2005**, *26*, 962–984. [CrossRef]
16. Zou, Q. GMRES algorithms over 35 years. *Appl. Math. Comput.* **2023**, *445*, 127869. [CrossRef]
17. Thomas, S.; Carson, E.; Rozložník, M. Iterated Gauss–Seidel GMRES. *SIAM J. Sci. Comput.* **2023**, *46*, S254–S279. [CrossRef]
18. Imakura, A.; Sogabe, T.; Zhang, S.L. An efficient variant of the GMRES(m) method based on the error equations. *East Asian J. Appl. Math.* **2012**, *2*, 19–32. [CrossRef]
19. Imakura, A.; Sogabe, T.; Zhang, S.L. A look-back-type restart for the restarted krylov subspace methods to solve non-Hermitian systems. *Jpn. J. Ind. Appl. Math.* **2018**, *35*, 835–859. [CrossRef]
20. Benzi, M.; Meyer, C.D.; Tuma, M. A sparse approximate inverse preconditioner for the conjugate gradient method. *SIAM J. Sci. Comput.* **1996**, *17*, 1135–1149. [CrossRef]
21. Benzi, M.; Tuma, M. A sparse approximate inverse preconditioner for nonsymmetric linear systems. *SIAM J. Sci. Comput.* **1998**, *19*, 968–994. [CrossRef]
22. Benzi, M.; Cullum, J.K.; Tuma, M. Robust approximate inverse preconditioning for the conjugate gradient method. *SIAM J. Sci. Comput.* **2000**, *22*, 1318–1332. [CrossRef]
23. Benzi, M.; Bertaccini, D. Approximate inverse preconditioning for shifted linear systems. *BIT Numer. Math.* **2003**, *43*, 231–244. [CrossRef]
24. Baglama, J.; Calvetti, D.; Golub, G.H.; Reichel, L. Adaptively preconditioned GMRES algorithms. *SIAM J. Sci. Comput.* **1998**, *20*, 243–269. [CrossRef]
25. Bakhos, T.; Kitanidis, P.; Ladenheim, S.; Saibaba, A.K.; Szyld, D.B. Multipreconditioned GMRES for shifted systems. *SIAM J. Sci. Comput.* **2017**, *39*, S222–S247. [CrossRef]
26. Liu, Q.; Morgan, R.B.; Wilcox, W. Polynomial preconditioned GMRES and GMRES-DR. *SIAM J. Sci. Comput.* **2015**, *37*, S407–S428. [CrossRef]
27. Morgan, R.B. A restarted GMRES method augmented with eigenvectors. *SIAM J. Matrix Anal. Appl.* **1995**, *16*, 1154–1171. [CrossRef]
28. Baglama, J.; Reichel, L. Augmented GMRES-type methods. *Numer. Linear Alg. Appl.* **2007**, *14*, 337–350. [CrossRef]
29. Dong, Y.; Garde, H.; Hansen, P.C. R^3GMRES: Including prior information in GMRES-type methods for discrete inverse problems. *Electron. Trans. Numer. Anal.* **2014**, *42*, 136–146.

30. Carson, E.; Liesen, J.; Strakoš, Z. Towards understanding CG and GMRES through examples. *Linear Alg. Appl.* **2024**, *692*, 241–291. [CrossRef]
31. Erhel, J.; Burrage, K.; Pohl, B. Restarted GMRES preconditioned by deflation. *J. Comput. Appl. Math.* **1996**, *69*, 303–318. [CrossRef]
32. Baker, A.H.; Jessup, E.R.; Kolev, T.Z. A simple strategy for varying the parameter in GMRES(m). *J. Comput. Appl. Math.* **2009**, *230*, 751–761. [CrossRef]
33. Moriya, K.; Nodera, T. The deflated-GMRES(m,k) method with switching the restart frequency dynamically. *Numer. Linear Alg. Appl.* **2000**, *7*, 569–584. [CrossRef]
34. Sosonkina, M.; Watson, L.T.; Kapania, R.K.; Walker, H.F. A new adaptive GMRES algorithm for achieving high accuracy. *Numer. Linear Alg. Appl.* **1998**, *5*, 275–297. [CrossRef]
35. Zhang, L.; Nodera, T. A new adaptive restart for GMRES(m) method. *Austr. N. Z. Ind. Appl. Math. J.* **2005**, *46*, 409–426. [CrossRef]

algorithms MDPI

Article

On Implementing a Two-Step Interior Point Method for Solving Linear Programs

Sajad Fathi Hafshejani *, Daya Gaur * and Robert Benkoczi

Department of Math and Computer Science, University of Lethbridge, Lethbridge, AB T1K 3M4, Canada; robert.benkoczi@uleth.ca
* Correspondence: sajad.fathihafshejan@uleth.ca (S.F.H.); daya.gaur@uleth.ca (D.G.)

Abstract: A new two-step interior point method for solving linear programs is presented. The technique uses a convex combination of the auxiliary and central points to compute the search direction. To update the central point, we find the best value for step size such that the feasibility condition is held. Since we use the information from the previous iteration to find the search direction, the inverse of the system is evaluated only once every iteration. A detailed empirical evaluation is performed on NETLIB instances, which compares two variants of the approach to the primal-dual log barrier interior point method. Results show that the proposed method is faster. The method reduces the number of iterations and CPU time(s) by 27% and 18%, respectively, on NETLIB instances tested compared to the classical interior point algorithm.

Keywords: linear programming; interior point method; Newton method

1. Introduction

1.1. Background

The following linear optimization (LO) problem is considered:

$$(P) \quad \min\{c^T x : Ax = b, x \geq 0\},$$

and the corresponding dual

$$(D) \quad \max\{b^T y : A^T y + s = c, s \geq 0\},$$

where $A \in \mathbb{R}^{m \times n}$, $x, c, s \in \mathbb{R}^n$ and $y, b \in \mathbb{R}^m$.

LO is an essential tool used widely in theoretical and practical science and engineering domains. Its applications extend to various fields such as operations research, engineering, economics and combinatorics. LO finds numerous applications across various fields. For instance, it is used in $\mathcal{L}_1$-regularized support vector machines (SVMs) [1], basis pursuit (BP) problems [2], sparse inverse covariance matrix estimation (SICE) [3], non-negative matrix factorization (NMF) [4] and maximum a posteriori (MAP) inference [5].

The Interior-Point Method (IPM) is a powerful optimization technique that has been widely used in the field of LO. It offers an alternative approach to solving LO problems compared to traditional simplex methods. Karmarkar [6] described the first efficient algorithm for the IPM. Nesterov and Nemirovskii [7] introduced a class of self-concordant barrier functions and extended the algorithm to a more general type of optimization problems, including convex programming, non-linear complementarity problem (NCP), semidefinite optimization (SDO) and second-order cone optimization (SOCO) problems.

1.2. Literature Review

IPMs are generally divided into two categories, that is, small-update methods and large-update methods [8]. In practice, large-update methods are significantly more efficient.

However, in theory, the best results are associated with small-update methods. From another perspective, IPMs are divided into feasible and infeasible methods [9]. Feasible methods start from a point that satisfies the constraints and seek to find the optimal solution within the feasible region. In contrast, infeasible methods begin with a point that does not meet the problem's constraints (infeasible). The algorithm first seeks to find a feasible point, and once a feasible point is found, it moves towards the optimal solution [9]. The IPM based on the kernel function is a popular method. The kernel function is used not only to find the search direction for the algorithm but also to estimate a lower bound for the step size [10]. In this type of IPM, the complexity bounds of the interior point algorithm are computed based on the kernel function. Terlaky et al. [10] proposed a new variant of IPMs for solving linear optimization problems in which the self-regular barrier function replaces the logarithmic barrier function and they showed that their method has the best-known complexity bounds. Following this, several different kernel functions with the best-known complexity bounds have been introduced, including the exponential kernel functions [11], trigonometric kernel functions [12,13] and hyperbolic kernel functions [14,15].

Predictor-corrector methods for interior point algorithms are an effective approach for solving both linear and nonlinear optimization problems [16,17]. These methods incorporate two main steps, i.e., the predictor step and the corrector step. In the predictor step, an approximate direction is determined by solving a linearized system of the Karush–Kuhn–Tucker (KKT) conditions [17]. In the subsequent corrector step, this direction is refined to ensure that the iterates remain within the feasible region and maintain the balance between the primal and dual variables. This refinement typically involves solving another system of linear equations that corrects any infeasibilities introduced in the predictor step, thereby enhancing the convergence properties of the algorithm. Mehrotra's algorithm [17] is a popular predictor-corrector algorithm used for solving optimization problems. In recent years, several new predictor-corrector algorithms have been introduced, further enhancing the effectiveness and efficiency of IPMs for various optimization problems [18,19].

The Newton method is a popular optimization technique that plays a vital role in interior-point methods. In the context of IPMs, the Newton method is used to solve the system of equations that arise from the Karush–Kuhn–Tucker (KKT) optimality conditions, which include both the primal and dual feasibility conditions, as well as the complementary slackness conditions. This approach is known for its fast convergence rate, often exhibiting quadratic convergence. A modification of Newton's method was proposed by McDougall and Wotherspoon [20] that can significantly reduce the number of iterations when used to the root of functions. Their approach has a convergence of the order of $1 + \sqrt{2}$. To compute the search direction, their algorithm uses an auxiliary point and computes the gradient of the objective function only once in each iteration.

McDougall and Wotherspoon [20] applied this method to solve the power flow problem, which is a multivariate problem. They used the improved method to find the roots of a system of multivariable equations and demonstrated that it can reduce the number of iterations compared to Newton's method. Additionally, Argyros et al. [21] present some theoretical results related to the convergence of the method. In Example 4 of [21], it is shown that the method is effective for solving multivariable problems and can converge to the optimal solution.

1.3. Contribution

Motivated by these ideas, this paper makes the following contributions:

- To solve optimization problems using the IPM, the search direction plays an important role. We use the idea proposed in [20] for determining the search direction. The main goal of this paper is to extend the strategy proposed in [20] to IPMs. This paper introduces a two-step method that calculates the inverse of the matrix only once to determine the search direction in each iteration.
  The next goal of this article is to demonstrate how this method can be integrated with the feasible interior-point algorithm for linear optimization problems to enhance the

efficiency of the IPM. We combine the technique described in [20] with the algorithm presented in [8], resulting in the development of two new algorithms.

- To demonstrate the efficiency of the new method, we implement the algorithms and solve a collection of NETLIB test problems. The results show that the proposed method can improve the algorithm's efficiency by 27% in terms of the number of iterations and 18% in terms of CPU time.
  The implementation on a Graphics Processing Unit (GPU), a specialized processor designed to accelerate graphics rendering and computational tasks, shows a reduction in the number of iterations and GPU time compared to the classical algorithm by 39% and 30%, respectively.

The paper is organized as follows: Section 2 briefly recalls the basics of the IPMs and the central path for solving linear programs. We explain the two-step algorithm proposed here next. Section 3 reports the numerical results on a collection of LP instances from NETLIB for CPU and GPU implementations of the algorithm. We end with concluding remarks in Section 4.

In this paper, we use the following notational conventions. The Euclidean norm of a vector is denoted by $\|.\|$. The non-negative and positive orthants are represented by $\mathbb{R}^n_+$ and $\mathbb{R}^n_{++}$ respectively. The vectors xs and $\frac{x}{s}$ indicate coordinate-wise operations on the vectors x and s, with components $x_i s_i$ and $\frac{x_i}{s_i}$ respectively. We also use diagonal matrices X and S with entries x and s respectively. For any variable v belonging to $\{x, y, s\}$, subscripts denote the nth iteration value. For instance, x_n refers to the value of x after n updates. If the current iteration is n, we use v instead of v_n and v_+ instead of v_{n+1} for $v \in \{x, y, s, \tilde{x}, \tilde{y}, \tilde{s}, \hat{x}, \hat{y}, \hat{s}\}$. Here, x refers to the current point and x_+ refers to the updated value of x.

2. Algorithm

In this paper, without loss of generality, we make the following assumptions:

- A is a full rank matrix, that is rank $A = m \leq n$.
- Both problems (P) and (D) satisfy the Interior Point Condition (IPC) [8], i.e., there exists $(x^0, s^0) > 0$ and y^0 so that:

$$\begin{aligned} Ax^0 &= b, & x^0 &> 0, \\ A^T y^0 + s^0 &= c, & s^0 &> 0. \end{aligned}$$

An optimal solution for (P) and (D) can be found by solving the following system [10]:

$$\begin{aligned} Ax &= b, & x &\geq 0, \\ A^T y + s &= c, & s &\geq 0, \\ xs &= 0, \end{aligned} \tag{1}$$

where xs stands for the coordinate-wise (Hadamard) product of the vectors x and s. Note that the first and second equations in (1) guarantee the feasibility of x and (y, s) for problems (P) and (D), respectively, and the last equation is satisfied only if the feasible solution is optimal (complementarity conditions).

The algorithm presented here is a two-step primal-dual IPM; therefore, we summarize the key idea behind the generic primal-dual IPMs for solving linear programming problems from the literature, see e.g., [8,10]. The key idea is to replace the complementarity conditions with parameterized equations $xs = \mu\mathbf{e}$, where $\mathbf{e}$ denotes the all-one vector of length n and μ is a real positive parameter. With these changes, the system of Equation (1) can be expressed as [10]:

$$\begin{aligned} Ax &= b, & x &\geq 0, \\ A^T y + s &= c, & s &\geq 0, \\ xs &= \mu\mathbf{e}. \end{aligned} \tag{2}$$

The parametric system (2) has a unique solution for any $\mu > 0$ if the IPC holds [10] (Theorem 1.2.4). For a given $\mu > 0$, denote the unique solution of (2) by $(x(\mu), y(\mu), s(\mu))$, where $x(\mu)$ is the μ-center of (P) and $(y(\mu), s(\mu))$ is called the μ-center of (D). The set of all μ-centers, for all $\mu > 0$, defines a path the so-called *central path* for (P) and (D). It has been shown that, as $\mu \to 0$, then the limit of the central path exists and converges to an analytic center of the optimal solutions set of (P) and (D) [22,23].

Given a current point (x, y, s), the goal is to find direction $(\Delta x, \Delta y, \Delta s)$ such that the point $(x + \Delta x, y + \Delta y, s + \Delta s)$ lies on the central path at μ. If it did lie exactly on the central path, then Equation (2) will be satisfied, i.e., [10]:

$$\begin{aligned} A(x + \Delta x) &= b \\ A^T(y + \Delta y) + (s + \Delta s) &= c \\ (X + \Delta x)(S + \Delta s) &= \mu \mathbf{e}, \end{aligned} \tag{3}$$

where:

$$X := \mathrm{diag}(x), \qquad S := \mathrm{diag}(s)$$

Assuming that the point (x, y, s) is feasible for the LO and dropping the non-linear term $\Delta x \Delta s$ [10], we get the following system of equations:

$$\begin{aligned} A\Delta x &= 0 \\ A^T \Delta y + \Delta s &= 0 \\ S\Delta x + X\Delta s &= \mu \mathbf{e} - XS\mathbf{e} \end{aligned} \tag{4}$$

System (4) has a unique solution, referred to as the search direction for IPMs. Many interior-point algorithms rely on the Newton method to solve the system (4) and find the search direction. Therefore, improving the speed of convergence of Newton's method significantly impacts the convergence speed and efficiency of IPMs. To calculate the search direction in Newton's method, we use the method proposed in [20], which has a convergence speed of $1 + \sqrt{2}$. Accordingly, we introduce a two-step method for the IPM based on the approach proposed in [20]. This change can significantly reduce the number of iterations needed to achieve the optimal solution.To apply the mentioned method, we present the system (4) as a problem of finding roots, followed by an explanation of the new approach.

2.1. Relationship with Root Finding

Define:

$$\xi = \begin{bmatrix} x \\ y \\ s \end{bmatrix} \text{ and } F(\xi) = \begin{bmatrix} Ax - b \\ A^T y + s - c \\ \mu \mathbf{e} - XS\mathbf{e} \end{bmatrix},$$

then a solution to the system (2) is ξ^* such that $F(\xi^*) = 0$. The root ξ^* can be computed using Newton's method (in the neighbourhood of the root), which, given ξ, finds $\Delta\xi$ such that $F(\xi + \Delta\xi) = 0$ [20]. Using Taylor series approximation (linear approximation) around ξ, we have [20]:

$$F(\xi) + F'(\xi)\Delta\xi \simeq 0$$

where:

$$\text{Jacobian } F' = \begin{bmatrix} \frac{\partial F_1}{\partial \xi_1} & \frac{\partial F_1}{\partial \xi_2} & \frac{\partial F_1}{\partial \xi_3} \\ \frac{\partial F_2}{\partial \xi_1} & \frac{\partial F_2}{\partial \xi_2} & \frac{\partial F_2}{\partial \xi_3} \\ \frac{\partial F_3}{\partial \xi_1} & \frac{\partial F_3}{\partial \xi_2} & \frac{\partial F_3}{\partial \xi_3} \end{bmatrix} = \begin{bmatrix} A & 0 & 0 \\ 0 & A^T & I \\ S & 0 & X \end{bmatrix} \text{ and } \Delta\xi = \begin{bmatrix} \Delta x \\ \Delta y \\ \Delta s \end{bmatrix}.$$

This is precisely the system that was derived earlier (4). This approximation yields an iterative algorithm, the Newton–Raphson method, for finding roots. Starting with an initial estimate ξ_0, the current estimate ξ_n of the root is updated using the following rule for some appropriate step size α:

$$\xi_{n+1} = \xi_n - \alpha(F'(\xi_n))^{-1}F(\xi_n)$$

The proposed method generalizes the Newton–Raphson method as follows (assume any F). In the first step, an auxiliary point $\tilde{\xi}_n$ is computed using a linear approximation to the function around the current estimate of the root ξ_n. The average $\hat{\xi}_{n+1}$ of the auxiliary point and the current estimate is computed in the second step. In the third step, the current estimate is updated using the linear approximation to F around ξ_n, but instead of using the Jacobian at point ξ_n, the Jacobian is evaluated at the average $\hat{\xi}_{n+1}$. The update rules used in n^{th} iteration can be summarized as [20]:

$$\begin{aligned}
\tilde{\xi}_{n+1} &= \xi_n - \alpha(F'(\hat{\xi}_n))^{-1}F(\xi_n) \\
\hat{\xi}_{n+1} &= \frac{1}{2}(\tilde{\xi}_{n+1} + \xi_n) \\
\xi_{n+1} &= \xi_n - \alpha(F'(\hat{\xi}_{n+1}))^{-1}F(\xi_n)
\end{aligned}$$

This sequence of updates requires two evaluations of the Jacobian, once at point $\hat{\xi}_n$ and the second time at point $\hat{\xi}_{n+1}$. This requirement doubles the computation burden. In practice, we use information regarding $F'(\hat{\xi}_{n+1})$ for the next iteration. Thus only a single evaluation of the Jacobian is needed in each iteration if we save the result from the previous iteration.

2.2. Two-Step IPMs

Here, we extend the idea presented in Section 2.1 to the interior-point algorithm. The extension involves two phases:

- Phase 1: We obtain a feasible starting point (x, y, s) using the method presented in [17]. Then, we create an auxiliary point $(\tilde{x}, \tilde{y}, \tilde{s})$, which initially has the same values as (x, y, s). The parameter μ is updated by a factor of $1-\theta$, and $(\tilde{x}, \tilde{y}, \tilde{s})$ is set to (x, y, s). After that, we set $(\hat{x}, \hat{y}, \hat{s}) = \frac{1}{2}((\tilde{x}, \tilde{y}, \tilde{s}) + (x, y, s))$ and solve the following system to determine the search direction:

$$\begin{aligned}
A\Delta x &= 0 \\
A^T\Delta y - \Delta s &= 0 \\
\hat{S}\Delta x + \hat{X}\Delta s &= \mu\mathbf{e} - XS\mathbf{e}
\end{aligned} \tag{5}$$

  To satisfy the feasibility condition, the algorithm performs a line search to find the maximum step size. To obtain the new point, the current point is updated in the following way:

$$(x, y, s) \leftarrow (x + \alpha\Delta x, y + \alpha\Delta y, s + \alpha\Delta s)$$

- Phase 2: In this phase, starting from $k \geq 1$, we begin by updating the parameter μ and constructing the auxiliary point. Then, we utilize it to create the principal point. As in Phase 1, we solve the following system to determine the search direction for the auxiliary point:

$$\begin{aligned}
A\Delta x &= 0 \\
A^T\Delta y - \Delta s &= 0 \\
\hat{S}\Delta x + \hat{X}\Delta s &= \mu\mathbf{e} - XS\mathbf{e}.
\end{aligned} \tag{6}$$

  Since the left-hand side of (6) is the same as the left-hand side of (5), so there is no need to compute the inverse. Next, a line search is performed to determine the maximum step size, and the auxiliary point is updated as follows:

$$(\tilde{x}, \tilde{y}, \tilde{s}) \leftarrow (x + \alpha\Delta x, y + \alpha\Delta y, s + \alpha\Delta s)$$

Next, the middle point is updated as:

$$(\hat{x},\hat{y},\hat{s}) \leftarrow \frac{1}{2}((\tilde{x},\tilde{y},\tilde{s}) + (x,y,s))$$

Finally, the algorithm updates the principal point using the search direction $(\Delta x, \Delta y, \Delta s)$ is obtained by solving the following system:

$$\begin{aligned} A\Delta x &= 0 \\ A^T\Delta y - \Delta s &= 0 \\ \hat{S}\Delta x + \hat{X}\Delta s &= \mu \mathbf{e} - XS\mathbf{e} \end{aligned} \tag{7}$$

As earlier, to ensure that the principal point satisfies the feasibility condition, a line search is performed to determine the maximum step size. The principal point is then updated as follows:

$$(x,y,s) \leftarrow (x + \alpha\Delta x, y + \alpha\Delta y, s + \alpha\Delta s).$$

In order to determine the maximum value for the step size α in each iteration, we utilize the method presented by Cai et al. [24]. This method initially calculates an upper bound based on the positivity constraints of the variables. If $(x + \alpha\Delta x, y + \alpha\Delta y, s + \alpha\Delta s)$ is not feasible, then α is decreased by a certain multiplicative factor until the point is feasible. This approach is more efficient than determining the optimal value for α, as it avoids an additional computational step. Algorithm 1 outlines the steps of this new approach.

Algorithm 1 A two-step feasible IPM algorithm for LOs

Input: x_0, y_0, s_0, μ, θ, ϵ
$(x,y,s) \leftarrow (x_0, y_0, s_0)$
$(\tilde{x},\tilde{y},\tilde{s}) \leftarrow (x,y,s)$
$(\hat{x},\hat{y},\hat{s}) \leftarrow \frac{1}{2}((\tilde{x},\tilde{y},\tilde{s}) + (x,y,s))$
$\mu \leftarrow (1-\theta)\mu$
Find search direction $(\Delta x, \Delta y, \Delta s)$ by solving (5).
Find the maximum value for the step size α.
Set $(x,y,s) \leftarrow (x + \alpha\Delta x, y + \alpha\Delta y, s + \alpha\Delta s)$
while *stopping criteria is not met* **do**
  $\mu \leftarrow \mu(1-\theta)$
  Solve (6) to find the search direction.
  Find the maximum value for the step size.
  Set $(\tilde{x},\tilde{y},\tilde{s}) \leftarrow (x + \alpha\Delta x, y + \Delta y, s + \Delta s)$.
  Set $(\hat{x},\hat{y},\hat{s}) \leftarrow \frac{1}{2}((\tilde{x},\tilde{y},\tilde{s}) + (x,y,s))$
  Find search direction $(\Delta x, \Delta y, \Delta s)$ by solving (7).
  Find the maximum value for the step size α.
  Set $(x,y,s) \leftarrow (x + \alpha\Delta x, y + \alpha\Delta y, s + \alpha\Delta s)$
end

The method used in Algorithm 1 stands out from existing interior point algorithms. It utilizes a two-step Newton method to determine the search direction, resulting in a higher convergence rate of $1 + \sqrt{2}$ compared to traditional methods. This improved rate leads to a significant reduction in the number of iterations needed by the algorithm. Additionally, unlike interior-point predictor-corrector methods that require computing the matrix's inverse twice per iteration, Algorithm 1 only needs to compute it once per iteration. Specifically, it uses the previous iteration's matrix inverse to calculate the search direction for the auxiliary point. This computational characteristic aligns Algorithm 1 with one-step methods but requires fewer iterations to converge.

Incorporating the idea proposed by Wang et al. [25], we will now implement Algorithm 2. To begin, they introduced a fresh variable v, defined below:

$$v := \sqrt{\frac{xs}{\mu}} \tag{8}$$

Next, they introduced a kernel function as $\psi(t) = \frac{(t-1)^2}{2}$. Based on this kernel function and the fact that $(x,s) > 0$, they showed that [25]:

$$v = \mathbf{e} \Leftrightarrow v^2 = \mathbf{e} \Leftrightarrow \frac{xs}{\mu} = \sqrt{\frac{xs}{\mu}}$$

Now, we use this idea and rewrite the (6) and (7) as [25]:

$$\begin{aligned} A\Delta x &= 0 \\ A^T\Delta y - \Delta s &= 0 \\ \hat{S}\Delta x + \hat{X}\Delta s &= \mu\mathbf{e} - \sqrt{\mu}\sqrt{XS\mathbf{e}} \end{aligned} \tag{9}$$

and

$$\begin{aligned} A\Delta x &= 0 \\ A^T\Delta y - \Delta s &= 0 \\ \hat{S}\Delta x + \hat{X}\Delta s &= \mu\mathbf{e} - \sqrt{\mu}\sqrt{XS\mathbf{e}} \end{aligned} \tag{10}$$

The resulting changes lead to Algorithm 2 which is described next:

Algorithm 2 A two-step feasible IPM algorithm for the algorithm proposed in [25]

Input: $x_0,\ y_0,\ s_0,\ \mu, \theta, \epsilon$
$(x,y,s) \leftarrow (x_0, y_0, s_0)$
$(\tilde{x},\tilde{y},\tilde{s}) \leftarrow (x,y,s)$
$(\hat{x},\hat{y},\hat{s}) \leftarrow \frac{1}{2}((\tilde{x},\tilde{y},\tilde{s}) + (x,y,s))$
$\mu \leftarrow (1-\theta)\mu$
Find search direction $(\Delta x, \Delta y, \Delta s)$ by solving (10).
Find the maximum value for the step size α.
Set $(x,y,s) \leftarrow (x+\alpha\Delta x, y+\alpha\Delta y, s+\alpha\Delta s)$
while *stopping criteria is not met* **do**
    $\mu \leftarrow \mu(1-\theta)$
    Solve (9) to find the search direction.
    Find the maximum value for the step size.
    Set $(\tilde{x},\tilde{y},\tilde{s}) \leftarrow (x+\alpha\Delta x, y+\Delta y, s+\Delta s)$.
    Set $(\hat{x},\hat{y},\hat{s}) \leftarrow \frac{1}{2}((\tilde{x},\tilde{y},\tilde{s}) + (x,y,s))$
    Find search direction $(\Delta x, \Delta y, \Delta s)$ by solving (10).
    Find the maximum value for the step size α.
    Set $(x,y,s) \leftarrow (x+\alpha\Delta x, y+\alpha\Delta y, s+\alpha\Delta s)$
end

2.3. Discussion

The idea in the Algorithm is also used in [20] for finding roots of a univariate function f using a modification to the Newton–Raphson method. The method requires two steps a predictor step and a corrector step. The predictor step computes the point $\tilde{x}_+$ based on a linear approximation of the function around the current point x. In the corrector step, the new point is calculated using the derivative at a point which is the average of the current point and the predictor. This can be expressed as [20]:

$$\tilde{x}_+ = x - \frac{f(x)}{f'(\frac{x+\tilde{x}}{2})}$$

If we call $1/2(x + \tilde{x}_+)$ as $\hat{x}$ and use the derivative at $\hat{x}$ instead at x in the predictor step, then we have the following scheme:

$$\begin{aligned} \hat{x} &= \frac{x + \tilde{x}_+}{2} \\ x_+ &= x - \frac{f(x)}{f'(\hat{x})} \end{aligned}$$

In particular applications $f'(\hat{x})$ is a good estimate of $f'(x)$. Therefore this reduces the number of derivate evaluations. If the predictor-corrector scheme was not used, then the new point is $\tilde{x}$, and the function value would be $f(\tilde{x}) = f(x - f(x)/f'(\hat{x}))$. Since a corrector is used, the value of f at the new point is $f(x - f(x)/f'(\hat{x}_+))$. If there is a favourable relationship between the derivatives of the predictor and the corrector points, then (assuming a minimization problem) we have:

$$f(x - f(x)/f'(\hat{x})) > f(x - f(x)/f'(\hat{x}_+)) \tag{11}$$

Suppose this inequality is valid in many iterative steps. In that case, there will be a reduction in the number of iterations compared to the Newton–Raphson method. The decrease in iterations required can be credited to the new approach's ability to attain the optimal point of the Newton method at a rate of $1 + \sqrt{2}$. This higher convergence rate means that the new method needs fewer iterations than the traditional approach.

In general, if the function is multivariate denoted F, then the updates can be written as:

$$\begin{aligned} \tilde{x}_+ &= x - \alpha(\nabla F(\hat{x}))^{-1}F(x) \\ \hat{x}_+ &= 1/2(x + \tilde{x}_+) \\ x_+ &= x - \alpha(\nabla F(\hat{x}_+))^{-1}F(x) \end{aligned}$$

The algorithm presented in this paper is fundamentally distinct from the predictor-corrector method proposed by Mehrotra [17]. Notably, one of the main differences lies in the approach to calculating the search direction. While both algorithms leverage the two-stage search direction concept, Algorithm 1 computes the search direction by utilizing an auxiliary point and taking the average of this auxiliary point and the main point. Another significant contrast arises in the computation of the matrix inverse. Algorithm 1 requires calculating the matrix inverse only once per iteration. Consequently, it falls into the category of one-step methods in terms of time consumption, while achieving efficiency akin to two-step algorithms. These distinctions set Algorithm 1 apart from the predictor-corrector method of [17], highlighting its unique and promising characteristics in solving optimization problems. Moreover, the method in [17] requires tuning many parameters in each iteration, which can be expensive and complex for some problems to find their optimal values. In contrast, our proposed method involves far fewer parameters to tune, simplifying the process and potentially reducing the computational burden.

3. Experimental Evaluation

We use linear programming instances from NETLIB [26]. There are a total of 138 linear programming problems with varying sizes. The smallest one has 22 non-zeros and the largest one has 1,408,073 non-zeros in the coefficient matrix. There are 30 instances that are infeasible. Thirty nine of the instances are not full rank. Further, four test problems have $b = 0$. If $b = 0$, the initialization scheme [27] not produce an initial feasible solution. We do not use these 73 instances in our experimentation. We define the absolute and relative gap as follows:

$$\text{absolute gap: } |c^T x - b^T y|$$
$$\text{relative gap: } \frac{|c^T x - b^T y|}{1 + |c^T x| + |b^T y|}.$$

For all experiments, we use a relative-gap of at most 10^{-6}, at most 700 iterations as the stopping criteria. Although the absolute gap is generally more stringent than the relative-gap, achieving the absolute gap for problems with large objective function values may require many iterations. In IPMs, significant reductions occur during the initial iterations, but only small reductions happen in the final iterations. Therefore, in this article, we use the relative-gap in our experiments.

3.1. Experiments on CPU

We demonstrate the effectiveness of the new approach by executing Algorithm 1 and comparing the results to the classical interior-point algorithm (the primal-dual logarithmic barrier method Chapter 7) [8]. Note that instead of calculating the search direction using Newton's method mentioned in the classical algorithm [8], Algorithm 1 finds the search direction using the modified Newton approach. This is the primary difference between Algorithm 1 and the classical algorithm. Specifically, in each iteration, Algorithm 1 first updates the auxiliary point. It then uses the auxiliary point, along with the primary principle, to find the search direction. In contrast, the classical algorithm relies solely on the information from the principal point to determine the search direction. In this regard, we coded the three algorithms (Algorithm 1, Algorithm 2 and Classical approach) in Python 3.10 using BLAS routines for the equation-solving step in all the algorithms. Of the remaining 65 instances the three algorithms solved 50 instances within a total time limit of 24 h on alliance canada cluster CEDAR, 2 × Intel E5-2683 v4 Broadwell CPU with 125 G RAM 2.1 GHz (https://alliancecan.ca/en, accessed on 27 August 2023) [28]. The remaining fifteen instances large-sized instances were not completed due to the time limit for at least one of the algorithms tested.

To demonstrate the efficacy of the new method, we applied Algorithms 1 and 2 to a test problem. For each step of the two-step method, we computed the objective function. Figure 1 illustrates the difference in the objective function between the two points, $\tilde{x}_+$ (auxiliary point) and x_+ (principal point), during each iteration. The plot shows that the objective function value in the second stage is consistently lower than in the first stage, indicated by positive values of $c^T\tilde{x}_+ - c^T x_+$. This validates the proposed method, demonstrating its potential to reduce the number of iterations required. Additionally, it indicates that the auxiliary point effectively guides the algorithm in correcting its search direction, leading to more efficient convergence.

Next, we want to see for the same instance `lp_scsd8`, (i) the rate at which the absolute gap decreases, (ii) the decrease in the relative gap between primal and dual solution values as a function of the iterations. Figure 2 shows the two plots. Algorithms 1 and 2 achieve a faster reduction in the relative gap and the absolute gap compared to the classical approach, i.e., a lesser number of iterations are needed overall. Also note a sharp reduction in the latter set of iterations for Algorithms 1 and 2 (this phenomenon is not universal though, sometimes there is a sharp reduction in the initial iteration for some instances).

To demonstrate the effectiveness of the two-step method we observe two quantities, (i) the speedup and (ii) the relative reduction in the number of iterations achieved by Algorithms 1 and 2 compared to the classical algorithm. If Algorithm i takes time $t(A_i, j)$ on instance j, classical algorithm C takes time $t(C, j)$ on the same instance j then the speedup of Algorithm i is defined as $t(C, j)/t(A_i, j)$. Similarly, we define the relative reduction in the number of iterations. If Algorithm i takes $n(A_i, j)$ iterations on instance j, classical algorithm C takes $n(C, j)$ iterations on the same instance j then the relative reduction of Algorithm i is defined as $n(C, j)/n(A_i, j)$.

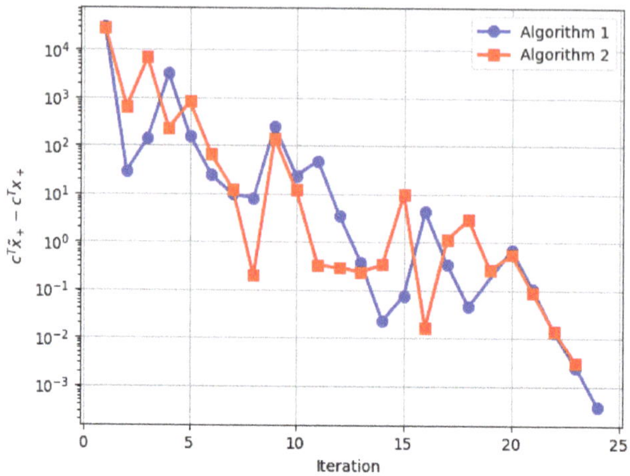

Figure 1. $c^T\tilde{x}_+ - c^T x_+$ for Algorithms 1 and 2 on instance `lp_scsd8`.

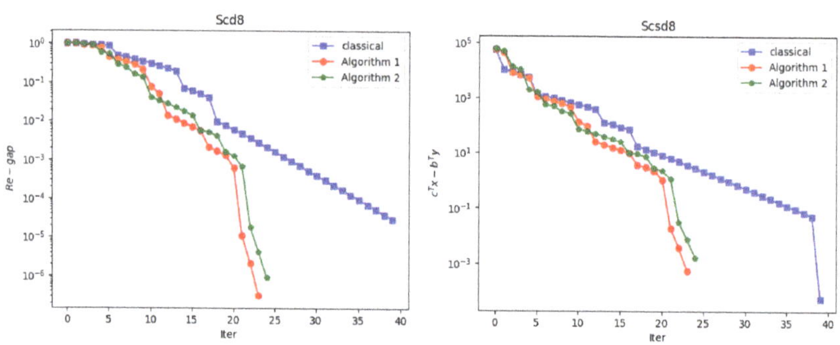

Figure 2. (**Left**) Relative gap, (**Right**) absolute gap vs. the iteration number.

Figures 3 and 4 show the relative reduction and the speedup in the number of iterations as a function of the number of instances. The instances are ordered according to the size in array. The average value of speedup, relative reduction if computed for each suffix of the array. If the x-coordinate is n then the y-coordinate is the average speedup, relative reduction for $50 - n$ largest instances. The average speedup over all 50 instances is between 1.3 and 1.4 for both Algorithms 1 and 2. The average relative reduction is between 1.20 and 1.25 for both Algorithms 1 and 2. If we focus only on 5 of the largest instances then the speedup and relative reduction are comparably higher. This indicates that the method scales well. Algorithm 1 has a greater relative reduction but it does not translate into a greater speedup. Algorithm 2 has a better speedup. The instances indexed 40–45 in the plots show anomalous behaviour, it appears that these are specific types of instances (staircase instances) and the performance of Algorithms 1 and 2 seems to be different (compared to the average) on these instances. One of the prominent features of the new proposed algorithm is its ability to find an appropriate search direction compared to traditional interior-point algorithms. As demonstrated in [20], the new modified Newton method exhibits a faster convergence rate, approximately $1 + \sqrt{2}$. Therefore, the new algorithm leverages this property, allowing it to reach the optimal solution in fewer iterations—a particularly valuable advantage when dealing with high-dimensional test problems. In scenarios where matrix multiplication and inversion can be computationally intensive, the reduced number of iterations provided by the new algorithm significantly decreases the execution time and enhances overall performance.

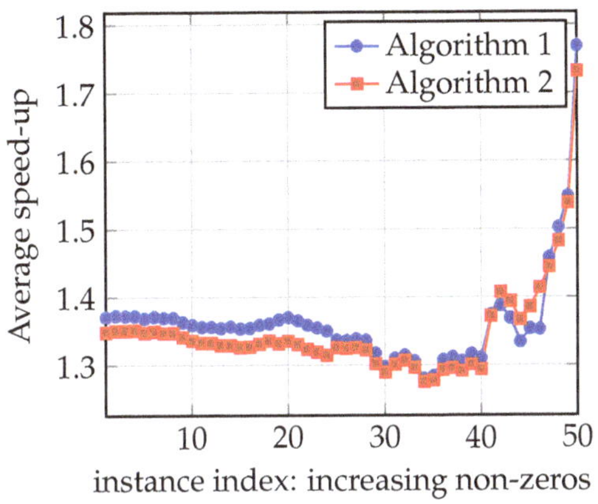

Figure 3. Relative reduction.

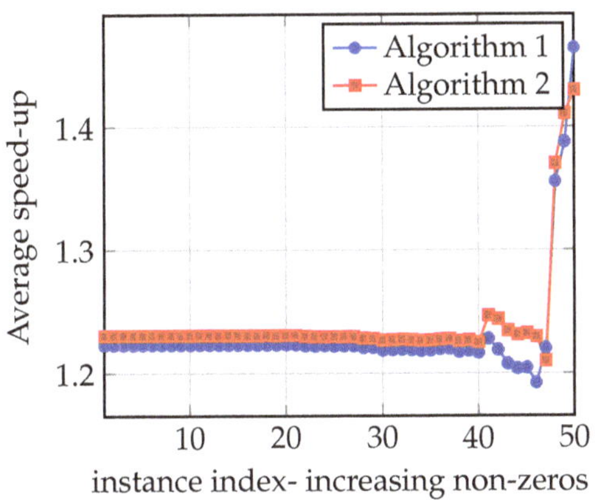

Figure 4. Speedup vs. number of instances.

In Table 1, the average number of iterations and CPU time (measured in seconds) for three different methods—Classical Algorithm, Algorithms 1 and 2—are presented. The Classical Algorithm required an average of 97.82 iterations and 166.71 s of CPU time. In contrast, Algorithm 1 demonstrated improved efficiency with an average of 71.34 iterations and 136.29 s of CPU time. Similarly, Algorithm 2 also exhibited promising results with 72.52 iterations and 135.46 s of CPU time. These findings indicate that both Algorithms 1 and 2 are more effective than the Classical Algorithm in reducing the number of iterations and CPU time. Notably, Algorithm 1 achieved the lowest number of iterations, while Algorithm 2 achieved the lowest CPU time. Moreover, the fourth column presents the standard deviation for the number of iterations for the three algorithms, indicating that Algorithm 1 has the lowest variability, followed by Algorithm 2, with the Classical Algorithm having the highest variability.

The details of the problem instances can be found in [29]. The raw data used to build these figures is in the Appendix A and is self-explanatory. The data in Appendix B is divided into blocks of three rows each. The first row in each block is data for the classical algorithm and the other two rows are the data for Algorithms 1 and 2. Table A2 lists the relative gap and the norms and shows the convergence of the three algorithms to an optimal solution on several instances from NETLIB.

Table 1. The average number of iterations, CPU time (s) and standard deviation.

Methods	Aver. Iter.	Aver. CPU (s)	Std-Iter.
Classical Algorithm	97.82	166.71	70.96
Algorithm 1	**71.34**	136.29	**55.64**
Algorithm 2	72.52	**135.46**	57.55

Note: Bold numbers indicate the lowest values in each column.

3.2. Experiments on GPU

A study by Świrydowicz et al. [30] investigated the use of GPU-accelerated linear solvers for optimizing large-scale power grid problems. Motivated by their study, we evaluate the performance of the algorithms on massively parallel hardware (GPU with tensor cores) using library routines to perform factoring.

We utilized the Pytorch library to develop a GPU-enabled version of our program. The library assumes that all tensors are located on the same device. During our experiments, we utilized the NVIDIA V100 Volta (on the GRAHAM Compute Canada cluster), which has a maximum size limitation due to its 16GB HBM2 memory. This card features tensor cores that perform certain tensor operations very efficiently in hardware. GPU arithmetic is typically conducted in low precision, ranging from 8 to 32 bits which account for the speed. Unfortunately, the algorithms failed to converge to an optimal solution when we employed low-precision arithmetic on V100. As a result, we opted to use double precision at 64 bits, even though this substantially increased the run time.

To utilize GPU acceleration, we convert all vectors and matrices into tensors, leveraging the PyTorch library instead of Numpy for computations. In contrast, when running on a CPU, we represent these structures as regular vectors and matrices. By performing essential operations like matrix multiplication, inverse calculation and step size computation in tensor mode. This approach enhances the efficiency of calculations and enables faster reaching the optimal point.

The computationally expensive steps in the algorithm proposed here (as in any IPM) are (i) the determination of α and (ii) the computation of the inverse of the Jacobian to determine the step direction. The first expensive step has quadratic complexity ($O((n+m)^2)$ when implemented naturally). The second step is computationally more costly $O((n+m)^3)$. In the approach here, gains are realized asymptotically in the second step. Therefore, we focus on the time the GPU takes to perform matrix inversions. The raw data is shown in Appendix C. The second and third columns are the iterations needed by the classical and the algorithm proposed here. The last two columns in the table are the times (in seconds) needed by the GPU to perform the matrix inversions.

Results for GPU

Table A3 presents the results for Classical Algorithm and Algorithm 1 on GPU. We report the number of iterations and GPU times for 41 test problems. Note that, Classical algorithm could not reach the optimal solution for 5 test problems and Algorithm 1 could not find the optimal solution for 1 test problem based on the stopping conditions. We remove the results for these 6 test problems and calculate the relative reduction and speed-up. Figure 5 presents the relative reduction and speed-up for 35 test problems. Moreover, we denote the average of the number of iterations and GPU time for these 35 test problems in Table 2. Based on this Table, we can conclude that the new proposed method reduced the number of iterations and GPU by 39% and 30% respectively. Moreover, the standard deviation for the number of iterations indicates that Algorithm 1 has the lowest variability at 60.19.

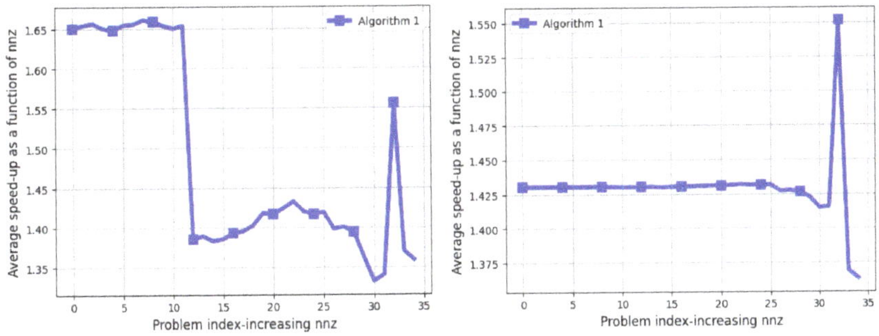

Figure 5. (**Left**) Relative reduction. (**Right**) Speedup vs. the number of instances.

Table 2. The average number of iterations, GPU time and standard deviation for the number of iterations.

Methods	Aver. Iter.	Aver. GPU (s)	Std-Iter.
Classical Algorithm	101.77	39.81	113.71
Algorithm 1	**61.68**	**27.83**	**60.19**

Note: Bold numbers indicate the lowest values in each column.

3.3. Discussion

Although it has not been theoretically proven that the convergence rate of the modified Newton method is exactly $1+\sqrt{2}$, numerical results show that this rate is achievable. This rate may vary slightly depending on the structure of the problem (see example in [20]). As shown in [20], the modified Newton method can reduce the number of iterations and computational time for solving a system of equations by a factor of 1.22, calculated as $(1+\sqrt{2})/2$, where 2 is the convergence rate of the Newton method and $1+\sqrt{2}$ is the convergence rate of the modified Newton method. However, the average results in Tables 1 and 2 demonstrate that when we apply the modified Newton method to find the search direction in IPMs, the convergence rate of the proposed method remains close to $1+\sqrt{2}$.

4. Conclusions

This paper presents a two-step method for optimizing linear programming problems using the interior point method. Building on a previous idea to solve the Karush–Kuhn–Tucker (KKT) conditions, we extend this approach to IPMs. Our method re-examines and enhances the classical algorithm proposed by [8] by incorporating a two-step process. In this method, the search direction is calculated using a modified Newton method that employs an auxiliary point. Like the standard Newton method, the modified Newton method computes the matrix inverse only once. We incorporate this approach into IPM and demonstrate its efficiency through implementations on both CPU and GPU, as well as by solving NETLIB test problems. The results show that the proposed method can improve efficiency by 27% in terms of iterations and 18% in terms of CPU time. These results confirm that the convergence rate of the modified Newton method is equal to $1+\sqrt{2}$. This work is pioneering in integrating the modified Newton method into IPM. This approach has the potential to be applied to other interior point algorithms, enhancing their performance as well.

One of the key areas for future work is proving the complexity bounds of the algorithm, which could enhance the overall complexity bounds of IPMs. Additionally, demonstrating the effectiveness of the algorithm on real-world problems modeled as linear optimization problems is another important avenue for exploration. Another focus will be evaluating the algorithm's efficiency on large datasets, which is particularly relevant given the increasing dimensions of real-world problems. Further, we plan to optimize the implementation on GPUs, which includes developing parallel implementations of the algorithm and leveraging parallel processing to calculate the inverse of the Jacobian matrix. This work can yield significant performance improvements and provide valuable insights into the practical applications of the algorithm.

Author Contributions: Conceptualization, S.F.H. and D.G.; methodology, S.F.H., D.G. and R.B.; visualization, S.F.H. and D.G.; investigation, D.G. and R.B.; writing—original draft preparation, S.F.H. and D.G.; writing—review and editing, D.G. and R.B.; supervision, D.G. and R.B. All authors have read and agreed to the published version of the manuscript.

Funding: This research received no external funding.

Institutional Review Board Statement: Not applicable.

Informed Consent Statement: Not applicable.

Data Availability Statement: The datasets used in this paper are available from the following link https://sparse.tamu.edu/LPnetlib, accessed on 25 August 2023.

Acknowledgments: This research was enabled in part by support provided by the Digital Research Alliance of Canada (https://alliancecan.ca, accessed on 27 August 2023).

Conflicts of Interest: The authors declare no conflicts of interest.

Abbreviations

The following abbreviations are used in this manuscript:

LO	Linear optimization
IPM	Interior-point method
SDO	Semidefinite optimization
NCP	Non-linear complementarity problem
SOCO	Second-order cone optimization
KKT	Karush–Kuhn–Tucke
IPC	Interior point condition

Appendix A. Raw Tables—CPU Data

Table A1. Iteration number and CPU times.

	Iteration			CPU Time (s)		
Name	**Classical**	**Algo 1**	**Algo 2**	**Classical**	**Algo 1**	**Algo 2**
adlittle	54	35	33	0.09	0.07	0.06
afiro	16	17	18	0.46	0.49	0.24
agg	91	66	66	6.20	4.64	4.63
agg2	95	60	58	7.99	5.24	5.03
agg3	124	60	66	10.51	5.31	5.85
bandm	105	69	71	3.97	2.67	2.74
beaconfd	105	48	46	1.39	0.76	0.64
blend	50	26	24	0.07	0.05	0.04
bnl2	203	183	153	1089.95	1000.31	827.93
capri	85	74	55	3.11	2.84	2.06
chemcom	58	45	45	3.69	2.97	2.94
czprob	186	137	146	434.42	313.03	342.57
e226	130	72	120	3.76	2.20	3.66
fffff800	118	121	112	22.88	24.03	22.13
fit1p	80	38	42	43.91	21.15	23.39
forest6	43	22	22	0.07	0.05	0.04
ganges	50	43	42	49.60	42.40	42.73

Table A1. *Cont.*

	Iteration			CPU Time (s)		
Name	Classical	Algo 1	Algo 2	Classical	Algo 1	Algo 2
gfrd_pnc	118	74	74	28.88	18.68	18.56
israel	150	100	102	1.31	0.98	0.97
lotfi	74	61	56	1.28	1.16	1.04
maros_r7	75	6	65	47.15	41.12	41.72
pilot87	329	186	190	842.15	575.14	589.04
pilot	168	135	133	472.34	371.98	342.67
pilot_we	357	325	371	612.35	558.41	632.02
sc50a	28	20	22	0.03	0.03	0.02
sc50b	29	20	22	0.03	0.03	0.02
sc105	24	24	23	0.06	0.16	0.13
sc205	42	24	24	0.308	0.30	0.27
scagr25	43	42	42	3.75	3.62	3.55
scfxm1	83	53	54	4.45	4.33	3.17
scfxm2	80	68	66	20.06	17.84	17.34
scfxm3	86	77	77	49.00	44.49	44.68
scrs8	149	108	106	85.76	63.90	61.95
scagr7	30	22	23	0.11	0.21	0.11
scsd1	26	27	28	1.94	2.08	2.12
scsd6	24	25	27	5.91	6.24	6.77
scsd8	40	25	24	42.27	25.68	24.58
sctap1	71	56	63	3.79	3.16	3.50
sctap2	82	56	58	99.29	68.78	71.00
sctap3	99	60	62	232.41	144.41	149.83
share1b	127	86	89	0.85	0.71	0.71
share2b	41	31	30	1.65	1.06	1.21
stair	102	97	97	4.82	4.81	4.71
standata	105	83	82	28.15	22.48	23.12
standmps	211	129	142	58.05	36.08	39.97
stocfor1	91	51	52	0.27	0.32	0.25
stocfor2	230	176	191	848.32	651.89	668.12
truss	70	59	58	2917.26	2517.39	2560.80
lp-vtp-base	64	42	39	0.89	0.59	0.57
woodw	50	43	46	239.01	198.67	201.92

Appendix B. Relative Gap and the Norm—CPU Data

Table A2. Relative gap and the norm.

Name	Regap	$\|\|Ax - b\|\|$	$\|\|c - s - A^T y\|\|$
adlittle	8.69×10^{-7}	5.62×10^{-14}	1.29×10^{-10}
adlittle	8.82×10^{-7}	3.32×10^{-12}	1.29×10^{-10}
adlittle	1.27×10^{-7}	1.31×10^{-13}	1.83×10^{-5}
afiro	8.37×10^{-7}	4.70×10^{-17}	1.61×10^{-4}
afiro	7.18×10^{-7}	2.13×10^{-9}	1.61×10^{-4}
afiro	9.54×10^{-7}	1.64×10^{-9}	3.69×10^{-4}
agg	1.41×10^{-7}	2.38×10^{-12}	1.78×10^{-5}
agg	7.57×10^{-7}	2.02×10^{-11}	1.78×10^{-5}
agg	7.57×10^{-7}	2.02×10^{-11}	1.78×10^{-5}

Table A2. *Cont.*

Name	Regap	$\|\|Ax-b\|\|$	$\|\|c-s-A^Ty\|\|$
agg2	3.23×10^{-8}	9.15×10^{-12}	3.34×10^{-6}
agg2	7.21×10^{-7}	5.27×10^{-13}	3.34×10^{-6}
agg2	2.03×10^{-9}	3.29×10^{-14}	3.37×10^{-6}
agg3	5.27×10^{-7}	1.13×10^{-10}	2.82×10^{-6}
agg3	5.77×10^{-7}	4.40×10^{-11}	2.82×10^{-6}
agg3	2.62×10^{-7}	1.01×10^{-11}	3.24×10^{-6}
bandm	8.64×10^{-7}	7.02×10^{-13}	1.76×10^{-3}
bandm	1.50×10^{-7}	4.43×10^{-15}	1.76×10^{-7}
bandm	7.66×10^{-7}	2.86×10^{-13}	8.31×10^{-9}
beaconfd	6.09×10^{-9}	7.13×10^{-5}	3.75×10^{-8}
beaconfd	4.26×10^{-8}	1.25×10^{-8}	3.75×10^{-8}
beaconfd	2.405×10^{-8}	6.97×10^{-9}	1.52×10^{-8}
blend	5.83×10^{-7}	2.87×10^{-16}	2.78×10^{-6}
blend	5.25×10^{-7}	6.43×10^{-10}	2.78×10^{-6}
blend	9.32×10^{-7}	7.54×10^{-14}	7.10×10^{-5}
bnl2	9.79×10^{-7}	1.65×10^{-16}	3.37×10^{-3}
bnl2	8.55×10^{-7}	2.27×10^{-16}	3.37×10^{-3}
bnl2	6.19×10^{-7}	1.97×10^{-16}	2.66×10^{-2}
capri	2.34×10^{-9}	2.57×10^{-16}	2.41×10^{-7}
capri	1.57×10^{-7}	1.45×10^{-10}	2.41×10^{-7}
capri	4.77×10^{-7}	6.37×10^{-10}	7.62×10^{-6}
chemcom	3.88×10^{-9}	1.48×10^{-16}	6.08×10^{-9}
chemcom	1.52×10^{-7}	1.66×10^{-12}	6.08×10^{-9}
chemcom	1.98×10^{-7}	1.72×10^{-12}	6.11×10^{-8}
czprob	9.42×10^{-7}	1.01×10^{-12}	1.59×10^{-3}
czprob	8.18×10^{-7}	8.94×10^{-13}	1.59×10^{-3}
czprob	7.07×10^{-9}	2.98×10^{-13}	3.81×10^{-5}
e226	2.03×10^{-7}	3.85×10^{-15}	5.40×10^{-4}
e226	2.95×10^{-7}	2.60×10^{-14}	5.40×10^{-4}
e226	7.65×10^{-7}	3.31×10^{-16}	4.22×10^{-3}
fffff800	9.51×10^{-7}	1.16×10^{-10}	3.45×10^{-7}
fffff800	3.96×10^{-7}	8.72×10^{-11}	3.45×10^{-7}
fffff800	1.31×10^{-8}	1.65×10^{-10}	6.25×10^{-7}
fit1p	8.80×10^{-7}	1.48×10^{-16}	1.02×10^{-5}
fit1p	1.25×10^{-7}	1.39×10^{-11}	1.02×10^{-5}
fit1p	1.01×10^{-7}	1.26×10^{-11}	7.49×10^{-10}
forest6	6.54×10^{-8}	6.95×10^{-16}	1.71×10^{-6}
forest6	8.14×10^{-7}	5.54×10^{-12}	1.71×10^{-6}
forest6	8.59×10^{-7}	5.57×10^{-12}	3.51×10^{-5}
ganges	7.87×10^{-7}	1.48×10^{-16}	2.40×10^{-7}
ganges	5.00×10^{-7}	6.77×10^{-12}	2.40×10^{-7}
ganges	4.25×10^{-7}	4.97×10^{-12}	3.41×10^{-8}
gfrd_pnc	1.88×10^{-8}	9.77×10^{-17}	1.24×10^{-6}
gfrd_pnc	2.99×10^{-7}	1.69×10^{-16}	1.24×10^{-6}
gfrd_pnc	2.11×10^{-7}	1.57×10^{-16}	3.98×10^{-6}
israel	8.63×10^{-7}	3.42×10^{-16}	4.00×10^{-6}
israel	2.37×10^{-7}	7.63×10^{-14}	4.00×10^{-6}
israel	2.33×10^{-7}	7.32×10^{-14}	3.99×10^{-6}

Table A2. *Cont.*

Name	Regap	$\|\|Ax-b\|\|$	$\|\|c-s-A^Ty\|\|$
lotfi	1.07×10^{-8}	2.34×10^{-14}	1.89×10^{-8}
lotfi	4.72×10^{-7}	4.05×10^{-12}	1.89×10^{-8}
lotfi	2.23×10^{-7}	1.84×10^{-12}	5.32×10^{-7}
maros_r7	2.36×10^{-10}	2.94×10^{-16}	7.53×10^{-9}
maros_r7	6.67×10^{-8}	2.92×10^{-12}	7.53×10^{-9}
maros_r7	5.15×10^{-8}	2.059×10^{-12}	1.82×10^{-9}
pilot87	2.10×10^{-5}	7.50×10^{-12}	3.58×10^{-7}
pilot87	3.86×10^{-5}	2.46×10^{-12}	3.58×10^{-7}
pilot87	1.35×10^{-5}	2.76×10^{-12}	2.85×10^{-7}
pilot	2.26×10^{-7}	7.30×10^{-13}	5.62×10^{-7}
pilot	1.62×10^{-6}	1.00×10^{-13}	5.62×10^{-7}
pilot	1.53×10^{-6}	7.56×10^{-14}	1.55×10^{-6}
pilot_we	9.70×10^{-7}	6.77×10^{-15}	9.06×10^{-4}
pilot_we	2.87×10^{-7}	4.01×10^{-15}	9.06×10^{-4}
pilot_we	8.19×10^{-7}	6.18×10^{-15}	1.27×10^{-3}
sc50a	9.73×10^{-7}	7.74×10^{-17}	7.90×10^{-3}
sc50a	6.08×10^{-7}	1.39×10^{-12}	7.901×10^{-3}
sc50a	7.08×10^{-7}	7.57×10^{-16}	3.92×10^{-2}
sc50b	9.94×10^{-7}	1.36×10^{-16}	5.45×10^{-3}
sc50b	6.46×10^{-7}	3.58×10^{-14}	5.45×10^{-3}
sc50b	4.58×10^{-7}	1.90×10^{-16}	2.71×10^{-3}
sc105	1.02×10^{-7}	1.02×10^{-16}	1.20×10^{-2}
sc105	4.64×10^{-7}	8.09×10^{-13}	1.20×10^{-2}
sc105	6.63×10^{-7}	2.12×10^{-16}	9.76×10^{-3}
sc205	5.80×10^{-9}	1.12×10^{-16}	3.80×10^{-6}
sc205	3.35×10^{-7}	9.02×10^{-13}	3.80×10^{-6}
sc205	5.83×10^{-7}	1.69×10^{-16}	9.60×10^{-3}
scagr25	6.25×10^{-7}	1.32×10^{-16}	2.79×10^{-6}
scagr25	2.11×10^{-7}	5.24×10^{-12}	2.79×10^{-6}
scagr25	2.80×10^{-7}	8.46×10^{-11}	6.52×10^{-5}
scfxm1	9.01×10^{-7}	1.82×10^{-10}	4.43×10^{-6}
scfxm1	1.74×10^{-7}	9.71×10^{-10}	4.43×10^{-6}
scfxm1	9.94×10^{-7}	1.45×10^{-9}	7.66×10^{-6}
scfxm2	7.87×10^{-7}	5.19×10^{-10}	1.62×10^{-6}
scfxm2	4.25×10^{-7}	1.41×10^{-10}	1.62×10^{-6}
scfxm2	9.46×10^{-7}	1.25×10^{-10}	3.06×10^{-6}
scfxm3	9.61×10^{-7}	8.06×10^{-9}	7.08×10^{-7}
scfxm3	1.57×10^{-7}	3.51×10^{-10}	7.08×10^{-7}
scfxm3	1.60×10^{-7}	3.86×10^{-10}	6.77×10^{-7}
scrs8	8.13×10^{-7}	5.25×10^{-13}	4.79×10^{-8}
scrs8	7.91×10^{-7}	1.84×10^{-13}	4.79×10^{-8}
scrs8	2.03×10^{-7}	1.82×10^{-14}	2.15×10^{-8}
scagr7	9.96×10^{-7}	1.33×10^{-16}	1.01×10^{-6}
scagr7	7.30×10^{-7}	1.71×10^{-16}	1.01×10^{-6}
scagr7	3.32×10^{-7}	4.26×10^{-14}	6.08×10^{-6}
scsd1	6.18×10^{-7}	1.18×10^{-16}	1.18×10^{-7}
scsd1	6.90×10^{-7}	1.04×10^{-16}	1.18×10^{-7}
scsd1	7.51×10^{-7}	8.19×10^{-17}	1.08×10^{-5}

Table A2. *Cont.*

Name	Regap	$\|\|Ax - b\|\|$	$\|\|c - s - A^T y\|\|$
scsd6	7.89×10^{-7}	1.56×10^{-16}	3.14×10^{-5}
scsd6	8.16×10^{-7}	2.03×10^{-16}	3.14×10^{-5}
scsd6	6.87×10^{-7}	1.83×10^{-16}	3.64×10^{-5}
scsd8	2.94×10^{-8}	2.52×10^{-16}	2.64×10^{-5}
scsd8	3.12×10^{-7}	2.76×10^{-16}	2.64×10^{-5}
scsd8	5.75×10^{-7}	2.41×10^{-16}	5.35×10^{-6}
sctap1	5.77×10^{-9}	2.84×10^{-16}	4.66×10^{-6}
sctap1	4.76×10^{-7}	1.28×10^{-9}	4.66×10^{-6}
sctap1	2.18×10^{-9}	4.64×10^{-12}	3.66×10^{-8}
sctap2	6.17×10^{-9}	5.70×10^{-11}	1.33×10^{-7}
sctap2	1.49×10^{-7}	4.38×10^{-12}	1.33×10^{-7}
sctap2	1.79×10^{-7}	1.94×10^{-11}	2.16×10^{-6}
sctap3	1.96×10^{-8}	8.318×10^{-11}	2.61×10^{-6}
sctap3	5.08×10^{-7}	2.86×10^{-10}	2.61×10^{-6}
sctap3	2.63×10^{-7}	1.41×10^{-10}	2.41×10^{-6}
share1b	9.68×10^{-7}	1.27×10^{-16}	1.40×10^{-4}
share1b	9.29×10^{-8}	3.31×10^{-15}	1.40×10^{-4}
share1b	8.53×10^{-8}	2.59×10^{-15}	1.38×10^{-5}
share2b	9.60×10^{-7}	6.11×10^{-15}	4.34×10^{-6}
share2b	8.65×10^{-7}	1.23×10^{-9}	4.34×10^{-6}
share2b	1.14×10^{-7}	3.10×10^{-10}	1.41×10^{-5}
stair	9.30×10^{-7}	1.90×10^{-11}	1.48×10^{-5}
stair	6.96×10^{-7}	2.04×10^{-10}	1.48×10^{-5}
stair	2.84×10^{-7}	1.22×10^{-10}	2.15×10^{-5}
standata	7.12×10^{-7}	2.66×10^{-12}	3.02×10^{-12}
standata	3.87×10^{-7}	1.22×10^{-13}	3.02×10^{-12}
standata	9.88×10^{-7}	2.35×10^{-15}	4.45×10^{-11}
standmps	9.68×10^{-7}	8.69×10^{-9}	5.74×10^{-7}
standmps	1.75×10^{-7}	1.10×10^{-9}	5.74×10^{-7}
standmps	3.92×10^{-8}	6.92×10^{-11}	4.53×10^{-7}
stocfor1	1.20×10^{-8}	2.06×10^{-16}	5.68×10^{-6}
stocfor1	1.30×10^{-7}	2.69×10^{-14}	5.68×10^{-6}
stocfor1	1.96×10^{-7}	1.98×10^{-15}	1.28×10^{-5}
stocfor2	1.39×10^{-10}	2.15×10^{-16}	6.40×10^{-7}
stocfor2	5.61×10^{-10}	6.68×10^{-16}	6.40×10^{-7}
stocfor2	2.08×10^{-8}	6.16×10^{-15}	5.51×10^{-6}
truss	5.72×10^{-11}	1.40×10^{-15}	6.37×10^{-7}
truss	2.12×10^{-9}	6.87×10^{-16}	6.37×10^{-7}
truss	1.99×10^{-9}	6.67×10^{-16}	1.22×10^{-5}
lp-vtp-base	9.02×10^{-9}	2.12×10^{-9}	3.55×10^{-6}
lp-vtp-base	3.82×10^{-7}	4.49×10^{-9}	3.55×10^{-6}
lp-vtp-base	5.09×10^{-7}	2.42×10^{-8}	4.26×10^{-6}
woodw	3.85×10^{-5}	2.01×10^{-4}	4.05×10^{-4}
woodw	4.29×10^{-5}	1.51×10^{-5}	4.05×10^{-4}
woodw	1.33×10^{-4}	7.42×10^{-6}	2.09×10^{-4}

Appendix C. Data Tables for GPU

Table A3. Results for GPU.

	Iteration	Iteration	GPU Time (s)	GPU Time (s)
Name	Classical	Algorithm 1	Classical	Algorithm 1
adlittle	49	31	0.151	0.099
afiro	19	17	0.028	0.027
agg	74	60	1.635	1.339
agg2	84	47	2.108	1.181
agg3	81	54	2.108	1.378
bandm	700	105	10.05	1.529
beaconfd	700	50	5.318	0.392
blend	53	26	0.148	0.073
bnl2	222	236	182.065	192.634
capri	67	51	0.963	0.733
chemcom	76	47	1.719	1.075
d2q06c	306	166	467.293	252.344
fit1p	41	35	3.137	2.670
forest6	42	22	0.125	0.071
ganges	700	41	71.770	4.190
israel	157	128	1.311	1.109
lotfi	68	54	0.642	0.522
pilot87	246	181	462.709	339.440
pilot	128	92	105.27	75.457
pilot_we	399	269	84.727	56.946
sc50a	46	20	0.089	0.042
sc50b	28	20	0.052	0.039
sc105	21	21	0.085	0.082
sc205	45	23	0.397	0.207
scagr25	47	37	1.133	0.890
scfxm1	75	51	1.399	0.960
scfxm2	72	53	3.480	2.551
scfxm3	76	56	7.496	5.498
scagr7	32	26	0.141	0.117
scsd1	26	27	0.526	0.551
scsd6	24	25	1.071	1.110
scsd8	37	22	6.320	3.732
sctap1	60	52	1.229	1.073
3sctap2	90	47	16.300	8.485
sctap3	105	61	36.785	21.257
share1b	700	69	4.096	0.417
share2b	38	27	0.144	0.104
stair	700	82	13.832	1.640
standmps	183	700	8.892	34.161
stocfor1	67	38	0.255	0.151
lp-vtp-base	558	37	0.521	0.354

References

1. Zhu, J.; Rosset, S.; Tibshirani, R.; Hastie, T. 1-norm support vector machines. *Adv. Neural Inf. Process. Syst.* **2003**, *16*, 49–56.
2. Yang, J.; Zhang, Y. Alternating direction algorithms for ℓ_1-problems in compressive sensing. *SIAM J. Sci. Comput.* **2011**, *33*, 250–278. [CrossRef]
3. Yuan, M. High dimensional inverse covariance matrix estimation via linear programming. *J. Mach. Learn. Res.* **2010**, *11*, 2261–2286.
4. Recht, B.; Re, C.; Tropp, J.; Bittorf, V. Factoring nonnegative matrices with linear programs. *Adv. Neural Inf. Process. Syst.* **2012**, *25*, 1214–1222.
5. Meshi, O.; Globerson, A. An alternating direction method for dual MAP LP relaxation. In Proceedings of the Machine Learning and Knowledge Discovery in Databases: European Conference, ECML PKDD 2011, Athens, Greece, 5–9 September 2011; Proceedings, Part II; Springer: Berlin/Heidelberg, Germany, 2011; Volume 22, pp. 470–483.

6. Karmarkar, N. A new polynomial-time algorithm for linear programming. In Proceedings of the Sixteenth Annual ACM Symposium on Theory of Computing, Washington, DC, USA, 30 April–2 May 1984; pp. 302–311.
7. Nesterov, Y.; Nemirovskii, A. *Interior-Point Polynomial Algorithms in Convex Programming*; SIAM: Philadelphia, PA, USA, 1994.
8. Roos, C.; Terlaky, T.; Vial, J.-P. *Theory and Algorithms for Linear Optimization: An Interior Point Approach*; Wiley: Chichester, UK, 1997.
9. Roos, C. A full-Newton step O(n) infeasible interior-point algorithm for linear optimization. *SIAM J. Optim.* **2006**, *16*, 1110–1136. [CrossRef]
10. Terlaky, T.; Roos, C.; Peng, J. *Self-Regularity: A New Paradigm for Primal-Dual Interior-Point Algorithms*; Princeton University Press: Princeton, NJ, USA, 2002.
11. Li, M.M.; Zhang, M.W.; Huang, Z.W. A primal-dual interior-point method for linear optimization based on a new parameterized kernel function. *J. Nonlinear Funct. Anal.* **2019**, *38*, 1–20.
12. Fathi-Hafshejani, S.; Peyghami, R.M. An interior point algorithm for solving linear optimization problems using a new trigonometric kernel function. *Filomat* **2020**, *34*, 1471–1486. [CrossRef]
13. Kheirfam, B.; Haghighi, M. A wide neighborhood interior-point algorithm based on the trigonometric kernel function. *J. Appl. Math. Comput.* **2020**, *64*, 119–135. [CrossRef]
14. Touil, I.; Chikouche, W. Polynomial-time algorithm for linear programming based on a kernel function with hyperbolic-logarithmic barrier term. *Palest. J. Math.* **2022**, *11*, 127–135.
15. Touil, I.; Chikouche, W. Novel kernel function with a hyperbolic barrier term to primal-dual interior point algorithm for SDP problems. *Acta Math. Appl. Sin. Engl. Ser.* **2022**, *38*, 44–67. [CrossRef]
16. Dexter, G.; Chowdhury, A.; Avron, H.; Drineas, P. On the convergence of inexact predictor-corrector methods for linear programming. In Proceedings of the International Conference on Machine Learning, Baltimore, MD, USA, 17–23 July 2022; pp. 5007–5038.
17. Mehrotra, S. On the implementation of a primal-dual interior point method. *SIAM J. Optim.* **1992**, *2*, 575–601. [CrossRef]
18. Darvay, Z.; Illés, T.; Kheirfam, B.; Rigó, P.R. A corrector–predictor interior-point method with new search direction for linear optimization. *Cent. Eur. J. Oper. Res.* **2020**, *28*, 1123–1140. [CrossRef]
19. Yang, X.; Yin, H. A Mizuno-Todd-Ye Predictor-Corrector Infeasible-Interior-Point Method with the l1-Norm Wide Neighborhood for Linear Programming. *Appl. Math. Sci.* **2022**, *16*, 511–528. [CrossRef]
20. McDougall, T.J.; Wotherspoon, S.J. A simple modification of Newton's method to achieve convergence of order $1 + \sqrt{2}$. *Appl. Math. Lett.* **2014**, *29*, 20–25. [CrossRef]
21. Argyros, I.K.; Deep, G.; Regmi, S. Extended Newton-like midpoint method for solving equations in Banach space. *Foundations* **2023**, *3*, 82–98. [CrossRef]
22. Monteiro, R.D.; Adler, I. Interior path following primal-dual algorithms. Part I: Linear programming. *Math. Program.* **1989**, *44*, 27–41. [CrossRef]
23. Sonnevend, G. An "analytical centre" for polyhedrons and new classes of global algorithms for linear (smooth, convex) programming. In *System Modelling and Optimization: Proceedings of 12th IFIP Conference, Budapest, Hungary, 2–6 September 1985*; Springer: Berlin/Heidelberg, Germany, 2006; pp. 866–875.
24. Cai, X.; Wang, G.; Zhang, Z. Complexity analysis and numerical implementation of primal-dual interior-point methods for convex quadratic optimization based on a finite barrier. *Numer. Algorithms* **2013**, *62*, 289–306. [CrossRef]
25. Wang, W.; Bi, H.; Liu, H. A full-newton step interior-point algorithm for linear optimization based on a finite barrier. *Oper. Res. Lett.* **2016**, *44*, 750–753. [CrossRef]
26. Gay, D.M. Electronic mail distribution of linear programming test problems. *Math. Program. Soc. Coal Newsl.* **1985**, *13*, 10–12.
27. Andersen, E.D.; Roos, C.; Terlaky, T. On implementing a primal-dual interior-point method for conic quadratic optimization. *Math. Program.* **2003**, *95*, 249–277. [CrossRef]
28. Baldwin, S. Compute Canada: Advancing computational research. *J. Phys. Conf. Ser.* **2012**, *341*, 012001. [CrossRef]
29. Koch, T. The Final NETLIB-LP Results. *Oper. Res. Lett.* **2003**, *32*, 138–142. [CrossRef]
30. Świrydowicz, K.; Darve, E.; Jones, W.; Maack, J.; Regev, S.; Saunders, M.A.; Thomas, S.J.; Peleš, S. Linear solvers for power grid optimization problems: A review of GPU-accelerated linear solvers. *Parallel Comput.* **2022**, *111*, 102870. [CrossRef]

algorithms

MDPI

Article

Algorithms for Various Trigonometric Power Sums

Victor Kowalenko

School of Mathematics and Statistics, The University of Melbourne, Parkville, VIC 3010, Australia;
vkowa@unimelb.edu.au

Abstract: In this paper, algorithms for different types of trigonometric power sums are developed and presented. Although interesting in their own right, these trigonometric power sums arise during the creation of an algorithm for the four types of twisted trigonometric power sums defined in the introduction. The primary aim in evaluating these sums is to obtain exact results in a rational form, as opposed to standard or direct evaluation, which often results in machine-dependent decimal values that can be affected by round-off errors. Moreover, since the variable, m, appearing in the denominators of the arguments of the trigonometric functions in these sums, can remain algebraic in the algorithms/codes, one can also obtain polynomial solutions in powers of m and the variable r that appears in the cosine factor accompanying the trigonometric power. The degrees of these polynomials are found to be dependent upon v, the value of the trigonometric power in the sum, which must always be specified.

Keywords: algorithm; alternating; basic trigonometric power sum; code; cosine; Dowker; Gardner–Fisher; module; summand; twisted trigonometric power sum

MSC: 05A15; 11B83; 33B10

1. Introduction

Over the last half-century or so, finite sums involving powers of trigonometric functions have fascinated mathematicians and physicists. Although there had been sporadic interest in these sums before that, the modern interest was sparked in 1966 when Quoniam [1] sought the solution of

$$\sum_{k=1}^{\lfloor n/2 \rfloor} 2^{2m} \cos^{2m}\left(\frac{k\pi}{n+1}\right) = (n+1)\binom{2m-1}{m-1} - 2^{2m-1}, \tag{1}$$

where $m < n+1$, with m and n being positive integers. The problem was solved shortly afterwards by Greening et al. [2]. Later, Gardner [3] observed that the finite sum of inverse powers of cosines given by

$$S_{v,2}(m) = \left(\frac{\pi}{2m}\right)^{2v} \sum_{k=1}^{m-1} \cos^{-2v}\left(\frac{k\pi}{2m}\right), \tag{2}$$

where both m and v are positive integers, appears in the calculation of the v-th cumulant of a certain quadratic form in m independent standardized normal variates. Although he was able to find that

$$\lim_{m\to\infty} S_{v(m)} = \zeta(2v), \tag{3}$$

where $\zeta(s)$ is the Riemann zeta function, he asked whether it was possible to obtain a "simpler" closed form expression for $S_v(m)$, for all m and v.

In solving this problem, Fisher [4,5] observed that the above sum could be expressed as

$$S_{v,2}(m) = \left(\frac{\pi}{2m}\right)^{2v} \sum_{k=1}^{m-1} \sin^{-2v}\left(\frac{k\pi}{2m}\right). \tag{4}$$

He then set about devising an ingenious generating function approach for the more general trigonometric power series given by

$$Q_{m,v}(\delta) = \sum_{k=1}^{m-1} \sin^{-2v}\left(\frac{k\pi+\delta}{2m}\right). \tag{5}$$

Eventually, he arrived at

$$S_{1,2}(m) = \frac{\pi^2}{6}\left(1 - \frac{1}{m^2}\right) \quad \text{and} \quad S_{2,2}(2) = \frac{\pi^4}{90}\left(1 + \frac{5}{2m^2} - \frac{7}{2m^4}\right), \tag{6}$$

while for large m, he found that

$$S_{v,2}(m) = \zeta(2v) + \frac{v}{12m^2}\zeta(2v-2) + O\left(m^{-4}\right). \tag{7}$$

A flurry of activity on trigonometric power sums occurred in the late 1980s due to their appearance in important physical applications, such as the work of Dowker [6–8] which dealt with the Casimir effect and heat kernel expansion on a generalized cone, Verlinde [9], whose formula predicts the dimension of the space of conformal blocks in a chiral conformal field theory (CFT), and Zagier [10], who developed Verlinde's results further. A decade later, studies of related sums appeared, such as by Berndt and Yeap [11] on the explicit evaluation and reciprocity theorems of specific trigonometric power sums, and Cvijović and Srivastava [12,13] on Dowker's and related sums. It should be mentioned here that Berndt and Yeap also presented an extensive literature survey on these sums in the introduction to their paper. At the same time, other authors [14–16], motivated by the intrinsic fascination of these sums, derived formulas where the summand was a power of secant, as in

$$\sum_{k=0}^{m-1} \sec^{2p}\left(\frac{k\pi}{m}\right) = m \sum_{k=1}^{2p-1} (-1)^{p+k} \binom{p-1+km}{2p-1} \sum_{j=k}^{2p-1} \binom{2p}{j+1}. \tag{8}$$

It should be noted that when both sides of this result are evaluated separately in Mathematica (Version 14.0) [17], it is found that they agree in decimal form, although for large values of n and p, one may encounter problems with complex infinities on the left-hand side (lhs) due to the summand becoming very small.

Afterwards, Dowker [18] concentrated only on sums involving powers of cosecant with a similar argument as in the above sum, since he noted that they also occur in Rényi entropy and Dedekind sums. More recently, however, these sums have appeared in a study of the entanglement entropy in string theory [19], where the argument inside the trigonometric powers are the fractions k/m multiplied by 2π, rather than π.

The most fascinating trigonometric power sums are those possessing inverse powers of the sine or cosine, including tangent and cotangent, because their evaluation generally involves the zeta function directly, or through related numbers such as the Bernoulli and Euler numbers. In these types of trigonometric power sums, the solution is number-theoretical in nature [10,12,13], whereas if it is a basic trigonometric power sum, i.e., possessing only even positive powers of sine or cosine, then it is combinatorial [20,21]. Nevertheless, this paper will present new developments concerning both positive and negative powers of the trigonometric function in the summand.

Specifically, this paper is concerned with developing an algorithm for evaluating general twisted trigonometric power sums, which are defined as

$$T^{\pm}_{v,l}(m,r) \doteqdot \sum_{k=1}^{lm-1} (\pm 1)^k \cos\Big(\frac{2rk\pi}{lm}\Big) \csc^{2v}\Big(\frac{k\pi}{lm}\Big). \tag{9}$$

In the above equation, r is an integer lying between 1 and $m-1$, while l can equal to 1 or 2. The major difference between these sums and the other sums mentioned above is the appearance of an extra cosine factor next to the trigonometric power. That is, most of the previous studies have been concerned with the untwisted or $r = 0$ forms of (9), with the exception being the papers of Cvijovic and Srivastava [12,13].

As indicated earlier, Dowker was the first to find physical applications of trigonometric power sums possessing an extra cosine factor next to the trigonometric power [6,8]. The sums were later referred to as twisted Dowker sums by Cvijovic and Srivastava [12,13], when developing Dowker's results further. Nevertheless, these twisted trigonometric power sums were concerned mainly with the non-alternating $l=1$ case of (9). In keeping with these papers, we shall refer to the sums with $l=1$ as the Dowker case. In addition, we shall generalize the works of these authors so that one can observe the underlying number-theoretical and combinatorial behavior emanating from the sums. Moreover, we shall study the corresponding $l=2$ forms, which will be referred to as the Gardner–Fisher case following from the seminal role of these two authors in this case. Our ultimate goal, however, will be to develop general algorithms capable of yielding: (1) purely rational values for the four types of trigonometric power sums defined by (9) when m and r are specified values, and (2) polynomials in powers of m and r when both variables are algebraic.

To complete this introduction, it should be stated that trigonometric power sums are not only affected by the power and type of trigonometric function in the summand, but also by the different sequences of rationals multiplying π in the arguments of these functions. As a result, it becomes a formidable task to devise a master theorem that can cover all kinds of trigonometric power sums. Nevertheless, we shall obtain deep and meaningful results for both the Dowker and Gardner–Fisher cases. In addition, whilst polynomial expressions in powers of m and r can be obtained, the same cannot be said for the power, v, in (9). That is, the value of the trigonometric power must always be specified before the solution to a particular sum can be determined.

2. Background

Before more intricate trigonometric power sums as discussed in [21] can be considered, one needs to study the untwisted Dowker and Gardner–Fisher sums, which can be represented in a general form as

$$S_{v,l}(m) \doteqdot \sum_{k=1}^{m-1} \csc^{2v}\Big(\frac{k\pi}{lm}\Big). \tag{10}$$

The untwisted Dowker sum is given by $l=1$ in the above definition, while the Gardner–Fisher sum corresponds to the $l=2$ case. Later in this work, we shall study more general alternating forms of the sums as given in (9), where r is an integer ranging from 1 to $m-1$. It should also be mentioned here that the Gardner–Fisher sum studied in [4,5] was defined with an extra normalization factor of $(\pi/2m)^{2v}$. Therefore, to make contact with the results presented in [21], the results presented here need to be multiplied by the appropriate normalization factor for each value of v.

General expressions for the two sums were obtained by employing a novel integral technique in Theorems 3.1 and 4.1 of [21]. In the case of the Gardner–Fisher sum, it was found to be given by

$$S_{v,2}(m) = \frac{1}{(2v-1)!}\left(\frac{2m}{\pi}\right)^{2v}\sum_{n=0}^{v-1}\left(\frac{\pi}{m}\right)^{2n} s(v,n)\Gamma(2v-2n)\,\zeta(2v-2n)\left(1-\frac{1}{m^{2v-2n}}\right), \tag{11}$$

where $n \leq v-1$ and $s(v,n)$ represents the n-th elementary symmetric polynomial obtained by summing over the entire sequence of quadratic powers or squares of integers from 1^2 to $(v-1)^2$. That is, the polynomials are given by

$$s(v,n) = \sum_{1\leq i_1<i_2<\cdots<i_n<v-1} x_{i_1}x_{i_2}\cdots x_{i_n}\ , \tag{12}$$

where $x_{i_1} < x_{i_2} < \cdots < x_{i_n}$ and each x_{i_j} is equal to at least one value in the set $\{1, 2^2, 3^2, \ldots, (v-1)^2\}$. It was also found in [21] that

$$s(v,n) = 2^{-2n}\,(2v-2n)_{2n}\, c_{2v,n}\ , \tag{13}$$

where $(x)_k$ denotes the Pochhammer notation for $\Gamma(k+x)/\Gamma(x)$ and the $c_{\rho,n}$, known as the generalized cosecant numbers, represent the polynomial coefficients in ρ of the generating function for $(x\csc x)^{\rho}$. The first ten orders are displayed in Table 1, while important applications and alternative methods of evaluating them are described extensively in [21,22]. For the special case of $\rho = 1$, the generalized cosecant numbers reduce to the cosecant numbers, c_k, which are given by

$$c_k = 2\left(1-2^{1-2k}\right)\frac{\zeta(2k)}{\pi^{2k}} = \frac{(-1)^{k+1}}{(2k)!}\left(2^{2k}-2\right)B_{2k}. \tag{14}$$

In the above equation, B_{2k} represent the Bernoulli numbers. The c_k have the advantage of converging quickly to zero as $k \to \infty$, whereas the Bernoulli numbers diverge.

Table 1. Generalized cosecant numbers $c_{\rho,n}$ up to $n = 10$.

n	$c_{\rho,n}$
0	1
1	$\frac{1}{3!}\,\rho$
2	$\frac{2}{6!}\,(2\rho+5\rho^2)$
3	$\frac{8}{9!}\,(16\rho+42\rho^2+35\rho^3)$
4	$\frac{2}{3\times 10!}\,(144\rho+404\rho^2+420\rho^3+175\rho^4)$
5	$\frac{4}{3\times 12!}\,(768\rho+2288\rho^2+2684\rho^3+1540\rho^4+385\rho^5)$
6	$\frac{2}{9\times 15!}\,(1061376\rho+3327594\rho^2+4252248\rho^3+2862860\rho^4+1051050\rho^5+175175\rho^6)$
7	$\frac{1}{27\times 15!}\,(552960\rho+1810176\rho^2+2471456\rho^3+1849848\rho^4+820820\rho^5+210210\rho^6$ $+25025\rho^7)$
8	$\frac{2}{45\times 18!}\,(200005632\rho+679395072\rho^2+978649472\rho^3+792548432\rho^4+397517120\rho^5$ $+125925800\rho^6+23823800\rho^7+2127125\rho^8)$
9	$\frac{4}{81\times 21!}\,(129369047040\rho+453757851648\rho^2+683526873856\rho^3+589153364352\rho^4$ $+323159810064\rho^5+117327450240\rho^6+27973905960\rho^7+4073869800\rho^8+282907625\rho^9)$
10	$\frac{2}{6075\times 22!}\,(38930128699392\rho+140441050828800\rho^2+219792161825280\rho^3+199416835425280\rho^4$ $+117302530691808\rho^5+47005085727600\rho^6+12995644662000\rho^7+2422012593000\rho^8$ $+280078548750\rho^9+15559919375\rho^{10})$

On the other hand, the untwisted Dowker sum was found in [21] to be given by

$$S_{v,1}(m) = \sum_{k=1}^{m-1} \csc^{2v}\left(\frac{k\pi}{m}\right) = \frac{2}{(2v-1)!}\left(\frac{m}{\pi}\right)^{2v} \sum_{n=0}^{v-1}\left(\frac{2\pi}{m}\right)^{2n} \times s(v,n)\Gamma(2v-2n)\zeta(2v-2n)\left(1-\frac{1}{m^{2v-2n}}\right). \tag{15}$$

Thus, it is seen that both sums are similar, with the differences occurring in the powers of the summands and the factors outside the sums.

By applying the SymmetricPolynomial, Gamma, Zeta, Sum, and Table routines in Mathematica [17], we were able to derive the first 15 values of v for both sums with only a few lines of code. To accomplish this, (11) is implemented in Mathematica as

```
S2[m_,v_]:= (1/((2v−1)!)) (2m/Pi)^(2v)Sum[(Pi/m)^(2n)
SymmetricPolynomial[n,Table[k^2,{k,1,v−1}]]
Gamma[2v−2n] Zeta[2(v−n)](1−1/(m^(2(v−n)))), {n,0,v−1}] .
```

The above line only prints out the coefficient of π^2 to the v-th power in $S_{v,2}$. For example, setting $m = 15$ and $v = 12$ yields

In[1]:= S2[15, 12]

$$\text{Out[1]} = \frac{286843846076469686166197007 2338432}{8303765625},$$

whilst putting $m = 15$ and $v = 3$ gives

In[2]:= S2[15, 3]

$$\text{Out[2]} = \frac{105368576}{135}.$$

For m, algebraic, the instruction prints out all the terms for the v-th power of π^2. To obtain the coefficient in terms of a polynomial in m, one needs to wrap the instruction in the FullSimplify routine. Therefore, for $v = 3$ one finds that

In[3]:= FullSimplify[S2[m, 3]]

$$\text{Out[3]} = \frac{8}{945}\left(-71 + 42m^2 + 21m^4 + 8m^6\right).$$

Moreover, by typing the following instruction in Mathematica:

```
F2[m_,n_] := Table[Simplify[S2[m, v]], {v, 1, n}],
```

one obtains a table of all the results of $S_{v,2}(m)$ for $v = 1$ to n or the first n values of the Gardner–Fisher sum. For example, setting $n = 3$ results in the following output:

In[4]:= F[m, 3]

$$\text{Out[4]} = \left\{\frac{2}{3}\left(-1+m^2\right), \frac{4}{45}\left(-7+5m^2+2m^4\right), \frac{8}{945}\left(-71+42m^2+21m^4+8m^6\right)\right\}.$$

The above results can be displayed as a table by wrapping F2[m_, n_] first inside the Partition routine and then using the Grid routine as in [21], but this is not necessary here. In addition, by wrapping it inside the Timing routine, one can obtain the amount of CPU time taken to generate the results. For example, the above results took only 0.00048 CPU seconds on a Venom Blackbook Zero laptop with 64 GB of RAM. For completeness, Table 2 displays the first ten values of $S_{v,2}(m)$.

The above statements can be easily adapted to yield values of the untwisted Dowker sum. All we need to do is modify the first instruction to

```
S1[v_, m_]:= (2^(2 v + 1)/(2 v − 1)!) Sum[(2 Pi/m)^(2 n −2 v)
SymmetricPolynomial[n, Table[k^ 2, {k, 1, v − 1}]] Gamma[2 (v − n)] Zeta[2 (v − n)]
(1 − 1/(m^(2 (v − n)))),{n, 0, v − 1}].
```

Consequently, for $n = 3$, we obtain the following output:

In[5]:= FullSimplify[S1[3, m]]

$$\text{Out[5]=}\ \frac{1}{945}\,(-1 + \mathrm{m})\,(1 + \mathrm{m})\,(191 + 23\,\mathrm{m}^2 + 2\,\mathrm{m}^4),$$

which represents $S_{3,1}(m)$.

Table 2. Values of $S_{v,2}(m)$ for v up to and including 10.

v	$S_{v,2}(m)$
1	$\frac{2}{3}(m^2-1)$
2	$\frac{4}{45}(2m^4+5m^2-7)$
3	$\frac{8}{945}(8m^6+21m^4+42m^2-71)$
4	$\frac{16}{14175}(24m^8+80m^6+147m^4+270m^2-521)$
5	$\frac{32}{93555}(32m^{10}+132m^8+286m^6+451m^4+792m^2-1693)$
6	$\frac{64}{638512875}(44224m^{12}+218400m^{10}+558558m^8+993850m^6+1438437m^4+2457000m^2$ $-5710468)$
7	$\frac{128}{273648375}(3840m^{14}+22112m^{12}+65520m^{10}+133419m^8+211640m^6+289926m^4$ $+486000m^2-1212457)$
8	$\frac{256}{488462349375}(1388928m^{16}+9139200m^{14}+30824128m^{12}+71073600m^{10}+126249123m^8$ $+185684200m^6+244813158m^4+404838000m^2-1074010337)$
9	$\frac{512}{194896477400625}(112299520m^{18}+831273408m^{16}+3145141440m^{14}+8109940848m^{12}$ $+16038878310m^{10}+25993141083m^8+36233012730m^6+46442359908m^4$ $+76014288000m^2-212920335247)$
10	$\frac{1024}{32157918771103125}(3754834944m^{20}+30882368000m^{18}+129540106080m^{16}$ $+369483576000m^{14}+806213290740m^{12}+1435972970250m^{10}+2178408236235m^8$ $+2917253715750m^6+3659175925020m^4+5941116720000m^2-17471801743019)$

To obtain all the results for the untwisted Dowker sum up to v, one must write the second instruction as

$$\mathrm{F1[n\_,m\_] := Table[Simplify[S1[v,m]],\{v,1,n\}]}.$$

Then, by setting $n=3$ in $\mathrm{F1[m\_,n\_]}$ and wrapping it around the Timing routine in Mathematica, one arrives at

In[6]:= Timing[F1[3, m]]

$$\text{Out[6]=}\ \Big\{0.000377, 1/3\,(-1 + \mathrm{m}^2), 1/45\,(-11 + 10\,\mathrm{m}^2 + \mathrm{m}^4), 1/945\,(-191 + 168\,\mathrm{m}^2 + 21\,\mathrm{m}^4 + 2\,\mathrm{m}^6)\Big\}.$$

This means it took 0.000377 s to generate the first three untwisted Dowker sums on the same Venom Blackbook Zero laptop mentioned previously. Note also that there are no powers of π as in the Gardner–Fisher sum. In fact, the factor outside the results in Table 3 is equal to $\zeta(2v)/\pi^{2v}$.

It should be mentioned here that the results obtained above and later on are exact due to the fact that they are computed using integer arithmetic, not machine-dependent decimal numbers. This means that there is no error involved in their determination. Furthermore, if m is set to a rational value, then the numerical value will be expressed as a rational subject to the memory of the computing system which, for the laptop used in this paper, is a few terabytes. In this work, we will be frequently generating rational values for the trigonometric power sums, but when we wish to compare them with those generated by the standard instructions in Mathematica, we will need to convert them to number or decimal form. For these situations, any number of decimal places can be specified by the user, but here they will be restricted to 16 decimal places, since those evaluated by standard means are limited by the byte size for the computing machine, especially when $m > 5$. Therefore, whilst it is possible to obtain much higher decimal precision from the computed values of the new results in this work, they will only be expressed to 16 decimal places in order to make a fair comparison with the machine-dependent numbers arising from the standard instructions in Mathematica.

Table 3. Values of $S_{v,1}(m)$ for v up to and including 10.

v	$S_{v,1}(m)$
1	$\frac{1}{3}(m^2-1)$
2	$\frac{1}{45}(m^4+10m^2-11)$
3	$\frac{1}{945}(2m^6+21m^4+168m^2-191)$
4	$\frac{1}{14175}(3m^8+40m^6+294m^4+21600m^2-2497)$
5	$\frac{1}{93555}(2m^{10}+33m^8+286m^6+1804m^4+12672m^2-14797)$
6	$\frac{1}{638512875}(1382m^{12}+27300m^{10}+279279m^8+1987700m^6+11507496m^4+78624000m^2$ $-92427157)$
7	$\frac{1}{273648375}(60m^{14}+1382m^{12}+16380m^{10}+133419m^8+846560m^6+4638816m^4$ $+31104000m^2-36740617)$
8	$\frac{1}{488462349375}(10851m^{16}+285600m^{14}+3853016m^{12}+35536800m^{10}+252498246m^8$ $+1485473600m^6+7834021056m^4+51819264000m^2-61430943169)$
9	$\frac{1}{194896477400625}(438760m^{18}+12988647m^{16}+196571340m^{14}+2027485212m^{12}$ $+16038878310m^{10}+103972564332m^8+579728203680m^6+2972311034112m^2$ $+19459657728000m^2-23133945892303)$
10	$\frac{1}{32157918771103125}(7333662m^{20}+241268500m^{18}+4048128315m^{16}+46185447000m^{14}$ $+403106645370m^{12}+2871945940500m^{10}+17427265889880m^8+93352118904000m^6$ $+468374518402560m^4+3041851760640000m^2-3624331198599787)$

To complete this section, it should also be mentioned that the untwisted Dowker sums can be obtained in Mathematica simply by typing in the $l = 1$ version of (10) as follows:

```
Sv1[v_, m_]:= Sum[Csc[k π/m]^(2 v), {k,1,m−1}].
```

With $v = 6$ and $m = m$ in the instruction, Mathematica prints out

```
In[7]:= Sv1[6, m]
```

Out[7] = (1/638512875)(−92427157 + 78624000 m^2 + 11507496 m^4 + 1987700 m^6 + 279279 m^8 + 27300 m^{10} + 1382 m^{12}).

The same approach, however, does not apply to the Gardner–Fisher case, but can be applied to the alternating version of the untwisted Dowker sum as described in the next section.

3. Alternating Sums

In the previous section, algorithms for both the untwisted Gardner–Fisher and Dowker sums were presented. As indicated in [21], these results formed the basis for determining more intricate trigonometric power sums, which will be discussed in later sections of this paper. For now, however, we can extend this basis by considering the alternating versions of the sums. That is, we are interested in solving

$$S^{\pm}_{v,l}(m) \div \sum_{k=1}^{m-1} (\pm 1)^k \csc^{2v}\left(\frac{k\pi}{lm}\right). \tag{16}$$

From here on, we shall denote $S_{v,l}(m)$ in the previous section as $S^+_{v,l}(m)$.

The solutions for $l=1$ and 2 of $S^-_{v,l}(m)$ can be determined by modifying the derivations for $l=1$ and 2 of $S^+_{v,l}(m)$, as can be seen by the following theorem.

Theorem 1. *General solutions of $S^-_{v,l}(m)$ for the Dowker ($l=1$) and Gardner–Fisher ($l=2$) cases are, respectively,*

$$S^-_{v,1}(m) = \begin{cases} 0, & m, \text{ odd}, \\ 2^{2v+1}\sum_{n=0}^{v-1}\frac{(2\pi)^{2n-2v}}{(2v-2n)_{2n}}s(v,n)\Big(\big(2^{2n-2v+1}-1\big)m^{2v-2n}-1\Big)\zeta(2v-2n), & m, \text{ even}, \end{cases} \tag{17}$$

and

$$S^-_{v,2}(m) = 2^{2v}\sum_{n=0}^{v-1}\frac{(2\pi)^{2n-2v}}{(2v-2n)_{2n}}s(v,n)\Big(2m^{2v-2n}-(2m)^{2v-2n} - (-1)^m\,2^{2v-2n}+(-1)^m-1\Big)\zeta(2v-2n). \tag{18}$$

Proof. In the proof of Theorem 3.1 in [21], the following result was obtained:

$$\csc^{2v}\left(\frac{k\pi}{2m}\right) = -\frac{4^v m^2}{\pi^2\Gamma(2v)}\int_0^1 \frac{\ln u}{1-u^{2m}}\prod_{n=1}^{v-1}\left(\frac{m^2}{\pi^2}\ln^2 u+n^2\right)\left(u^k+u^{2m-k}\right)\frac{du}{u}. \tag{19}$$

This can be generalized to $\csc^{2v}(k\pi/lm)$ by replacing m with $lm/2$. Then, we have

$$\csc^{2v}\left(\frac{k\pi}{lm}\right) = -\frac{4^{v-1}l^2m^2}{\pi^2\Gamma(2v)}\int_0^1 \frac{\ln u}{1-u^{lm}}\prod_{n=1}^{v-1}\left(\frac{l^2m^2}{4\pi^2}\ln^2 u+n^2\right)\left(u^k+u^{lm-k}\right)\frac{du}{u}. \tag{20}$$

Multiplying the above result by $(-1)^k$ and summing over k from 1 to $m-1$, we obtain $S^-_{v,l}(m)$, which is given by

$$S^-_{v,l}(m) = -\frac{4^{v-1}l^2m^2}{\pi^2\Gamma(2v)}\int_0^1 \frac{\ln u}{1-u^{lm}}\prod_{n=1}^{v-1}\left(\frac{l^2m^2}{4\pi^2}\ln^2 u+n^2\right)\left(\sum_{k=1}^{m-1}(-u)^k+\sum_{k=1}^{m-1}(-1)^k u^{lm-k}\right)\frac{du}{u}. \tag{21}$$

The sums over k in (21) are simply the finite geometric series, which yield

$$\frac{1}{u}\sum_{k=1}^{m-1}\left((-u)^k+(-1)^k u^{lm-k}\right) = -\left(1+(-1)^m u^{(l-1)m}\right)\left(\frac{1-(-u)^{m-1}}{1+u}\right). \tag{22}$$

In addition, from [21], the product in (21) can be expressed via Newton's identities as

$$\prod_{n=1}^{v-1}\left(\frac{l^2m^2}{4\pi^2}\ln^2 u+n^2\right) = \sum_{n=0}^{v-1}\left(\frac{l^2m^2}{4\pi^2}\ln^2 u\right)^n s(v,n). \tag{23}$$

By introducing (22) and (23) into (21), one is able to write the integrals over u in terms of derivative of digamma function via No. 4.271(16) in [23]. Thus, after some algebra, one arrives at

$$S^-_{v,l}(m) = \frac{2^{2v}}{\Gamma(2v)}\sum_{n=0}^{v-1}(2\pi)^{2n-2v}s(v,n)\sum_{k=1}^{m-1}\Big((-1)^k\psi^{(2v-2n-1)}\left(\frac{k}{lm}\right) + (-1)^{m-k}\psi^{(2v-2n-1)}\left(\frac{k+(l-1)m}{lm}\right)\Big), \tag{24}$$

where $s(v,n)$ is given by (12). Noting that $\psi^{(n)}(x) = (-1)^{n+1} n! \zeta(n+1,x)$, where $\zeta(n,x)$ represents the Hurwitz zeta function, we can simplify (24) to

$$S^{-}_{v,l}(m) = \frac{2^{2v}\Gamma(2v-2n)}{\Gamma(2v)} \sum_{n=0}^{v-1} (2\pi)^{2n-2v} s(v,n) \sum_{k=1}^{m-1} \Big((-1)^k \zeta\Big(2v-2n, \frac{k}{lm}\Big) + (-1)^{m-k} \zeta\Big(2v-2n, \frac{k+(l-1)m}{lm}\Big)\Big). \quad (25)$$

Now, we examine the $l=1$ and $l=2$ cases of (24) separately. For $l=1$ or the Dowker case, we find that the summation over k reduces to

$$\sum_{k=1}^{m-1} \Big((-1)^k \zeta\Big(2v-2n, \frac{k}{m}\Big) + (-1)^{m-k} \zeta\Big(2v-2n, \frac{k}{m}\Big)\Big) = \Big(1+(-1)^m\Big) \sum_{k=1}^{m-1} (-1)^k \zeta\Big(2v-2n, \frac{k}{m}\Big). \quad (26)$$

This result indicates that when m is an odd integer, $S^{-}_{v,1}(m)$ vanishes. Hence, the trigonometric power sum will only yield non-zero solutions for even integer values of m. Thus, we arrive at

$$S^{-}_{v,1}(m) = \frac{2^{2v+1}\Gamma(2v-2n)}{\Gamma(2v)} \sum_{n=0}^{v-1} (2\pi)^{2n-2v} s(v,n) \sum_{k=1}^{m-1} (-1)^k \zeta\Big(2v-2n, \frac{k}{m}\Big). \quad (27)$$

The sum over k can be expressed as

$$\sum_{k=1}^{m-1} (-1)^k \zeta(v, k/m) = m^v \sum_{k=1}^{m-1} (-1)^k \sum_{j=0}^{\infty} \frac{1}{(jm+k)^v}. \quad (28)$$

Since m can only be an even integer, we can combine the two dimensional sum into one dimensional sums by noting that

$$\sum_{k=1}^{m-1} (-1)^k \zeta(v, k/m) = m^v \Big(\sum_{j=1}^{\infty} \frac{(-1)^j}{j^v} - \sum_{j=1}^{\infty} \frac{1}{(jm)^v} \Big) = \Big(2^{1-v} m^v - m^v - 1\Big) \zeta(v). \quad (29)$$

Introducing (29) into (25) yields the first result displayed in the theorem.

For $l=2$ or the Gardner–Fisher case, the summation over k in (25) for any positive integer v becomes

$$\sum_{k=1}^{m-1} \Big((-1)^k \zeta\Big(v, \frac{k}{2m}\Big) + (-1)^{m+k} \zeta\Big(v, \frac{k+m}{2m}\Big)\Big) = \sum_{\substack{k=1 \\ k\neq m}}^{2m-1} (-1)^k \zeta\Big(v, \frac{k}{2m}\Big). \quad (30)$$

By inserting the $k=m$ term into the sum over k on the right-hand side (rhs) and subtracting it separately, one eventually arrives at

$$\sum_{\substack{k=1 \\ k\neq m}}^{2m-1} (-1)^k \zeta\Big(v, \frac{k}{2m}\Big) = \Big(2m^v - (2m)^v - (-1)^m 2^v + (-1)^m - 1\Big) \zeta(v). \quad (31)$$

Introducing (31) into (25) gives the second result in the theorem. □

We can use the material in the previous section to create an algorithm/program of these results. First, we begin by programming the alternating sum directly. This will show that Mathematica is able to provide the polynomials in powers of m^2 for the Dowker case, but can only provide numerical values in the Gardner–Fisher case. Furthermore, we can combine both cases into one instruction as follows:

Sminus[v_, l_, m_]: = Sum[(−1)^k Csc[k Pi/(l m)]^(2 v), {k, 1, m − 1}].

Obviously, if one puts $l=1$, then one obtains the Dowker case, while if it equals 2, then results for the Gardner–Fisher case are printed out. Putting $v=4$, $l=1$ and $m=m$ yields

In[8]:= Sminus[4, 1, m]

Out[8] = $(-319616 - 138240\,m^2 - 32928\,m^4 - 4960\,m^6 - 381\,m^8)/1814400$.

However, if $l=2$, then the following occurs:

In[9]:= Sminus[4, 2, m]

Out[9] = $\sum_{k=1}^{-1+m}(-1)^k \mathrm{Csc}\left[\frac{k\pi}{2m}\right]^8$.

Therefore, Mathematica is unable to generate algebraic forms in the Gardner–Fisher case. In this case, only numerical values can be obtained by specifying values of m.

If we examine (17) and (18) closely, then we see that main difference, aside from a factor of 2 outside the Dowker case, is that the expressions inside the large parentheses are different. That is, the summands are identical except for the terms inside the large brackets. Consequently, we can create two different factors corresponding to the $l=1$ and $l=2$ cases that need to be introduced into the summand. Let us call this the summand factor, which can be specified by the following instructions:

Smdfctr[v_, m_, n_, 1]:= 2 ((2^(2 n − 2 v + 1) − 1) m^(2 v− 2n)− 1)

Smdfctr[v_, m_, n_, 2]:= (2 m^(2 v − 2 n) − (2 m)^(2 v − 2 n) − (−1)^m 2^(2 v − 2 n) + (−1)^m − 1).

Moreover, both cases can be combined into one instruction by introducing l as a variable, which would need to be specified by the user as follows:

SMinus[v_, m_, l_]:= 2^(2 v) Sum[((2 π)^(2 n − 2 v) Zeta[2 v − 2 n]/Pochhammer [2 v − 2 n, 2 n]) Smdfctr[v, m, n,l] SymmetricPolynomial[n, Table[k^2, {k, 1, v − 1}]], {n, 0,v − 1}].

To generate the results in the same manner as in the previous section, one simply types

F1[n_, m_, l_]:= Table[Simplify[SMinus[v, m, l]], {v, 1, n}].

If one puts $n=3$, $m=m$, and $l=1$ in the above instruction, then Mathematica prints out

In[10]:= F1[3, m, 1]

Out[10] = $\{1/6\,(-2 - m^2), 1/360\,(-88 - 40\,m^2 - 7\,m^4), (-3056 - 1344\,m^2 - 294\,m^4 - 31\,m^6)/15120\}$.

Thus, the first three results for the alternating version of the Dowker sum are obtained. Table 4 presents the first 10 results for the sum generated by putting n = 10 in the above instruction. Wrapping the instruction inside the Timing routine gives a CPU time of 0.030429 s using the same Venom laptop mentioned in the previous section.

Table 4. Values of $S^-_{v,1}(m)$ for v up to and including 10.

v	$S^-_{v,1}(m)$
1	$-\frac{1}{6}(m^2+2)$
2	$-\frac{1}{360}(7m^4+40m^2+88)$
3	$-\frac{1}{15120}(31m^6+294m^4+1344m^2+3056)$
4	$-\frac{1}{1814400}(381m^8+4960m^6+32928m^4+138240m^2+319616)$
5	$-\frac{1}{23950080}(511m^{10}+8382m^8+70928m^6+404096m^4+1622016m^2+3788032)$
6	$-\frac{1}{653837184000}(1414477m^{12}+27900600m^{10}+283747464m^8+1971798400m^6$ $+10310716416m^4+40255488000m^2+94645408768)$
7	$-\frac{1}{560431872000}(122865m^{14}+2828954m^{12}+33480720m^{10}+271107408m^8$ $+1679575040m^6+8312758272m^4+31850496000m^2+75244783616)$
8	$-\frac{1}{16005934264320000}(355554717m^{16}+9357398400m^{14}+126193980032m^{12}$ $+7513929603048960000m^{10}+45327621488942284800m^8+237069572999086080000m^6$ $+1074338735331056025600m^4+3987015939706060800000m^2+9500966777257425633280)$
9	$-\frac{1}{1277273554292736000}(28748457785m^{18}+851197992498m^{16}+12880926767520m^{14}$ $+132808391326848m^{12}+1049070952500480m^{10}+6760712023123968m^8$ $+36805784195235840m^6+170444203940118528m^4+637654064431104000m^2$ $+1516106277997969408)$
10	$-\frac{1}{8430005458332057600000}(1922471824497m^{20}+63246607127000m^{18}+1061160163980840m^{16}$ $+12105759884064000m^{14}+105620390793265920m^{12}+7513929603048960000m^{10}$ $+45327621488942284800m^8+237069572999086080000m^6+1074338735331056025600m^4$ $+3987015939706060800000m^2+9500966777257425633280)$

To obtain the alternating version of the Gardner–Fisher sum, all one needs to do is to put $l=2$. Therefore, typing

In[11]:= Timing[F1[3, m, 2]],

yields

Out[11] = {0.000618, {1/6 (−1 −3 $(-1)^m$ − 2 m^2), 1/90 (−11 − 45 $(-1)^m$ − 20 m^2 − 14 m^4), (−191 − 945 $(-1)^m$ − 336 m^2 − 294 m^4 − 124 m^6)/1890}}.

This indicates that it took 0.000618 s to produce the first three results of the alternating version of the Gardner–Fisher sum.

Table 5 displays the first ten results of the alternating version of the Gardner–Fisher sum, which took only 0.004516 s to produce. As for $S^-_{v,1}(m)$ and the non-alternating versions of both types of sum, the result yields polynomials in powers of m^2 of degree $2v$. There is, however, one interesting property of $S^-_{v,2}(m)$ that the other sums do not possess. This is that the constant in the polynomials is different when m is odd and even, and might explain why Mathematica is unable to provide the results in Table 5 as it does for Table 4. Nevertheless, all four sums yield rational values for any value of m.

Table 5. Values of $S^-_{v,2}(m)$ for v up to and including 10.

v	$S^-_{v,2}(m)$
1	$-\frac{1}{6}(2m^2+1+(-1)^m 3)$
2	$-\frac{1}{90}(14m^4+20m^2+11+(-1)^m 45)$
3	$-\frac{1}{1890}(124m^6+294m^4+336m^2+191+(-1)^m 945)$
4	$-\frac{1}{28350}(762m^8+2480m^6+4116m^4+4320m^2+2497+(-1)^m 14175)$
5	$-\frac{1}{187110}(2044m^{10}+8382m^8+17732m^6+25256m^4+25344m^2++14797+(-1)^m 555)$
6	$-\frac{1}{1277025750}(5657908m^{12}+27900600m^{10}+70936866m^8+123237400m^6+161104944m^4$ $+157248000m^2+94645408768+(-1)^m 638512875$
7	$-\frac{1}{547296750}(982920m^{14}+5657908m^{12}+16740360m^{10}+33888426m^8+52486720m^6$ $+64943424m^4+62208000m^2+36740617+(-1)^m 273648375)$
8	$-\frac{1}{976924698750}(711109434m^{16}+4678699200m^{14}+15774247504m^{12}+36318609600m^{10}$ $+64134554484m^8+92099363200m^6+109676294784m^4+103638528000m^2$ $+61430943169+(-1)^m 488462349375)$
9	$-\frac{1}{389792954801250}(114993831140m^{18}+851197992498m^{16}+3220231691880m^{14}$ $+8300524457928m^{12}+16391733632820m^{10}+26409031340328m^8+35943148628160m^6$ $+41612354477568m^4+38919315456000m^2+23133945892303+(-1)^m 194896477400625)$
10	$-\frac{1}{643158375422062 50}(7689887297988m^{20}+63246607127000m^{18}+265290040995210m^{16}$ $+756609992754000m^{14}+1650318606144780m^{12}+2935128751191000m^{10}$ $+4426525536029520m^8+5787831372048000m^6+6557243257635840m^4$ $+6083703521280000m^2+16399688681447+(-1)^m 145510944665625)$

To conclude this section, it should be noted that if we modify the upper limit of $S^-_{v,2}(m)$ to $2m-1$, as will be the case in Section 5, then we find that

$$T^-_{v,2}(m,0) = \sum_{k=1}^{2m-1} (-1)^k \csc^{2v}\Big(\frac{k\pi}{2m}\Big) = 2S^-_{v,2}(m) + (-1)^m, \tag{32}$$

where the general definition of $T^{\pm}_{v,l}(m,r)$ appears in the introduction. In any case, we can use the results in Table 5 to obtain $T^-_{v,2}(m,0)$ for specific values of v. Furthermore, from (18), we arrive at

$$\begin{aligned} T^-_{v,2}(m,0) = 2^{2v+1} \sum_{n=0}^{v-1} \frac{(2\pi)^{2n-2v}}{(2v-2n)_{2n}} s(v,n) \Big(2m^{2v-2n} - (2m)^{2v-2n} \\ - (-1)^m\, 2^{2v-2n} + (-1)^m - 1 \Big) \zeta(2v-2n) + (-1)^m. \end{aligned} \tag{33}$$

4. Basic Trigonometric Power Sums

Basic trigonometric power sums are defined as finite trigonometric power sums of the form

$$\left\{ \begin{matrix} S(m,n) \\ C(m,n) \end{matrix} \right\} \doteqdot \sum_{k=0}^{g(n)} (\pm 1)^k f(k) \left\{ \begin{matrix} \cos^{2m} \\ \sin^{2m} \end{matrix} \right\} \Big(\frac{qk\pi}{n}\Big), \tag{34}$$

where m, q and n are positive integers, $g(n)$ is dependent upon n such as $n-1$ or even $\lfloor m/n \rfloor$, and $f(k)$ is a relatively simple function of k, such as unity or $\cos(k\pi/p)$ with p, an integer. Besides the derivation of new results for interesting versions of these sums in [20], important applications were studied, such as the determination of the number of closed random walks of length $2m$ with $n-1$ vertices. The key results of this work are

$$C(m,n) = \begin{cases} 2^{1-2m} n\left(\binom{2m-1}{m-1} + \sum_{p=1}^{\lfloor m/n \rfloor} \binom{2m}{m-pn}\right), & \text{if } m \geqslant n, \\ 2^{1-2m} n\binom{2m-1}{m-1}, & \text{if } m < n, \end{cases} \tag{35}$$

and

$$S(m,n) = \begin{cases} 2^{1-2m} n\left(\binom{2m-1}{m-1} + \sum_{p=1}^{\lfloor m/n \rfloor} (-1)^{pn}\binom{2m}{m-pn}\right), & \text{if } m \geqslant n, \\ 2^{1-2m} n\binom{2m-1}{m-1}, & \text{if } m < n, \end{cases} \tag{36}$$

where

$$C(m,n) := \sum_{k=0}^{n-1} \cos^{2m}\left(\frac{k\pi}{n}\right), \tag{37}$$

and

$$S(m,n) := \sum_{k=0}^{n-1} \sin^{2m}\left(\frac{k\pi}{n}\right). \tag{38}$$

Although the second result in each case is not necessary, since it is covered by the first result, they were displayed separately (1) because of their simplicity and (2) to make contact with the results on p. 640 of [24]. There, they are given as

$$\sum_{k=1}^{\lfloor n/2 \rfloor} \cos^{2m}\left(\frac{k\pi}{n}\right) = 2^{1-2m} n\binom{2m}{m} - \frac{1}{2}, \quad m < n, \tag{39}$$

and

$$\sum_{k=1}^{\lfloor n \rfloor} \sin^{2m}\left(\frac{k\pi}{2n}\right) = 2^{-2m} n\binom{2m}{m} + \frac{1}{2}, \quad m < 2n. \tag{40}$$

The above results can be programmed easily in Mathematica by typing the following instructions:

```
CosSum[m_, n_]:= Sum[Cos[k Pi/n]^(2 m), {k, 0, n − 1}]
CosSum2[m_, n_]:= 2^(1 − 2 m) n (Binomial[2 m − 1, m − 1] +
Sum[Binomial[2 m, m − p n], {p, 1, Floor[m/n] }]).
```

If we put $m = 15$ and $n = 17$, then Mathematica prints out

In[12]:= CosSum[15, 17]

$$\text{Out[12]} = 1 + 2\,\text{Cos}\left[\frac{\pi}{17}\right]^{30} + 2\,\text{Cos}\left[\frac{2\pi}{17}\right]^{30} + 2\,\text{Cos}\left[\frac{3\pi}{17}\right]^{30} + 2\,\text{Cos}\left[\frac{4\pi}{17}\right]^{30} + 2\,\text{Sin}\left[\frac{\pi}{34}\right]^{30} + 2\,\text{Sin}\left[\frac{3\pi}{34}\right]^{30} + 2\,\text{Sin}\left[\frac{5\pi}{34}\right]^{30} + 2\,\text{Sin}\left[\frac{7\pi}{34}\right]^{30}.$$

In number form, this gives a value of 2.455895617604257. On the other hand, typing in

In[13]:= CosSum2[15, 17]

yields

Out[13] = 164812365/67108864.

The decimal value of the above result yields the same decimal value as CosSum[15, 17] except a 6 appears as the last figure, and not a 7. This means that the first result is limited by machine precision, and will not be accurate beyond 16 decimal places. Moreover, one cannot observe whether the results are rational or not, which is clearly the case when employing CosSum2.

The other basic trigonometric power sums studied in [20] were written in terms of $C(m,n)$ and $S(m,n)$. Nearly all involved combinations of $C(m,n)$ or $S(m,n)$ with different arguments. A typical example is

$$\sum_{k=0}^{3n-1} \cos\Big(\frac{k\pi}{3}\Big) \cos^{2m}\Big(\frac{k\pi}{3n}\Big) = 3\,C(m,n/2) - 3\,C(m,n)/2 + C(m,3n)/2 - C(m,3n/2), \tag{41}$$

where n is a positive even integer. Based on the preceding material we can program this basic trigonometric power sum as

CosSum3[m_, n_]:= Sum[Cos[k Pi/3] Cos[k Pi/(3 n)]^(2 m), k, 0, 3 n−1]

CosSum4[m_, n_]:= 3 CosSum2[m, n/2] − 3 CosSum2[m, n]/2 + CosSum2[m, 3 n]/2 − CosSum2[m, 3 n/2].

By putting $m=15$ and $n=18$ in both instructions, one obtains

In[14]:= CosSum3[15, 18]

$$\text{Out[14]} = 261261003/268435456 + \text{Cos}\Big[\frac{\pi}{54}\Big]^{30} - \text{Cos}\Big[\frac{\pi}{27}\Big]^{30} - 2\,\text{Cos}\Big[\frac{\pi}{18}\Big]^{30} - \text{Cos}\Big[\frac{2\pi}{27}\Big]^{30} + \text{Cos}\Big[\frac{5\pi}{54}\Big]^{30} + 2\,\text{Cos}\Big[\frac{\pi}{9}\Big]^{30} + \text{Cos}\Big[\frac{7\pi}{54}\Big]^{30} - \text{Cos}\Big[\frac{4\pi}{27}\Big]^{30} - \text{Cos}\Big[\frac{5\pi}{27}\Big]^{30} + \text{Cos}\Big[\frac{11\pi}{54}\Big]^{30} + 2\,\text{Cos}\Big[\frac{2\pi}{9}\Big]^{30} + \text{Cos}\Big[\frac{13\pi}{54}\Big]^{30} - \text{Sin}\Big[\frac{\pi}{54}\Big]^{30} + \text{Sin}\Big[\frac{\pi}{27}\Big]^{30} + 2\,\text{Sin}\Big[\frac{\pi}{18}\Big]^{30} + \text{Sin}\Big[\frac{2\pi}{27}\Big]^{30} - \text{Sin}\Big[\frac{11\pi}{54}\Big]^{30} - 2\,\text{Sin}\Big[\frac{\pi}{9}\Big]^{30} - \text{Sin}\Big[\frac{7\pi}{54}\Big]^{30} + \text{Sin}\Big[\frac{4\pi}{54}\Big]^{30} + \text{Sin}\Big[\frac{5\pi}{27}\Big]^{30} - \text{Sin}\Big[\frac{11\pi}{54}\Big]^{30} - 2\,\text{Sin}\Big[\frac{2\pi}{9}\Big]^{30} - \text{Sin}\Big[\frac{13\pi}{54}\Big]^{30}.$$

In number form, this yields 0.02986178733408444. However, by applying the second form we find that

In[15]:= CosSum4[15, 18]

Out[15] = 16031925/536870912.

This gives the same value in number form as the above value except the last two digits are 51, not 44. Therefore, programming (35) and (36) into Mathematica yields rational values for all the basic trigonometric power sums studied in [20].

More recently, Du et al. [25] have developed basic trigonometric power sums further in two directions. The first involves the introduction of an offset in the argument of the power of the trigonometric function so that (37) and (38) are re-defined as

$$C_\alpha(m,n) := \sum_{k=0}^{n-1} \cos^{2m}\Big(\Big(\frac{k}{n} + \alpha\Big)\pi\Big), \tag{42}$$

and

$$S_\alpha(m,n) := \sum_{k=0}^{n-1} \sin^{2m}\Big(\Big(\frac{k}{n} + \alpha\Big)\pi\Big). \tag{43}$$

Consequently, it was found that

$$C_\alpha(m,n) = 2^{1-2m} n \binom{2m-1}{m} + 2^{1-2m} n \sum_{p=1}^{\lfloor m/n \rfloor} \binom{2m}{m-pn} \cos(2\alpha pn\pi). \tag{44}$$

and

$$S_\alpha(m,n) = 2^{1-2m} n \binom{2m-1}{m} + 2^{1-2m} n \sum_{p=1}^{\lfloor m/n \rfloor} (-1)^{pn} \binom{2m}{m-pn} \cos(2\alpha pn\pi). \tag{45}$$

Here, the $m < n$ forms have been dropped, since they are included in (44) and (45).

The above results can be checked by implementing them in Mathematica [17]. For example, (44) can be implemented via the following instructions:

CosSum[m_, n_, α_]:= Sum[Cos[(k/n + α) π]^(2 m), {k, 0, n − 1 }]

CosSum2[m_, n_, α_]:= 2^(1 − 2 m) n (Binomial[2 m − 1, m − 1] + Sum[Binomial[2 m, m − p n] Cos[2 α p n π], {p, 1, Floor[m/n]}]).

Putting $m = 12$, $n = 15$ and $\alpha = 1/3$ in the first instruction yields

In[16]:= CosSum[12, 15, 1/3]

Out[16] = 8388609/8388608 + (−1 − √5)^24/140737488355328 + (1 −√5)^24/281474976710656 + (−1 + √5)^24/281474976710656 + 2 Cos[π/15]^24 + 2 Cos[(2 π)/15]^24 + 2 Sin[π/30]^24 + 2 Sin[(7 π)/30]^24.

In number form, Mathematica prints out a value of 2.41770386695862. On the other hand, typing CosSum2[12,15,1/3] generates the rational value of 10140585/4194304, which yields the same decimal value except the final digit of 2 is replaced by 18.

The second direction taken in [25] deals with the introduction of a twisted cosine phase factor, e.g., $\cos(2lk\pi/q)$, in the summand with the cosine power. This will become become important when we study twisted forms of Dowker-like sums in the next section. It should be mentioned that twisted basic trigonometric power sums were also studied in [20], but only in limited cases. In [25], the following result was obtained:

$$\sum_{k=0}^{qn-1} \cos\Big(\frac{2\ell k\pi}{q}\Big)\cos^{2m}\Big(\frac{k\pi}{qn}\Big) = 2^{2\ell n-1}C(m+\ell n, qn) + \ell n\sum_{j=0}^{\ell n-1}\frac{(-1)^{j+1}}{j+1}\binom{2\ell n-j-2}{j} \times 2^{2\ell n-2j-2}C(m+\ell n-j-1, qn). \tag{46}$$

The above result was obtained by adapting No. I.1.10 in [24] and then multiplying by $\cos^{2m}(k\pi/q)$. By summing over k from zero to $qn-1$, one finds that

$$\sum_{k=0}^{qn-1} \cos\Big(\frac{2\ell k\pi}{q}\Big)\cos^{2m}\Big(\frac{k\pi}{qn}\Big) = 2^{2\ell n-1}\sum_{k=0}^{qn-1}\cos^{2m+2\ell n}\Big(\frac{k\pi}{qn}\Big) + \ell n\sum_{j=0}^{\ell n-1}\frac{(-1)^{j+1}}{j+1}\binom{2\ell n-j-2}{j} \times 2^{2\ell n-2j-2}\sum_{k=0}^{qn-1}\cos^{2m+2\ell n-2j-2}\Big(\frac{k\pi}{qn}\Big). \tag{47}$$

The sums over k on the rhs in (47) are simply versions of the basic cosine power sum given by (35). As a consequence, their introduction results in (46). However, it should be stressed that (47) is only valid for $\ell = 1$ and 2.

Since $C(m,n)$ was observed in [21] to yield rational values, it follows that the cosine power sum given on the lhs of (46) will also yield rational rational values. As an example, consider $q = 5$, which yields

$$\sum_{k=0}^{5n-1} \cos\Big(\frac{2\ell k\pi}{5}\Big)\cos^{2m}\Big(\frac{k\pi}{5n}\Big) = 2^{2\ell n-1}C(\ell n+m, 5n) + \ell n\sum_{j=0}^{\ell n-1}\frac{(-1)^{j+1}}{j+1}\binom{2\ell n-j-2}{j} \times 2^{2\ell n-2j-2}C(\ell n+m-j-1, 5n). \tag{48}$$

Both the $l = 1$ and $l = 2$ results were not checked numerically in [25]. However, we can check them here, since a knowledge of how they are coded will be required in the following section.

The definition of a basic trigonometric power sum also includes those with alternating phase factors in the summands. Surprisingly, these have not been studied yet. Moreover, we shall require these sums in the next section. Therefore, the following lemma needs to be proved.

Lemma 1. *The alternating basic sine power sum, $S^-_{\alpha,l}(m,n)$, stemming from the definition, $S^{\pm}_{\alpha,l}(m,n) \doteqdot \sum_{k=1}^{n}(\pm 1)^k \sin^{2m}((k+\alpha)\pi/ln)$, has the following solutions for $\alpha = 0$. For the Dowker case or $l=1$, we have*

$$S^-_{0,1}(m,n) = \begin{cases} 0, & n, \text{ odd}, \\ (-1)^m 2^{1-2m} \sum_{j=0}^{m}(-1)^j \binom{2m}{j} H_2(m,n,j) - 2^{-2m}\binom{2m}{m}, & n, \text{ even}, \end{cases} \tag{49}$$

where

$$H_2(m,n,j) = \begin{cases} n+1, & \exp(2i(m-j)\pi/n) = -1, \\ (1+(-1)^n)/2, & \text{for all other values of } j, m \text{ and } n. \end{cases} \tag{50}$$

For the Gardner–Fisher case or $l=2$, the sum has the following solution:

$$S^-_{0,2}(m,n) = \begin{cases} (-1)^m 2^{1-2m} \sum_{j=0}^{m}(-1)^j \binom{2m}{j} H_1(m,n,j), & n, \text{ odd}, \\ (-1)^m 2^{1-2m} \sum_{j=0}^{m}(-1)^j \binom{2m}{j} H_1(m,n,j) - 2^{-2m}\binom{2m}{m}, & n, \text{ even}, \end{cases} \tag{51}$$

where

$$H_1(m,n,j) = \begin{cases} n+1, & \exp(i(m-j)\pi/n) = -1, \\ (1+(-1)^{m+n-j})/2, & \text{for all other values of } j,\ m \text{ and } n. \end{cases} \tag{52}$$

Remark 1. *The conditions on $H_1(m,n,j)$ and $H_2(m,n,j)$ in the lemma mean that $H_1(m,n,j)$ and $H_2(m,n,j)$ are equal to $n+1$, when $2(m-j)/n$ and $(m-j)/n$, respectively, are odd numbers. Consequently, we shall find that the above results can only be implemented as algorithms/programs yielding rational values.*

Proof. To prove the Dowker case for odd values of n, we set $n = 2p+1$, where p is a non-negative integer, and write $S^-_{0,1}(m,n)$ as

$$S^-_{0,1}(m,n) = \sum_{k=1}^{p}(-1)^k \sin^{2m}\Big(\frac{k\pi}{2p+1}\Big) + \sum_{k=p+1}^{2p-1}(-1)^k \sin^{2m}\Big(\frac{k\pi}{2p+1}\Big). \tag{53}$$

In the second sum, we replace k by $2p+1-k$. Then, $(-1)^k$ becomes $(-1)^{2p+1-k}$ or $(-1)^{k+1}$, while $\sin(\pi - k\pi/(2p+1)) = \sin(k\pi/(2p+1))$. Accordingly, the second sum becomes the negative of the first sum on the rhs (rhs) of (53), thereby yielding zero or the first result of (49). To prove the remaining results in the lemma, we use No. 17.4.5 of [26], which can be expressed as

$$\sin^{2m}\Big(\frac{x}{2}\Big) = \frac{(-1)^m}{2^{2m-1}} \sum_{j=0}^{m} \frac{(-2m)_j}{j!} \cos((m-j)x) - \frac{1}{2^{2m}}\binom{2m}{m}. \tag{54}$$

For the Dowker case, we need to put $x/2$ equal to $k\pi/n$, while for the Gardner–Fisher case, we put $x=k\pi$. Then, we multiply both sides of each resulting equation by $(-1)^k$, sum over k from zero to n and interchange the summations. Hence, for the Dowker case, we arrive at

$$\sum_{k=0}^{n}(-1)^k \sin^{2m}\Big(\frac{k\pi}{n}\Big) = \frac{(-1)^m}{2^{2m-1}} \sum_{j=0}^{m}(-1)^j \frac{(-2m)_j}{j!} \sum_{k=0}^{n}(-1)^k \cos\Big(\frac{2(m-j)k\pi}{n}\Big)$$
$$- \frac{1}{2^{2m}}\binom{2m}{m}\sum_{k=0}^{n}(-1)^k, \tag{55}$$

while for the Gardner–Fisher case, we obtain

$$\sum_{k=0}^{n}(-1)^k \sin^{2m}\Big(\frac{k\pi}{2n}\Big) = \frac{(-1)^m}{2^{2m-1}} \sum_{j=0}^{m}(-1)^j \frac{(-2m)_j}{j!} \sum_{k=0}^{n}(-1)^k \cos\Big(\frac{(m-j)k\pi}{n}\Big)$$
$$- \frac{1}{2^{2m}}\binom{2m}{m}\sum_{k=0}^{n}(-1)^k. \tag{56}$$

The second sum appearing on the rhs of both (55) and (56) simply yields $(1+(-1)^n)/2$, which means, in turn, that it vanishes for odd values of n and gives unity when n is even. It provides the common term in the n even cases of (51) and (52).

The first sum on the rhs of either (55) or (56) can be evaluated either in Mathematica or using the formula for a geometric series. We shall define them by

$$H_r(m,n,j,x) \doteqdot \sum_{k=0}^{n}(-1)^k \cos\big(rk(m-j)x\big). \tag{57}$$

This is the general form for the sum. When x is set equal to π/n shortly, the sum will be denoted as $H_r(m,n,j)$. For the Dowker or $l=1$ case, we require $r=2$ in (57), which becomes

$$H_2(m,n,j,x) = \sum_{k=0}^{n}(-1)^k \cos\big(2k(m-j)x\big). \tag{58}$$

The above sum can be determined using Mathematica or splitting the cosine into exponentials of imaginary arguments and applying the geometric series. In adopting the first approach, one simply types

```
H2[m_, n_, j_, x_]:= Sum[(−1)^k Cos[2 k (m − j) x], {k, 0, n }].
```

Next, one types

```
In[17]:= H2[m, n, j, x].
```

Then, one obtains the following output:

```
Out[17] = 1/2 (Cos[j x − m x] Sec[(j − m) x] + Cos[n π − j x + m x − 2 j n x + 2 m n x]
Sec[(j − m) x]).
```

Putting $x = k\pi/n$ and dropping the fourth variable, since it is summed out eventually, one arrives at

$$H_2(m,n,j) = \frac{1}{2}\Big(\Big(\cos\Big(\frac{(j-m)\pi}{n}\Big) + \cos\Big(2(j-m)\pi + \frac{(j-m)\pi}{n} - n\pi\Big)\Big)$$
$$\times \sec\Big(\frac{(j-m)\pi}{n}\Big). \tag{59}$$

There is, however, a problem with using this result in the summand, due to the fact that the secant can yield infinity. This problem can be circumvented by simplifying (59) with the aid of trigonometric identities. Thus, we find that

$$H_2(m,n,j) = \frac{1}{2}\Big(1+(-1)^n \frac{\cos\big((m-j)\pi(2+1/n)\big)}{\cos\big((j-m)\pi/n\big)}\Big)$$
$$= \frac{1}{2}\Big(1+(-1)^n\Big). \tag{60}$$

The above equation is beginning to approach the second result of (50). Unfortunately, this is not the end of the matter, as (59) cannot be valid for all values of m, n and j. If we return to (58) and put $x = k\pi/n$, then we observe that whenever $2(m-j)/n = (2p+1)$, where p is a non-negative integer, $H_2(m,n,j) = \sum_{k=0}^{n} 1 = n+1$, not the result obtained from (60). As j is summed from 0 to m, this condition can be met occasionally. That is, there will be at least one value of p where $H_2(m,n,j) = n+1$. Hence, we arrive at (50). If we insert (50) and the result below (56) into (55), then it becomes

$$\sum_{k=0}^{n} (-1)^k \sin^{2m}\Big(\frac{k\pi}{n}\Big) = \frac{(-1)^m}{2^{2m-1}} \sum_{j=0}^{m} (-1)^j \frac{(-2m)_j}{j!} H_2(m,n,j) - \frac{1}{2^{2m+1}} \binom{2m}{m} \big(1+(-1)^n\big). \tag{61}$$

From the reflection formula for the gamma function, one obtains

$$\frac{(-2m)_j}{j!} = (-1)^j \binom{2m}{j}. \tag{62}$$

Introducing the above result into (61) yields (49). In the Gardner–Fisher case given by (56), we observe that the summation over k can be represented by $H_1(m,n,j,x)$ according to (57). This, too, can be summed in Mathematica or by employing the geometric series. Thus, by typing the instruction as

```
H1[m_, n_, j_, x_]:= Sum[(−1)^k Cos[k (m − j) x], {k, 0, n }],
```

followed by

```
In[18]:= H1[m, n, j, x],
```

one obtains following output:

```
Out[18] = 1/2 (−Cos[1/2(−2 π + j x − m x)] Sec[1/2(j − m) x] + Cos[1/2 (2n π − j x + m x −
2 j n x + 2 m n x)] Sec[1/2(j − m) x]).
```

Putting $x = k\pi/n$ in the above result yields $H_1(m,n,j)$, which is given by

$$H_1(m,n,j) = \frac{1}{2}\bigg(\Big(-\cos\Big(\frac{(j-m-2n)\pi}{2n}\Big) + \cos\Big(\frac{(j-m+2jn-2mn-2n^2)\pi}{2n}\Big)\Big) \times \sec\Big(\frac{(j-m)\pi}{2n}\Big). \tag{63}$$

As in the case of $H_2(m,n,j)$, the above result can be simplified further by the application of trigonometric identities. Therefore, after a bit of manipulation, one finds that

$$H_1(m,n,j) = \frac{1}{2}\Big(1+(-1)^{m+n-j}\Big). \tag{64}$$

However, (64) is not valid when $(m-j)/n = 2p+1$ and p is a non-negative integer. Thus, $H_1(m,n,j) = (n+1)$, and we arrive at (52). Finally, if we replace the sum over k by $H_1(m,n,j)$ and introduce (62) into (56), then we obtain (51). □

The results in Lemma 1 indicate that the values of $S^-_{0,1}(m,n)$ and $S^-_{0,2}(m,n)$ are rational. Even though they have been expressed as formulas, in reality, they represent algorithms when one wishes to determine values from them. Consequently, Mathematica modules need to be created.

As an example, let us consider the second result in (49). This basic trigonometric sum can be programmed as

```
Szero1minus[m_, n_]:= Sum[(−1)^k Sin[k π/n]^(2 m), {k, 0, n}].
```

For relatively small values of n, typically less than or equal to 6, the above instruction generates rational values. For example, for $m=16$ and $n=6$, Mathematica prints out

In[19]:= Timing[Szero1minus[16, 6]]

Out[19] = {0.000195, −(32881827/33554432)}.

However, putting $m=16$ and $n=8$ yields

In[20]:= Timing[Szero1minus1[16, 8]]

Out[20] = {0.00016, 32769/32768 − 2 Cos[π/8]^{32} − 2 Sin[π/8]^{32}}.

In number form, this yields 0.841277837753296. Unfortunately, one cannot tell whether the value is rational or not, unless the trigonometric powers are expressed as surds as in Section 5.

To create a module that is able to calculate rational values of $S^-_{0,1}(m,n)$ for even integer values of n, we require a Do loop to determine each of the contributions from the summand of the second result of (49). Before entering the Do loop, we calculate the second term on the rhs first, which will be denoted as e1. To obtain the correct value of the summand, we introduce an If statement with the condition being whether $\exp((2(m-j)i\pi/n)$ is equal to -1 or not. If it is, then $H2(m,n,j)$ or H2 as it is denoted in the module below equals $n+1$. Otherwise, it will equal $(1+(-1)^n)/2$. Then, the summand, which will be denoted as e2, can be evaluated and summed with e1 to yield a new value for e1. On exiting the Do loop, the final value will become the value of $S^-_{0,1}(m,n)$, which can then be printed out as in the following module:

```
Szero1minusA[m_, n_]:= Module[{}, e1 = −Binomial[2 m, m] (1 + (−1)^n)/2^(2 m +
1); Print["The value of e1 is ", e1];
Do[If[ Exp[2 (m − j) I π/n] == −1, H2 = (n + 1), H2 = (1 + (−1)^n)/2];
e2 = (−1)^(m + j) Binomial[2 m, j] H2/2^(2 m − 1);
Print["For j=", j, " the value of H2 is ", H2, " while the value of e2 is ", e2];
e1 = e1 + e2, {j, 0, m}]; Print["The value of Szero1minus is ", e1]].
```

Running the module for $m=16$ and $n=8$ gives

```
Szero1minusA[16, 8]
The value of e1 is −(300540195/2147483648)
For j = 0 the value of H2 is 1 while the value of e2 is 1/2147483648
For j = 1 the value of H2 is 1 while the value of e2 is −(1/67108864)
For j = 2 the value of H2 is 1 while the value of e2 is 31/134217728
For j = 3 the value of H2 is 1 while the value of e2 is −(155/67108864)
For j = 4 the value of H2 is 9 while the value of e2 is 40455/268435456
For j = 5 the value of H2 is 1 while the value of e2 is −(6293/67108864)
For j = 6 the value of H2 is 1 while the value of e2 is 56637/134217728
For j = 7 the value of H2 is 1 while the value of e2 is −(105183/67108864)
For j = 8 the value of H2 is 1 while the value of e2 is 2629575/536870912
For j = 9 the value of H2 is 1 while the value of e2 is −(876525/67108864)
For j = 10 the value of H2 is 1 while the value of e2 is 4032015/134217728
For j = 11 the value of H2 is 1 while the value of e2 is −(4032015/67108864)
For j = 12 the value of H2 is 9 while the value of e2 is 254016945/268435456
For j = 13 the value of H2 is 1 while the value of e2 is −(10855425/67108864)
For j = 14 the value of H2 is 1 while the value of e2 is 29464725/134217728
For j = 15 the value of H2 is 1 while the value of e2 is −(17678835/67108864)
For j = 16 the value of H2 is 1 while the value of e2 is 300540195/1073741824
The value of Szero1minusA is 3528575/4194304.
```

Finally, the decimal value is found by typing

In[21]:= N[3528575/4194304, 16]

Out[21] = 0.8412778377532959,

which agrees with the previously given value. Of course, the above process can be sped up by removing the Print instruction in the Do loop. From the output, we observe also that there are two instances where $H_2(m,n,j)$ equal to $n+1$ or 9, namely $j=4$ and $j=12$. These values of j will change when m and n are assigned different values. Finally, if we wish to evaluate $S^-_{0,2}(m,n)$, all that is required is to alter the first part of the Do loop to

```
Do[If[ Exp[(m − j) I π/n] == −1, H1 = (n + 1), H1 = (1 + (−1)^(m + n − j))/2];
e2 = (−1)^(m + j) Binomial[2 m, j] H1/2^(2 m − 1);.
```

In fact, we can combine both modules into one module by introducing the variable, l, which is set equal to 1 or 2 corresponding to the Dowker and Gardner–Fisher cases. This can be accomplished by using the Which instruction in Mathematica in the following manner:

```
Sminuszerol[−1, l_, i_, m_]:= Module[{H1, H2}, H1 = 0; H2 = 0;
e1 = −Binomial[2 i, i] (1 + (−1)^m)/2^(2 i + 1);
Which[l == 1, Do[If[ Exp[2 (i − j) I π/m] == −1, H2 = (m + 1), H2 = (1 + (−1)^m)/2];
e2 = (−1)^(i + j) Binomial[2 i, j] H2/2^(2 i − 1); e1 = e1 + e2, {j, 0, i}],
l == 2, Do[If[ Exp[(i − j) I π/m] == −1, H1 = (m + 1), H1 = (1 + (−1)^(i + m − j))/2];
e2 = (−1)^(i + j) Binomial[2 i, j] H1/2^(2 i − 1);
e1 = e1 + e2, {j, 0, i}]]; Print[e1]].
```

Therefore, by setting $l=1$ and 2 in the above module, one obtains the values for $S^-_{0,1}(i,m)$ and $S^-_{0,2}(i,m)$, respectively.

5. Twisted Trigonometric Power Sums

As indicated in the introduction, a twisted trigonometric power sum is one whose trigonometric power is multiplied by an extra cosine factor. Originally, this name was applied to the $l=1$ case of (10) with $\cos(2rk\pi/m)$ inserted in the sum, and r was any integer from 1 to $m-1$ [6,7,12]. In actual fact, r can be much greater m, but then it must be replaced by the modulo value with respect to m, viz., $\mathrm{mod}(r,m)$. This issue will be discussed in more detail in Section 6. Moreover, since the trigonometric power sum in (10) has lm in the denominator, we shall consider twisted trigonometric power sums where the cosine factor appears as $\cos(2rk\pi/lm)$ as in (9). Hence, we will be able to study both the Dowker and Gardner–Fisher cases of (9). Note, however, the upper limit is now $lm-1$, whereas in the previous sections, we considered $m-1$ as the upper limit in both cases. Obviously, this has no effect in the Dowker case, but in the Gardner–Fisher case, it will play a role. For $l=2$, we shall be required to express the sums in (9) in terms of those appearing in previous sections. It should also be mentioned that the case of the cosine factor accompanying the basic trigonometric power has already been studied in [25].

When one considers twisted trigonometric power sums, one frequently ends up having to study other trigonometric power sums, denoted here by $V^{\pm}_{v,j,l}(m)$, and defined as

$$V^{\pm}_{v,j,l}(m) \doteqdot \sum_{k=1}^{lm-1} (\pm 1)^k \csc^{2v}\left(\frac{k\pi}{lm}\right) \cos^{2j}\left(\frac{k\pi}{lm}\right), \tag{65}$$

where j ranges from 0 to r. These sums arise when the factor of $\cos(2rk\pi/lm)$ in a twisted trigonometric power sum is expressed as a sum in powers of $\cos^2(k\pi/lm)$, as will be seen later. Nevertheless, these sums are not only interesting in their own right, but are also formidable. By expressing the cosine powers in terms of power of sines via the binomial theorem, one finds that

$$V^{\pm}_{v,j,l}(m) = \sum_{i=0}^{j} (-1)^i \binom{j}{i} \sum_{k=1}^{lm-1} (\pm 1)^k \sin^{2i-2v}\left(\frac{k\pi}{lm}\right). \tag{66}$$

For the case of $j < v$, the inner sum becomes a pure cosecant power sum. Consequently, it will yield polynomials in powers of m^2 with number-theoretical coefficients or coefficients expressible in terms of the Riemann zeta function. However, when $j > v$, there will also be sums involving basic trigonometric power sums. In particular, when $i \geq v$ and the summand does not alternate, the sum over k simply becomes $S(i-v,m)$ given by (36). On the other hand, for the alternating case, it can be expressed in terms of $S^{-}_{0,l}(i-v,m)$ in Lemma 1. As we have seen in the previous section, such sums are combinatorial in nature. Hence the second case is more complicated than the first case. Nevertheless, both cases can be considered together in one lemma as follows.

Lemma 2. *For $j < v$, the trigonometric power sum defined by* (65) *can be expressed as*

$$V^{\pm}_{v,j,l}(m) = \sum_{i=0}^{j}(-1)^i\binom{j}{i}\sum_{n=0}^{v-i-1}\frac{s(v-i,n)}{(2v-2i-2n)_{2n}}\,\frac{\zeta(2v-2i-2n)}{\pi^{2v-2i-2n}}\,Sf(v-i,n,\pm 1,l), \quad (67)$$

where the summand factors, $Sf(v-i,n,\pm,l)$ are displayed in Table 6 for both signs in the Dowker and Gardner–Fisher cases.

Table 6. Summand factors for the inverse powers of sine in $V^{\pm}_{v,j,l}(m)$.

sgn	*l*	$Sf(v,n,sgn,l)$
1	1	$2^{2n+1}(m^{2v-2n}-1)$
1	2	$2^{2v+1}(m^{2v-2n}-1)$
−1	1	$2^{2n}((2^{2n-2v+1}-1)m^{2v-2n}-1)(1+(-1)^m)$
−1	2	$(2^{2n+2}-2^{2v+1})m^{2v-2n}-(-1)^m 2^{2v+1}+((-1)^m-1)2^{2n+1}$

For $j \geq v$, $V^{\pm}_{v,j,l}(m)$ is given by

$$\begin{aligned} V^{\pm}_{v,j,l}(m) &= \sum_{i=0}^{v-1}(-1)^i\binom{j}{i}\sum_{n=0}^{v-i-1}\frac{s(v-i,n)}{(2v-2i-2n)_{2n}}\,\frac{\zeta(2v-2i-2n)}{\pi^{2v-2i-2n}}\,Sf(v-i,n,\pm 1,l) \\ &\quad + l\sum_{i=v+1}^{j}(-1)^i\binom{j}{i}S^{\pm}_{0,l}(i-v,m) - \frac{(\pm 1)^m}{2}(-1)^v\left(1+(-1)^l\right)\binom{j-1}{v} \\ &\quad + (-1)^v P(\pm 1,l)\binom{j}{v}, \end{aligned} \quad (68)$$

where $S^{+}_{0,1}(i-v,m)$, $S^{+}_{0,2}(i-v,m)$, $S^{-}_{0,1}(i-v,m)$ and $S^{-}_{0,2}(i-v,m)$ are, respectively, given by (81), (82), (49) *and* (51), *while the various m-dependent values of $P(\pm 1,l)$ are displayed in Table 7. In addition, $\delta_{l,2}$ denotes the Kronecker delta, which equals unity only when $l=2$ or in the Gardner–Fisher case.*

Table 7. Factors $P(\pm 1,l)$ in the $j=v$ term of $V^{\pm}_{v,j,l}(m)$.

sgn	*l*	$P(sgn,l)$
1	1	$m-1$
1	2	$2m-1$
−1	1	$\begin{cases}-1, & m \text{ even} \\ 0, & m \text{ odd}\end{cases}$
−1	2	−1

Proof. When $j < v$, all the sine powers in (66) are negative. Consequently, the results of Sections 2 and 3 can be used to evaluate $V^{\pm}_{v,j,l}$. For $l=1$ and the non-alternating case, we introduce (15) into (66), which gives

$$V^{+}_{v,j,1}(m) = 2\sum_{i=0}^{j}(-1)^{i}\binom{j}{i}\sum_{n=0}^{v-i-1}\frac{s(v-i,n)}{(2v-2i-2n)_{2n}}\frac{\zeta(2v-2i-2n)}{\pi^{2v-2i-2n}}\left(m^{2v-2i-2n}-1\right). \quad (69)$$

On the other hand, for the Gardner–Fisher case, where $l=2$, we note that

$$\sum_{k=1}^{2m-1}\sin^{2i-2v}\left(\frac{k\pi}{2m}\right) = 2\sum_{k=1}^{m-1}\sin^{2i-2v}\left(\frac{k\pi}{2m}\right)+1. \quad (70)$$

Therefore, we can replace the sum on the rhs of (66) by $2S^{+}_{v,2}(m)+1$. Since $S^{+}_{v,2}(m)$ is given by (11), we arrive at

$$V^{+}_{v,j,2}(m) = \sum_{i=0}^{j}(-1)^{i}\binom{j}{i}\sum_{n=0}^{v-i-1}\frac{s(v-i,n)}{(2v-2i-2n)_{2n}}\frac{\zeta(2v-2i-2n)}{\pi^{2v-2i-2n}}\left(\frac{m^{2v-2i-2n}-1}{2^{2v-2i-2n-1}}\right), \quad (71)$$

where the contribution from the term of unity on the rhs of (70) vanishes courtesy of the binomial theorem. When the summand alternates in sign, i.e., for $V^{-}_{v,j,l}(m)$, we still have a purely cosecant power sum, but now the results in Section 3 apply. For the Dowker case, we introduce (17) into (66). Thus, we arrive at

$$V^{-}_{v,j,1}(m) = \sum_{i=0}^{j}(-1)^{i}\binom{j}{i}\begin{cases}0, & m,\ \text{odd},\\ \displaystyle\sum_{n=0}^{v-i-1}\frac{s(v-i,n)}{(2v-2i-2n)_{2n}}\frac{\zeta(2v-2i-2n)}{\pi^{2v-2i-2n}} & \\ \times\ 2^{2n+1}\left(\left(2^{2n+2i-2v+1}-1\right)m^{2v-2i-2n}-1\right), & m,\ \text{even}.\end{cases} \quad (72)$$

For $V^{-}_{v,j,2}(m)$, however, we need to take note of the fact that

$$\sum_{k=1}^{2m-1}(-1)^{k}\sin^{2i-2v}\left(\frac{k\pi}{2m}\right) = 2S^{-}_{v-i,2}(m)+(-1)^{m}, \quad (73)$$

where $S^{-}_{v-i,2}(m)$ is defined by (16). Now, by introducing (18) into (66), one obtains

$$V^{-}_{v,j,2}(m) = \sum_{i=0}^{j}(-1)^{i}\binom{j}{i}\sum_{n=0}^{v-i-1}\frac{s(v-i,n)}{(2v-2i-2n)_{2n}}\frac{\zeta(2v-2i-2n)}{\pi^{2v-2i-2n}}$$
$$\times\left(\left(2^{2n+2}-2^{2v-2i+1}\right)m^{2v-2i-2n}-(-1)^{m}\,2^{2v-2i+1}+((-1)^{m}-1)2^{2n+1}\right). \quad (74)$$

Similarly to the term of unity in the non-alternating case, the resulting sum arising from the constant of $(-1)^m$ in (73) vanishes due to the binomial theorem. From these results, we see that the sums over n have a common term of $s(v-i,n)\zeta(2v-2i-2n)/(2v-2i-2n)_{2n}\pi^{2v-2i-2n}$. Hence, the sums can be expressed as this term multiplied by a summand factor, $Sf(v,n,sgn,l)$, for each value of sgn in both the Dowker and Gardner–Fisher cases. By adopting this approach, we arrive at (67), where the summand factors are displayed in Table 6. These results have been obtained from (11), (15), (17) and (18), bearing in mind that, in the Gardner–Fisher cases, there is an extra factor of 2, as indicated by (70) and (73). Furthermore, since the above results for $V^{\pm}_{v,j,l}(m)$ depend on the Riemann zeta function, they are purely number-theoretical. When $j\geq v$, however, the situation changes. Then, (66) becomes

$$V^{\pm}_{v,j,l}(m) = \sum_{i=0}^{v-1}(-1)^{i}\binom{j}{i}\sum_{k=1}^{lm-1}(\pm1)^{k}\csc^{2v-2i}\left(\frac{k\pi}{lm}\right)+(-1)^{v}\binom{j}{v}\sum_{k=1}^{lm-1}(\pm1)^{k}$$
$$+\sum_{i=v+1}^{j}(-1)^{i}\binom{j}{i}\sum_{k=1}^{lm-1}(\pm1)^{k}\sin^{2i-2v}\left(\frac{k\pi}{lm}\right). \quad (75)$$

The first term on the rhs of (75) is number-theoretical as in the $j < v$ case, except the upper limit is $v-1$, not j. Consequently, the earlier results in the proof can be applied here except the summation over i has an upper limit of $v-1$. The next term on the rhs is the $j = v$ term, which has been isolated or separated to avoid any potential problem concerning the determinacy of the zeroth power of the sine in Mathematica. For $l=1$ and the positive-sign or non-alternating case, the sum over k equals $m-1$. However, for $l=1$ and the negative-sign or alternating case, the sum over k vanishes when m is odd, but equals -1 when m is even. For the Gardner–Fisher case, the sum over k equals $2m-1$ for $sgn = +1$ and -1 for $sgn = -1$. We shall represent the latter sums over k for the various values of the sign and l by $P(\pm 1, l)$. Hence, the second term on the rhs of (75) can be represented by $(-1)^v P(\pm 1, l)\binom{j}{v}$, where the various values of $P(\pm 1, l)$ appear in Table 7.

The final term on the rhs of (75) represents a sum over basic sine power sums. It also vanishes when $j = v$. From the previous section, we have seen that these sums yield combinatorial results. For j close to v, the first sum on the rhs of (75) dominates. Hence, we expect that the overall results will be mainly number-theoretical in nature when j is approximately equal to v. This applies especially to the alternating Dowker case because, as we shall see, the sum over the basic sine power sums vanishes when coded as a program. For the Dowker or $l=1$ case, the final term on the rhs of (66) can be expressed as

$$\sum_{i=v+1}^{j} (-1)^i \binom{j}{i} \sum_{k=1}^{m-1} (\pm 1)^k \sin^{2i-2v}\left(\frac{k\pi}{m}\right) = \sum_{i=v+1}^{j} (-1)^i \binom{j}{i} S_{0,1}^{\pm}(i-v, m). \tag{76}$$

On the other hand, in the Gardner–Fisher case, we have to split the sum over k into two separate sums, since the upper limit is $2m-1$ for $V_{v,j,2}^{\pm}(m)$. In terms of $S_{0,2}^{\pm}(i, m)$, the sum can be expressed as

$$\sum_{k=1}^{2m-1} (\pm 1)^k \sin^{2i}\left(\frac{k\pi}{2m}\right) = 2S_{0,2}^{\pm}(i, m) + (\pm 1)^m, \tag{77}$$

which is identical to the inverse power cases given by (70) and (73). As a consequence, for the Gardner–Fisher case, the third term on the rhs of (75) becomes

$$\begin{aligned} \sum_{i=v+1}^{j} (-1)^i \binom{j}{i} \sum_{k=1}^{2m-1} (\pm 1)^k \sin^{2i-2v}\left(\frac{k\pi}{2m}\right) &= 2 \sum_{i=v+1}^{j} (-1)^i \binom{j}{i} S_{0,2}^{\pm}(i-v, m) \\ &+ \sum_{i=v+1}^{j} (-1)^i \binom{j}{i} (\pm 1)^m. \end{aligned} \tag{78}$$

As a result of the preceding discussion, (75) becomes

$$\begin{aligned} V_{v,j,l}^{\pm}(m) = &\sum_{i=0}^{v-1} (-1)^i \binom{j}{i} \sum_{n=0}^{v-1} \frac{s(v-i, n)}{(2v-2i-2n)_{2n}} \frac{\zeta(2v-2i-2n)}{\pi^{2v-2i-2n}} Sf(v-i, n, \pm 1, l) \\ &+ l \sum_{i=v+1}^{j} (-1)^i \binom{j}{i} S_{0,l}^{\pm}(i-v, m) + \frac{(\pm 1)^m}{2}\left(1 + (-1)^l\right) \sum_{i=v+1}^{j} (-1)^i \binom{j}{i} \\ &+ (-1)^v P(\pm 1, l) \binom{j}{v}. \end{aligned} \tag{79}$$

The sum in the penultimate term on the rhs of (79) can be evaluated with the aid of No. 4.2.1.5 in Ref. [24]. Then, we find that

$$\sum_{i=v+1}^{j} (-1)^i \binom{j}{i} = (-1)^{v+1} \binom{j-1}{v}. \tag{80}$$

By introducing (80) into (79), we arrive at the second result in Lemma 2. □

From the proof, we observe that the powers of m in the final results are governed by the summand factors. In fact, since each summand factor possesses a term with m^{2v-2n}, the degree of the polynomials in m will be $2v$ corresponding to $n = 0$. In other words, the highest power of $V^{\pm}_{v,j,l}(m)$ is m^{2v}, where $s(v-i,0) = 1$.

We are now in a position to consider the development of an algorithm that will evaluate $V^{\pm}_{v,j,l}(m)$ as defined by (65). Of the terms comprising $V^{\pm}_{v,j,l}(m)$, the most formidable is the second term on the rhs of (68), namely the term involving the alternating basic sine power sums since, from (36), the positive sign or non-alternating case can be reduced to

$$S^{+}_{0,1}(i,m) = 2^{1-2i} m \sum_{p=0}^{\lfloor i/m \rfloor} (-1)^{pm} c_p \binom{2i}{i-pm}, \tag{81}$$

and

$$S^{+}_{0,2}(i,m) = 2^{1-2i} m \sum_{p=0}^{\lfloor i/2m \rfloor} c_p \binom{2i}{i-2pm} - \frac{1}{2}, \tag{82}$$

where

$$c_p = \begin{cases} 1/2, & p = 0, \\ 1, & p \geq 1. \end{cases} \tag{83}$$

These forms are easily implemented in Mathematica. However, when the summands are alternating, i.e., for $S^{-}_{0,l}(i-v,m)$, we have to deal with $H_r(m,n,j)$. Previously, it was shown how this function can be implemented by itself in Mathematica, but now it appears in another sum involving binomial coefficients.

To simplify the problem, we combine the results appearing in Lemma 2 into one expression as

$$\begin{aligned} S^{-}_{0,l}(i-v,m) = & \Big(\frac{(-1)^{lm}+1}{2^{2i-2v}}\Big) \sum_{j=0}^{i-v} (-1)^j \binom{2i-2v}{j} H_{3-l}(i-v,m,j) \\ & - \frac{(1+(-1)^m)}{2^{2i-2v+1}} \binom{2i-2v}{i-v}. \end{aligned} \tag{84}$$

Therefore, when $l=1$ and m is odd, (84) vanishes as demonstrated by the top result of (49), while when $l=2$ or m is even, the first sum contributes as demonstrated by the other results in (49) and (50) with the major difference being that $H_2(i-v,m,j)$ and $H_1(i-v,m,j)$ are implemented when $l=1$ and $l=2$, respectively. In accordance with Lemma 1, the second or final term on the rhs of (84) only contributes when m is even.

When it comes to the creation of a module in Mathematica, we do not really need to implement H_2 and H_1. Instead, we employ a condition to evaluate H, which is different according to whether $l=1$ or $l=2$ when the sign or *sgn* equals -1. On the other hand, we can introduce another conditional statement to cover the case when the sign is equal to +1. Moreover, the $l=1$ and $l=2$ cases can be combined into one statement by separating the $i=v$ term in the second term of (68) and expressing it as

$$\begin{aligned} l \sum_{i=v+1}^{j} (-1)^i \binom{j}{i} S^{+}_{0,l}(i-v,m) = & \, l \sum_{i=v+1}^{j} (-1)^i \binom{j}{i} \Bigg(2^{2v-2i} m \binom{2i-2v}{i-v} + 2^{1+2v-2i} m \\ & \times \sum_{p=1}^{\lfloor (i-v)/(lm) \rfloor} (-1)^{plm} \binom{2i-2v}{i-v-pml} - \frac{1+(-1)^l}{4} \Bigg). \end{aligned} \tag{85}$$

Although the above equation is elegant from a mathematical point of view, it makes a module less readable rather than considering the $l=1$ and $l=2$ cases separately, which can be handled by employing a conditional If statement. Consequently, we shall create a module that will calculate the entire last term on the rhs of (75) for both the Dowker and

Gardner–Fisher cases in addition to both their non-alternating (sgn = +1) and alternating versions (sgn = −1). Thus, we shall evaluate

$$U^{\pm}_{l,v,j}(m) = \sum_{i=v+1}^{j} (-1)^i \binom{j}{i} \sum_{k=1}^{lm-1} (\pm 1)^k \sin^{2i-2v}\left(\frac{k\pi}{lm}\right). \tag{86}$$

The above result can be regarded as a different type of trigonometric power sum. A Mathematica module that is capable of calculating $U^{\pm}_{l,v,j}(m)$ appears below:

```
U[sgn_,l_,v_,j_,m_]:= Module[{H}, H = 0; totval = 0;
Do[e1 = 0;Which[sgn == 1&&(l == 1),
e1 = 2^(1−2(i−v))m Sum[(−1)^(p m) Binomial[2 (i−v),i−v−p m], {p,1,Floor[(i−v)/m]}]
+2^(−2(i−v))m Binomial[2(i−v),i−v];
(sgn == 1)&&(l == 2),
e1 = 2^(1−2(i−v)) m Sum[Binomial[2 (i−v),i−v−2p m],{p,1,Floor[(i−v)/(2m)]}]
+2^(−2 (i−v))m Binomial[2 (i−v),i−v]−1/2,
(sgn == −1)&&(l == 1),
e1 = −Binomial[2 (i−v),i−v](1 + (−1)^m)/2^(2 (i−v) + 1);
Do[If[Exp[2(i−v−k)I π/m] == −1,H = (m + 1),H = (1 + (−1)^m)/2];
e2 = (−1)^(i−v + k) Binomial[2 (i−v),k]H/2^(2(i−v)−1); e1 = e1 + e2,{k,0,(i−v)}];
(sgn == −1)&& (l == 2),
e1 = −Binomial[2 (i−v),i−v](1 + (−1)^m)/2∧(2 (i−v) + 1);
Do[If[ Exp[(i−v−k)I π/m] == −1,H = (m + 1),H = (1 + (−1)^(i−v + m−k))/2];
e2 = (−1)^(i−v + k) Binomial[2 (i−v),k]H/2^(2(i−v)−1); e1 = e1 + e2,{k,0,i−v}]];
If[l == 2,e1 = (−1)^i Binomial[j,i](2 e1−(−1)^m),
e1 = (−1)^i Binomial[j,i] e1] e1;totval = totval + e1,{i,v + 1,j}];
Print["For l=",l,", sgn=",sgn,", j= ",j,"v=",v," and m=",m,", the sum of the basic
trigonometric power sums in U±_{v,j,l} is equal to ",totval]]].
```

After setting both the values of H, given by (57), and $U^{\pm}_{l,v,j}(m)$ or totval to zero initially, the code encounters a Do loop, whose purpose is to evaluate the summation over i in (86). Within the Do loop, the module encounters a Which instruction, which is required to handle all four types of basic trigonometric power sums according to the values of sgn and l. The first case in the Which instruction corresponds to the non-alternating Dowker case, where the value of $S^{\pm}_{0,1}(i-v,m)$ given by e1 is calculated in accordance with the upper signed version of (76). The second case is the non-alternating Gardner–Fisher sum, where e1 is evaluated according to (82). The next two cases in the Which instruction handle both types of alternating sums. This is where H figures in the evaluation of e1. As indicated previously, both these cases require another Do loop, since H is conditional on the value of $\exp(i(2-l)\pi/m)$. After e1 has been calculated for the particular values of sgn and l, the code continues with the first Do loop by testing the upper limit of the inner sum. That is, for the $l=2$ case, the code takes account of the fact that the upper limit is $lm-1$ as in (77). Then, each e1 is multiplied by the binomial factor of $(-1)^i\binom{j}{i}$ to give each value of the summand in the sum over i in (86). Finally, all the e1 values are summed within the first Do loop to become the current value of totval. When the Do loop has been completed, totval becomes the value of $U^{\pm}_{l,v,j}(m)$, which is the only value printed out by the code.

Table 8 displays the values obtained from the above Mathematica module of all four cases of $U^{\pm}_{l,v,j}(m)$, where v and j are varied for fixed values of m. All told, there are 100 values that have been evaluated using the Mathematica module. As expected, the tabulated values are rational. The values were then converted into decimal notation so that they could be checked against the values calculated by programming each basic trigonometric sum in Mathematica. For example, $U^{-}_{2,v,j}(m)$ or the alternating Gardner–Fisher case can also be programmed as

```
TRGSM4[v_,j_,m_]:= Sum[(−1)∧i Binomial[j,i] Sum[(−1)∧k Sin[k Pi/(2 m)]∧ (2 i−2 v),
{k,1,2m−1}],{i,v + 1,j}].
```

Obviously, in this module, the $(-1)^k$ phase factor has to be removed for the non-alternating cases, while when dealing with a Dowker case sum, $2m$ is replaced by m.

It was found that the decimal values of all the results in Table 8 agreed with the results obtained from the alternative modules, except for the final value in the $l = 2$ and $sgn = -1$ column. The decimal value of 4250495/8589934592 is 0.0004948227..., whereas TRGSM4[20,40,16] gives a value of 0.000369022... As indicated previously, a closer examination of $U^{-}_{2,20,40}(16)$ can be conducted by inserting a Print[e1] instruction before totval is evaluated in the module. This means that each term in the sum over i in (86) is now printed out. In doing so, it is found that e1 vanishes for i ranging from $i = 21$ to $i = 36$. That is, only the last five values in the sum contribute to the value of $U^{-}_{2,20,40}(16)$. Specifically, the last five or non-zero values in the sum are

$$\frac{45695}{33554432}, -\frac{20995}{16777216}, \frac{61425}{134217728}, -\frac{10545}{134217728}, \text{ and } \frac{45695}{8589934592}.$$

Summing these values yields the final value in Table 8 or 4250495/8589934592.

Table 8. Values of the Four Basic Trigonometric Power Sums, $U^{sgn}_{l,v,j}(m)$, for various values of v, j and m.

sgn	v	j	$l=1$ $+$	$l=2$ $+$	$l=1$ $-$	$l=2$ $-$
	2	9	$-\frac{422343}{8192}$	$-\frac{223545}{2048}$	0	$\frac{24747}{4096}$
	6	12	$-\frac{1598103}{2048}$	$-\frac{1620819}{1024}$	0	$\frac{5679}{256}$
$m=3$	9	15	$\frac{6744267}{2048}$	$\frac{6798747}{1024}$	0	$-\frac{3045}{64}$
	12	19	$-\frac{248613615}{8192}$	$-\frac{125215101}{2048}$	0	$\frac{1810587}{4096}$
	15	30	$\frac{70980258098522691}{536870912}$	$\frac{36217263345658047}{134217728}$	0	$-\frac{1454268592793403}{268435456}$
	3	15	$\frac{5103981565}{4194304}$	$\frac{5130885535}{2097152}$	0	$-\frac{13451985}{1048576}$
	5	10	$\frac{183225}{512}$	$\frac{91615}{128}$	0	$-\frac{5}{256}$
$m=5$	10	20	$-\frac{139679654025}{524288}$	$-\frac{139710446585}{262144}$	0	$-\frac{1924535}{16384}$
	20	30	$-\frac{14318455371625}{524288}$	$-\frac{14318864559225}{262144}$	0	$\frac{25574225}{16384}$
	25	40	$\frac{11226079196963413825}{268435456}$	$\frac{11226967864447527165}{134217728}$	0	$-\frac{222166871028335}{33554432}$
	4	13	$-\frac{5074887}{2048}$	$-\frac{20299565}{4096}$	$\frac{40919}{2048}$	$\frac{17}{4096}$
	8	16	$-\frac{30295991}{1024}$	$-\frac{121183965}{2048}$	$\frac{53191}{1024}$	$\frac{1}{2048}$
$m=8$	11	19	$\frac{145793675}{1024}$	$\frac{583174701}{2048}$	$-\frac{131707}{1024}$	$\frac{1}{2048}$
	14	30	$-\frac{1530284095489007}{4194304}$	$-\frac{24484565631151767}{33554432}$	$\frac{5626959414511}{4194304}$	$\frac{20103327655}{33554432}$
	18	35	$-\frac{85397153404618323}{8388608}$	$-\frac{1366354983263679667}{67108864}$	$\frac{221267848529747}{8388608}$	$\frac{528789786499}{67108864}$
	5	14	$\frac{154581141}{16384}$	$\frac{154581141}{8192}$	$-\frac{1653}{2048}$	0
	7	16	$\frac{746494545}{16384}$	$\frac{746494545}{8192}$	$-\frac{2967}{2048}$	0
$m=12$	12	20	$-\frac{2751221655}{8192}$	$-\frac{2751221655}{4096}$	$\frac{765}{1024}$	0
	14	25	$-\frac{1743775392027}{131072}$	$-\frac{1743775392027}{65536}$	$\frac{34909329}{262144}$	0
	16	32	$-\frac{1123494714551097135}{536870912}$	$-\frac{1123494714561502359}{268435456}$	$\frac{3006142168365}{33554432}$	$\frac{1300653}{33554432}$
	6	13	$-\frac{1099943}{128}$	$-\frac{1099943}{64}$	0	0
	9	17	$\frac{213188525}{2048}$	$\frac{213188525}{1024}$	$-\frac{1}{2048}$	0
$m=16$	11	20	$\frac{2802276035}{4096}$	$\frac{2802276035}{2048}$	$-\frac{31}{4096}$	0
	15	26	$\frac{963117586951}{32768}$	$\frac{963117586951}{16384}$	$-\frac{22755}{32768}$	0
	20	40	$-\frac{27535407845508314400045}{4294967296}$	$-\frac{110141631382033300010675}{8589934592}$	$\frac{1989694564057805}{4294967296}$	$\frac{4250495}{8589934592}$

On the other hand, if one examines the individual terms in the sum over i in TRGSM4[40, 20,16], which is achieved by the following instruction:

```
Term[j_,v_,i_,m_]:= (−1)^i Binomial[j,i] Sum[(−1)^ Sin[k Pi/(2 m)]^(2 i−2 v),
{k,1,2m−1}],
```

then one finds that the five values given above are obtained in decimal form for i ranging from 36 to 40. Thus, both Mathematica modules calculate the correct components for $U^{-}_{2,20,40}(16)$. The problem, however, lies with the vanishing terms. These are found to be zero using the first module, but are rounded off when using TRGSM4. These round-off errors become sizeable because of the binomial factor multiplying the round-off value obtained from the sum over k in TRGSM4. Normally, round-off errors remain small due since they often oscillate in sign and remain negligible. However, when the terms from $i = 21$ to 35 are summed by typing the following instruction:

```
Sum[Term[40,20,i,16],{i,21,35}],
```

then Mathematica prints out a value of -0.0001258. If this value is added to 4250495/8589934592, then one obtains 0.00036902..., which agrees with the value of TRGSM4[20,40,16] given above. From this study, we observe that utilizing the machine dependent approach to obtain values of $U^{\pm}_{l,v,j}(m)$ will not necessarily guarantee a correct answer, especially when the binomial factor, $\binom{j}{i}$, becomes very large. Nonetheless, we see that such trigonometric power sums are rational, which is also not noticeable when utilizing the machine dependent approach.

Having produced a Mathematica module for the term involving basic trigonometric power sums or the second term in (68), we turn our attention to the last two terms in the same equation. The first of these can be handled by one instruction, such as

```
BinomTerm[sgn_,j_,v_,m_]:= sgn^m (−1)^(v + j) Binom[j−1,v](1 + (−1)^l)/2.
```

This term only makes a contribution in the Gardner–Fisher or $l = 2$ case. It also changes when the summand of $V^{\pm}_{v,j,l}(m)$ alternates in sign. Nevertheless, it is rather simple to program by employing the Binomial instruction in Mathematica. On the other hand, the final term in (68) is dependent upon both l and sgn, since $P(sgn, l)$ is different for all cases of $V^{\pm}_{v,j,l}(m)$. This term can be programmed as a Mathematica module as follows:

```
PSum[sgn_,l_,j_,v_,m_]:= Module[, Which[ (sgn == 1) &&(l == 1),e1 = m−1,
(sgn == 1) &&(l == 2),e1 = 2m−1,(sgn == −1) &&(l == 1), e1 = −(1 + (−1)^m)/2,
(sgn == −1) &&(l == 2), e1 = −1];
PSumval = (−1)^v Binomial[j,v] e1; Print[" The value of the sum with P(sgn,l) is ",
PSumval]].
```

From this module we see that the same conditions in the Which instruction for U[sgn_, l_,v_,j_,m_] apply. Therefore, one could, perhaps, incorporate the above module into U[sgn_,l_,v_,j_,m_]. However, this is hampered by the Do loop corresponding to the summation over i in (68) even though having a module with one Which instruction would result in a more elegant code. Therefore, a simpler option is to insert the module after totval has been evaluated, assign a new variable e2 to determine PSumval and then add PSumval and BinomTerm to determine totval. A code combining the modules appears below:

```
U1[sgn_,l_,v_,j_,m_]:= Module[{H}, H = 0; totval = 0;
Do[e1 = 0; Which[sgn == 1&&(l == 1),
e1 = 2^(1−2(i−v))m Sum[(−1)^(p m) Binomial[2 (i−v),i−v−p m], {p,1,Floor[(i−v)/m]}]
+2^(−2(i−v))m Binomial[2(i−v),i−v];
(sgn == 1)&&(l == 2),
e1 = 2^(1−2(i−v)) m Sum[Binomial[2 (i−v),i−v−2p m],{p,1,Floor[(i−v)/(2m)]}]
+2^(−2 (i−v))m Binomial[2 (i−v),i−v]−1/2,
(sgn == −1)&&(l == 1),
e1 = −Binomial[2 (i−v),i−v](1 + (−1)^m)/2^(2 (i−v) + 1);
```

```
Do[If[Exp[2(i−v−k)I π/m] == −1,H = (m + 1),H = (1 + (−1)^m)/2];
e2 = (−1)^(i−v + k) Binomial[2 (i−v),k]H/2^(2(i−v)−1); e1 = e1 + e2,{k,0,(i−v)}];
(sgn == −1)&& (l == 2),
e1 = −Binomial[2 (i−v),i−v](1 + (−1)^m)/2∧(2 (i−v) + 1);
Do[If[ Exp[(i−v−k)I π/m] == −1,H = (m + 1),H = (1 + (−1)^(i−v + m−k))/2];
e2 = (−1)^(i−v + k) Binomial[2 (i−v),k]H/2^(2(i−v)−1); e1 = e1 + e2,{k,0,i−v}]];
If[l == 2,e1 = (−1)^i Binomial[j,i](2 e1−(−1)^m),
e1 = (−1)^i Binomial[j,i] e1] e1;totval = totval + e1,{i,v + 1,j}];
Which[(sgn == 1)&&(l == 1),e3 = m−1, (sgn == 1)&&(l == 2),e3 = 2m−1,
(sgn == −1)&&(l == 1),e3 = −(1 + (−1)^m)/2, (sgn == −1)&&(l == 2),e3 = −1];
scndval = (−1)^ Binomial[j,v] e3;totval = totval + scndval;
totval = totval + (−1)^v Binomial[j,v] Sum[(sgn)^k,{k,1,lm−1}];
Print["For l=",l,", sgn=",sgn,", j= ",j,"v=",v," and m=",m,", the sum of the basic
trigonometric power sums in U±_{v,j,l} is equal to ",totval]]].
```

This interim module is essentially the same as the previous one, except it calculates the last two terms in (68) and adds them to totval. Therefore, to produce a code that determines values of $V^{\pm}_{v,j,l}(m)$, we only require the code to evaluate the first term on the rhs of (68) when $j \geq v$. Moreover, this term is identical to the rhs of (67), except for the upper limit of the outer sum. Hence, one can use the same code to calculate (67), except for a change in the upper limit. To handle this problem, a conditional If instruction has been inserted near the beginning of the program. This instruction tests whether the upper limit of the outer sum equals j when $j < v$ or equals $v - 1$ for all other values of j. It then assigns a variable called upplmt to whichever case applies. Moreover, the same instructions can be used to evaluate the double sum in either case. As a result of the introduction of upplmt into the code, there needs to be another conditional If instruction or statement following the calculation of the double sum, which calculates the remaining terms in (68) for $v \leq j$.

In Section 3, we introduced a summand factor to cover both the $l=1$ and $l=2$ cases of $S^{-}_{v,l}(m)$. Here, we adopt this approach further by noting that the four cases of $S^{\pm}_{v,l}(m)$ all include $\zeta(2v-2n)s(v,n)/\pi^{2v-2n}(2v-2n)_{2n}$ in their summands. Therefore, we can employ the Which instruction again to determine the appropriate summation factor for the case of $V^{\pm}_{v,j,l}(m)$ we are studying. This is achieved by implementing the following instruction:

```
Which[sgn == 1 &&(l == 1), Sfctr = 2^(2n + 1) (m^(2 v−2 n) −1),
sgn == 1 &&(l == 2), Sfctr = 2^(2v + 1) (m^(2v−2n)−1),
sgn == −1 &&(l == 1), Sfctr = 2^(2n)(1 + (−1)^m)(((2^(2 n−2 v + 1)−1) m^(2 v−2n) −1),
sgn == −1 &&(l == 2), Sfctr = (2^(2n + 2)− 2^(2v + 1)) m^(2v−2n)− (−1)^m 2^(2v + 1)
+((−1)^m −1)2^(2n + 1)].
```

Now, we are in a position to present an algorithm for determining rational values of $V^{\pm}_{v,j,l}(m)$ by combining the preceding material into one module called Vpm[sgn_,l_,v_, j_,m_]. To execute the module, one needs to input the values of sgn, l, v, j and m, although the value of m can be algebraic. The module begins by initializing the value of the trigonometric power sum given by the variable totval to zero in addition to H, which is used in evaluating the term involving the basic trigonometric power sum or the second term on the rhs of (68). Next, the summand factor for either the rhs of (67) or the first term on the rhs of (68) is determined from the input values of sgn and l. The summand factor is printed out via the first Print instruction, but this diagnostic can be commented out, if desired. In fact, there are other Print instructions in the module that can also be removed, since they only appear to observe that the module is behaving properly. Hence, only the last Print instruction, which prints out the final value of totval, needs to be retained.

Once the appropriate summand factor has been determined, the module determines whether the upper limit of the number-theoretical term is j or $v-1$ corresponding to $j < v$ or $j \geq v$. If it is the former value, then upplmt is set equal to j and the variable totval becomes the value of the double sum in (67). Otherwise, upplmt is set equal $v-1$ and totval becomes the value of the first term on the rhs of (68). It is then printed out as the current value of totval. In the first case, totval becomes the value of $V^{\pm}_{v,j,l}(m)$. The rest of the module is discarded because of the conditional If instruction after the second Print instruction. Hence, totval becomes the final value of $V^{\pm}_{v,j,l}(m)$, and is printed out again in the final Print instruction. However, if $v \leq j$, then the module calculates the remaining terms on the rhs of (68).

As stated earlier, the second term on the rhs of (68) represents a finite sum of basic trigonometric power sums, which can be evaluated by employing the module presented for $U^{\pm}_{l,v,j}(m)$ below (32). It was also stated that this is the most difficult term to evaluate because in the case of an alternating sum H changes according to the conditions of the If instruction as it is being tested for each iteration in a Do loop. Therefore, this term is left until after the third and fourth terms on the rhs of (68) are evaluated in the module below. After the number-theoretical term has been calculated, the module evaluates the fourth term on the rhs of (68), which only contributes to the Gardner–Fisher or $l=2$ trigonometric power sums. This calculation is represented by e1 in the module.

```
Vpm[sgn_, l_, v_, j_, m_] := Module[{H}, H = 0; totval = 0;
Which[(sgn == 1 && l == 1), Sfctr = 2^(2 n + 1) (m^(2 v − 2 i − 2 n) − 1),
(sgn == 1 && l == 2), Sfctr = 2^(2 v − 2 i + 1) (m^(2 v −2 i − 2 n) − 1),
(sgn == −1 && l == 1), Sfctr = 2^(2 n) (1 + (−1)^m) ((2^(2 n − 2 v + 2 i + 1)−1)
m^(2 v−2 i− 2 n)− 1),
(sgn == −1 && l == 2), Sfctr = (2^(2 n + 2) − 2^(2 v − 2 i + 1)) m^(2 v − 2 i − 2 n) −
(−1)^ m 2^(2 v − 2 i + 1) + ((−1)^m − 1) 2^(2 n + 1)];
Print["The value of the summand factor is ", Sfctr];
If[j < v, upplmt = j, upplmt = v − 1];
totval = Expand[FullSimplify[ Sum[(−1)^i Binomial[j, i] Sum[ SymmetricPolynomial[n,
Table[k^2, k, 1, v − i − 1]] Zeta[2 v−2 i−2 n] Sfctr/(Pi^(2 v− 2 i −2 n) Pochhammer[2 v−
2 i−2 n, 2 n]), {n, 0, v − i − 1}], {i, 0, upplmt}]]];
Print["The current value of totval is ", totval];
If[v <= j, If[l == 2 && j != v, e1 = −sgn^m (−1)^v Binomial[j−1, v], e1 = 0];
totval = totval + e1;
Print["The value of the l = 2 extra term is ", e1];
Which[(sgn == 1 && l == 1), e2 = m − 1, (sgn == 1 && l == 2), e2 = 2 m − 1, (sgn
== −1 && l == 1), e2 = −(1 + (−1)^m)/2, (sgn == −1 && l == 2), e2 = −1];
Print["The value of P(sgn,l) is ", e2];
e3 = (−1)^v Binomial[j, v] e2;
Print["The value of the extra term is ", e3];
totval = totval + e3];
If[v<j, Do[e4 = 0; Which[(sgn == 1) && (l == 1),
e4 = 2^(1 −2 (i −v)) m Sum[(−1)^(p m)
Binomial[2 (i −v), i−v−p m], {p, 1, Floor[(i − v)/m]}] + 2^(−2 (i− v)) m Binomial[2 (i−
v),
i− v], (sgn == 1) && (l ==2), e4 = 2^(1 −2 (i −v)) m Sum[Binomial[2 (i−v), i−v −2 p m],
{p, 0, Floor[(i − v)/(2 m)]}] − 2^(−2 (i−v)) m Binomial[2 (i− v), i− v] − 1/2, (sgn == −1)
&& (l == 1), e4 = −Binomial[2 (i−v), i−v] (1 + (−1)^m)/2^(2 (i−v) + 1);
Do[If[ Exp[2 (i− v− k) I Pi/m] == −1, H = (m + 1), H = (1 + (−1)^m)/2]; e5 = (−1)^(i−
v + k) Binomial[2 (i−v), k] H/2^ (2 (i− v) − 1); e4 = e4 + e5, {k, 0, (i− v)}], (sgn == −1)
&& (l == 2), e4 = −Binomial[2 (i− v), i− v] (1 + (−1)^m)/2^ (2 (i− v) + 1);
Do[If[ Exp[(i − v − k) I Pi/m] == −1, H = (m + 1), H = (1 + (−1)^(i− v+ m− k))/2];
e5 = (−1)^(i −v+ k) Binomial[2 (i−v), k] H/2^(2 (i − v) − 1); e4 = e4 + e5, {k, 0, i − v}]];
```

```
If[l == 2, e4 = (−1)^i Binomial[j, i] (2 e4 + (sgn)^(m)), e4 = (−1)^i Binomial[j, i] e4];
Print["The value of e4 is ", e4];
totval = totval + e4, {i, v + 1, j}]];
Print["The final value of the sum is ", totval]].
```

The fourth term on the rhs of (68) is the next term that is evaluated in the module. Before this term can be evaluated, the appropriate value of $P(\pm 1, l)$ given in Table 7 must be determined. Like the summand factor for the number-theoretical terms, a Which instruction is invoked, and the value is given by e2. A diagnostic Print statement that prints out the appropriate value of $P(\pm, l)$ follows. Then, this value is multiplied by the binomial factor to give the value of e3, which is subsequently added to totval. Finally, e4 representing the value for the sum of the basic trigonometric power sums is obtained by adapting the Mathematica module for $U^{\pm}_{l,v,j}(m)$, and is added to totval to give the final value for $V^{\pm}_{v,j,l}(m)$. Before this value is added to totval, the module checks whether a doubling up is required due to the fact that for $l=2$, the summation over k has an upper limit of $2m-1$, rather than $m-1$.

For the Dowker case, we introduce (17) and (49) into (77). Therefore, we observe that for m odd, $V^{-}_{v,j,1}(m)$ vanishes, as in (72). However, there is no need to display the result here, as we can use the algorithms for $S^{-}_{v-i}(m)$ and $S^{-}_{0,1}(i-v,m)$ to obtain the values of $V^{-}_{v,j,1}(m)$.

Let us now examine the output generated from the above module for two examples. In the first example, we consider $j=v$, while in the second example, we consider $j>v$. That is, in both examples, we shall require the more complicated (68) rather than (67). Moreover, we shall consider the alternating or $sgn=-1$ forms of $V^{\pm}_{v,j,l}(m)$, since they require the results of Lemma 1. Specifically, we shall set $l=1$ or consider the Dowker case for $j=v$ in the first example, and then set $l=2$, or consider the Gardner–Fisher case for $j>v$. For the first example, we do not require (49) because j is not greater than v, but in the second example, we require (50), since there will be basic trigonometric power sums for all those values between $j-v+1$ and j. Finally, we shall set m to a specific value, viz. 12 in the first example, while it will remain algebraic in the second example.

Below is the output generated by Vpm when $j=v=9$ and $m=12$ for the alternating form of the Dowker case. That is, the trigonometric power sum reduces to

$$V^{-}_{9,9,1}(12) = \sum_{k=1}^{11} (-1)^k \cos^{18}\left(\frac{k\pi}{12}\right) \csc^{18}\left(\frac{k\pi}{12}\right). \tag{87}$$

Typing Vpm[−1, 1, 9, 9, 12] into a Mathematica notebook yields

Vpm[−1, 1, 9, 9, 12]
The value of the summand factor is $2^{1+2n}\,(-1+12^{18-2i-2n}\,(-1+2^{-17+2i+2n}))$
The current value of totval is $-\dfrac{776563028730601}{19683}$
The value of the l = 2 extra term is 0
The value of P(sgn,l) is −1
The value of the extra term is 1
The final value of the sum is $-\dfrac{776563028710918}{19683}$.

In decimal form, the final value is given as

$N\left[-\dfrac{776563028710918}{19683}\right]$

-3.94535×10^{10}.

By clicking on the NumberForm of this result, one can extend the decimal value to 16 places. For example,

NumberForm[-3.94535×10^{10}, 16]
NumberForm= $-3.94534892399999\times10^{10}$.

The above output begins by displaying the value of the summand factor, which is a simplification of the $j=v=9$ and $m=12$ form of the third result in Table 6. Next, the value of the first term on the rhs of (68) is given. As expected, because $l=1$, the third term on the rhs of (68) does not contribute. Thus, the $l=2$ term is given as zero. Then, the last term on the rhs of (68) is calculated, which requires $P(-1,1)$ from Table 7. Since m is even, the entire term is non-zero. In fact, it yields unity due to the fact that $j=v$ and v is odd. Finally, the term involving basic trigonometric powers sums, viz. the second term on the rhs of (68) is evaluated. However, as $j=v$ in the example, there is no contribution from this term.

On the other hand, one can evaluate $V^{-}_{9,9,1}(12)$ by using the standard instructions in Mathematica [17], which means expressing it as

Sum[(−1)^k Csc[k Pi/12]^(18) Cos[k Pi/12]^(18), {k, 1, 11}].

The above instruction generates the following output:

$$\frac{774801614}{19683}-\frac{(1-\sqrt{3})^{18}(-1+\sqrt{3})^{18}}{262144}-\frac{(-1+\sqrt{3})^{36}}{262144}-\frac{(1+\sqrt{3})^{36}}{131072}.$$

Once again, the above expression can be evaluated via the N[] instruction in Mathematica as indicated below:

$$\text{In[22]:= N}\left[\frac{774801614}{19683}-\frac{(1-\sqrt{3})^{18}(-1+\sqrt{3})^{18}}{262144}-\frac{(-1+\sqrt{3})^{36}}{262144}-\frac{(1+\sqrt{3})^{36}}{131072}\right].$$

Out[22] = -3.94535×10^{10}.

As before, we can extend the number of decimal places using NumberForm. Therefore, we find that

In[23]:= NumberForm[-3.94535×10^{10}, 16]

Out[23] = $-3.945348923999989\times10^{10}$.

In this example, we observe that the standard approach yields an almost identical result to the algorithmic approach employing $V^{\pm}_{v,j,l}(m)$ which, as we have seen already, is not always the case.

Let us now consider the second example. As indicated above, this example has the same values for the arguments as the first example except that $l=2$ and m remains algebraic. In this instance, the sum becomes $V^{-}_{9,12,2}(m)$, which is given by

$$V^{-}_{9,12,2}(m)=\sum_{k=1}^{2m-1}(-1)^k\cos^{24}\left(\frac{k\pi}{2m}\right)\csc^{18}\left(\frac{k\pi}{2m}\right). \tag{88}$$

Then, typing Vpm[−1, 2, 9, 9, m] into a Mathematica notebook generates the following output:

In[24]:= Vpm[−1, 2, 9, 12, m]

The value of the summand factor is

$-(-1)^m\ 2^{19-2i}+2^{1+2n}\ (-1+(-1)^m)+(-2^{19-2i}+2^{2+2n})\ \mathrm{m}^{18-2i-2n}$

The current value of totval is

$$-\frac{14141033520614911}{194896477400625}-165(-1)^m-\frac{51766106m^2}{328185}+\frac{4946109824m^4}{50675625}-\frac{1794850075988m^6}{40226311125}+$$
$$\frac{559016662m^8}{35083125}-\frac{10363664m^{10}}{2338875}+\frac{187005175216m^{12}}{201131555625}-\frac{12646904m^{14}}{91216125}+\frac{33862354m^{16}}{2584456875}-$$

$$\frac{22998766228m^{18}}{38979295480125}$$

The value of the l = 2 extra term is 55 $(-1)^m$
The value of P(sgn,l) is -1
The value of the extra term is 220
The value of e4 is 66 $\left((-1)^m+2\left(\frac{1}{4}(-1-(-1)^m)+\frac{1}{2}(1+(-1)^m)\right)\right)$
The value of e4 is $-12\left((-1)^m+2\left(\frac{3}{8}(-1-(-1)^m)+\frac{3}{8}(1+(-1)^m)\right)\right)$
The value of e4 is $\left((-1)^m+2\left(\frac{5}{64}(-1-(-1)^m)+\frac{5}{64}(1+(-1)^m)\right)\right)$
The final value of the sum is

$$\frac{28736191507522589}{194896477400625}-109(-1)^m+2\left(\frac{5}{64}(-1-(-1)^m)+\frac{5}{64}(1+(-1)^m)\right)-12\left((-1)^m+2\left(\frac{3}{8}(-1-(-1)^m)+\frac{3}{8}(1+(-1)^m)\right)\right)+66\left((-1)^m+2\left(\frac{1}{4}(-1-(-1)^m)+\frac{1}{2}(1+(-1)^m)\right)\right)-\frac{51766106m^2}{328185}+\frac{4946109824m^4}{50675625}-\frac{1794850075988m^6}{40226311125}+\frac{559016662m^8}{35083125}-\frac{10363664m^{10}}{2338875}+\frac{187005175216m^{12}}{201131555625}-\frac{12646904m^{14}}{91216125}+\frac{33862354m^{16}}{2584456875}-\frac{22998766228m^{18}}{38979295480125}.$$

The output begins by printing out the summand factor, $Sf(9-i,n.-1,2)$, for the first sum on the rhs of (68). Then, the module prints out the value of the number-theoretical term or the first term on the rhs of (68). Next the third term on the rhs of (68), which only contributes when $l=2$, is calculated and printed out. After this, the term containing $P(-1,2)$ is calculated and printed out followed by the value of the last term on the rhs of (68). The last term handled by the module is the sum over the basic trigonometric power sums or the second term on the rhs of (68). As expected, there are three distinct sums corresponding to $j-v$. These combinatorial terms only affect the constant term in the polynomials derived from summing the number-theoretical terms represented by the first value of totval. Finally, all these terms are added to totval to give the value of $V^-_{9,12,2}(m)$. From the output, we observe that the behavior of the trigonometric power sum is mainly dominated by the number-theoretical term in (68).

One can check the above result for $V^-_{9,12,2}(m)$ by setting m equal to specific values. Before this can be accomplished, one must first create a function out of the final value as follows:

$$\text{Q[m\_]:=}\ 28736191507522589/194896477400625-219\,(-1)^m+2\left(\frac{5}{64}(-1-(-1)^m)+\frac{5}{64}(1+(-1)^m)\right)-12\left((-1)^m+2\left(\frac{3}{8}\left(-1-(-1)^m\right)+\frac{3}{8}\left(1+(-1)^m\right)\right)\right)+66\left((-1)^m+2\left(\frac{1}{4}(-1-(-1)^m)+\frac{1}{2}(1+(-1)^m)\right)\right)-\frac{51766106m^2}{328185}+\frac{4946109824m^4}{50675625}-\frac{1794850075988m^6}{40226311125}+\frac{559016662m^8}{35083125}-\frac{10363664m^{10}}{2338875}+\frac{187005175216m^{12}}{201131555625}-\frac{12646904m^{14}}{91216125}+\frac{33862354m^{16}}{2584456875}-\frac{22998766228m^{18}}{38979295480125}.$$

By putting $m = 10$ in the above polynomial, one finds that

In[25]:= Q[10]

$$\text{Out[25]} = -\frac{294987411105763646}{625}.$$

In number form or to machine precision, Mathematica prints out

In[26]:= NumberForm[-4.7198×10^{14}, 16]

Out[26]//NumberForm = $-4.719798577692218\times 10^{14}$.

On the other hand, for m equal to an odd value, say 11, one finds that $Q[11]$ gives a value of

In[27]:= Q[11]

$$\text{Out[27]} = -\frac{30011175790744604}{11}.$$

In decimal form to 16 places, the above value becomes $-2.7282887082495095\times 10^{15}$.

Since the above trigonometric power sum can be expressed as

$$V^{-}_{9,12,2}(m) = \sum_{k=1}^{2m-1} (-1)^k \cos^{24}\Big(\frac{k\pi}{2m}\Big)\csc^{18}\Big(\frac{k\pi}{2m}\Big), \tag{89}$$

it can also be expressed in terms of the standard instructions in Mathematica as

VSum[m_]:= Sum[(−1)^k Cos[k Pi/(2 m)]^(24) Csc[k Pi/(2 m)]^(18), {k, 1, 2 m − 1}].

Putting $m = 10$ in the above form for $V^{-}_{9,12,2}(m)$ yields

In[28]:= VSum[10]

$$-\frac{1}{4} + \frac{(-1-\sqrt{5})^{24}}{281474976710656(5/8-\sqrt{5}/8)^9} + \frac{(1-\sqrt{5})^{24}}{281474976710656(5/8+\sqrt{5}/8)^9} +$$
$$2\Big(\frac{5}{8}-\frac{\sqrt{5}}{8}\Big)^{12}(-1+\sqrt{5})^{18} + \frac{(-1+\sqrt{5})^{24}}{281474976710656(5/8+\sqrt{5}/8)^9} +$$
$$2\Big(\frac{5}{8}+\frac{\sqrt{5}}{8}\Big)^{12}(1+\sqrt{5})^{18} + \frac{(1+\sqrt{5})^{24}}{281474976710656(5/8-\sqrt{5}/8)^9} -$$
$$2\cos^6\Big(\frac{\pi}{20}\Big)\cot^{18}\Big(\frac{\pi}{20}\Big) - 2\cos^6\Big(\frac{3\pi}{20}\Big)\cot^{18}\Big(\frac{3\pi}{20}\Big) - 2\sin^6\Big(\frac{\pi}{20}\Big)\tan^{18}\Big(\frac{\pi}{20}\Big) -$$
$$2\sin^6\Big(\tfrac{3\pi}{20}\Big)\tan^{18}\Big(\tfrac{3\pi}{20}\Big).$$

If one asks for the number form value of the above expression to 16 decimal places, then Mathematica yields the following result:

In[29]:= NumberForm[-4.7198×10^{14}, 16]

Out[29]//NumberForm = $-4.719798577691239\times 10^{14}$.

Therefore, we obtain the same numerical value as the algorithm for $V^{-}_{9,12,2}(10)$ except for the last few decimal places, which is due to the fact that the above method has implemented far more machine precision. Moreover, if we set $m=11$ in VSum, then we obtain a similar expression in products of power of trigonometric functions as displayed above, but with no algebraic terms. In this instance, the denominators of the arguments of the trigonometric functions are either 11 or 22. In number form to 16 decimal places, we find that VSum[11] equals $-2.7282887082496685\times 10^{15}$, which equals Q[11] to 13 decimal places.

Now that we have created an algorithm for $V^{\pm}(v,j,l)(m)$, we can turn our attention to the creation of an algorithm for the general twisted trigonometric power sum defined in the introduction. This is accomplished in the following theorem.

Theorem 2. *For r, an integer lying in range of $[1,\ldots,m-1]$, the general twisted trigonometric power sum, $T^{\pm}_{v,l}(m,r)$, as defined by (9), can be expressed as a sum of various forms of the trigonometric power sum, $V^{\pm}_{v,j,l}(m)$, given by (66) as follows:*

$$T^{\pm}_{v,l}(m,r) = 2^{2r-1} V^{\pm}_{v,r,l}(m) + \sum_{j=0}^{r-1} C(r,j) V^{\pm}_{v,r-j-1,l}(m), \tag{90}$$

where

$$C(r,j) = \frac{(-1)^{j+1} r}{j+1} \binom{2r-j-2}{j} 2^{2r-2j-2}. \tag{91}$$

Proof. To prove the above theorem, we require No. I.1.10 in [24] which, as mentioned previously, was used in the derivation of (46). On this occasion, however, we replace n and x by $2r$ and $k\pi/lm$, respectively. Hence, we arrive at

$$\cos\left(\frac{2rk\pi}{lm}\right) = 2^{2r-1}\cos^{2r}\left(\frac{k\pi}{lm}\right) + r\sum_{j=0}^{r-2} \frac{(-1)^{j+1}}{j+1}\binom{2r-j-2}{j} 2^{2r-2j-2} \times \cos^{2r-2j-2}\left(\frac{k\pi}{lm}\right), \tag{92}$$

where, without loss of generality, r is an integer lying between 1 and $m-1$. For $r=0$ or the untwisted case, one needs to insert an additional term of $1/2$ into (92) as a result of No. 4.2.1.8 in [24].

If one introduces (92) into (2), then one obtains

$$T^{\pm}_{v,l}(m,r) = 2^{r-1}\sum_{k=1}^{lm-1} (\pm 1)^k \csc^{2v}\left(\frac{k\pi}{lm}\right)\cos^{2r}\left(\frac{k\pi}{lm}\right) + \sum_{j=0}^{r-1} C(r,j) \times \sum_{k=1}^{lm-1} (\pm 1)^k \csc^{2v}\left(\frac{k\pi}{lm}\right)\cos^{2r-2j-2}\left(\frac{k\pi}{lm}\right), \tag{93}$$

where $C(r,j)$ is given by (91). Now, it is readily observed that the summations over k represent different forms of (66). Therefore, by replacing the summations by the corresponding form of $V^{\pm}_{v,j,l}(m)$, we arrive at (90). □

From (90), we see that the general twisted trigonometric power sum reduces to a finite sum involving different forms of $V^{\pm}_{v,j,l}(m)$. Therefore, we can call the Mathematica module for $V^{\pm}_{v,j,l}(m)$, given earlier in this section, inside another module that handles only (92). Such a module appears below:

```
Tpm[sgn_, l_, v_, m_, r_]:= Module[{}, Vpm[sgn, l, v, r, m];
sumval = 2^(2 r − 1) totval;
Print["The value of the trigonometric power sum Vpm is ", sumval];
Do[Crj = (−1)^(j + 1) r Binomial[2 r − j − 2, j] 2^(2 r − 2 j − 2)/(j + 1);
Print["The value of the coefficient Crj is ", Crj];
Vpm[sgn, l, v, r − j − 1, m]; e5 = totval Crj;
Print["The value of e5 is ", e5]; sumval = sumval + e5, {j, 0, r − 1}];
Print["The value of Tpm[", sgn,",", l, ",", v, ",", m, ",", r"] is ", Expand[
FullSimplify[sumval]]]].
```

For this module to yield values, one needs to specify values for all the input variable except m, which can remain algebraic.

The module begins with the evaluation of the first term on the rhs of (90). This involves the $j=v$ form of $V^{\pm}_{v,j,l}(m)$ multiplied by 2^{2r-1}, and is set equal to the variable sumval, which will ultimately yield the value of the entire twisted trigonometric power sum. Next, the summation over j on the rhs of (90) is evaluated in a Do loop. For each value of j, the coefficient, $C(r,j)$ is calculated first and printed out. Then, a call to $V^{\pm}_{v,r-j-1,l}(m)$ is made. With each call a new value of totval is calculated. In each instance, the variable is multiplied by the appropriate value of $C(r,j)$, since it is not local in the module, Vpm[sgn_,l_,v_,j−,m_]. The resulting value denoted by e5 is printed out before being added to sumval. After exiting the Do loop, the final value of sumval is printed out yielding the value of $T^{\pm}_{v,l}(m,r)$.

Now, we confirm that the module does give correct values of $T^{\pm}_{v,l}(m,r)$ by considering a few examples. To facilitate this study, we shall remove all diagnostic Print statements or instructions in both modules. That is, we shall only maintain the output provided by the last Print instruction in Tpm[sgn_, l_, v_, m_, r_]. In the first example, we shall consider twisted trigonometric power sums in the Dowker case. Specifically, we shall study

$$T^{\pm}_{v,1}(m,r) = \sum_{k=1}^{m-1} (\pm 1)^k \cos\left(\frac{2kr\pi}{m}\right) \csc^{2v}\left(\frac{k\pi}{m}\right). \tag{94}$$

In terms of standard Mathematica instructions, this sum can be expressed as

TSumP[m_, v_, r_]:= Sum[Cos[2 k r Pi/m] Csc[k Pi/m]^(2 v), {k, 1, m − 1}].

The corresponding sum for the alternating case becomes TSumM[m_, v_, r_], and has an extra phase factor of $(-1)^k$ in the summand.

If we put $v = 9$ and $r = 3$, then we find that Mathematica prints out the following result for TSumP[m,9,3]:

In[30]:= TSumP[m, 9, 3]

$$\text{Out[30]} = \frac{1}{38979295480125}(12964140545731 - 11467298304000\, m^2 - 1421455464576\, m^4 - 112614327264\, m^6 + 24754314312\, m^8 + 10826441262\, m^{10} + 1620381180\, m^{12} + 39314268\, m^{14} - 12988647\, m^{16} + 87734\, m^{18}).$$

On the other hand, if we put sgn = 1, $l = 1$, $v = 9$ and $r = 3$ in the module Tpm[sgn_, l_, v_, m_, r_], then Mathematica prints out

In[31]:= Tpm[1, 1, 9, m, 3]

The value of Tpm[1,1,9,m,3] is

$$\frac{12964140545731}{38979295480125} - \frac{4096m^2}{13923} - \frac{7761536m^4}{212837625} - \frac{116217056m^6}{40226311125} + \frac{4456m^8}{7016625} + \frac{13642m^{10}}{49116375} + \frac{30404m^{12}}{731387475} + \frac{92m^{14}}{91216125} - \frac{3617m^{16}}{10854718875} + \frac{87734m^{18}}{38979295480125}.$$

Both forms for $T^{+}_{9,1}(m,3)$ are identical to each other, though they appear to be different due to the FullSimplify instruction being implemented in one of the algorithms.

An interesting situation occurs when 3 is replaced by r in the command for TSum. That is, typing

In[32]:= TSumP[m,9,r]

yields

$$\frac{1}{194896477400625}(-23133945892303 + 19459657728000m^2 + 2972311034112m^4 + 579728203680m^6 + 103972564332m^8 + 16038878310m^{10} + 2027485212m^{12} + 196571340m^{14} + 12988647m^{16} + 438670m^{18} - 116757946368000mr + 116757946368000r^2 - 89169331023360m^2r^2 - 12174292277280m^4r^2 -$$

$2079451286640m^6r^2 - 317569790538m^8r^2 - 40050901800m^{10}r^2 - 3880879884m^{12}r^2 - 256397400m^{14}r^2 - 8659098m^{16}r^2 + 178338662046720mr^3 - 89169331023360r^4 + 60871461386400m^2r^4 + 7278079503240m^4r^4 + 1058565968460m^6r^4 + 132167975940m^8r^4 + 12777137100m^{10}r^4 + 843669540m^{12}r^4 + 28488600m^{14}r^4 - 73045753663680mr^5 + 24348584554560r^6 - 14556159006480m^2r^6 - 1481992355844m^4r^6 - 176223967920m^6r^6 - 16865820972m^8r^6 - 1111055400m^{10}r^6 + 37496424m^{12}r^6 + 12476707719840mr^7 - 3119176929960r^8 + 1587848952690m^2r^8 + 132167975940m^4r^8 + 12047014980m^6r^8 + 785674890m^8r^8 + 26453700m^{10}r^8 - 1058565968460mr^9 + 211713193692r^{10} - 88111983960m^2r^{10} - 5621940324m^4r^{10} - 349188840m^6r^{10} - 11639628m^8r^{10} + 48061082160mr^{11} - 8010180360r^{12} + 2555427420m^2r^{12} + 111105540m^4r^{12} + 3527160m^6r^{12} - 1179428040mr^{13} + 168489720r^{14} - 36628200m^2r^{14} - 813960m^4r^{14} + 14651280mr^{15} - 1831410r^{16} + 203490m^2r^{16} - 71820mr^{17} + 7980r^{18})$.

Therefore, we see that Mathematica is able to produce not only the polynomials of degree $2v$ in m as the module Tpm[sgn_, l_, v_, m_, r_] does, but also polynomials of degree $2v$ in r. In fact, the polynomials are composed of terms of the form m^ir^j, where i and j are less than or equal to $2v$. This issue forms the basis of the next section where the integral approach in [21] is extended to $T^{\pm}_{v,l}(m,r)$. For now, we are concerned with how the module Vpm[sgn_,l_,v_,m_,r_] can be used to obtain values for the various twisted trigonometric power sums.

Before discussing an example of the Gardner–Fisher case, it should be mentioned that there is another representation of (90). Instead of using (92), one can use the intermediate form of No. I.1.10 in [24], which can be expressed as

$$\cos 2nx = \sum_{k=0}^{n} (-1)^{n-k} \binom{2n}{2k} \sin^{2n-2k} x \cos^{2k} x. \tag{95}$$

Introducing (95) into (2) yields

$$T^{\pm}_{v,l}(m,r) = (-1)^r \sum_{k=1}^{lm-1} (\pm 1)^k \sum_{j=0}^{r} (-1)^j \binom{2r}{2j} \sin^{2r-2j-2v}\left(\frac{k\pi}{lm}\right) \cos^{2j}\left(\frac{k\pi}{lm}\right). \tag{96}$$

By interchanging the order of the summations, we can substitute (66), thereby obtaining

$$T^{\pm}_{v,l}(m,r) = (-1)^r \sum_{j=0}^{r} (-1)^j \binom{2r}{2j} V^{\pm}_{v+j-r,j,l}(m). \tag{97}$$

This can be programmed in a slightly shorter Mathematica module as follows:

```
Tpm3[sgn_, l_, v_, m_, r_]:= Module[{}, e1 = 0;
Do[Vpm[sgn, l, v + j − r, j, m]; e2 = (−1)^(r + j) totval Binomial[2 r, 2 j];
e1 = e1 + e2, j, 0, r];
Print["The value of Tpm3[",sgn,",",l,",",v,",",m,",",r,"] is ", Expand[FullSimplify[e1]]]].
```

As expected, this module gives the same values for Tp[1,1,9,m,3] and TSum[m,9,3] presented above. Hence, we can use this form when we study the Gardner–Fisher case.

For $l = 2$, we shall consider the alternating case ($sgn = -1$) with v and r again equal to 9 and 3, respectively. If one types in Tpm[−1, 2, 9, m, 3], then Mathematica prints out

The final value of the sum is $-\frac{23133945892303}{194896477400625} - (-1)^m -$

$\frac{65536m^2}{328185} - \frac{32459264m^4}{152026875} - \frac{7418606528m^6}{40226311125} - \frac{4753864m^8}{35083125} - \frac{590132m^{10}}{7016625} -$

$\frac{8566072712m^{12}}{201131555625} - \frac{1507144m^{14}}{91216125} - \frac{33862354m^{16}}{7753370625} - \frac{22998766228m^{18}}{38979295480125}.$

In this case, we need to modify TSumM[v_, m_, r_] to

TSumM2[v_, m_, r_]:= Sum[(−1)^k Cos[k r Pi/m] Csc[k Pi/(2 m)]^(2 v), {k, 1, m − 1}].

If one types TSumM2[9,m,3] into Mathematica, then all that is generated is the original sum itself. That is, Mathematica is unable to generate the polynomials of degree $2v$ or 18 in m as above. On the other hand, if one types in TSumM2[9,11,3], then the package prints out

$$1486025307131/177147 - \mathrm{Csc}\left[\frac{\pi}{18}\right]^{22} - \mathrm{Csc}\left[\frac{\pi}{9}\right]^{22} - \mathrm{Csc}\left[\frac{2\pi}{9}\right]^{22} - \mathrm{Sec}\left[\frac{\pi}{18}\right]^{22} - \mathrm{Sec}\left[\frac{\pi}{9}\right]^{22} -$$

$$\mathrm{Sec}\left[\frac{2\pi}{9}\right]^{22}.$$

In NumberForm this gives a value of $-2.283894674697865 \times 10^{15}$ compared with $-2.283894674697867 \times 10^{15}$ when m is set equal to 11 in the above polynomial expression.

6. Integral Approach Revisited

The previous section was aimed at constructing an algorithm for twisted trigonometric power sums, $T^{\pm}_{v,l}(m,r)$, by using the general results for another important trigonometric power sum, $V^{\pm}_{v,j,l}(m)$, given in Lemma 1. Two different forms were determined, the first appearing in Theorem 2, and the second given by (97). Consequently, we obtained polynomials in m of degree $2v$, provided r was set to fixed values. These results agreed with the values obtained by programming the power sums directly in Mathematica. From the direct approach, it was found in the Gardner–Fisher case that both m and r had to be specified or set to numerical values, which meant only decimal values for the trigonometric power sum could be compared with each other, even though rational values were obtained from the new algorithm. In addition, m did not need to be specified to obtain expressions for $T^{\pm}_{v,l}(m,r)$ via the new algorithm. Surprisingly, in the Dowker case, Mathematica was not only able to provide the polynomials in m for the values of the twisted trigonometric power sums, but also combined with powers of r, when r was algebraic. This is not possible with the new algorithm since r has to be specified. Consequently, this section aims to adapt the integral approach in [21] in order to generate results possessing powers of m and r like those generated by Mathematica in the Dowker case. In doing so, we shall be able to obtain the results for the Gardner–Fisher case, where Mathematica fails. Thus, the results presented here will be more general than any of the results presented in previous sections.

Theorem 3. *For r, a positive integer less than lm, the twisted trigonometric power sum, $T^{+}_{v,l}(m,r)$, as defined by* (9) *is given by*

$$T^{+}_{v,l}(m,r) = \frac{2^{2v}}{\Gamma(2v)} \sum_{n=0}^{v-1} \frac{(-1)^n}{2n+2} s(v, v-n-1)\Big((lm)^{2n+2} B_{2n+2}(r/lm) - B_{2n+2}\Big), \tag{98}$$

where, as before, $B_{2n+2}(x)$ represents the Bernoulli polynomial of order $2n+2$.

Proof. The proof begins with (20), where we assume $m = lm$ for the time being. Multiplying this equation by $\cos(2\pi kr/m)$ yields

$$\cos\left(\frac{2\pi kr}{m}\right)\csc^{2v}\left(\frac{k\pi}{m}\right) = -\frac{2^{2v-2}m^2}{2\pi^2\Gamma(2v)}\int_0^1 \frac{\ln u}{1-u^m}\prod_{n=1}^{v-1}\left(\frac{m^2}{4\pi^2}\ln^2 u + n^2\right) \times \left(u^k + u^{m-k}\right)\left(e^{2i\pi kr/m} + e^{-2i\pi kr/m}\right)\frac{du}{u}. \tag{99}$$

Therefore, the summation over k becomes

$$T^{\pm}_{v,l}(m,r) = -\frac{2^{2v-2}m^2}{2\pi^2\Gamma(2v)}\int_0^1 \frac{\ln u}{1-u^m}\prod_{n=1}^{v-1}\left(\frac{m^2}{4\pi^2}\ln^2 u + n^2\right) \times \sum_{k=1}^{m-1}(\pm 1)^k\left(u^k + u^{m-k}\right)\left(e^{2i\pi kr/m} + e^{-2i\pi kr/m}\right)\frac{du}{u}. \tag{100}$$

For the upper-signed case, one finds, by substituting $m-k$ by k in the sums with u^{m-k} and r equal to an integer, that

$$\sum_{k=1}^{m-1} u^{m-k}e^{\pm 2i\pi kr/m} = \sum_{k=1}^{m-1}\left(ue^{\mp 2i\pi r/m}\right)^k. \tag{101}$$

In other words, we obtain the same sums involving u^k, which means that we can multiply the sums involving u^k by a factor of 2 and discard the complex conjugate terms. The upper-signed case of (100) can be expressed as

$$T^{+}_{v,l}(m,r) = -\frac{2^{2v-1}m^2}{2\pi^2\Gamma(2v)}\sum_{n=0}^{v-1}\left(\frac{m}{2\pi}\right)^{2n} s(v, v-n-1)\int_0^1 \frac{\ln^{2n+1} u}{1-u^m} \times \sum_{k=1}^{m-1}\left(u^k e^{2i\pi kr/m} + \text{c.c.}\right)\frac{du}{u}, \tag{102}$$

where c.c. denotes the complex conjugate of the preceding sum over k. The summation over k can be easily evaluated, yielding

$$\sum_{k=1}^{m-1} u^k e^{2i\pi kr/m} = \frac{u^m e^{2i\pi r/m} - u}{e^{2i\pi r/m} - u} = \frac{u^2 - ue^{-2i\pi r/m} + u^m - u^{m+1}e^{2i\pi r/m}}{1 - 2u\cos(2\pi r/m) + u^2}. \tag{103}$$

If we introduce (103) into (102) and carry out a little algebra, then we find that

$$T^{+}_{v,l}(m,r) = -\frac{2^{2v-1}m^2}{\pi^2\Gamma(2v)}\sum_{n=0}^{v-1}\left(\frac{m}{2\pi}\right)^{2n} s(v, v-n-1)\int_0^1 \frac{\ln^{2n+1} u}{1-u^m} \times \frac{\left((u^2 - 2u\cos(2\pi r/m) + 1) + (u^m - 1)(1 - 2u\cos(2\pi r/m))\right)}{1 - 2u\cos(2\pi r/m) + u^2}\frac{du}{u}. \tag{104}$$

Expressed in this form, one can cancel terms in the denominator. Consequently, (102) reduces to

$$T^{+}_{v,l}(m,r) = -\frac{2^{2v-1}m^2}{\pi^2\Gamma(2v)}\sum_{n=0}^{v-1}\left(\frac{m}{2\pi}\right)^{2n} s(v, v-n-1)\int_0^1 \ln^{2n+1} u \times \left(\frac{u^{-1}}{1-u^m} - \frac{u^{-1} - \cos(2\pi r/m)}{1 - 2u\cos(2\pi r/m) + u^2}\right)du. \tag{105}$$

The integrals in (105) are divergent at the lower limit, despite the fact that the lhs or $T^{+}_{v,j,l}(m,r)$ is finite. This means that the divergences must cancel. To observe this cancellation, we re-write (105) as follows:

$$T^{+}_{v,l}(m,r) = -\frac{2^{2v-1}m^2}{\pi^2\Gamma(2v)}\sum_{n=0}^{v-1}\left(\frac{m}{2\pi}\right)^{2n} s(v,v-n-1)\lim_{\epsilon\to 0^+}\int_0^1 \ln^{2n+1} u \times \left(\frac{u^{\epsilon-1}}{1-u^m} - \frac{u^{\epsilon-1}(1+u\cos(2\pi r/m))}{1-2u\cos(2\pi r/m)+u^2}\right)du. \tag{106}$$

According to No. 4.271(15) in [23], we have

$$\int_0^1 \frac{u^{\epsilon-1}\ln^{2n+1} u}{1-u^m}\,du = -m^{-2n-2}\psi^{(2n+1)}\left(\frac{\epsilon}{m}\right), \tag{107}$$

where $\psi^{(n)}(x)$ represents the digamma function differentiated n times. From No. 8.363(8) in [23] and separating the $k=0$ term, we find that

$$-\frac{1}{m^{2n+2}}\psi^{(2n+1)}\left(\frac{\epsilon}{m}\right) = -\frac{\Gamma(2n+2)}{m^{2n+2}}\left(\sum_{k=1}^{\infty}\frac{1}{k^{2n+2}} + \frac{m^{2n+2}}{\epsilon^{2n+2}}\right). \tag{108}$$

Thus, we observe that only one term is responsible for the divergence in the first integral.

The second integral in (105) is evaluated by using No. 4.272(3) in [23] subject to the condition that $r < m/2$. Then, we find that

$$\int_0^1 \ln^{2n+1} u\left(\frac{u^{\epsilon-1}-\cos(2\pi r/m)}{1-2u\cos(2\pi r/m)+u^2}\right)du = \frac{\Gamma(2n+2)}{\sin(2\pi r/m)}\left(\sum_{k=1}^{\infty}\frac{\sin(2\pi kr/m)}{(k+\epsilon-1)^{2n+2}} - \cos(2\pi r/m)\sum_{k=1}^{\infty}\frac{\sin(2\pi kr/m)}{k^{2n+2}}\right). \tag{109}$$

Again, only the $k=1$ term in the first summation on the rhs of (109) is divergent. As expected, this term cancels the divergent term in (108).

Now, we can take the limit as $\epsilon \to 0^+$ without any issues. Consequently, (106) becomes

$$T^{+}_{v,l}(m,r) = -\frac{2^{2v-1}m^2}{\pi^2\Gamma(2v)}\sum_{n=0}^{v-1}\left(\frac{m}{2\pi}\right)^{2n} s(v,v-n-1)\Bigg[-\frac{\Gamma(2n+2)}{m^{2n+2}}\zeta(2n+2) + \frac{\Gamma(2n+2)}{\sin(2\pi r/m)}\left(\sum_{k=1}^{\infty}\frac{\sin(2\pi(k+1)r/m)}{k^{2n+2}} - \cos(2\pi r/m)\sum_{k=1}^{\infty}\frac{\sin(2\pi kr/m)}{k^{2n+2}}\right)\Bigg]. \tag{110}$$

Expanding the sine in the numerator of the first summation over k results in the cancellation of the second summation over k in (110). So, (110) reduces to

$$T^{+}_{v,l}(m,r) = -\frac{2^{2v-1}m^2}{\pi^2\Gamma(2v)}\sum_{n=0}^{v-1}\left(\frac{m}{2\pi}\right)^{2n} s(v,v-n-1)\left[\frac{\Gamma(2n+2)}{m^{2n+2}}\zeta(2n+2) - \Gamma(2n+2)\sum_{k=1}^{\infty}\frac{\cos(2\pi kr/m)}{k^{2n+2}}\right]. \tag{111}$$

From No. 9.622(3) in [23], we can identify the remaining summation over k in (111) as being related to the Bernoulli polynomial, $B_{2n+2}(r/m)$. Moreover, as the divergences have been removed, we can analytically continue the result from $r < m/2$ to $r < m$, since the summation over k is valid for these values of r. Then, we obtain

$$T^{+}_{v,l}(m,r) = -\frac{2^{2v-1}m^2}{\pi^2\Gamma(2v)}\sum_{n=0}^{v-1}\Big(\frac{m}{2\pi}\Big)^{2n} s(v,v-n-1)\Big[\frac{\Gamma(2n+2)}{m^{2n+2}}\,\zeta(2n+2)$$
$$-\quad \frac{(-1)^n}{2}\,(2\pi)^{2n+2}B_{2n+2}(r/m)\Big]. \tag{112}$$

From No. 9.616 in [23], we can replace the first term in the parentheses by the Bernoulli number, B_{2n+2}, which results in (98) with $l=1$. For $l=2$ or the Gardner–Fisher case, all that is required is to replace m by $2m$ in (112). Thus, we end up with (98) for both cases. □

To verify (98), all we need to do is construct a code and compare it with the output from the instruction for TSumP below (94). Of course, this can only be performed in the Dowker or $l=1$ case. A simple code for (98) in Mathematica is

```
Tplus[m_, r_, v_]:= Expand[FullSimplify[(2^(2 v)/Gamma[2 v])
Sum[(−1)^n SymmetricPolynomial[v−n−1,Table[k^2, {k,1,v−1}]]
(m^(2 n + 2) BernoulliB[2n + 2,r/m] − BernoulliB[2n + 2])/(2 n + 2), {n,0,v−1}]]].
```

Note that the first argument in the SymmetricPolynomial instruction is now $v-n-1$ instead of n. This has occurred because previously when Newton's identities were applied to the product term inside (21) to yield (23), the ordering over n was reversed.

Let us consider an example by putting $m=m$, $r=r$ and $v=4$ in the above compound instruction. Then, we find that

In[33]:= Tplus[m, r, 4]

$$\text{Out[33]} = -\frac{2497}{14175}+\frac{16m^2}{105}+\frac{14m^4}{675}+\frac{8m^6}{2835}+\frac{m^8}{4725}-\frac{32mr}{35}+\frac{32r^2}{35}-\frac{28m^2r^2}{45}-$$
$$\frac{8m^4r^2}{135}-\frac{4m^6r^2}{45}+\frac{56mr^3}{45}-\frac{28r^4}{45}+\frac{8m^2r^4}{27}+\frac{2m^4r^4}{135}-\frac{16mr^5}{45}+\frac{16r^6}{135}-\frac{4m^2r^6}{135}+$$
$$\frac{8mr^7}{315}-\frac{2r^8}{315}.$$

This is the same result obtained when one types TSumP[m,4,r] into Mathematica. Nevertheless, it should be pointed out that (98) applies to the Gardner–Fisher case ($l=2$), which cannot be determined by a simple Mathematica instruction as in TSumP.

On the other hand, it should be pointed out that the results for the Dowker case are more general in Mathematica, in the sense that r is not restricted to being an integer less than m. Whilst one obtains polynomial results via (98) when m and v are specified, this does not apply to TSumP[m,v,r]. For example, typing

In[34]:= TSumP[5, r, 4]

produces the following result:

$$\text{Out[34]} = \frac{\cos\big[2\pi r/5\big]}{\big(5/8-\sqrt{5}/8\big)^4}+\frac{\cos\big[4\pi r/5\big]}{\big(5/8+\sqrt{5}/8\big)^4}+\frac{\cos\big[6\pi r/5\big]}{\big(5/8+\sqrt{5}/8\big)^4}+\frac{\cos\big[8\pi r/5\big]}{\big(5/8-\sqrt{5}/8\big)^4}.$$

The above result, of course, is simply an expression totaling the individual summands in $T^{+}_{4,1}(5,r)$, and can be expressed in reduced form as

$$T^{+}_{4,1}(5,r) = \frac{32}{25}\Big[\cos\Big(\frac{2\pi r}{5}\Big)-\cos\Big(\frac{\pi r}{5}\Big)\Big]. \tag{113}$$

An expression for the second cosine in (113) appears in Appendix I.11.5 of [24], but it is misleading. The more correct version, which appears as (3.15) in [25], is given by

$$\cos\Big(\frac{\pi r}{5}\Big) = (-1)^{\lfloor (r+2)/5\rfloor}\frac{\sqrt{5}}{4}+\frac{(-1)^{r+1}}{4}. \tag{114}$$

The first cosine is simply obtained by replacing r by $2r$, which gives (3.16) in [24]. In this case, the second term on the rhs of (114) equals $-1/4$. Introducing (114) into (112) and carrying out some algebra, one obtains

$$T^+_{4,1}(5,r) = \frac{64}{25}\Big(15p_-(r) + 7\sqrt{5}p_+(r) - 14\Big), \tag{115}$$

where

$$p_\pm(r) = (-1)^{\lfloor(2r+2)/5\rfloor)} \pm (-1)^{r+\lfloor(r+2)/5\rfloor)}. \tag{116}$$

Since both $p_-(r)$ and $p_+(r)$ are periodic or modulo 5, $T^+_{4,1}(5,r)$ is modulo 5. Moreover, $p_+(r)$ is zero for all values of r greater than zero and less than 5, where Theorem 3 applies. The modulo 5 values of r equal to zero can be handled with the results in Section 2. Consequently, the $\sqrt{5}$-term in (115) vanishes, and the values of $T^+_{4,1}(5,r)$ are rational, as expected. In fact, for r equal to 1, 2, 3 and 4, $T^+_{4,1}(5,r)$ equals $1024/25$, $-2816/25$, $-2816/25$ and $1024/25$, respectively. The above example can be extended to other values of v and m using other results in Appendix I.11 of [24].

We now turn our attention to the alternating cases of the results in Theorem 3, which are presented in the following theorem.

Theorem 4. *For r, a positive integer less than lm, the alternating twisted trigonometric power sum in the Dowker case, $T^-_{v,1}(m,r)$, is given by*

$$T^-_{v,1}(m,r) = \begin{cases} 0\ , & m, \quad \text{odd}, \\ \dfrac{2^{2v}}{\Gamma(2v)} \displaystyle\sum_{n=0}^{v-1} \frac{(-1)^n}{2n+2} s(v, v-n-1)\Big(2(m/2)^{2n+2} B_{2n+2}(2r/m) & \\ -m^{2n+2} B_{2n+2}(r/m) - B_{2n+2}\Big), & m, \quad \text{even}, \end{cases} \tag{117}$$

while the corresponding alternating twisted trigonometric power sum in the Gardner–Fisher case is given by

$$T^-_{v,2}(m,r) = \frac{2^{2v}}{\Gamma(2v)} \sum_{n=0}^{v-1} \frac{(-1)^n}{2n+2} s(v, v-n-1)\Big(2m^{2n+2} B_{2n+2}(r/m) - (2m)^{2n+2} B_{2n+2}(r/2m) - B_{2n+2}\Big). \tag{118}$$

Proof. The second result in the theorem will be proved first, since it will be used in the proof of the first result. As a consequence, we express $T^-_{v,2}(m,r)$ as

$$\sum_{k=1}^{2m-1} (-1)^k \cos\Big(\frac{k\pi r}{m}\Big) \csc^{2v}\Big(\frac{k\pi}{2m}\Big) = -\sum_{k=1}^{m-1} \cos\Big(\frac{k\pi r}{m}\Big) \csc^{2v}\Big(\frac{k\pi}{2m}\Big) + 2 \sum_{k=2,4}^{2\lfloor(m-1)/2\rfloor} \cos\Big(\frac{k\pi r}{m}\Big) \csc^{2v}\Big(\frac{k\pi}{2m}\Big). \tag{119}$$

The first sum on the rhs of (119) is the $l=2$ or Gardner–Fisher version of (98), while the second sum on the rhs of (119) can also be expressed as

$$\sum_{k=2,4}^{2\lfloor(m-1)/2\rfloor} \cos\Big(\frac{k\pi r}{m}\Big) \csc^{2v}\Big(\frac{k\pi}{2m}\Big) = \sum_{k=1}^{m-1} \cos\Big(\frac{2k\pi r}{m}\Big) \csc^{2v}\Big(\frac{k\pi}{m}\Big). \tag{120}$$

Hence, the second sum becomes the $l=1$ or Dowker version of (98). Introducing the two versions of (98) into (119) yields (118) after some algebra.

If one assumes that m is even, then we can put $m = m/2$ in $T^{-}_{v,2}(m,r)$, thereby obtaining

$$T^{-}_{v,2}(m/2,r) = \sum_{k=1}^{m-1} (-1)^k \cos\left(\frac{2\pi kr}{m}\right) \csc^{2v}\left(\frac{k\pi}{m}\right). \tag{121}$$

In other words, we arrive at $T^{-}_{v,1}(m,r)$. Therefore, substituting m by $m/2$ in (118) yields the second result in (117).

Finally, we consider the situation when m is odd. Already, we have seen that $T^{-}_{v,1}(m,0)$ vanishes as demonstrated in the proof of Theorem 1. Here, we let $m = 2N+1$, where N is a non-negative integer, and write $T^{-}_{v,1}(2N+1,r)$ as

$$T^{-}_{v,1}(2N+1,r) = -\sum_{k=1,3,\ldots}^{2N-1} \cos\left(\frac{2\pi kr}{2N+1}\right) \csc^{2v}\left(\frac{k\pi}{2N+1}\right) + \sum_{k=2,4,\ldots}^{2N} \cos\left(\frac{2\pi kr}{2N+1}\right) \times \csc^{2v}\left(\frac{k\pi}{2N+1}\right). \tag{122}$$

In the second sum we replace k by $2N+1-k$. This gives

$$\sum_{k=2,4,\ldots}^{2N} \cos\left(\frac{2\pi kr}{2N+1}\right) \csc^{2v}\left(\frac{k\pi}{2N+1}\right) = \sum_{k=1,3,\ldots}^{2N-1} \cos\left(\frac{2\pi kr}{2N+1}\right) \csc^{2v}\left(\frac{k\pi}{2N+1}\right). \tag{123}$$

The above result cancels the first sum on the rhs of (122), leaving us with zero for odd values of m in the alternating twisted trigonometric power sum for the Dowker case. □

From the results in Theorems 3 and 4, it can be seen that the solutions to twisted trigonometric power sums in the Dowker case are only valid for integer values of r from zero to $m-1$. However, it was mentioned at the beginning of Section 5 that, for r larger than m, solutions of twisted trigonometric power sums could be obtained for the Dowker case by assigning r to become the modulo of r with respect to m. Thus, there are m distinct values of this twisted trigonometric power sum. In other words, the twisted trigonometric power sum is cyclic over the integers.

In the Gardner–Fisher case, there are $2m$ distinct values for the twisted trigonometric power sum, but the upper m values of the non-alternating version become the lowest m values of the alternating version or vice versa. To see this more clearly, let $r = pm + q$, where p is an integer and $q = \mathrm{mod}(r,m)$. Introducing this value of r into (9) yields

$$T^{\pm}_{v,2}(m,pm+q) = \sum_{k=1}^{lm-1} (\pm 1)^k \cos\left(k\pi\left(p + \frac{q}{m}\right)\right) \csc^{2v}\left(\frac{k\pi}{2m}\right). \tag{124}$$

This equation can be simplified further, giving

$$T^{\pm}_{v,2}(m,pm+q) = \sum_{k=1}^{lm-1} (\pm 1)^k (-1)^{pk} \cos\left(\frac{qk\pi}{m}\right) \csc^{2v}\left(\frac{k\pi}{2m}\right). \tag{125}$$

Let us assume that $sgn = +1$, which means the summand has only a factor of $(-1)^{pk}$. Therefore, whenever p is even, $T^{+}_{v,2}(m,pm+q) = T^{+}_{v,2}(m,q)$, while if p is odd, then $T^{+}_{v,2}(m,pm+q) = T^{-}_{v,2}(m,q)$. On the other hand, if $sgn = -1$, we find that when p is even, $T^{-}_{v,2}(m,pm+q) = T^{-}_{v,2}(m,q)$, while for p, odd, $T^{-}_{v,2}(m,pm+q) = T^{+}_{v,2}(m,q)$.

Following the descriptions around previous codes throughout this work, it is now a straightforward matter to combine the results in Theorems 3 and 4 into one Mathematica module/program. First, we note that the results in both theorems except the vanishing odd m result in the alternating version of the Dowker case are very similar. The main difference lies in the Bernoulli polynomial expression accompanying the symmetric polynomial term, $s(v, v-n-1)$. As we observed with the summand factor of the previous section, the Which instruction can be used to determine the appropriate Bernoulli polynomial expression, which will be called BrnTrm. Then, the Sum instruction can be utilized to sum over the symmetric polynomials multiplied by BrnTrm for the various values of n. Finally, we use the Print instruction to print out the value of the particular twisted trigonometric power sum requested in the input. An implementation of this algorithm appears below:

```
Ttwstd[sgn_, l_, m_, v_, r_]:= Module[{}, flg = 1;
If[1 + (−1)^m == 0 || 1 + (−1)^m,flg = −1]; Which[(sgn == +1),
BrnTrm = (l m)^(2 n + 2) BernoulliB[2 n + 2, r/(l m)] − BernoulliB[2 n + 2],
(sgn == −1 && l == 1 && flg> 0), Print["The result below is not valid for odd
values of m."];
BrnTrm = 2 (m/2)^(2 n + 2) BernoulliB[2 n + 2, 2 r/m]− m^(2 n + 2)
BernoulliB[2 n + 2, r/m] − BernoulliB[2 n + 2],
(sgn == −1 && l == 1 && 1 + (−1)^m == 0 && flg< 0), BrnTrm = 0,
(sgn == −1 && l == 1 && 1 + (−1)^m ! = 0 && flg< 0), BrnTrm = 2 (m/2)^(2 n + 2)
BernoulliB[2 n + 2, 2 r/m] − m^(2 n + 2) BernoulliB[2 n + 2, r/m]
− BernoulliB[2 n + 2],
(sgn == −1 && l == 2), BrnTrm = 2 m^(2 n + 2) BernoulliB[2 n + 2,r/m]
− (2 m)^(2 n + 2) BernoulliB[2 n + 2, r/(2 m)] − BernoulliB[2 n + 2] ];
e1 = (2^(2 v)/Gamma[2 v]) Sum[((−1)^n/(2 n + 2)) BrnTrm
SymmetricPolynomial[v−n−1, Table[k^2,{k, 1, v−1}]], {n, 0, v−1}];
Print["The value of T^(", sgn, ")_{", v, ",", l, "}(", m, ",", r, ")=", Expand[e1]]].
```

Whilst the above module largely follows in a straightforward manner from the other modules presented earlier, there is one new feature that has been inserted above. This is the variable flg, which represents a flag that is initially set equal to 1. Its primary purpose is to determine whether m is algebraic or has been set equal to an integer. If m is algebraic, then flg remains positive and the module continues treating m as an algebraic quantity. Whether m is algebraic or not does not affect the Gardner–Fisher forms. However, it does affect the alternating Dowker results. If one should input $sgn = -1$, $l = 1$ and $m = m$, then flg will remain greater than zero and the module will print out a statement that the result for the twisted trigonometric power sums is not valid for odd values of m. On the other hand, if m has been given an integer value, then flg is set equal to -1 in the If statement at the beginning of the module. In this instance, the module prints out zero for the twisted trigonometric power sum if m is odd or its value for even values of m.

We now consider some examples of the above module before concluding this paper. Let us consider a relatively small value of m initially, say $m = 5$. By putting $sgn = +1$, $l = 2$, $v = 9$ and $r = 2$, Ttwstd prints out

In[35]:= Ttwstd[+1, 2, 5, 9, 2]

Out[35] = The value of T^(1)_{9,2}(5,2) = $\dfrac{117011644541}{125}$

In[36]:= N[(117011644541/125)]

Out[36] = -9.36093×10^8

In[37]:= NumberForm[-9.36093×10^8, 16]

Out[37] = $-9.36093156328 \times 10^8$.

On the other hand, if we consider the same values except set $sgn = -1$, then we obtain

In[38]:= Ttwstd[−1, 2, 5, 9, 2]

Out[38] = The value of T^(−1)_{9,2}(5,7) = $-\dfrac{117017411709}{125}$

In[39]:= N[−117017411709/125]

Out[39] = -9.36139×10^8

In[40]:= NumberForm[-9.36139×10^8, 16]

Out[40]:= $-9.36139293672 \times 10^8$.

It was mentioned above that when $r = pm + q$ in the Gardner–Fisher case, where p is odd, the values of $T^{+}_{v,2}(m, pm+q)$ equal $T^{-}_{v,2}(m,q)$. If we replace $r = 2$ by $r = 7$, in the first calculation, then we expect that calculated value will equal Ttwstd[−1,2,5,9,2]. Typing in the values of the variables, we find that :

In[41]:= Ttwstd[1, 2, 5, 9, 7]

The value of T^(1)_{9,2}(5,7) = $-\dfrac{117017411709}{125}$,

which is indeed the value of $T^{-}_{9,2}(5,2)$.

To compare the above results with a direct evaluation in Mathematica, one requires the following instruction:

TSum[sgn_, l_, m_, v_, r_]:= Sum[sgn^ k Cos[2 k r Pi/(l m)] Csc[k Pi/(l m)]^(2 v), {k, 1, l m − 1}].

Since m is relatively small in this case, we can expect to obtain a solution in terms of powers of $\sqrt{5}$ as occurred earlier when evaluating the non-alternating Dowker case directly. Typing in

In[42]:= TSum[1, 2, 5, 9, 7]

$$\text{Out[42]} = -1 + \frac{-1-\sqrt{5}}{2(5/8-\sqrt{5}/8)^9} + \frac{-1+\sqrt{5}}{2(5/8+\sqrt{5}/8)^9} + \frac{1}{2}(-1+\sqrt{5})^{18}(1+\sqrt{5}) + \frac{1}{2}(1-\sqrt{5})(1+\sqrt{5})^{18},$$

while for $sgn = -1$, one finds

In[43]:= TSum[−1, 2, 5, 9, 7]

$$\text{Out[43]} = 1 + \frac{-1-\sqrt{5}}{2(5/8-\sqrt{5}/8)^9} + \frac{-1+\sqrt{5}}{2(5/8+\sqrt{5}/8)^9} - \frac{1}{2}(-1+\sqrt{5})^{18}(1+\sqrt{5}) - \frac{1}{2}(1-\sqrt{5})(1+\sqrt{5})^{18}.$$

As expected, the terms in powers of $\sqrt{5}$ cancel and we are left with the rational values given above for the two types of Gardner–Fisher twisted trigonometric power sums.

Let us now consider the alternating Dowker case. First, we shall let m remain algebraic. For $v=9$ and $r=7$, we obtain

Ttwstd[−1, 1, m, 9, 7]

The result below is not valid for odd values of m.

The value of T^(−1)_{9,1}(m,7)=

$$\frac{2503953954305731}{38979295480125} + \frac{2048m^2}{51} - \frac{33163664m^4}{3378375} - \frac{29761296203m^6}{4469590125} + \frac{151231981m^8}{67359600} - \frac{1394041799m^{10}}{5388768000} + \frac{24715156621m^{12}}{1872351936000} - \frac{12392983m^{14}}{43110144000} + \frac{321692363m^{16}}{152437469184000} - \frac{5749691557m^{18}}{2554547108585472000}.$$

With the same input values, TSum yields

In[44]:= TSum[−1, 1, m, 9, 7]

Out[44] = (16409912634938038 6816 + 10258259761535385600 0 m^2 −

$2507659510305993523 2\,m^4 - 17009755041717485568\,m^6 + 5735325325405777920\,m^8 - 660845938604656896\,m^{10} + 33720173366174592\,m^{12} - 734362169827104\,m^{14} + 5390920619154\,m^{16} - 5749691557\,m^{18})/2554547108585472000.$

This result is identical to the corresponding result immediately above it, since subtracting the results from each other yields zero. Because both forms are in a different format, this implies the methods used to determine them are also different.

If we set $m = 5$, then we find that Ttwstd prints out

Ttwstd[−1, 1, 5, 9, 2]

The value of T^(−1)_{9,1}(5,2) = 0,

while if $m = 6$, then the module prints out

In[45]:= Ttwstd[−1, 1, 6, 9, 2]

The value of T^(−1)_{9,1}(6,2) = $-\dfrac{5160022813}{19683}$.

On the other hand, putting $m = 5$ and $m = 6$ in TSum[−1,1,m,9,2], one obtains zero and $-5160022813/19683$, respectively, confirming the results generated by Ttwstd.

7. Conclusions

This paper has been concerned with developing algorithms aimed at evaluating rational values and polynomials in powers of m and r for the general twisted trigonometric power sum, $T^{\pm}_{v,l}(m,r)$, as defined by (9). In the process, we have encountered other interesting trigonometric power sums for which algorithms were also developed. It is hoped that the material in this work in conjunction with the results in [21] can be used to study more complicated trigonometric power sums in the future. Although the descriptions of the codes or algorithms have been converted into Mathematica modules here, it should be possible to create alternative programs using other software packages such as Maple and MatLab.

Funding: This research has received no external funding.

Data Availability Statement: There are no external databases where the results in this work can be analyzed.

Acknowledgments: The author thanks Carlos M. da Fonseca and M. Lawrence Glasser for their encouragement of this work and also for their earlier contributions on trigonometric power sums without which this work would not have been possible.

Conflicts of Interest: The author declares no conflict of interest.

References

1. Quoniam, J.M. Problem E1937. *Am. Math. Mon.* **1966**, *73*, 1122.
2. Greening, M.G.; Carlitz, L.; Gootkind, D.; Hoffman, S.; Reich, S.; Scheinok, P.; Quoniam, J.M. A trigonometric summation—Solution to Problem E1937. *Am. Math. Mon.* **1968**, *75*, 405–406.
3. Gardner, L.A., Jr. Sum of inverse powers of cosines. *Siam Rev.* **1969**, *11*, 1621. [CrossRef]
4. Fisher, M.E. Sum of inverse powers of cosines (L.A. Gardner, Jr.). *Siam Rev.* **1971**, *13*, 116–119. [CrossRef]
5. Klamkin, M.S. (Ed.) *Problems in Applied Mathematics—Selections from SIAM Review*; SIAM: Philadelphia, PA, USA, 1990; pp. 157–160.
6. Dowker, J.S. Casimir effect around a cone. *Phys. Rev. D* **1987**, *36*, 3095–3101. [CrossRef] [PubMed]
7. Dowker, J.S. Heat kernel expansion on a generalized cone. *J. Math. Phys.* **1989**, *30*, 770–773. [CrossRef]
8. Dowker, J.S. On Verlinde's formula for the dimensions of vector bundles on moduli spaces. *J. Phys. A Math. Gen.* **1992**, *25*, 2641–2648. [CrossRef]
9. Verlinde, E. Fusion rules and modular transformations in 2D conformal field theory. *Nucl. Phys. B* **1988**, *300*, 360–376. [CrossRef]
10. Zagier, D. Elementary aspects of the Verlinde formula and of the Harder-Narasimhan-Bott formula. *Israel Math. Conf. Proc.* **1996**, *9*, 445–462.
11. Berndt, B.C.; Yeap, B.P. Explicit evaluations and reciprocity theorems for finite trigonometric sums. *Adv. Appl. Math.* **2002**, *29*, 358–385. [CrossRef]

12. Cvijovic, D.; Srivastava, H.M. Closed-form summation of Dowker and related sums. *J. Math. Phys.* **2007**, *48*, 043507. [CrossRef]
13. Cvijovic, D.; Srivastava, H.M. Closed-form summations of Dowker's and related trigonometric sums. *J. Phys. A Math. Theor.* **2012**, *45*, 1–10. [CrossRef]
14. Chu, W.; Marini, A. Partial fractions and trigonometric identities. *Adv. Appl. Math.* **1999**, *23*, 115–175. [CrossRef]
15. Chu, W. Summations on trigonometric functions. *Appl. Math. Comput.* **2003**, *141*, 161–176. [CrossRef]
16. Wang, X.; Zheng, D.Y. Summation formulae on trigonometric functions. *J. Math. Anal. Appl.* **2007**, *335*, 1020–1037. [CrossRef]
17. Wolfram, S. *Mathematica-A System for Doing Mathematics by Computer*; Addison-Wesley: Reading, PA, USA, 1992.
18. Dowker, J.S. On sums of powers of cosecs. *arXiv* **2015**, arXiv:1507.01848.
19. He, S.; Numasawa, T.; Takayanagi, T.; Watanabe, K. Notes on entanglement entropy in string theory. *J. High Energy Phys.* **2015**, *106*, 1–17. [CrossRef]
20. da Fonseca, C.M.; Glasser, M.L.; Kowalenko, V. Basic trigonometric power sums with applications. *Ramanujan J.* **2017**, *42*, 401–428. [CrossRef]
21. da Fonseca, C.M.; Glasser, M.L.; Kowalenko, V. Generalized Cosecant Numbers and Trigonometric Inverse Power Series. *Appl. Anal. Discrete Math.* **2018**, *12*, 70–109.
22. Kowalenko, V. Alternative Formulations of the Generalized Cosecant and Secant Numbers. *Integers* **2021**, *21*, A1.
23. Gradshteyn, I.S.; Ryzhik, I.M.; Jeffrey, A. (Eds.) *Table of Integrals, Series, and Products*, 5th ed.; Academic Press, Inc.: Boston, DC, USA, 1994.
24. Prudnikov, A.P.; Brychkov, Y.A.; Marichev, O.I. *Integrals and Series: Vol. I Elementary Functions*; Gordon & Breach: New York, NY, USA, 2003.
25. Du, Z.; da Fonseca, C.M.; Kowalenko, V. Further Developments of Basic Trigonometric Power Sums. *Rev. Real Acad. Cienc. Exactas Fis. Nat. Ser. A-Mat* **2023**, *117*, 112. [CrossRef]
26. Hansen, E.R. *A Table of Series and Products*; Prentice-Hall, Inc.: Englewood Hills, OH, USA, 1975.

algorithms

MDPI

Article

Algorithm for Option Number Selection in Stochastic Paired Comparison Models

László Gyarmati †, Csaba Mihálykó *,† and Éva Orbán-Mihálykó †

Department of Mathematics, University of Pannonia, Egyetem u. 10, 8200 Veszprém, Hungary; gyarmati.laszlo@phd.mik.uni-pannon.hu (L.G.); orban.eva@mik.uni-pannon.hu (É.O.-M.)
* Correspondence: mihalyko.csaba@mik.uni-pannon.hu
† These authors contributed equally to this work.

Abstract: In this paper, paired comparison models with a stochastic background are investigated and compared from the perspective of the option numbers allowed. As two-option and three-option models are the ones most frequently used, we mainly focus on the relationships between two-option and four-option models and three-option and five-option models, and then we turn to the general s- and $(s+2)$-option models. We compare them from both theoretical and practical perspectives; the latter are based on computer simulations. We examine, when it is possible, mandatory, or advisable how to convert four-, five-, and $(s+2)$-option models into two-, three-, and s-option models, respectively. The problem also exists in reverse: when is it advisable to use four-, five-, and $(s+2)$-option models instead of two-, three-, and s-option models? As a result of these investigations, we set up an algorithm to perform the decision process.

Keywords: paired comparisons; Bradley–Terry model; maximum likelihood estimation; number of options; algorithm for model choice

1. Introduction

Comparisons in pairs are frequently used in a broad spectrum of problems, across all areas of life. Without claiming to be exhaustive, we mention education [1], management [2], psychology [3], marketing [4], construction [5], social science [6], group decisions [7], engineering [8], medicine [9], economics [10], resource allocation [11], and sports [12] as areas of their application. A paper by Vaidya and Kumar [13] presents more than 150 of the applications that they had been used for before 2003.

Paired comparisons play a significant role in decision-making [14,15]. This method allows for a detailed analysis of the differences and similarities between objects, properties, and aspects, aiding in the selection of the best choice. It is particularly useful in situations where multiple factors must be considered and a clear ranking is necessary [13,16]. By applying paired comparisons, complex problems can be simplified, leading to more objective decisions.

Paired comparisons means that, instead of using a scale, the objects to be evaluated are compared directly to each other, and the decision of the observer (evaluator) is based on the relative merits of the two objects. These opinions are generally more reliable than evaluations based on scales, as it is easier to determine which of two objects, properties, aspects, etc., is 'better', 'more appropriate', 'more important', etc., than to describe them with a single number. This is especially true for evaluations based on subjective criteria. In the case of paired comparisons, their data represent the relations between the observed objects. These types of data require methods that are different from conventional statistical procedures.

The most popular method based on comparisons in pairs is the Analytic Hierarchy Process (AHP) associated with Saaty [17]. This single paper has more than 10 thousand citations; some of these articles provide criticism, see for example [18]. The starting point of the model is a pairwise comparison matrix (PCM), the elements of which show how many times one object is stronger than another. The matrix $A = (a_{i,j})$ is reciprocal–symmetric, that is $a_{i,j} = \frac{1}{a_{i,j}}$. According to Saaty's original model [17], $a_{i,j} \in \{1/9, 1/8, \ldots, 1/2, 1, 2, \ldots, 8, 9\}$. The observers usually express their decisions verbally by voting for 'equally preferred', 'moderately preferred', 'strongly preferred', 'very strongly preferred', or 'extremely preferred'. These verbal expressions help decision-makers articulate their preferences in a qualitative manner, which are converted to quantitative values; 1, 3, 5, 7, or 9, respectively. Intermediate values are used for comparisons that fall between these levels. We mention that the number of options may be reduced in light of the problem under consideration. Their quantitative results are analyzed using different methods. Usually, every object is compared to every other object; that is, the PCM does not contain missing elements. In this case, the number of pairs under comparison increases quadratically with the number of objects examined. If the number of objects is 10, then the number of pairs equals $\frac{10 \cdot 9}{2} = 45$, which can be quite overwhelming for an observer. The reliability of their opinions and the measure of the contradictions between their decisions are checked by the inconsistency index of the PCM. The definition of consistency is given in [17] for complete cases; it was recently generalized to incomplete cases in [19]. There are numerous ways to quantify the degree of inconsistency seen [20,21]. Some research papers present procedures for reducing the level of inconsistency [22–25]. Most measurements are for complete cases, but they can also be generalized to incomplete cases [26]. The most frequently used measurement is related to the eigenvector method [17,24], which is the most common evaluation method, but the logarithmic least squares method (LLSM) is also readily used for evaluations, although mainly in incomplete cases [19,27,28].

Not only can the evaluation method change, but so can the way the PCM is constructed. Maintaining its reciprocal–symmetric property, the matrix can also be formed in another way than how Saaty constructed it. In [29], the authors define the elements of the PCM as the ratio of the number of wins to losses. In a socio-economic study [30], the author used the ratio of certain economic characteristics as the elements of the matrix. In this case, the number of possible options that require a decision is infinitely many. Both the eigenvector method (with some modification) and LLSM work for these PCM cases under the same conditions: the PCM matrix has to be irreducible, i.e., the representing graph must be connected [27,28].

We note that other methods, such as the Preference Ranking Organization Method for Enrichments Evaluations (PROMETHEE) and Elimination and Choice Translating Reality (ELECTREE, ELECTRE) are often applied, mainly when focusing on multi-criteria decision making [31–33].

Another branch of the models connected to comparisons in pairs are the Thurstone motivated models, which have stochastic background. Based on Thurstone's brilliant idea, these models envision the latent probabilistic variables behind each object [34]. The starting points were psychological problems but, based on Thurstone motivated models, it is also possible to solve the ranking and rating of objects related to other research questions [35–38]. Thurstone assumed the latent variables had a Gaussian distribution, while Bradley and Terry assumed a logistic distribution [39]. In [34,39], two options were allowed for the decisions of the observers, but the models were generalized to three options, allowing ties [40,41]. Models allowing more than three options are presented in [42]. The ranking and rating of these objects are determined by the expectation of latent random variables. There are numerous possibilities for estimating their parameters. In the case of a Gaussian distribution, least squares methods are usually applied, but in cases of logistic distribution, maximum likelihood estimations are widespread. One publication [42] also applied least squares estimations based on relative frequencies and a cumulative distribution function. In [43], a general model for applying a Gaussian distribution and allowing more than two options is presented. The parameter estimation method is the maximum likelihood method, as it contains many inherent possibilities.

The maximum likelihood estimation (MLE) method is based on the maximization of a multi-variate function, and it requires the clarification of the conditions under which the maximum exists and is unique. In [43], the authors provide a sufficient condition for its existence and uniqueness in the case of a Gaussian distribution, and in [44] they generalize this for the case of general strictly log-concave distributions. Previously known conditions are used in [45], assuming a logistic distribution in the case of the two-option model, and in [46] they are used to investigate a modified model in the case of three choices. This condition, in the case of two options, is a necessary and sufficient condition. The condition given by Davidson is a sufficient condition but not necessary. It has been significantly generalized in [47] for the case of two and three options within choices, and the new system of conditions is a generalization of the set of conditions in [44]. The general conditions are met much more frequently [47], making them more useful for both theoretical research and practical applications. This set of conditions and its generalization for more than three options are used in this paper during comparisons of option numbers from a theoretical perspective.

During this research, the applied number of options is usually two or four, but there are cases when further numbers of options are allowed: four options are used in [43] for evaluating women tennis players, five options are used in [48], six options are used in [38], and seven options are used in [49]. In these studies, the numbers of options were fixed before data collection, based on preliminary empirical observations. If we fix the number of options, then the data will be available to us in a corresponding structure. It seems natural to use a fixed number of options during an evaluation. But data can also be transformed. In the past, ties were transformed into half a win and half a loss. Which option number is the best choice for an evaluation? Is this method better than an evaluation using a three-option model? Is it possible that by using a different number of options and transforming the data, we could obtain a model that fits the data better? In which cases is this possible? What can be said about the choice of the number of options? After studying the literature, we did not find an answer to this question. We intend this paper to fill this gap. As a result, we develop an algorithm by which the appropriate number of options can be selected. This algorithm relies on both theoretical and experimental aspects. Theoretical aspects concern evaluability considerations, while experimental aspects concern more accurate estimations and are supported by a large number of computer simulations.

This paper is structured as follows: In Section 2 we introduce a general model allowing for s options with an MLE parameter estimation. Then, we turn to the theoretical aspects of evaluability. First we investigate the relationship between two- and four-option models, and then the general case of s and $s+2$ follows. The preliminary results for the case $s=3$ and $s+2=5$ are contained in the conference paper [50]. We can see that if data are evaluable in the s-option model and the data's transformation satisfies a natural condition, then the data can also be evaluated in the $(s+2)$-option model. After these theoretical considerations, in Section 4 we present the results of a large number of computer simulations. The runs investigate how often the s-option model ($s=2,3$) and the $(s+2)$-option model ($s+2=4,5$) are evaluable. If both are evaluable, we investigate which one fits better with the data, and that is the one worth using for evaluations. We define a measurement for fitting which helps us decide on the choice of the number of options. In the subsequent section, we elaborate on and present an algorithm for performing the decision process, and the paper then concludes with our final remarks.

2. The Investigated Models

Let the objects to be rated and ranked be denoted by the numbers $1,2,\ldots,n$. The models with a stochastic background assume that the current performances of the objects are random variables denoted by ξ_i, $i=1,2,\ldots,n$. Let the expectations be $E(\xi_i)=m_i$, which expresses the expected strengths of the object i. Decisions comparing two objects i and j are related to the difference between these latent random variables, i.e., $\xi_i-\xi_j$. We can separate the expectations as follows:

$$\xi_i-\xi_j=m_i-m_j+\eta_{i,j} \tag{1}$$

where $\eta_{i,j}$ are independent identically distributed random variables with the cumulative distribution function F. F is supposed to be three times continuously differentiable, $0<F<1$, its probability density function is symmetric to zero, i.e., $f(-x)=f(x)$, $x\in\mathbb{R}$, and its logarithm is strictly concave. If $\eta_{i,j}$ are normally distributed, then we use the nomination of the Thurstone model, if their distribution is logistic, we use the Bradley–Terry model. These models were originally defined for two-option choices, 'worse' or 'better', and then were generalized for three options: 'worse', 'better', or 'equal'. In the case of the two-option model, the differences are compared to zero: the decision 'better'/'worse' indicates whether the difference is positive/negative, respectively. In the three-option model, the value of the difference is compared to the parameter $0 < d$; if it is in $[-d, d]$, we can consider it a tie. More than three, (s), options can be allowed too, as follows: The options within choices are denoted by C_k, $k = 1, 2, \ldots, s$ ($s = 5$: 'much worse', 'slightly worse', 'equal', 'slightly better', 'much better'). The real line $\mathbb{R}$ is divided into disjoint sub-intervals $I_1, I_2, \ldots, I_s$. $I_1 = (-\infty, d_1)$, $I_2 = [d_1, d_2), \ldots, I_s = [d_{s-1}, \infty)$. Decisions have a kind of symmetry, i.e., if i is 'better'/'much better'/...than j, then j is 'worse'/'much worse'/...than i. This means that $d_{s-1} = -d_1$ and, in general, $d_{s-i} = -d_i$, $i = 1, 2, \ldots, [s/2]$. The figure belonging to the model allowing s options can be seen in Figure 1.

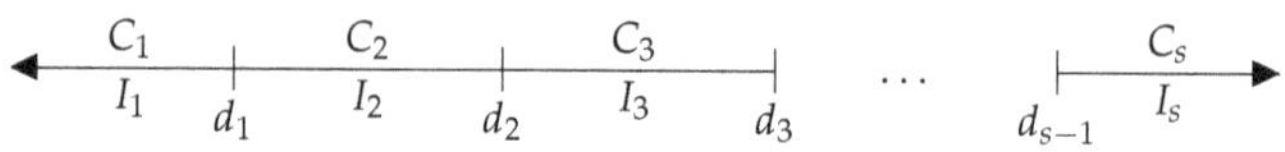

Figure 1. The options and the intervals belonging to them in an s-option model.

If $s = 2$, then no parameter is needed to assign the endpoints of the intervals. If $s = 3, 4$, then we need one positive parameter; we will denote it using $0 < d = d_{s-1}$. If $s = 5, 6$, we need two parameters, denoted by $0 < d = d_{s-2} < D = d_{s-1}$, to assign the intervals, etc. These parameters will be estimated on the basis of the comparison results, which are called data.

The data are included into a three-dimensional $(nxnxs)$ data matrix $A^{(s)}$. Its elements $A^{(s)}_{i,j,k}$ are the number of comparisons in which decision C_k is the result when we compare object i and j. Of course, $A^{(s)}_{i,i,k} = 0$, $i = 1, \ldots, n$, $k = 1, \ldots, s$ and $A^{(s)}_{i,j,k} = A^{(s)}_{j,i,s+1-k}$, $i = 1, \ldots, n$, $j = 1, \ldots, n$, $k = 1, \ldots, s$. The upper index $^{(s)}$ indicates the number of options.

The probability that the difference between the random variables belonging to the objects i and j is in the interval I_k can be expressed as follows, in the case of $2 < s$:

$$p_{i,j,1} = P(\xi_i - \xi_j \in I_1) = F(d_1 - (m_i - m_j)) \tag{2}$$

$$p_{i,j,k} = P(\xi_i - \xi_j \in I_k) = F(d_k - (m_i - m_j)) - F(d_{k-1} - (m_i - m_j)), k = 2, \ldots, s-1 \tag{3}$$

and

$$p_{i,j,s} = P(\xi_i - \xi_j \in I_s) = 1 - F(d_{s-1} - (m_i - m_j)). \tag{4}$$

If $s = 2$, $d_1 = 0$. In this case, instead of (2) and (4), the appropriate probabilities can be expressed by

$$p_{i,j,1} = P(\xi_i - \xi_j \in I_1) = F(0 - (m_i - m_j)) \tag{5}$$

and

$$p_{i,j,2} = P(\xi_i - \xi_j \in I_2) = 1 - F(0 - (m_i - m_j)). \tag{6}$$

Assuming independent decisions, the probability of the data, i.e., the likelihood function, is

$$L(A^{(s)} | m_1, \ldots, m_n, d_1, \ldots, d_{[(s-1)/2]}) = \prod_{k=1}^{s} \prod_{i=1}^{n-1} \prod_{j=i+1}^{n} p_{i,j,k}^{A^{(s)}_{i,j,k}}. \tag{7}$$

We estimate the parameters by maximizing the likelihood function (7), or, equivalently, its logarithm:

$$logL(A^{(s)}|m_1, m_2, \ldots, m_n, d_1, \ldots, d_{[(s-1)/2]}) = \sum_{k=1}^{s} \sum_{i=1}^{n-1} \sum_{j=i+1}^{n} A_{i,j,k}^{(s)} \cdot log(p_{i,j,k}). \tag{8}$$

The maximum likelihood estimation of the parameters $\underline{m} = (m_1, \ldots, m_n)$ and $\underline{dv} = (-d_1, \ldots, -d_{[(s-1)/2]})$ is the argument for the maximal value of (8); that is

$$(\widehat{\underline{m}}^{(s)}, \widehat{\underline{dv}}^{(s)}) = \underset{\underline{m} \in \mathbb{R}^n,\, 0 < -d_{[\frac{s-1}{2}]} < \ldots < -d_1}{\arg\max} L(A^{(s)}|\underline{m}, \underline{dv}). \tag{9}$$

We note that the probabilities (2), (3), (4), (5), and (6) depend only on the differences between their expectations; therefore, one coordinate of the parameter vector $\underline{m}$, for example m_1, can be fixed at zero or the constraint $\sum_{i=1}^{n} m_i = 0$ can be assumed. Naturally, the maximum value is not necessarily always attained, but some conditions for the data can guarantee the existence and uniqueness of a maximizer. These conditions are different in the case of different options. Hereinafter, data will be referred to as evaluable in an s-option model if the maximum likelihood estimate of the parameters (9) exists and is unique.

3. Comparison of Models Allowing Different Options in Choices from a Theoretical Aspect

3.1. The Case of $s = 2$ and $s + 2 = 4$

The early models of paired comparisons allowed for only two options: 'worse' or 'better' [34,39]. Thurstone himself argued that, based on experience, people are able to make decisions between these two options and there is not necessarily a need for more. In some sports, such as tennis or basketball, matches can only end in a win or a loss. However, knowing the result, we can sometimes add nuance: for example, the victory was decisive, or one side barely defeated the other. In other words, there may be justification for additional decision options, such as a narrow victory ('slightly better'/'slightly defeated') or a decisive victory ('much better'/'much worse'). These models, which allow for and process more information, could yield more nuanced results.

The data's structure is contained in the data matrices but the data matrices $A^{(4)}$ and $A^{(2)}$ can be transformed into each other. We follow the following natural transformation rules:

$$A_{i,j,1}^{(2)} = A_{i,j,1}^{(4)} + A_{i,j,2}^{(4)} \tag{10}$$

$$A_{i,j,2}^{(2)} = A_{i,j,3}^{(4)} + A_{i,j,4}^{(4)} \tag{11}$$

This means that if we want to transform the four-option model into a two-option model, we merge the options 'slightly better' and 'much better' into the 'better' option. In the opposite case, we split the option 'better' into two parts, and the numbers of decisions will be divided into two parts. We require that $0 < A_{i_1,j_1,3}^{(4)}$ and $0 < A_{i_2,j_2,4}^{(4)}$ for at least one pair of the indices (i_1, j_1) and (i_2, j_2), respectively. This requirement ensures that the data $A_{i,j,k}^{(4)}$ actually belong to the four-option model.

Now, we compare these models based on their evaluability. First, we track which theoretical cases are possible:

(A) Data are evaluable only in the two-option model.
(B) Data are evaluable only in the four-option model.
(C) Data are evaluable in both models.
(D) Data are not evaluable in either model.

(Case A) We can prove that this theoretical case may not happen. For this, we summarize the conditions for evaluability in the case of the two-option and the four-option models. For this, we need two graph definitions.

Definition 1 (The graph belonging to $A^{(2)}$). *The nodes are the objects. There is a directed edge from i to j (denoted by $\rightarrow$) if $0 < A^{(2)}_{i,j,2}$, i.e., there is a decision according to which object i is 'better' than object j. This graph will be denoted by $GC^{(A^{(2)})}$.*

Definition 2 (The graph belonging to $A^{(4)}$). *The nodes are the objects. Two types of directed edges are contained in the graph. There is a directed edge from i to j (denoted by $\rightarrow$ and called the 'slightly better' edge) if $0 < A^{(4)}_{i,j,3}$, i.e., there is a decision according to which object i is 'slightly better' than object j. The other type of directed edge is defined as follows: there is a double directed edge from i to j, denoted by $\twoheadrightarrow$ and called the 'much better edge', if $0 < A^{(4)}_{i,j,4}$. In other words, there exists a decision according to which object i is 'much better' than object j. This graph will be denoted by $GC^{(A^{(4)})}$.*

In the case of the two-option model, the necessary and sufficient condition for evaluability, given by Ford, for logistic distributions and generalized for strictly log-concave distributions in [47] is the following:

- C.2.1. For every nonempty partition of S and $\overline{S}$ of the set {1, 2, ..., n} ($S \cup \overline{S} = \{1,2,\ldots,n\}$, $S \cap \overline{S} = \emptyset$), there are two (not necessarily different) index pairs (i_1, j_1) and (i_2, j_2), $i_1, i_2 \in S$, $j_1, j_2 \in \overline{S}$, for which $0 < A^{(2)}_{i_1,j_1,1}$ and $0 < A^{(2)}_{i_2,j_2,2}$. In other words, these subsets are connected in both directions.

These types of graphs are known as strongly connected graphs in graph theory. This property is equivalent to the existence of a directed path between any two nodes.

In the case of the four-option model, the following set of conditions has been proven sufficient for evaluability:

- C.4.1. Both types of edges are included in the graph $GC^{(A^{(4)})}$.
- C.4.2. For every nonempty partition of S and $\overline{S}$ ($S \cup \overline{S} = \{1,2,\ldots,n\}$, $S \cap \overline{S} = \emptyset$), there are two (not necessarily different) index pairs (i_1, j_1) and (i_2, j_2), $i_1, i_2 \in S$, $j_1, j_2 \in \overline{S}$, for which
$$0 < A^{(4)}_{i_1,j_1,4} \text{ and } 0 < A^{(4)}_{i_2,j_2,1} \tag{12}$$
are satisfied or there exists such an index pair (i_3, j_3) for which
$$0 < A^{(4)}_{i_3,j_3,2} \text{ or } 0 < A^{(4)}_{i_3,j_3,3} \tag{13}$$
is satisfied.
- C.4.3. There exists a cycle in the graph $GC^{(A^{(4)})}$ along the 'slightly better' and 'much better' edges which contains at least one 'much better' edge. In other words, there exists a cycle $(i_1, i_2, \ldots, i_k, i_1)$, $i_l, l = 1, 2, \ldots, k$ for some $k = 2, \ldots, n$, where $i_1, i_2, \ldots, i_k$ are distinct, for which $0 < A^{(4)}_{i_l,i_{l+1},3}$ or $0 < A^{(4)}_{i_l,i_{l+1},4}$, and there exists at least one index pair in the cycle for which $0 < A^{(4)}_{i_l,i_{l+1},4}$. We note that this cycle might contain only two different nodes.

Using the above set of conditions for evaluability, C.2.1., as well as C.4.1., C.4.2., and C.4.3., we can prove the following theorem:

Theorem 1. *Suppose that there exists at least one index pair (i_1, j_1) for which $0 < A^{(4)}_{i_1,j_1,3}$ is satisfied and there exists an index pair (i_2, j_2) for which $0 < A^{(4)}_{i_2,j_2,4}$ holds. Redefine the data in the two-option model using* (10) *and* (11)*. If condition C.2.1.is satisfied, then $A^{(4)}$ is evaluable in the four-option model.*

Proof. We check the conditions C.4.1., C.4.2., and C.4.3.

C.4.1. is already required in the assumptions.

C.4.2. Consider an arbitrary partition $S, \overline{S}$. As there exists an index pair (i_1, j_1) $i_1 \in S$ $j_1 \in \overline{S}$ for which $0 < A^{(2)}_{i_1,j_1,2}$, there exists an index pair for which $0 < A^{(4)}_{i_1,j_1,3}$ or $0 < A^{(4)}_{i_1,j_1,4}$. Similarly, as there exists an index pair (i_2, j_2) $i_2 \in S$ $j_2 \in \overline{S}$ for which $0 < A^{(2)}_{i_2,j_2,1}$, there exists an index pair for which $0 < A^{(4)}_{i_2,j_2,2}$ or $0 < A^{(4)}_{i_2,j_2,1}$. These mean that (12) or (13) is satisfied.

C.4.3. Let us choose a node from which a 'much better' edge starts (i_1). It goes to the object i_2. C.2.1. implies that there is a path along the directed edges from i_2 to i_1 in $GC^{(A^{(2)})}$. We denote it by saying that $(i_2, i_3, \ldots, i_1)$ and i_j are different. These edges are also included as 'slightly better' or 'much better' edges in $GC^{(A^{(4)})}$. Therefore, the cycle $(i_1, i_2, \ldots, i_1)$ forms a cycle along the 'slightly better' and 'much better' edges in $GC^{(A^{(4)})}$ and it contains at least one 'much better' edge.

We proved that all three conditions, C.4.1., C.4.2., and C.4.3., hold; consequently, data matrix $A^{(4)}$ is evaluable. □

(Case B) This case may occur. A simple example can be seen in Figure 2. The justification is as follows:

Data are evaluable in the four-option model as

C.4.1. is obviously satisfied; $A^{(4)}_{1,2,4} = 1$ and $A^{(4)}_{2,3,2} = 1$.

C.4.2. In the case of every partition, the subsets are connected either by 'much better' edges, both forward and backward, or a 'slightly better' edge. More exactly, if $S = \{1\}$, $\overline{S} = \{2,3\}$, or $S = \{1,3\}$, $\overline{S} = \{2\}$, then $A^{(4)}_{1,2,4} = 1$, $A^{(4)}_{2,1,4} = A^{(4)}_{1,2,1} = 1$. If $S = \{1,2\}$ and $\overline{S} = \{3\}$, then $A^{(4)}_{3,2,3} = A^{(4)}_{2,3,1} = 1$. When S and $\overline{S}$ are changed, the situation remains the same.

C.4.3. The required cycle is (1,2,1).

Data cannot be evaluated using the two-option model as C.2.1. does not hold. See $S = \{1,2\}$ and $\overline{S} = \{3\}$; moreover, take into the consideration that $A^{(2)}_{1,2,1} = A^{(2)}_{1,2,2} = 2$, $A^{(2)}_{2,3,1} = A^{(2)}_{3,2,2} = 1$, $A^{(2)}_{2,3,2} = A^{(2)}_{3,2,1} = 0$, $A^{(2)}_{1,3,1} = A^{(2)}_{3,1,2} = A^{(2)}_{1,3,2} = A^{(2)}_{3,1,1} = 0$.

In this case, the situation is clear; there is no choice.

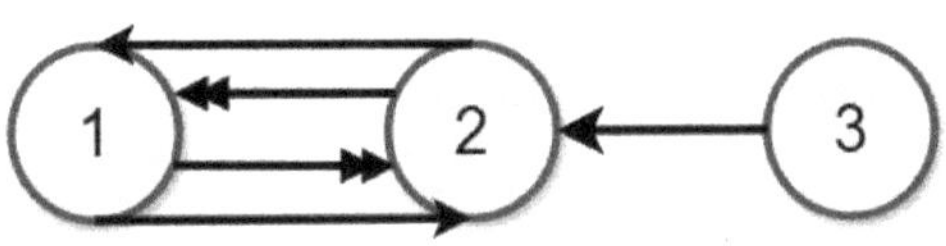

Figure 2. Comparison results which can be evaluated using the four-option model but not in the two-option model (the objects to be compared are denoted by 1, 2, and 3).

(Case C) The data given by $A^{(4)}$ and its transformed version $A^{(2)}$ can be evaluated in the four- and two-option model, respectively; see Figure 3. This is a very common case. This simple example is a modification of the data presented in Figure 2. The comparison results in Figure 2 are supplemented by a 'slightly better' edge from object 2 to object 3; see Figure 3. The data are obviously evaluable in the four-option model. Moreover, C.2.1. is also satisfied; hence, data can be evaluated in the two-option model too.

In this case, we can decide which model will be used for the data evaluation. In Section 4, we investigate these cases' practical aspects using computer simulations.

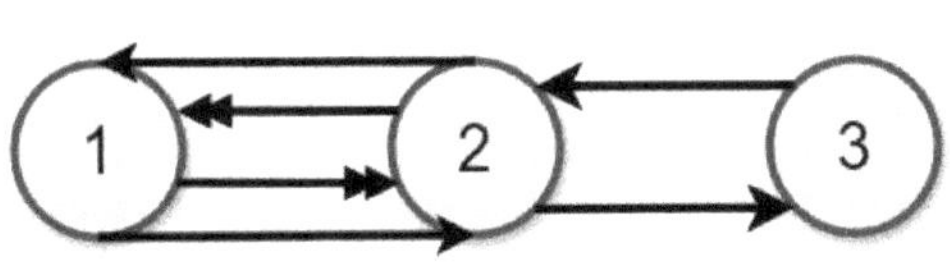

Figure 3. Comparison results which can be evaluated in both models (objects to be compared are denoted by 1, 2, and 3).

(Case D) Data cannot be evaluated in either the two-option or the four-option model.

Let $A^{(4)}_{1,2,1} = A^{(4)}_{1,2,2} = A^{(4)}_{1,2,3} = A^{(4)}_{1,2,4} = 1$, $A^{(4)}_{2,3,1} = A^{(4)}_{2,3,2} = A^{(4)}_{2,3,3} = 0$, $A^{(4)}_{2,3,4} = 1$, $A^{(4)}_{1,3,k} = 0$, and $k = 1,2,3,4$. The graph of the results belonging to the data matrix $A^{(4)}$ can be seen in Figure 4.

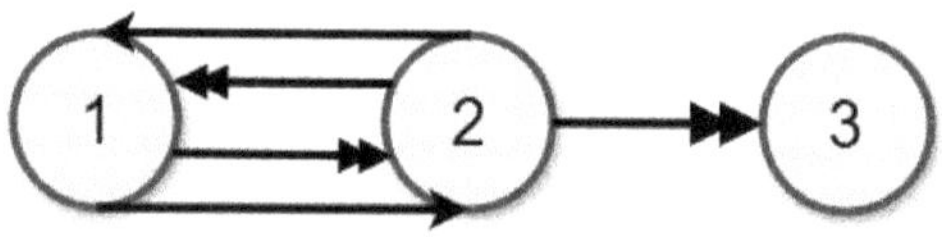

Figure 4. Comparison results which can be evaluated in neither model (objects to be compared are denoted by 1, 2, and 3).

The reason why the data are not evaluable is clear: object 3 is 'worse' than objects 2 and 1, but we do not know by how much they differ. Formally, conditions C.2.1. and C.4.2. do not hold: they take the partition $S = \{1,2\}$ and $\overline{S} = \{3\}$ in both cases.

3.2. Comparison of Models Allowing Different Options within Choices: The General Case

The comparison of three-option and five-option cases is included in the conference paper [50]; therefore, we do not provide details here. Instead, in this section, we turn to the general case, which includes the analysis in Section 3.1 and the analysis of the cases $s = 3$ and $s + 2 = 5$.

Data will be denoted by $A^{(s)}_{i,j,k}$, $(k = 1,2,\ldots,s)$ and $A^{(s+2)}_{i,j,k}$, $(k = 1,2,\ldots,s+2)$ in the s and (s + 2)-option models, respectively. Data conversion is performed using the following formulas:

$$A^{(s)}_{i,j,1} = A^{(s+2)}_{i,j,1} + A^{(s+2)}_{i,j,2} \tag{14}$$

$$A^{(s)}_{i,j,s} = A^{(s+2)}_{i,j,s+1} + A^{(s+2)}_{i,j,s+2} \tag{15}$$

and

$$A^{(s)}_{i,j,k} = A^{(s+2)}_{i,j,k+1}, k = 2,3,\ldots,s-1 \tag{16}$$

This means that the two outermost decisions are merged and the 'internal' decisions remain unchanged. The appropriate graph definition is the following:

Definition 3 (The graph belonging to $A^{(s^*)}$). *The nodes are the objects. Next to the edges belonging to ties are two types of directed edge: The first one is a 'simple' directed edge from i to j (denoted by $\rightarrow$ and called the 'better to some extent' edge) that exists if $0 < A^{(s^*)}_{i,j,k}$ $k = [\frac{s^*+1}{2}] + 1,\ldots,s^*-1$; i.e., there is a decision according to which object i is 'better to some extent' than object j. The other is an 'extremely better' directed edge: it exists from i towards j if $0 < A^{(s^*)}_{i,j,s^*}$ and it is denoted by $\twoheadrightarrow$. Expressed in words, there is an extreme decision according to which i exceeds j. This graph will be denoted by $GC^{(A^{(s^*)})}$. We substitute $s^* = s$ and $s^* = s+2$.*

The set of conditions that guarantees data evaluability in the model allowing s^* options is as follows:

- C.s*.1. For every value of $k = 1,2,\ldots,s^*$ there exists an index pair i_k, j_k for which $0 < A^{(s^*)}_{i_k,j_k,k}$.
  We note that this condition is necessary. If it does not hold, the number of options has to be reduced.
- C.s*.2. For every nonempty partition of S and $\overline{S}$ ($S \cup \overline{S} = \{1,2,\ldots,n\}$, $S \cap \overline{S} = \emptyset$), there are two (not necessarily different) index pairs (i_1, j_1) and (i_2, j_2) $i_1, i_2 \in S$, $j_1, j_2 \in \overline{S}$ for which

$$0 < A^{(s^*)}_{i_1,j_1,1} \text{ and } 0 < A^{(s^*)}_{i_2,j_2,s^*} \tag{17}$$

  are satisfied or there exists such an index pair (i_3, j_3) for which

$$0 < A^{(s^*)}_{i_3,j_3,k}, \text{ for some } k = 2,3,\ldots,s^*-1 \quad (18)$$

is satisfied. Expressed in words, the S and $\overline{S}$ are connected either by a non-extreme decision or by two extreme decisions made back and forth.
We note that this condition is also a necessary condition, not just sufficient.

- C.s*.3. There exists a cycle in the graph $GC^{(A^{(s^*)})}$ along the 'better to some extent' and 'extremely better' edges which contains at least one 'extremely better' edge. In other words, there exists a cycle $(i_1, i_2, \ldots, i_h, i_1)$ for which $i_1, i_2, \ldots, i_h$ are different, and $0 < A^{(s^*)}_{i_l,i_{l+1},k_l}$, where $k_l \in \{[\frac{s^*+1}{2}]+1, \ldots, s^*\}$, and there exists at least one index pair in the cycle for which $k_l = s^*$. We note that it may be the case that this cycle contains only two different nodes.

As a generalization of Theorem 4 in [47], one can prove that conditions C.s*.1., C.s*.2., and C.s*.3. guarantee the evaluability of the data matrix $A^{(s^*)}$ in the s^*-option model. We will apply this statement for $s^* = s$ and $s^* = s + 2$.

Comparison Based on Evaluability

(A) Data are evaluable only in the s-option model.
(B) Data are evaluable only in the (s+2)-option model.
(C) Data are evaluable in both models.
(D) Data are not evaluable in either model.

(Case A) This case cannot occur. We can state the following Theorem:

Theorem 2. *Suppose that there exists at least one index pair (i_1, j_1) for which $0 < A^{(s+2)}_{i_1,j_1,s+1}$ is satisfied and there exists an index pair (i_2, j_2) for which $0 < A^{(s+2)}_{i_2,j_2,s+2}$ holds. Redefine the data in the s-option model using (14), (15), and (16). If conditions C.s.1., C.s.2, and C.s.3. are satisfied, then $A^{(s+2)}$ is evaluable in the $(s+2)$-option model.*

The proof of this can be easily carried out by following the steps of $s = 2$ cases step by step in Section 3.1.

(Case B) This case can happen; for example, take the following data matrix $A^{(s+2)}$, which is a modification of the comparison results presented in Figure 2: $A^{(s+2)}_{1,2,k} = A^{(s+2)}_{2,1,k} = 1, k = 1,2,\ldots,s+2$, $A^{(s+2)}_{2,3,2} = A^{(s+2)}_{3,2,s+1} = 1$, and the other values are zero. This implies that $0 < A^{(s)}_{1,2,k}$, $k = 1,2,\ldots,s$ and $A^{(s)}_{2,3,1} = A^{(s)}_{3,2,s} = 1$. Similarly to the case $s = 2$, in the general s-option model, object 3 exceeds object 2 but we do not know by how much. By increasing its expectation, its likelihood function increases. C.s.2. does not hold if $S = \{1,2\}$ and $\overline{S} = \{3\}$.

In this case, it is clear which model should be used for the evaluation; there is no choice between them.

(Case C) Similar to the case where $s = 2$, it may be the case that the data can be evaluated using both numbers of options. As an example, consider the previous data matrix A, and supplement it with an 'extremely better' edge between the objects 2 and 3. That is, let $A^{(s+2)}_{1,2,k} = A^{(s+2)}_{2,1,k} = 1, k = 1,2,\ldots,s+2$, $A^{(s+2)}_{2,3,2} = A^{(s+2)}_{3,2,s+1} = A^{(s+2)}_{2,3,s+2} = A^{(s+2)}_{3,2,1} = 1$, and the other values be zero. One can easily check that C.s.1, C.s.2., C.s,3., C.s + 2.1., C.s + 2.2., and C.s + 2.3. are all satisfied. Consequently, the data can be evaluated by both models.

In this case, we can choose which model to use for evaluating the data.

(Case D) The data cannot be evaluated using either the s-option or the $(s+2)$-option model.

An easily understandable example is a modification of the data presented in Figure 4. Let $A^{(s+2)}_{1,2,k} = A^{(s+2)}_{2,1,k} = 1, k = 1,2,\ldots,s+2$, and $A^{(s+2)}_{2,3,s+2} = 1$, while the other values are zero. Now, $0 < A^{(s)}_{1,2,k} = A^{(s)}_{2,1,s+1-k}$ $k = 1,2,\ldots,s$ and $A^{(s)}_{2,3,s} = A^{(s)}_{3,2,1} = 1$, while the other values are zero. In this case, neither C.s.2. nor C.s + 2.2. holds, as $S = \{1,2\}$ and $\overline{S} = \{3\}$.

In this case, further data have to be collected; that is, new comparisons have to be made.

4. Simulation Results

In this chapter, we deal with the two most commonly used option numbers: $s = 2$ and $s = 3$. First, we present the relationship between the two- and four-option models, and then we turn to the relationship between the three- and five-option models. We analyze these relationships by applying computer simulations. We point out when it is more appropriate to use which model in order to generate the best possible retrieval of the strength of the objects.

In these cases, we predefined the strength (expectation) of each object within certain bounds and the separation parameters for each object, including $0 < d$ for the three-, four- and five-option models and $d < D$ for the five-option model. We then generated comparison results for different comparison numbers using these fixed strengths (m_i) and values d, D. We checked whether the data set could be evaluated using the s- and $(s+2)$-option models. If it was possible, we evaluated the data sets using both the s- and $(s+2)$-option models and compared the estimated strengths obtained by the models with the original strengths $m_i, i = 1, 2, \ldots, n$. We used a large number of simulations (10^5) in our work, applying logistic distribution. The simulation program was implemented in C#.

4.1. The Cases s = 2 and s + 2 = 4

The two-option and four-option models are compared in this subsection. Table 1 shows an example for the simulation results of the two models. We present a case in Table 1 where the probabilities of an object falling into the options available are approximately equal. In the presented situation, there are eight objects to compare, their expectations are between 0 and 2, and $d = 1.35$. In the case of the four-option model, the 'slightly better' and 'much better' options can be distinguished: a 'slightly worse' or 'much worse' decision is a 'slightly better' or 'much better' decision from the perspective of the other object. In this case, the ratio of 'much better' decisions to all decisions was 0.5186, while the ratio of 'slightly better' was 0.4814.

The first column in Table 1 shows the number of comparisons denoted by $|A|$. The second column shows the number of cases in which we were able to use both models. The third column shows the number of cases for which we were only able to evaluate objects using the four-option model and the fourth column shows the number of cases for which we were not able to make an evaluation using either model. The fifth and sixth columns contain the average sum of the squared differences between the exact and the estimated expectations found by the two-option model and the four-option model, respectively. The quantities

$$e^{(s*)} = \sum_{i=1}^{n} (\widehat{m}_i^{(s*)} - m_i)^2 \tag{19}$$

measure how close the estimated expected values are to the exact expected values in the model allowing $s*$ options. They are zero if, and only if, the estimated parameters coincide with the exact parameters. If we take the average of these values, we get a measure of the model's accuracy, denoted by $\overline{e^{(s*)}}$.

The seventh (last) column shows their relative relation, which is defined as follows:

$$R^{(2),(4)} = \frac{\overline{e^{(2)}} - \overline{e^{(4)}}}{\overline{e^{(4)}}}. \tag{20}$$

In these formulas, as previously, the upper index refers to the option number of the model used for the evaluation. As the quantities (19) are non-negative, the sign of (20) characterizes which model is better: if (20) is positive, then the four-option model is better; if it is negative, then the two-option model is. It can clearly be seen that in the case of few comparisons—i.e., when the number of comparisons conducted is less than 21—the two-option model is able to approximate the exact expectations more accurately. Meanwhile, in the case of more comparisons, the more complex four-option model performs better. It can also clearly be seen that as more comparisons are included, we can make better evaluations with both models and the average errors $\overline{e^{(2)}}$ and $\overline{e^{(4)}}$ become smaller and smaller. Due to space limitations, we have only presented the case m_i: $0-2, d=1.35$, but in other cases the results are similar: if the number of comparisons reaches 21, the more complex four-option model performs better than for 8 compared objects. We conducted simulations with different numbers of objects to be evaluated and we observed a linear relationship between the number of objects evaluated and the number of comparisons when the model with more options performs better, on average. The critical number is the twice of the number of the objects plus 5 in the case of $s = 2$. This means that if $2 \cdot n + 5 \leq |A|$, then the complicated model is advisable.

Table 1. Comparison of the simulation results of the two- and four-option models using the parameters m_i: $0-2, d=1.35$.

$\|A\|$	2 and 4	Only 4	Neither	$\overline{e^{(2)}}$	$\overline{e^{(4)}}$	$R^{(2),(4)}$
10	23	4006	95,971	3.887	6.059	−0.359
11	85	8611	91,304	4.566	6.405	−0.287
12	279	14,819	84,902	4.880	6.823	−0.285
13	708	21,872	77,420	5.168	7.231	−0.285
14	1473	29,086	69,441	5.313	6.917	−0.232
15	2646	35,962	61,392	5.421	6.736	−0.195
16	4195	42,121	53,684	5.576	6.655	−0.162
17	6401	47,080	46,519	5.709	6.465	−0.117
18	8981	50,959	40,060	5.839	6.332	−0.078
19	11,999	53,683	34,318	5.924	6.179	−0.041
20	15,499	55,244	29,257	5.946	5.970	−0.004
21	19,284	55,878	24,838	5.971	5.811	0.027
22	23,220	55,720	21,060	5.939	5.643	0.052
23	27,247	54,962	17,791	5.921	5.456	0.085
24	31,453	53,502	15,045	5.882	5.284	0.113
25	35,552	51,612	12,836	5.808	5.090	0.141
30	54,845	39,613	5542	5.424	4.296	0.263
40	79,977	18,949	1074	4.441	3.151	0.410
50	91,265	8514	221	3.569	2.417	0.477
60	96,142	3797	61	2.898	1.929	0.503
70	98,224	1763	13	2.388	1.590	0.502
80	99,167	831	2	2.011	1.343	0.497
90	99,587	412	1	1.729	1.163	0.487
100	99,781	219	0	1.510	1.023	0.476
200	99,999	1	0	0.650	0.460	0.412
300	100,000	0	0	0.414	0.297	0.394
400	100,000	0	0	0.304	0.219	0.387
500	100,000	0	0	0.240	0.174	0.382
600	100,000	0	0	0.199	0.144	0.381
700	100,000	0	0	0.169	0.123	0.378
800	100,000	0	0	0.148	0.107	0.376
900	100,000	0	0	0.131	0.095	0.374
1000	100,000	0	0	0.117	0.085	0.373

This phenomenon can be explained as follows: if there are few comparisons, randomness plays a significant role. The more options there are, the greater the variability. As the number of comparisons increases, the generated data increasingly follow the probabilities derived by the model. The parameters become easier to estimate accurately. Models with more options contain more information, allowing for more precise estimates.

4.2. The Cases $s = 3$ and $s + 2 = 5$

Now we compare the three-option and five-option models. The reason for choosing $s = 3$ as the general case is that it is often used, and particularly in sports applications. In numerous sports, the possible outcomes of matches are a draw, a victory, or a defeat. Examples of such sports include football and handball. However, if we want to evaluate the results using a three-option model, we cannot do so. One possible reason for this could be that one team defeats every other team. This happened in the 2024 UEFA European Championship: Spain won all of its matches. If we evaluate these results using the five-option model, the evaluation can be carried out. Many other examples can be given of this phenomenon. These also highlight the importance of examining the problem carefully.

We conducted computer simulations. In the presented example, the number of compared objects was eight. The expectations were chosen from the interval $[0,2]$ and d and D were fixed: $0 < d < D$, $d = 0.1$, $D = 1.35$. In this case, we wanted to show an extreme scenario in which there are almost no 'equal' decisions, while the number of 'slightly better' and 'much better' decisions is approximately equal. The ratios of the decisions 'equal', 'slightly better', and 'much better' are 0.0540/0.4768/0.4692, respectively. The simulation results are shown in Table 2.

Table 2. Comparison of the simulation results of the three- and five-option models using the parameters m_i: $0 - 2, d = 0.1, D = 1.35$.

$\|A\|$	3 and 5	Only 5	Neither	$\overline{e^{(3)}}$	$\overline{e^{(5)}}$	$R^{(3),(5)}$
10	39	2283	97,678	5.715	7.765	−0.264
11	168	4899	94,933	6.203	9.178	−0.324
12	454	8313	91,233	6.508	8.666	−0.249
13	936	12,437	86,627	6.838	8.495	−0.195
14	1788	16,617	81,595	6.870	8.342	−0.177
15	3041	20,513	76,446	6.959	8.112	−0.142
16	4738	23,837	71,425	7.091	7.820	−0.093
17	6766	26,470	66,764	7.074	7.563	−0.065
18	9171	28,704	62,125	7.045	7.260	−0.030
19	11,901	30,146	57,953	6.913	6.949	−0.005
20	14,931	30,912	54,157	6.847	6.706	0.021
30	46,295	21,860	31,845	5.583	4.466	0.250
40	66,621	10,441	22,938	4.310	3.181	0.355
50	77,156	4594	18,250	3.370	2.404	0.402
60	82,770	1964	15,266	2.688	1.896	0.417
70	86,060	887	13,053	2.206	1.559	0.415
80	88,263	412	11,325	1.857	1.316	0.411
90	89,784	191	10,025	1.593	1.137	0.401
100	90,892	92	9016	1.391	1.001	0.390
200	95,497	0	4503	0.607	0.452	0.341
300	96,976	0	3024	0.387	0.292	0.327
400	97,762	0	2238	0.285	0.215	0.323

Table 2. *Cont.*

$\|A\|$	3 and 5	Only 5	Neither	$\overline{e^{(3)}}$	$\overline{e^{(5)}}$	$R^{(3),(5)}$
500	98,232	0	1768	0.225	0.171	0.319
600	98,530	0	1470	0.186	0.142	0.317
700	98,714	0	1286	0.159	0.121	0.315
800	98,867	0	1133	0.138	0.105	0.314
900	98,991	0	1009	0.122	0.093	0.314
1000	99,105	0	895	0.110	0.084	0.314

The structure of Table 2 follows the structure of Table 1; the columns' names and reports change according to the option numbers of the models. Here, as in the two-option model, the accuracy of estimating expected values is better for the simpler model if the numbers of comparisons are small, while, for larger numbers of comparisons, the more complex model performs better. As more and more comparisons are made between random objects, the errors become smaller and smaller. Based on lots of further experiments, we can ascertain that, in the case of 8 objects and at least 20 comparisons, the five-option model estimates the expectations m_i $(i = 1, 2, \ldots, n)$ better, on average; therefore, it is recommended.

Of course, the number of comparisons depends on the number of objects to be evaluated when the model that allows more options is able to use its additional information. By conducting a large number of simulations with different object numbers, we can conclude that five-option model performs better than the three-option model if $2 \cdot n + 4 \leq |A|$.

5. Algorithm for Model Choice

In this section, based on the results of the previous sections, we present an algorithm that allows us to select which model to use; see Algorithm 1.

The function $Ev(A, om)$ specifies, for a given A (data set), whether A can be evaluated with an om-option model. As previously, we denote the number of comparisons as $|A|$. The number of 'equal' choices is denoted by A_{equal}. The auxiliary variable k that we use takes only positive integer values. The function $Calc(A, om)$ evaluates the data set A in an om-option model. $Mdc()$ represents additional data collection. oc denotes the number of objects in the comparisons, while $Cm(om)$ is the constant multiplier in each case where $om = s + 2$. For $s + 2 = 4$ and $s + 2 = 5$, the $Cm(om)$ is 2. The od is a constant value for each model; it equals 5 for $s + 2 = 4$ and it equals 4 for $s + 2 = 5$.

We present the algorithm described below in the flowchart in Figure 5.

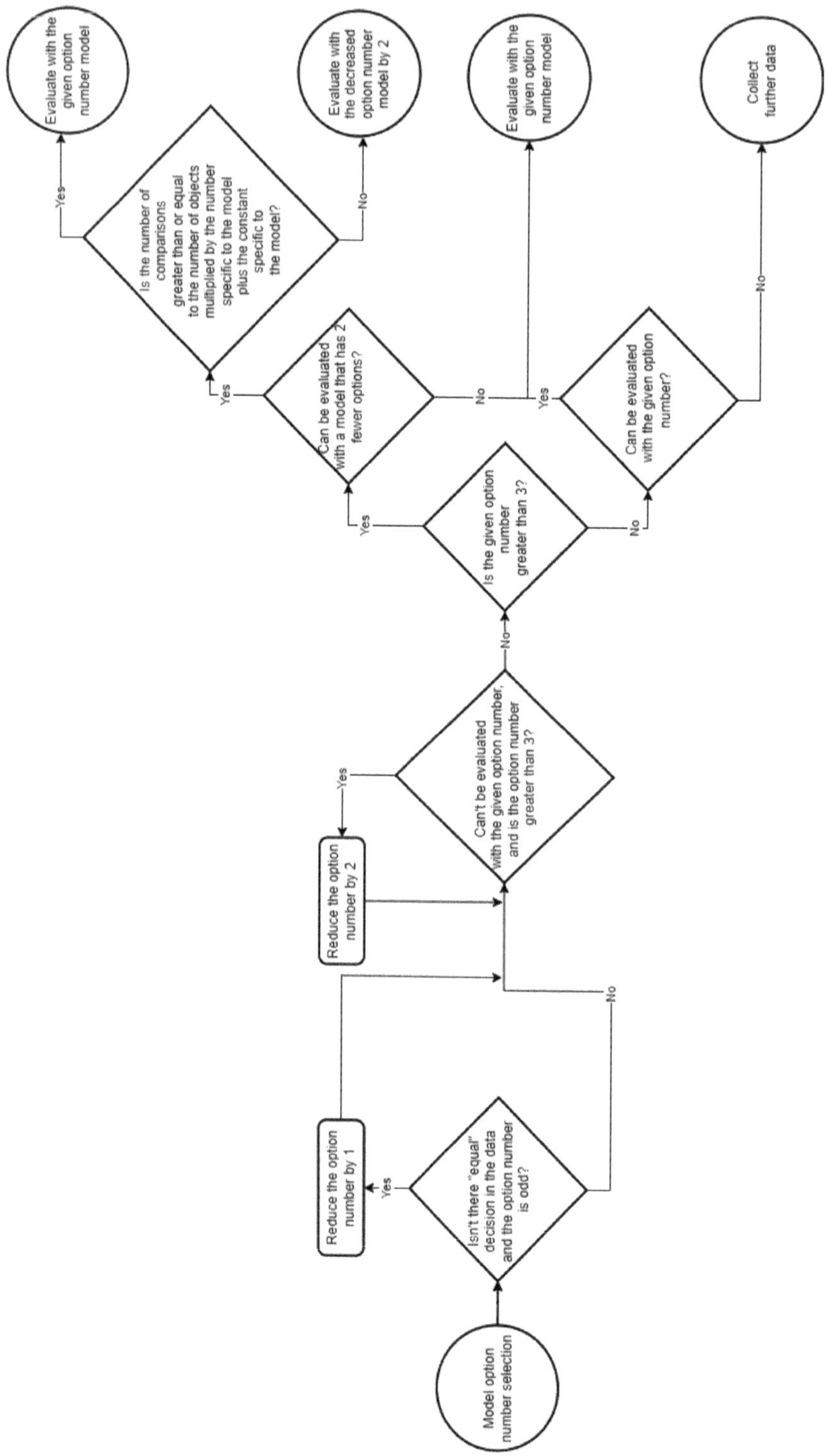

Figure 5. Flowchart of the algorithm for the choice of option numbers.

Algorithm 1 Option Number Selection

```
if s_2 = 2k + 1 and A_equal = 0 then
    s_2 := s_2 − 1
end if
while not Ev(A, s_2) and s_2 > 3 do
    s_2 := s_2 − 2
end while
if s_2 > 3 then
    s := s_2 − 2
    if Ev(A, s) then
        if |A| ≥ Cm(s_2) · oc + od then
            Calc(A, s_2)
        else
            Calc(A, s)
        end if
    else
        Calc(A, s_2)
    end if
else
    if Ev(A, s_2) then
        Calc(A, s_2)
    else
        Mdc()
    end if
end if
```

6. Conclusions

The selection of the number of allowed options was studied to improve evaluability and provide a better approximation of the strength of the objects compared. An algorithm was created to decide which option number is best to use for data evaluations. The algorithm includes theoretical results for evaluability and empirical observations based on simulations. Based on theoretical considerations, we can assert that if the data can be evaluated in an s-option model, then it can also be evaluated in an $(s+2)$-option model, provided that there are comparison results for every option. However, the statement is not true in the reverse scenario: there are situations where only the model with a higher number of options is suitable for the evaluation. If both models are suitable for the evaluation, we have provided a metric to determine which model allows for a more accurate estimation of the objects' strengths. We can conclude that, in the case of a small number of comparisons, the simpler model is more suitable. However, beyond a certain number of comparisons, the model that allows for more options is better, as it can utilize its additional information. This critical number is approximately twice the number of objects to be evaluated plus five for the $s + 2 = 4$-option model and four for the $s + 2 = 5$-option model. Further investigations are needed for cases where $s + 2 = 6, 7$, and beyond.

Author Contributions: Conceptualization, L.G., C.M. and É.O.-M.; methodology, L.G., C.M. and É.O.-M.; software, L.G.; validation, L.G. and C.M.; writing, É.O.-M. and C.M.; review and editing, C.M., É.O.-M. and L.G.; visualization, L.G.; supervision, C.M. All authors have read and agreed to the published version of the manuscript.

Funding: This research received no external funding.

Data Availability Statement: Most of the data are computer-generated. Little public data were used.

Acknowledgments: This research was partly supported by the ÚNKP-23-3 New National Excellence Program of the Ministry for Culture and Innovation from the source of the National Research, Development and Innovation Fund. L. Gyarmati thanks them for the support. This work has been implemented by the TKP2021-NVA-10 project with the support provided by the Ministry of Culture and Innovation of Hungary from the National Research, Development and Innovation Fund, financed under the 2021 Thematic Excellence Programme funding scheme.

Conflicts of Interest: The authors declare no conflicts of interest.

References

1. Erdil, Ü.D.; Tümer, M.; Nadiri, H.; Aghaei, I. Prioritizing information sources and requirements in students' choice of higher education destination: Using AHP analysis. *Sage Open* **2021**, *11*. [CrossRef]
2. Esangbedo, M.O.; Bai, S.; Mirjalili, S.; Wang, Z. Evaluation of human resource information systems using grey ordinal pairwise comparison MCDM methods. *Expert Syst. Appl.* **2021**, *182*, 115151. [CrossRef]
3. Kulak, A.; Selvi, H. Scaling the Psychological Variables that Affect Performance of Athletes by Means of Pairwise Comparison Method. *Int. J. Recreat. Sports Sci.* **2019**, *3*, 16–24. [CrossRef]
4. Krivulin, N.; Prinkov, A.; Gladkikh, I. Using pairwise comparisons to determine consumer preferences in hotel selection. *Mathematics* **2022**, *10*, 730. [CrossRef]
5. Darko, A.; Chan, A.P.C.; Ameyaw, E.E.; Owusu, E.K.; Pärn, E.; Edwards, D.J. Review of application of Analytic Hierarchy Process (AHP) in construction. *Int. J. Constr. Manag.* **2019**, *19*, 436–452. [CrossRef]
6. Lee, S.E.; Choi, M.; Kim, S. How and what to study about IoT: Research trends and future directions from the perspective of social science. *Telecommun. Policy* **2017**, *41*, 1056–1067. [CrossRef]
7. Sasaki, Y. Strategic manipulation in group decisions with pairwise comparisons: A game theoretical perspective. *Eur. J. Oper. Res.* **2023**, *304*, 1133–1139. [CrossRef]
8. Dym, C.L.; Wood, W.H.; Scott, M.J. Rank ordering engineering designs: Pairwise comparison charts and Borda counts. *Res. Eng. Des.* **2002**, *13*, 236–242. [CrossRef]
9. Amlani, A.M.; Schafer, E.C. Application of paired-comparison methods to hearing aids. *Trends Amplif.* **2009**, *13*, 241–259. [CrossRef]
10. Rosenberger, R.S.; Peterson, G.L.; Loomis, J.B. Applying a method of paired comparisons to measure economic values for multiple goods sets. *J. Agric. Appl. Econ.* **2002**, *34*, 215–229. [CrossRef]
11. Huang, J.J.; Chen, C.Y. Resource Allocation of Cooperative Alternatives Using the Analytic Hierarchy Process and Analytic Network Process with Shapley Values. *Algorithms* **2024**, *17*, 152. [CrossRef]
12. Temesi, J.; Szádoczki, Z.; Bozóki, S. Incomplete pairwise comparison matrices: Ranking top women tennis players. *J. Oper. Res. Soc.* **2024**, *75*, 145–157. [CrossRef]
13. Vaidya, O.S.; Kumar, S. Analytic hierarchy process: An overview of applications. *Eur. J. Oper. Res.* **2006**, *169*, 1–29. [CrossRef]
14. Weernink, M.G.; Janus, S.I.; van Til, J.A.; Raisch, D.W.; van Manen, J.G.; IJzerman, M.J. A systematic review to identify the use of preference elicitation methods in healthcare decision making. *Pharm. Med.* **2014**, *28*, 175–185. [CrossRef]
15. Cheng, K.E.; McHugh, J.A.; Deek, F.P. On the use of paired comparisons to construct group preference scales for decision making. *Group Decis. Negot.* **2013**, *22*, 519–540. [CrossRef]
16. Zhang, G.; Lu, J.; Gao, Y. *Multi-Level Decision Making*, 1st ed.; Springer: Berlin/Heidelberg, Germany, 2015; ISBN 978-3-662-46059-7
17. Saaty, T.L. A scaling method for priorities in hierarchical structures. *J. Math. Psychol.* **1977**, *15*, 234–281. [CrossRef]
18. Dyer, J.S. Remarks on the Analytic Hierarchy Process. *Manag. Sci.* **1990**, *36*, 249–258. [CrossRef]
19. Bozóki, S.; Tsyganok, V. The (logarithmic) least squares optimality of the arithmetic (geometric) mean of weight vectors calculated from all spanning trees for incomplete additive (multiplicative) pairwise comparison matrices. *Int. J. Gen. Syst.* **2019**, *48*, 362–381. [CrossRef]
20. Brunelli, M. A survey of inconsistency indices for pairwise comparisons. *Int. J. Gen. Syst.* **2018**, *47*, 751–771. [CrossRef]
21. Brunelli, M.; Fedrizzi, M. Inconsistency indices for pairwise comparisons and the Pareto dominance principle. *Eur. J. Oper. Res.* **2024**, *312*, 273–282. [CrossRef]
22. Sato, Y.; Tan, K.H. Inconsistency indices in pairwise comparisons: An improvement of the consistency index. *Ann. Oper. Res.* **2023**, *326*, 809–830. [CrossRef]
23. Ágoston, K.C.; Csató, L. A lexicographically optimal completion for pairwise comparison matrices with missing entries. *Eur. J. Oper. Res.* **2024**, *314*, 1078–1086. [CrossRef]
24. Tekile, H.A.; Fedrizzi, M.; Brunelli, M. Constrained eigenvalue minimization of incomplete pairwise comparison matrices by Nelder-Mead algorithm. *Algorithms* **2021**, *14*, 222. [CrossRef]
25. Pascoe, S. A simplified algorithm for dealing with inconsistencies using the Analytic Hierarchy Process. *Algorithms* **2022**, 15, 442. [CrossRef]
26. Ágoston, K.C.; Csató, L. Inconsistency thresholds for incomplete pairwise comparison matrices. *Omega* **2022**, *108*, 102576. [CrossRef]
27. Harker, P.T. Incomplete pairwise comparisons in the analytic hierarchy process. *Math. Modell.* **1987**, *9*, 837–848. [CrossRef]
28. Bozóki, S.; Fülöp, J.; Rónyai, L. On optimal completion of incomplete pairwise comparison matrices. *Math. Comput. Modell.* **2010**, *52*, 318–333. [CrossRef]
29. Bozóki, S.; Csató, L.; Temesi, J. An application of incomplete pairwise comparison matrices for ranking top tennis players. *Eur. J. Oper. Res.* **2016**, 248, 211–218. [CrossRef]

30. Petróczy, D.G. An alternative quality of life ranking on the basis of remittances. *Socio-Econ. Plann. Sci.* **2021**, *78*, 101042. [CrossRef]
31. Brans, J.P.; Vincke, P.; Mareschal, B. How to select and how to rank projects: The PROMETHEE method. *Eur. J. Oper. Res.* **1986**, *24*, 228–238. [CrossRef]
32. Behzadian, M.; Kazemzadeh, R.B.; Albadvi, A.; Aghdasi, M. PROMETHEE: A comprehensive literature review on methodologies and applications. *Eur. J. Oper. Res.* **2010**, *200*, 198–215. [CrossRef]
33. Aherdoost, H.; Madanchian, M. A Comprehensive Overview of the ELECTRE Method in Multi-Criteria Decision-Making. *J. Manag. Sci. Eng. Res.* **2023**, *6*, 5–16. [CrossRef]
34. Thurstone, L.L. A law of comparative judgment. *Psychol. Rev.* **1927**, *34*, 273–286. [CrossRef]
35. Tutz, G.; Schauberger, G. Extended ordered paired comparison models with application to football data from German Bundesliga. *AStA Adv. Stat. Anal.* **2015**, *99*, 209–227. [CrossRef]
36. Jeon, J.J.; Kim, Y. Revisiting the Bradley-Terry model and its application to information retrieval. *J. Korean Data Inf. Sci. Soc.* **2013**, *24*, 1089–1099. [CrossRef]
37. Stein, A.; Aryal, J.; Gort, G. Use of the Bradley-Terry model to quantify association in remotely sensed images. *IEEE Trans. Geosci. Remote Sens.* **2005**, *43*, 852–856. [CrossRef]
38. Kosztyán, Z.T.; Orbán-Mihálykó, É.; Mihálykó, C.; Csányi, V.V.; Telcs, A. Analyzing and clustering students' application preferences in higher education. *J. Appl. Stat.* **2020**, *47*, 2961–2983. [CrossRef]
39. Bradley, R.A.; Terry, M.E. Rank analysis of incomplete block designs: I. The method of paired comparisons. *Biometrika* **1952**, *39*, 324–345. [CrossRef]
40. Glenn, W.A.; David, H.A. Ties in paired-comparison experiments using a modified Thurstone-Mosteller model. *Biometrics* **1960**, *16*, 86–109. [CrossRef]
41. Rao, P.V.; Kupper, L.L. Ties in paired-comparison experiments: A generalization of the Bradley-Terry model. *J. Am. Stat. Assoc.* **1967**, *62*, 194–204. [CrossRef]
42. Agresti, A. Analysis of ordinal paired comparison data. *J. R. Stat. Soc. Ser. C Appl. Stat.* **1992**, *41*, 287–297. [CrossRef]
43. Orbán-Mihálykó, É.; Mihálykó, C.; Koltay, L. A generalization of the Thurstone method for multiple choice and incomplete paired comparisons. *Cent. Eur. J. Oper. Res.* **2019**, *27*, 133–159. [CrossRef]
44. Orbán-Mihálykó, É.; Mihálykó, C.; Koltay, L. Incomplete paired comparisons in case of multiple choice and general log-concave probability density functions. *Cent. Eur. J. Oper. Res.* **2019**, *27*, 515–532. [CrossRef]
45. Ford, L.R., Jr. Solution of a ranking problem from binary comparisons. *Am. Math. Mon.* **1957**, *64*, 28–33. [CrossRef]
46. Davidson, R.R. On Extending the Bradley-Terry Model to Accommodate Ties in Paired Comparison Experiments. *J. Am. Stat. Assoc.* **1970**, *65*, 317–328. [CrossRef]
47. Gyarmati, L.; Orbán-Mihálykó, É.; Mihálykó, C. Comparative analysis of the existence and uniqueness conditions of parameter estimation in paired comparison models. *Axioms* **2023**, *12*, 575. [CrossRef]
48. Orbán-Mihálykó, É.; Koltay, L.; Szabó, F.; Csuti, P.; Kéri, R.; Schanda, J. A new statistical method for ranking of light sources based on subjective points of view. *Acta Polytech. Hung.* **2015**, *12*, 195–214.
49. Gyarmati, L.; Edvy, L.; Mihálykó, C.; Orbán-Mihálykó, É. Decision support for sports selection based on performance measurement applying the generalized Thurstone method. *Int. J. Sports Sci. Coach.* **2024**, 17479541241240609. [CrossRef]
50. Gyarmati, L.; Mihálykó, C.; Orbán-Mihálykó, É. Application of different option numbers in Thurstone motivated models. In Proceedings of the 3rd Conference on Information Technology and Data Science, Debrecen, Hungary, 26–28 August 2024.

algorithms

MDPI

Article

Meshfree Variational-Physics-Informed Neural Networks (MF-VPINN): An Adaptive Training Strategy

Stefano Berrone [1] and Moreno Pintore [2,3,*]

[1] Dipartimento di Scienze Matematiche, Politecnico di Torino, Corso Duca degli Abruzzi 24, 10129 Torino, Italy; stefano.berrone@polito.it
[2] MEGAVOLT Team, Inria, 48 Rue Barrault, 75013 Paris, France
[3] Laboratoire Jacques-Louis Lions, Sorbonne Center for Artificial Intelligence, Sorbonne Université, 4 Place Jussieu, 75005 Paris, France
* Correspondence: moreno.pintore@inria.fr

Abstract: In this paper, we introduce a Meshfree Variational-Physics-Informed Neural Network. It is a Variational-Physics-Informed Neural Network that does not require the generation of the triangulation of the entire domain and that can be trained with an adaptive set of test functions. In order to generate the test space, we exploit an a posteriori error indicator and add test functions only where the error is higher. Four training strategies are proposed and compared. Numerical results show that the accuracy is higher than the one of a Variational-Physics-Informed Neural Network trained with the same number of test functions but defined on a quasi-uniform mesh.

Keywords: VPINN; meshfree; Physics-Informed Neural Networks; error estimator; patches

MSC: 65N12; 65N15; 65N50; 68T05; 92B20

1. Introduction

Physics-Informed Neural Networks (PINNs) are a rapidly emerging numerical technique used to solve partial Differential equations (PDEs) by means of a deep neural network. The first idea can be traced back to the works of Lagaris et al. [1–3], but, thanks to the hardware advancements and the existence of deep learning packages like Tensorflow [4], Pytorch [5] and JAX [6], they have recently became popular since the works of Raissi et al. [7,8], published in [9]. In its original formulation, the approximate solution is computed as the output of a neural network trained to minimize the PDE residual on a set of collocation points inside the domain and on its boundary.

The growing interest in PINNs is strictly related to their flexibility. In fact, with minor changes to the implementation, it is possible to solve a huge variety of problems. For example, exploiting the nonlinear nature of the involved neural network, nonlinear [10,11] and high-dimensional PDEs [12] can be solved without the need for globalization methods or additional nonlinear solvers. Moreover, by changing the neural network's input dimensions or suitably adapting the loss function, it is possible to solve parametric [13,14] or inverse [15,16] problems. When external data are available, they can also be used to guide the optimization phase and improve the PINN accuracy [17].

In order to improve the original PINN proposed in [9] and to adapt it to solve specific problems, several generalizations have been proposed. For example, the deep Ritz method (DRM) [18–20] looks for a minimizer of the PDE energy functional and, in the deep Galerkin method (DGM) [21–23], an approximation of the L^2 norm of the PDE residual is minimized. It is also possible to exploit domain decomposition strategies [24,25] as in the conservative PINN (CPINN) [26], in the parallel PINN [27], in the extended PINN (XPINN) [28], or in the Finite Basis PINN (FBPINN) [29]. Moreover, it is even possible to change the neural network architecture or the training strategy as in [14,30–35]; between the methods based on different architectures, we highlight some works based on the novel Kolmogorov–Arnold Network (KAN) [36] architecture [37,38] and on a Large Language Model (LLM) [39]. More extensive overviews of the existing approaches can be found in [40–43]. In the context of the current work, an important extension is the Variational-Physics-Informed Neural Network (VPINN) [44,45], where the weak formulation of the problem is used to construct the loss function.

In this work, we focus on VPINNs. As discussed in [44–47], in order to train a VPINN, one needs to choose a suitable space of test functions, compute the variational residuals against all the test functions on the basis of such a space, and minimize a linear combination of these residuals. Since a spatial mesh is required to define the test functions, the VPINN cannot be considered a meshfree method, even though it is an extension of the PINN, which is meshfree. In this work, we present an adaptive Meshfree VPINN (MF-VPINN) that does not require a global triangulation of the domain but is trained with the same loss function and neural network architecture of a standard VPINN. Note that the MF-VPINN and the original VPINN can solve the same differential problems because the neural network is trained with the same loss functions. We also highlight that they can solve problems where the solution has low regularity that cannot be solved with standard PINNs, for example, in the presence of singular forcing terms, thanks to the weak formulation of the PDE without introducing further approximations or regularizations. However, one of the VPINN's limitations is that a triangulation of the entire domain is required to define the test functions. Generating it may be very expensive or even impractical for very complex geometries (like, for example, the ones in [48]) and in moderate- or high-dimensional problems, for which automatic mesh-generation algorithms do not exist. For such domains, it is therefore highly advisable or computationally necessary to use a meshfree method such as the original PINN or the proposed MF-VPINN. Moreover, when dealing with complex geometries for which a mesh can be hardly generated, the refinement of the mesh for adaptive methods can be very difficult. In this paper, we describe an algorithm that solves the problem and provides a reliable solution.

The paper is organized as follows. In Section 2, we introduce the problem we are interested in. In particular, we focus on the problem discretization in Section 2.1 and on the MF-VPINN loss function in Section 2.2. Then, an a posteriori error estimator is presented in Section 2.3 and used in Section 2.4 to iteratively generate the required test functions. Numerical results are presented in Section 3. In Section 3.1, we describe the model implementation and some strategies to improve the model efficiency, in Section 3.2 we compare different approaches to generate the test functions and compare their performance and, in Section 3.3, we analyze the role of the error estimator introduced in Section 2.3. Similar tests are performed on a different problem in Section 3.4 to describe possible extensions on more complex domains. Finally, we conclude the paper in Section 4 and discuss future perspectives and ideas.

2. Problem Formulation

Let us consider the following second-order elliptic problem, defined on a polygonal or polyhedral domain $\Omega \subset \mathbb{R}^n$ with a Lipshitz boundary $\Gamma = \partial\Omega$:

$$\begin{cases} Lu := -\nabla \cdot (\mu \nabla u) + \boldsymbol{\beta} \cdot \nabla u + \sigma u = f & \text{in } \Omega\,, \\ u = g & \text{on } \Gamma\,, \end{cases} \tag{1}$$

where $\mu, \sigma \in \mathrm{L}^\infty(\Omega)$, $\boldsymbol{\beta} \in (\mathrm{W}^{1,\infty}(\Omega))^n$ satisfy $\mu \geq \mu_0$, $\sigma - \frac{1}{2}\nabla \cdot \boldsymbol{\beta} \geq 0$ in Ω for some constant $\mu_0 > 0$, whereas $f \in L^2(\Omega)$ and $g = \bar{u}_{|\Gamma}$ for some $\bar{u} \in \mathrm{H}^1(\Omega)$.

In order to derive the corresponding variational formulation, we define the bilinear form a and the linear form F as

$$a : V \times V \to \mathbb{R}\,, \qquad a(w,v) = \int_\Omega \mu \nabla w \cdot \nabla v + \boldsymbol{\beta} \cdot \nabla w\, v + \sigma w\, v\,, \tag{2}$$

$$F : V \to \mathbb{R}, \qquad F(v) = \int_\Omega f\, v; \tag{3}$$

where V is the function space $V = H_0^1(\Omega)$. We denote by $\alpha \geq \mu_0$ the coercivity constant of a and by $\|a\|$ and $\|F\|$ the continuity constants of a and F. Then, the variational formulation of Problem (1) reads as follows: Find $u \in \bar{u} + V$ such that

$$a(u, v) = F(v) \qquad \forall v \in V. \tag{4}$$

2.1. Problem Discretization

In order to numerically solve Problem (4), one needs to choose suitable finite-dimensional approximations of the trial space $\bar{u} + V$ and of the test space V. A Galerkin formulation is considered when we consider a finite-dimensional space V_h^{trial} for the trial space $\bar{u} + V_h^{\text{trial}}$ and a finite-dimensional test space V_h^{test}, with $V_h^{\text{trial}} = V_h^{\text{test}}$; whereas a Petrov–Galerkin formulation is considered otherwise. In this work, we consider a Petrov–Galerkin formulation in which the trial space is approximated by a set of functions $V^{\mathcal{NN}}$ of the form $V^{\mathcal{NN}} = \bar{u} + V_h^{\text{trial}}$, with V_h^{trial} represented by a neural network suitably modified to enforce the Dirichlet boundary conditions, and the test space is a space V_h of piecewise linear functions.

The neural network considered in the following is a standard fully connected feed-forward neural network. Given the number L of layers and a set of matrices $A_\ell \in \mathbb{R}^{N_\ell \times N_{\ell-1}}$ and vectors $b_\ell \in \mathbb{R}^{N_\ell}$, $\ell = 1, \dots, L$ containing the neural network's trainable weights, the function $w : \mathbb{R}^n \to \mathbb{R}$ associated with the considered neural network architecture is:

$$\begin{aligned} x_0 &= \boldsymbol{x}, \\ x_\ell &= \rho(A_\ell x_{\ell-1} + b_\ell), \qquad \ell = 1, \dots, L-1, \\ w(\boldsymbol{x}) &= A_L x_{L-1} + b_L. \end{aligned} \tag{5}$$

where $\rho : \mathbb{R} \to \mathbb{R}$ is a nonlinear function applied element-wise to the vector $A_\ell x_{\ell-1} + b_\ell$. In this section, we use $\rho(x) = \tanh(x)$; other common choices include, but are not limited to, $\rho(x) = \text{ReLU}(x) = \max\{0, x\}$, $\rho(x) = \text{RePU}(x) = \max\{0, x^p\}$ for $1 < p \in \mathbb{N}$, $\rho(x) = 1/(1 + e^{-x})$ and $\rho(x) = \log(1 + e^x)$. Note that, in order to represent a function $w : \mathbb{R}^n \to \mathbb{R}$, the layer widths N_ℓ of the first and last layers are chosen as $N_0 = n$ and $N_L = 1$. We denote by $W^{\mathcal{NN}}$ the set of functions that can be represented as in (5) for any combination of the neural network weights and by $\boldsymbol{w}^{\mathcal{NN}}$ the vector containing all the trainable weights of the neural network.

The function w defined in (5) is independent of the differential problem that has to be solved and is, in most papers on PINNs or related models, trained to minimize both the residual of the equation and a term penalizing the discrepancy between $w_{|\Gamma}$ and g. Instead, we add a non-trainable layer B to the neural network architecture in order to automatically enforce the required boundary conditions without the need to learn them during the training. As described in [49], the operator B acts on the neural network output as

$$Bw = \phi w + \bar{g}, \tag{6}$$

where $\phi : \Omega \to \mathbb{R}$ is a function vanishing on Γ and strictly positive inside Ω, and $\bar{g} : \Omega \to \mathbb{R}$ is a suitable extension of $g : \Gamma \to \mathbb{R}$. The advantages of such an approach are also described in [50]. Then, the discrete trial space approximating $\bar{u} + V$ can be defined as

$$V^{\mathcal{NN}} = \{v^{\mathcal{NN}} \in \bar{u} + V : v^{\mathcal{NN}} = Bw \text{ for some } w \in W^{\mathcal{NN}}\}.$$

On the other hand, the discrete test space V_h is not associated with the neural network and only contains known test functions. In standard VPINNs, one generates a triangulation $\mathcal{T}$ of the domain Ω and then defines V_h as the space of functions that coincide with a polynomial of order $p \in \mathbb{N}$ inside each element of $\mathcal{T}$. Instead, we want to construct a discrete space V_h of functions independent from a global triangulation $\mathcal{T}$. Moreover, since in [47] it has been proven that the VPINN convergence rate with respect to mesh refinement decreases when the order of the test functions is increased, we are interested in a space V_h that only contains piecewise linear functions. For the sake of simplicity, we only consider the case where $n = 2$; the discussion can be directly generalized to the more general case $n \in \mathbb{N}$.

Let $\hat{P} \subset \mathbb{R}^n$ be a reference patch. In the following discussion, $\hat{P}$ can be any arbitrary star-shaped polygon with $N_{\hat{P}}$ vertices and the dimension of its kernel strictly greater than zero. Nevertheless, in the numerical experiments, we only consider the reference patch $\hat{P} = [0,1]^2$ to avoid any unnecessary computational overhead. Let $\mathcal{M} = \{M_i\}_{i=1}^{n_{\text{patches}}}$ be a set of affine mappings such that $M_i : \hat{P} \to P_i \subset \Omega$, where we denote as P_i the patch obtained transforming the reference patch $\hat{P}$ through the map M_i. We assume that $\mathcal{P} = \{P_i\}_{i=1}^{n_{\text{patches}}}$ is a cover of Ω, i.e., $\cup_{i=1}^{n_{\text{patches}}} P_i = \Omega$, and we admit overlapping patches.

Let us consider the triangulation $\hat{\mathcal{T}} = \{\hat{T}_j : 1 \le j \le N_{\hat{P}}\}$ of $\hat{P}$ obtained by connecting each vertex with a single point $\boldsymbol{c}_{\hat{P}}$ in its kernel. It is then possible to define a piecewise linear function $\hat{\varphi}$ vanishing on the border of $\hat{P}$ such that $\hat{\varphi}(\boldsymbol{c}_{\hat{P}}) = 1$ and $\hat{\varphi}_{|\hat{T}_j} \in \mathbb{P}_1(\hat{T}_j)$, for any $j = 1, \ldots, N_{\hat{P}}$. Then, we define the discrete test space V_h as $V_h = \text{span}\{\varphi_i : i = 1, \ldots, n_{\text{patches}}\}$, where $\varphi_i \in V$ is the piecewise linear function:

$$\varphi_i(\boldsymbol{x}) = \begin{cases} \hat{\varphi}(M_i^{-1}(\boldsymbol{x})), & \boldsymbol{x} \in P_i, \\ 0, & \boldsymbol{x} \notin P_i. \end{cases} \tag{7}$$

We remark that the only required triangulation is $\hat{\mathcal{T}}$, which contains only $N_{\hat{P}}$ triangles (in the numerical tests in this paper, $N_{\hat{P}} = 4$). Instead, there exists no mesh on Ω and the test functions φ_i and their supports P_i are all independent. Therefore, the proposed method is said to be meshfree. A simple example of a set of patches $\mathcal{P}$ with $n_{\text{patches}} = 7$ on the domain $\Omega = [0,1]^2$ is shown in Figure 1. For the sake of simplicity, in this work, we consider a squared reference patch $\hat{P}$ with $\boldsymbol{c}_{\hat{P}}$ coinciding with its center, and we let each mapping M_i represent a combination of scalings and translations.

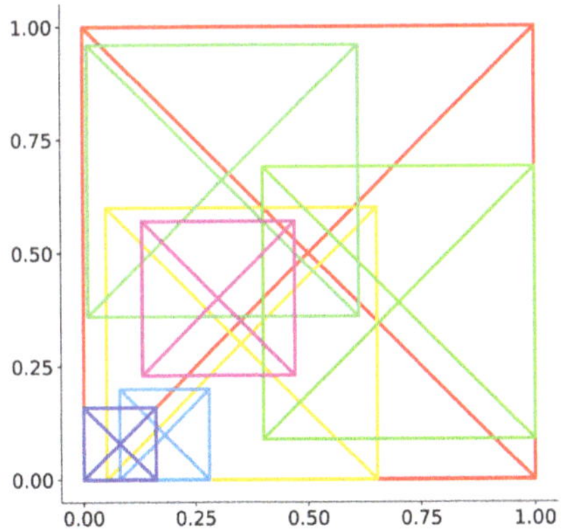

Figure 1. Graphical representation of a set $\{P_i\}_{i=1}^{n_{\text{patches}}}$ obtained from a squared reference patch $\hat{P}$ with $\boldsymbol{c}_{\hat{P}}$ in its center covering the domain $\Omega = (0,1)^2$.

Using the introduced finite-dimensional set of functions $V^{\mathcal{NN}}$ and V_h, it is possible to discretize Problem (4) as follows: Find $u^{\mathcal{NN}} \in V^{\mathcal{NN}}$ such that

$$a(u^{\mathcal{NN}}, v) = F(v) \qquad \forall v \in V_h\,. \tag{8}$$

2.2. Loss Function

In this section, we derive the loss function used to train the neural network. It has to be computable, and its minimizer has to be an approximate solution of Problem (1). We highlight that, when a standard PINN is used, the loss function can be seen as a discrete cost penalizing the residual of (1) directly. In this context, instead, the loss function penalizes the variational residuals of (4) as in standard VPINNs. This is the key difference that differentiates the VPINNs (and its extension proposed in this manuscript) from the other generalizations of the original PINN introduced in Section 1.

Let us consider a quadrature rule of order $q \geq 2$ on each triangle $T_j \in \hat{\mathcal{T}}, j = 1, \ldots, N_{\hat{P}}$, uniquely identified by a set of nodes and weights $\{(\tilde{\xi}^j_\ell, \tilde{\omega}^j_\ell) : \ell \in I^{T_j}\}$. The nodes and weights of a composite quadrature formula of order q on $\hat{P}$ can be obtained as

$$\{(\hat{\xi}_\ell, \hat{\omega}_\ell) : \ell \in I^{\hat{P}}\} = \bigcup_{j=1}^{N_{\hat{P}}} \{(\tilde{\xi}^j_\ell, \tilde{\omega}^j_\ell) : \ell \in I^{T_j}\}.$$

Then, the corresponding quadrature rule of order q of an arbitrary patch P_i is defined as

$$\left\{(\xi^i_\ell, \omega^i_\ell) : \ell \in I^{\hat{P}} | \xi^i_\ell = M_i(\hat{\xi}_\ell), \omega^i_\ell = \hat{\omega}_\ell \frac{\text{area}(P_i)}{\text{area}(\hat{P})}\right\}. \tag{9}$$

Using the quadrature rule in (9), it is possible to define an approximate restriction on each patch of the forms a and F as follows:

$$a^i_h(w, v) = \sum_{\ell \in I^{\hat{P}}} [\mu \nabla w \cdot \nabla v + \boldsymbol{\beta} \cdot \nabla w\, v + \sigma w v](\xi^i_\ell)\, \omega^i_\ell \approx a_{P_i}(w, v), \tag{10}$$

$$F^i_h(v) = \sum_{\ell \in I^{\hat{P}}} [f v](\xi^i_\ell)\, \omega^i_\ell \approx F_{P_i}(v), \tag{11}$$

where $a_{P_i}(w, v)$ and $F_{P_i}(v)$ are defined as in (2) and (3) but restricting the supports of the integrals to P_i. We remark that, since it is not possible to compute integrals involving a neural network exactly, we can only use the forms a^i_h and F^i_h in the loss function. Exploiting the linearity of $a(w, v)$ and $F(v)$ with respect to v to consider only the basis $\{\varphi_i\}_{i=1}^{n_{\text{patches}}}$ of V_h as set of test functions, we approximate Problem (8) as follows: Find $u^{\mathcal{NN}} \in V^{\mathcal{NN}}$ such that

$$a^i_h(u^{\mathcal{NN}}, \varphi_i) = F^i_h(\varphi_i) \qquad \forall i = 1, \ldots, n_{\text{patches}}. \tag{12}$$

Then, in order to cast Problem (12) into an optimization problem, we define the residuals

$$r_{h,i}(w) = F^i_h(\varphi_i) - a^i_h(w, \varphi_i), \qquad i = 1, \ldots, n_{\text{patches}} \tag{13}$$

and the loss function

$$R^2_h(w; \mathcal{P}) = \frac{1}{n_{\text{patches}}} \sum_{i=1}^{n_{\text{patches}}} \gamma_i r^2_{h,i}(w), \tag{14}$$

where γ_i are suitable positive scaling coefficients. In this work, we use $\gamma_i = \text{area}(P_i)^{-1}$ to give the same importance to each patch. Note that this is equivalent to normalizing the quadrature rules involved in (10) and (11); this way, each residual $r_{h,i}$ can be regarded as a linear combination of the MF-VPINN value and derivatives independent of the size of the support of the patch P_i. We also highlight that the loss function depends on the choice of $\mathcal{M}$ since all the used test functions are generated starting from the corresponding mappings $M_i \in \mathcal{M}$. We are now interested in a practical procedure to obtain a set $\widetilde{\mathcal{P}}$ such that the approximate solution computed minimizing $R^2_h(\cdot; \widetilde{\mathcal{P}})$ is as accurate as possible with $\widetilde{\mathcal{P}}$ being as small as possible.

2.3. The a Posteriori Error Estimator

The goal of this section is to derive an error estimator associated with an arbitrary patch P_i, with $i \in \{1, \dots, n_{\text{patches}}\}$. To do so, we rely on the a posteriori error estimator proposed in [46]. It has been proven to be efficient and reliable; therefore, such an estimator allows us to know where the error is larger without knowing the exact solution of the PDE. Let us consider the patch P_i, formed by the triangles $T_{i,1}, \dots, T_{i,N_{\hat{p}}}$ and a triangulation $\mathcal{T}_i$ of Ω such that $T_{i,j} \in \mathcal{T}_i$, for every $j = 1, \dots, N_{\hat{p}}$. We remark that the triangulation $\mathcal{T}_i$ does not have to be explicitly generated; it is only used to properly define all the quantities introduced in [46] required to derive the proposed error estimator.

Let $V_h^i = \text{span}\{\psi_j^i : 1 \le j \le \dim(V_h^i)\}$ be the space of piecewise linear functions defined on $\mathcal{T}_i$. Where $\{\psi_j^i : 1 \le j \le \dim(V_h^i)\}$ is a Lagrange basis of V_h^i. It is then possible to define two constants c_h^i and C_h^i, with $0 < c_h^i < C_h^i$, such that

$$c_h^i |v|_{1,\Omega} \le \|\boldsymbol{v}\|_2 \le C_h^i |v|_{1,\Omega} \qquad \forall v \in V_h^i\,, \tag{15}$$

where $v = \sum_{j=1}^{\dim(V_h^i)} v_j \psi_j^i$ is an arbitrary element of V_h^i associated with the expansion coefficients $\boldsymbol{v} = \left\{v_1, \dots, v_{\dim(V_h^i)}\right\}$ and $\|\boldsymbol{v}\|_2 = \left(\sum_{j=1}^{\dim(V_h^i)} v_i^2\right)^{1/2}$.

Then, given an integer $k \ge 0$, for any element $E \in \mathcal{T}_i$, we define the projection operator $\Pi_{E,k} : L^2(E) \to \mathbb{P}_k(E)$ such that

$$\int_E \Pi_{E,k}\phi = \int_E \phi \qquad \forall \phi \in L^2(E)\,. \tag{16}$$

We also denote by $\{(\xi_\ell^E, \omega_\ell^E) : \ell \in I^E\}$ a quadrature formula of order q on E and define the quadrature-based discrete seminorm:

$$\|v\|_{0,E,\omega} = \left(\sum_{\ell \in I^E} v^2(\xi_\ell^E)\, \omega_\ell^E\right)^{1/2}. \tag{17}$$

We require the weights and nodes of this quadrature rule to coincide with the ones introduced in (9) when E is a triangle included in P_i (i.e., when $E \in \{T_{i,1}, \dots, T_{i,N_{\hat{p}}}\}$). We can now introduce all the terms involved in the a posteriori error estimator.

Let $\eta_{\text{rhs},1}(E)$ and $\eta_{\text{rhs},2}(E)$ be the quantities:

$$\begin{aligned} \eta_{\text{rhs},1}(E) &= h_E \|f - \Pi_{E,q-1} f\|_{0,E}\,, \\ \eta_{\text{rhs},2}(E) &= h_E \|f - \Pi_{E,q-1} f\|_{0,E,\omega} + \|f - \Pi_{E,q} f\|_{0,E,\omega}\,. \end{aligned} \tag{18}$$

They measure the oscillations of the forcing term with respect to its polynomial projections in various norms. Similar oscillations are also measured for the diffusion, convection and reaction terms by the terms $\eta_{\text{coef},i}(E)$ for $i = 1, \dots, 6$:

$$\begin{aligned} \eta_{\text{coef},1}(E) &= \|\mu \nabla u^{\mathcal{NN}} - \Pi_{E,q}(\mu \nabla u^{\mathcal{NN}})\|_{0,E}\,, \\ \eta_{\text{coef},2}(E) &= h_E \|\boldsymbol{\beta} \cdot \nabla u^{\mathcal{NN}} - \Pi_{E,q-1}(\boldsymbol{\beta} \cdot \nabla u^{\mathcal{NN}})\|_{0,E}\,, \\ \eta_{\text{coef},3}(E) &= h_E \|\sigma u^{\mathcal{NN}} - \Pi_{E,q-1}(\sigma u^{\mathcal{NN}})\|_{0,E}\,, \\ \eta_{\text{coef},4}(E) &= \|\mu \nabla u^{\mathcal{NN}} - \Pi_{E,q}(\mu \nabla u^{\mathcal{NN}})\|_{0,E,\omega}\,, \\ \eta_{\text{coef},5}(E) &= h_E \|\boldsymbol{\beta} \cdot \nabla u^{\mathcal{NN}} - \Pi_{E,q-1}(\boldsymbol{\beta} \cdot \nabla u^{\mathcal{NN}})\|_{0,E,\omega}\,, \\ &\qquad + \|\boldsymbol{\beta} \cdot \nabla u^{\mathcal{NN}} - \Pi_{E,q}(\boldsymbol{\beta} \cdot \nabla u^{\mathcal{NN}})\|_{0,E,\omega} \\ \eta_{\text{coef},6}(E) &= h_E \|\sigma u^{\mathcal{NN}} - \Pi_{E,q-1}(\sigma u^{\mathcal{NN}})\|_{0,E,\omega} \\ &\qquad + \|\sigma u^{\mathcal{NN}} - \Pi_{E,q}(\sigma u^{\mathcal{NN}})\|_{0,E,\omega}\,, \end{aligned} \tag{19}$$

where $u^{\mathcal{NN}}$ is the output of the neural network after the enforcement of the Dirichlet boundary conditions through the operator B and h_E is the diameter of E. Then, let us define the term $\eta_{res}(E)$, which measures how well the equation is satisfied, as

$$\eta_{\text{res}}(E) = h_E \| \operatorname{bulk}_E(u^{\mathcal{NN}}) \|_{0,E} + h_E^{1/2} \sum_{e \subset \partial E} \| \operatorname{jump}_e(u^{\mathcal{NN}}) \|_{0,e} , \tag{20}$$

where

$$\operatorname{bulk}_E(u^{\mathcal{NN}}) = \Pi_{E,q-1} f + \nabla \cdot \Pi_{E,q}(\mu \nabla u^{\mathcal{NN}}) - \Pi_{E,q-1}(\boldsymbol{\beta} \cdot \nabla u^{\mathcal{NN}} + \sigma u^{\mathcal{NN}})$$

$$\operatorname{jump}_e(u^{\mathcal{NN}}) = \Pi_{E_1,q}(\mu \nabla u^{\mathcal{NN}}) \cdot \boldsymbol{n} - \Pi_{E_2,q}(\mu \nabla u^{\mathcal{NN}}) \cdot \boldsymbol{n} .$$

Note that $\operatorname{jump}_e(u^{\mathcal{NN}})$ measures the interelemental jumps of $\Pi_{E,q}(\mu \nabla u^{\mathcal{NN}})$ across the edge e with normal unit vector $\boldsymbol{n}$ shared by the elements E_1 and E_2.

Finally, we introduce the approximate elemental forms:

$$a_h^{i,E}(w,v) = \sum_{\ell \in I^E} [\mu \nabla w \cdot \nabla v + \boldsymbol{\beta} \cdot \nabla w \, v + \sigma w v](\boldsymbol{\xi}_\ell^E) \, \omega_\ell^E , \tag{21}$$

$$F_h^{i,E}(v) = \sum_{\ell \in I^E} [f v](\boldsymbol{\xi}_\ell^E) \, \omega_\ell^E , \tag{22}$$

where $\boldsymbol{\xi}_\ell^E$ and ω_ℓ^E, $\ell \in I^E$, are the nodes and weights used in Equation (17). With such forms, it is possible to define the residuals

$$r_{h,i,j}(w) = \sum_{E \in \mathcal{T}_i} F_h^{i,E}(\psi_j^i) - a_h^{i,E}(w, \psi_j^i), \qquad j = 1, \ldots, \dim\left(V_h^i\right)$$

and the quantity $\eta_{\text{loss}}(E)$ as

$$\eta_{\text{loss}}(E) = C_h \sqrt{\sum_{j \in I_h^E} r_{h,i,j}^2(u^{\mathcal{NN}})} . \tag{23}$$

Here, denoting the support of the function $\psi_j^i \in V_h^i$ by $\operatorname{supp} \psi_j^i$, the elemental index set

$$I_h^E = \{ j \in I_h : E \subset \operatorname{supp} \psi_j^i \}$$

is the set containing the indices of the functions whose support contains E. It is then possible to estimate the error between the unknown exact solution u and its MF-VPINN approximation $u^{\mathcal{NN}}$ by means of the computable quantities in Equations (18)–(20) and (23) as

$$|u - u^{\mathcal{NN}}|_{1,E} \lesssim \left(\eta_{\text{res}}^2(E) + \eta_{\text{loss}}^2(E) + \sum_{i=1}^{6} \eta_{\text{coef},i}^2(E) + \sum_{i=1}^{2} \eta_{\text{rhs},i}^2(E) \right)^{1/2} . \tag{24}$$

Once more, we refer to [46] for the proof of such a statement.

We recall that our goal is to obtain a computable error estimator associated with a single patch P_i. When evaluated on an element $E \in P_i$, the quantity on the right-hand side of Equation (24) implicitly depends on several elements in V_h^i that do not belong to P_i because of the presence of $\eta_{\text{res}}^2(E)$ and $\eta_{\text{loss}}^2(E)$. Therefore, such an estimator is not computable without generating the triangulation $\mathcal{T}_i$ and the corresponding space V_h^i. Instead, we look for an error estimator that does not control the error on the entire patch but only in a neighborhood $\mathcal{N}_i$ of its center $\boldsymbol{c}_{P_i} = M_i(\boldsymbol{c}_{\hat{P}})$. This can be carried out by considering only the terms whose computation involves geometric elements containing $\boldsymbol{c}_{P_i}$ and the only function ψ_j^i that does not vanish on $\boldsymbol{c}_{P_i}$. Note that such a function is the function φ_i defined in (7). Therefore, the error estimator η_i that controls the error in $\mathcal{N}_i$ can be computed as

$$\eta_i = \left[\eta_{\text{res},i}^2 + C_h^2 r_{h,i}^2(u^{\mathcal{NN}}) + \sum_{j=1}^{N_{\hat{p}}} \left(\sum_{k=1}^{6} \eta_{\text{coef},k}^2(T_{i,j}) + \sum_{k=1}^{2} \eta_{\text{rhs},k}^2(T_{i,j}) \right) \right]^{1/2}, \tag{25}$$

where $\eta_{\text{res},i}$ is defined as

$$\eta_{\text{res},i} = \sum_{j=1}^{N_{\hat{p}}} \left(h_{T_{i,j}} \| \text{bulk}_{T_{i,j}}(u^{\mathcal{NN}}) \|_{0,T_{i,j}} + h_{P_i}^{1/2} \| \text{jump}_{e_{i,j}}(u^{\mathcal{NN}}) \|_{0,e_{i,j}} \right). \tag{26}$$

In (26), we denote by h_{P_i} the diameter of the patch P_i and by $e_{i,j}, j = 1, \dots, N_{\hat{p}}$ the edges connecting its vertices with c_{P_i}.

Since η_i can be seen as an approximation of the right-hand side of (24), we use it as an indicator of the error $|u - u^{\mathcal{NN}}|_{1,\mathcal{N}_i}$. It is important to remark that η_i can be computed without generating $\mathcal{T}_i$ and V_h^i. In fact, its computation involves only the function φ_i, the triangles partitioning P_i and the edges connecting its vertices with its center.

2.4. The Choice of $\mathcal{M}$ and $\mathcal{P}$

In this section, the procedure adopted to generate the set of test functions used to train the MF-VPINN is described. We propose an iterative approach, in which the MF-VPINN is initially trained with very few test functions, and then other test functions are added in the regions of the domain in which the H_1 norm of the error is larger. We anticipate that, as shown in Section 3.3, generating test functions in regions where $r_{h,i}^2$ is large may not lead to accurate solutions because $r_{h,i}^2$ is not proportional to the H_1 error. Therefore, such a choice may increase the density of test functions where they are not required while maintaining only a few test functions in regions in which the error is large. Instead, we use the error indicator η_i defined in (25).

Let us initially consider a cover $\mathcal{P}_0 = \{P_i\}_{i=1}^{n_{\text{patches}}}$ of Ω comprising a few patches (i.e., n_{patches} is a small integer) and the corresponding set of mappings $\mathcal{M}_0 = \{M_i\}_{i=1}^{n_{\text{patches}}}$ and test functions $\{\varphi_i\}_{i=1}^{n_{\text{patches}}}$. These sets induce a loss function $R_h^2(w; \mathcal{P}_0)$ as defined in (14), which is used to train an MF-VPINN. After this initial training, one computes $\eta_i^\gamma = \gamma_i \eta_i$ for each patch $P_i \in \mathcal{P}_0$ and stores the result in the array $\boldsymbol{\eta} = \left[\eta_1^\gamma, \dots, \eta_{n_{\text{patches}}}^\gamma\right]$. Note that η_i^γ is a suitable rescaling of η_i to get rid of dependence from the size of P_i. Let us choose a threshold $1 \le \tau_0 \le n_{\text{patches}}$, sort $\boldsymbol{\eta}$ in descending order obtaining $\boldsymbol{\eta}_{\text{sort}} = \left[\eta_{s_1}^\gamma, \dots, \eta_{s_{n_{\text{patches}}}}^\gamma\right]$ (where we denote by $[s_1, \dots, s_{n_{\text{patches}}}]$ the index set corresponding to a suitable permutation of $[1, \dots, n_{\text{patches}}]$) and consider the vector $\overline{\boldsymbol{\eta}}_0 = \left[\eta_{s_1}^\gamma, \dots, \eta_{s_{\tau_0}}^\gamma\right]$. It is possible to note that $\overline{\boldsymbol{\eta}}_0$ contains only the τ_0 worst values of the indicator; it thus allows us to understand where the error is higher and where additional test functions are required to increase the model accuracy.

It is then possible to move forward with the second iteration of the iterative training. For each patch P_i such that $\eta_i^\gamma \in \overline{\boldsymbol{\eta}}_0$, we generate k_{new} new patches $P_i^k, k = 1, \dots, k_{\text{new}}$ with centers inside P_i and areas such that $\text{area}(P_i) < \cup_{k=1}^{k_{\text{new}}} \text{area}(P_i^k) < c \cdot \text{area}(P_i)$, where $c > 1$ is a tunable parameter. In the numerical experiments, we use $c = 1.25$. There exist different strategies to choose the number, the dimension, and the position of the centers of the new patches. Such strategies are described in Section 3 with particular attention to the effects of these choices on the MF-VPINN accuracy.

Let us denote by $\mathcal{P}_1$ the set $\mathcal{P}_1 = \mathcal{P}_0 \cup \{P_{s_1}^k\}_{k=1}^{k_{\text{new}}} \cup \dots \cup \{P_{s_{\tau_0}}^k\}_{k=1}^{k_{\text{new}}}$ and by $\mathcal{M}_1$ the corresponding set of mappings. Then, it is possible to define the loss function $R_h^2(w; \mathcal{P}_1)$, continue the training of the previously trained MF-VPINN, compute the error indicator η_i^γ for each patch $P_i \in \mathcal{P}_1$, and obtain the vector $\overline{\boldsymbol{\eta}}_1$ used to decide where to insert the new patches to generate $\mathcal{P}_2$. In general, iterating this procedure, it is possible to compute a set of patches $\mathcal{P}_m$ and of mappings $\mathcal{M}_m$ from the previously obtained sets $\mathcal{P}_{m-1}$ and $\mathcal{M}_{m-1}$. Technical optimization details are discussed in Section 3.1.

3. Numerical Results

In this section, we provide several numerical results to show the performance of the training strategy described in Section 2.4. In Section 3.1, we describe the structure of the MF-VPINN implementation and highlight some details that have to be taken into account in order to increase the efficiency of the training phase. Different strategies to choose the position of the new patches are discussed in Section 3.2. The importance of the use of the error indicator is remarked in Section 3.3 with additional numerical examples. An example on a more complex domain is shown in Section 3.4 to discuss some ideas to adapt the proposed strategies in more complex domains.

3.1. Implementation Details

The computer code used to perform the experiment is implemented in Python using the Python package Tensorflow [4] to generate the neural network architecture and train the MF-VPINN. Using the notation introduced in Section 2.1, the used neural network consists of $L = 5$ layers with $N_\ell = 50$ neurons in each hidden layer (i.e., for $\ell = 1, \ldots, L-1$); the activation function is the hyperbolic tangent in each hidden layer. For the first iteration of the iterative training, the neural network weights in the ℓ-th layer are initialized with a glorot normal distribution, i.e., a truncated normal distribution with mean 0 and standard deviation equal to $\sqrt{2/(N_{\ell-1}+N_\ell)}$. Then, for the subsequent iterations, their are initialized with the weights obtained at the end of the previous one.

During the first iteration of the training (during the minimization of $R_h^2(\cdot;\mathcal{P}_0)$), the optimization is carried out by exploiting the ADAM optimizer [51] with an exponentially decaying learning rate from 10^{-2} to 10^{-4} and with the second-order L-BFGS optimizer [52]. Then, from the second training iteration, we only use the L-BFGS optimizer. We remark that L-BFGS allows a very fast convergence but only if the initial starting point is close enough to the problem's solution. Therefore, in the first training iteration, we use ADAM to obtain a first approximation of the solution that is then improved via L-BFGS. Then, since the m-th training iteration starts from the solution computed during the $(m-1)$-th one, we assume that the starting point is close enough to the solution of the new optimization problem (associated with a difference loss function with more patches) and we only use L-BFGS to increase the training efficiency.

During the m-th iteration of the training, the training set consists of all the quadrature nodes ξ_ℓ^i, for any $\ell \in I^{\hat{P}}$ and for any patch $P_i \in \mathcal{P}_m$ as defined in (9). The order of the chosen quadrature rule is $q = 3$ inside each triangle. The Dirichlet boundary conditions are imposed by means of the operator B defined in (6). In this operator, for our first numerical test, the function ϕ is a polynomial bubble vanishing on Γ and $\overline{g}$ is the output of a neural network trained to interpolate the boundary data. For the numerical test in Section 3.4, instead, ϕ is computed as in [50] and $g = 0$. To decrease the training time, the functions ϕ, $\nabla\phi$, $\overline{g}$ and $\nabla\overline{g}$ are evaluated only once at the beginning of the m-th training iteration and they are then combined to evaluate $Bu^{\mathcal{NN}}$ and its gradient (where $u^{\mathcal{NN}}$ is the output of the last layer of the neural network). The derivatives of $u^{\mathcal{NN}}$ and $\overline{g}$ are computed via automatic differentiation [53] due to the complexity of their analytical expressions.

The output of the model is the value of the function $Bu^{\mathcal{NN}}$ and its gradient evaluated at the input points. Such values are then suitably combined using sparse and dense tensors to compute the quantity $R_h^2(Bu^{\mathcal{NN}};\mathcal{P}_m)$. The sparse tensors contain the evaluation of φ_i and $\nabla\varphi_i$ at each input point, whereas the dense ones store the quadrature weights, the vector $\gamma = \{\gamma_i\}_{i=1}^{n_{\text{patches}}}$ and the evaluation of μ, β, σ and f at the input points. We highlight that all these tensors have to be computed once at the beginning of the m-th training iteration (updating the ones of the $(m-1)$-th iteration) to significantly decrease the training computational cost.

As discussed in Section 2.1, we assume that all the patches and test functions can be generated from a reference patch $\hat{P}$. For each patch $P_i \in \mathcal{P}_m$, one has to generate all the data structures required to assemble the loss function and the error indicator η_i. To do so, it is possible to explicitly construct all the tensors required to assemble the term $\hat{a}_h(w,\hat{\varphi})$ and all the terms involved in the computation of the reference error indicator $\hat{\eta}$ only once, at the beginning of the first iteration of the training. Then, all these tensors can be suitably rescaled to obtain the ones corresponding to the patches and test functions involved in the loss function and error indicators computations.

To stabilize the MF-VPINN, we introduce the L^2 regularization term

$$\mathcal{L}_{\text{reg}}(u^{\mathcal{NN}}) = \lambda_{\text{reg}} \|\mathbf{u}^{\mathcal{NN}}\|_2^2,$$

where $\mathbf{u}^{\mathcal{NN}}$ is the set of weights of the neural network introduced in Section 2.1. In our numerical experiments, we use $\lambda_{\text{reg}} = 10^{-5}$. During the m-th iteration of the training, such a quantity is added to $R_h^2(Bu^{\mathcal{NN}}; \mathcal{P}_m)$ to obtain the training loss function

$$\mathcal{L}_m(u^{\mathcal{NN}}) = R_h^2(Bu^{\mathcal{NN}}; \mathcal{P}_m) + \mathcal{L}_{\text{reg}}(\mathbf{w}^{\mathcal{NN}}), \tag{27}$$

which has to be minimized accurately enough. Indeed, if $\mathcal{L}_m$ is minimized poorly, the new patches $\mathcal{P}_{m+1} \backslash \mathcal{P}_m$ may be added in regions where they are not necessary because the accuracy of $Bu^{\mathcal{NN}}$ may still improve during the training and may not be inserted in areas where they are required. Note that, in order to compute the numerical solution, the MF-VPINN has to be trained multiple times with a different set of patches $\mathcal{P}_m$ to minimize the losses $\{\mathcal{L}_m\}$. Since such an iterative training may be expensive, we propose an early stopping strategy [54] based on the discussed error indicator to reduce its computational cost. In its basic version, early stopping consists of evaluating a chosen metric on a validation set in order to know when the neural network accuracy on data that are not present in the training set start worsening. Interrupting the training there prevents overfitting and improves generalization. In our context, instead, we can directly track the behavior of the MF-VPINN H^1 error on each patch through the corresponding error indicator to understand when it stops decreasing. Therefore, given the set of patches $\mathcal{P}_m$, the chosen metric is the linear combination $ES_m = \sum_{i=1}^{\dim(\mathcal{P}_m)} \eta_i^{\gamma}$. Numerical results showing the performance of this strategy are presented in Sections 3.2 and 3.4.

3.2. Adaptive Training Strategies

Let us consider the Poisson problem:

$$\begin{cases} -\Delta u = f & \text{in } \Omega\,, \\ \quad u = g & \text{on } \Gamma\,, \end{cases} \tag{28}$$

defined on the unit square $\Omega = (0,1)^2$. The forcing term f and the boundary condition g are chosen such that the exact solution is, in polar coordinates,

$$u(r,\theta) = r^{\frac{2}{3}} \sin\left(\frac{2}{3}\left(\theta + \frac{\pi}{2}\right)\right). \tag{29}$$

We use this function, represented in Figure 2, because the solution u is such that $u \in H^{5/3-\varepsilon}(\Omega)$ but $u \in C^\infty(\Omega \backslash \mathcal{N}_0)$, where we denote by $\mathcal{N}_0$ a neighborhood of the origin. Therefore, we know that an efficient distribution of patches has to be characterized by a high density only near the origin.

Below, we propose, in order of complexity, three alternatives to construct the new patches after having marked the ones with the higher error indicator. The first strategy is the most simple and intuitive, and the new patches are randomly generated with centers inside the marked patches, whereas the second strategy and third one place the new centers on a small local cartesian grid to ensure a more regular distribution. The difference between the second and the third strategies is that the marked patches are removed to increase the efficiency and we add a constraint to the marking procedure to ensure more regular distributions of the new patches.

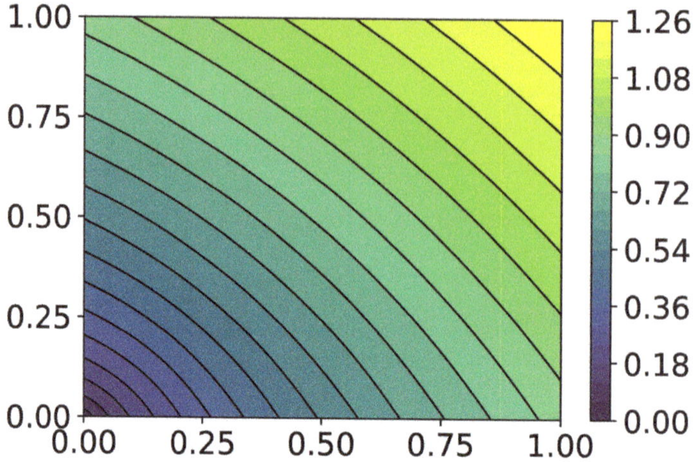

Figure 2. Graphical representation of the solution u in (29).

Strategy #1: Random patch centers with uniform distribution

To solve Problem (28), as a first strategy, we consider the reference patch $\hat{P} = (0,1)^2$ and generate a sequence of sets of patches. During the first training iteration, we use $\mathcal{P}_0 = \{\hat{P}\}$ since this is already a cover of Ω. During the second iteration, we enrich the set of patches as $\mathcal{P}_1 = \mathcal{P}_0 \cup \{P_1, P_2, P_3, P_4\}$ where P_1, P_2, P_3 and P_4 are squared patches with edge $h_i = 0.6$, $i = 1, \ldots, 4$ and centers

$$c_{P_1} = (0.3, 0.3), \qquad c_{P_2} = (0.7, 0.3),$$

$$c_{P_3} = (0.3, 0.7), \qquad c_{P_4} = (0.7, 0.7).$$

This allows us to start from a homogeneous distribution of patches before utilizing the error indicator to choose the location of the new patches. Then, to decide how many patches have to be added to $\mathcal{P}_{m-1}$ to generate $\mathcal{P}_m$, we choose $\widetilde{\tau}_m$ such that

$$\widetilde{\tau}_m = \dim\left(\left\{\widetilde{\tau} \in \{1, \ldots, \dim(\mathcal{P}_{m-1}\} : \frac{\sum_{i=1}^{\widetilde{\tau}} \eta_{s_i}^{\gamma}}{\sum_{i=1}^{\dim(\mathcal{P}_{m-1})} \eta_i^{\gamma}} < 0.75\right\}\right) + 1 \tag{30}$$

and fix

$$\tau_m = \min(\lceil 0.3 \cdot \dim(\mathcal{P}_{m-1}) \rceil, \widetilde{\tau}_m). \tag{31}$$

Note that (30) allows us to consider the smallest set of patches such that the corresponding error indicators contribute at least 75% of the global error indicator ES_{m-1}, whereas (31) is considered to limit the maximum number of patches that can be added for efficiency reasons.

Then, to generate the generic set of patches $\mathcal{P}_m$, we fix a multiplication factor C_M to decide how many new patches have to be inserted inside each patch P_i such that $\eta_i^{\gamma} \in \overline{\boldsymbol{\eta}}_{m-1}$. Inside each chosen patch P_i, C_M centers $\widetilde{\boldsymbol{c}}_{P_i^k} = (\widetilde{x}_i^k, \widetilde{y}_i^k)$, $k = 1, \ldots, C_M$, are randomly generated with a uniform distribution and the new patches' edges' lengths are chosen as $h_i^k = \lambda \frac{A_{\text{ratio}}}{\sqrt{C_M}} h_i$. Here, λ is a random real value from the uniform distribution $U\left(\left[\frac{9}{10}, \frac{10}{9}\right]\right)$, and the scaling coefficient $\frac{A_{\text{ratio}}}{\sqrt{C_M}}$ is chosen such that the sum of the areas of the new patches is A_{ratio} times the area of the original patch P_i. In the numerical experiments, we use $A_{\text{ratio}} = 1.25$. This way, it is possible to allow the new patches to overlap and keep the area of the region $P_i \setminus \left(\cup_{k=1}^{C_M} P_i^k\right)$ reasonably small.

We remark that, with this strategy, it may happen that some patches are outside Ω. In order to avoid this risk, we move the centers $\widetilde{c}_{P_i^k}$ to obtain the actual patches centers $c_{P_i^k}$ as follows:

$$c_{P_i^k} = (x_i^k, y_i^k) \leftarrow \left(\max\left\{ \min\left\{ \widetilde{x}_i^k, 1 - \frac{h_i^k}{2} \right\}, \frac{h_i^k}{2} \right\}, \max\left\{ \min\left\{ \widetilde{y}_i^k, 1 - \frac{h_i^k}{2} \right\}, \frac{h_i^k}{2} \right\} \right). \quad (32)$$

We remark that, when the patch P_i is very close to a vertex of the domain, it is possible that multiple original centers $\widetilde{c}_{P_i^k}$ are such that the distance of both $\widetilde{x}_i^k$ and $\widetilde{y}_i^k$ from the x and y coordinates of the domain vertex is smaller than $h_i^k/2$. In this case, it is important to consider the random coefficient λ in the definition of h_i^k to avoid updating all these centers with the same point; otherwise, multiple new patches would coincide (because they would share the same center and size).

For the numerical test, we consider $C_M = 4$ and $C_M = 9$. Using significantly more accurate quadrature rules, we compare the approximate solution with the exact one defined in (29) and compute the relative H^1 error $\|u - u^{\mathcal{NN}}\|_1/\|u\|_1$ at the end of each training iteration. The obtained errors are shown as blue circles ($C_M = 4$) and red triangles ($C_M = 9$) in Figure 3. It can be noted that, with both values of C_M, when more patches are used, the error is smaller, even though the convergence rate is limited by the low regularity of the solution. It is also interesting to observe the positions and sizes of the used patches; such information is summarized in Figures 4 and 5. In such figures, each dot is in the center of a patch P_i, and its size and color represent the size h_i^2 and the scaled indicator η_i^γ associated with P_i. It can be noted that, even if the new centers are chosen randomly in the few selected patches, the final distribution is the expected one. In fact, most of the patches cluster around the origin, whereas the rest of the domain is covered by fewer patches. Nevertheless, we highlight that, when $C_M = 9$, there are more small and medium patches far from the origin, yielding a more uniform covering of the areas far from the singular point and a slightly better accuracy.

Strategy #2: Fixed patch centers

From the results discussed in *Strategy #1*, it can be observed that choosing the position of the new centers randomly may lead to non-uniform patch distribution in regions far from the singular point. In order to obtain better distributions, let us fix a priori the position of the new centers. Let us consider the reference patch $\hat{P} = (0,1)^2$ and the points

$$\begin{aligned} \hat{c}_1 &= (0.25, 0.25), & \hat{c}_2 &= (0.75, 0.25), \\ \hat{c}_3 &= (0.25, 0.75), & \hat{c}_4 &= (0.75, 0.75), \end{aligned} \quad (33)$$

when $C_M = 4$ and

$$\begin{aligned} \hat{c}_1 &= (0.2, 0.2), & \hat{c}_2 &= (0.2, 0.5), & \hat{c}_3 &= (0.2, 0.8), \\ \hat{c}_4 &= (0.5, 0.2), & \hat{c}_5 &= (0.5, 0.5), & \hat{c}_6 &= (0.5, 0.8), \\ \hat{c}_7 &= (0.8, 0.2), & \hat{c}_8 &= (0.8, 0.5), & \hat{c}_9 &= (0.8, 0.8), \end{aligned} \quad (34)$$

when $C_M = 9$. At the end of the $(m-1)$-th training iteration, if $\eta_i^\gamma \in \overline{\boldsymbol{\eta}}_{m-1}$, the C_M centers inside P_i are chosen as $c_{P_i^k} = M_i(\hat{c}_k)$, $k = 1, \ldots, C_M$. Once more, to avoid patches partially outside Ω, we update such centers as in (32). We highlight that defining the new centers as in (33) and in (34) and the length h_i^k of the edges of the new patches as in *Strategy #1*, then the new patches with centers inside P_i form a cover of P_i, i.e., $P_i \subsetneq \cup_{k=1}^{C_M} P_i^k$. Such a property does not hold if the new centers are randomly chosen.

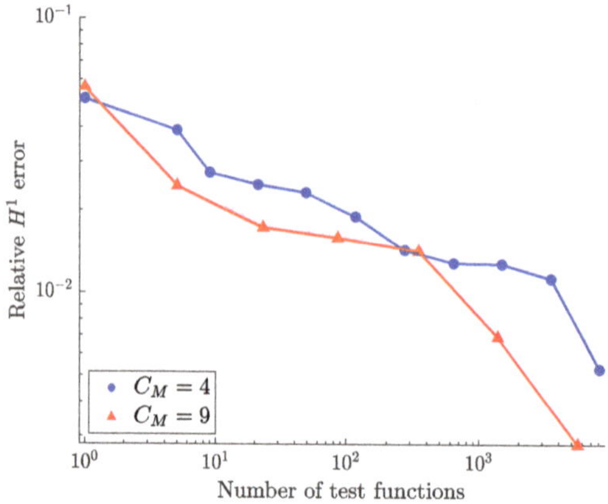

Figure 3. *Strategy #1*: Relative H^1 errors obtained at the end of each training iteration for $C_M = 4$ (blue circles) and $C_M = 9$ (red triangles).

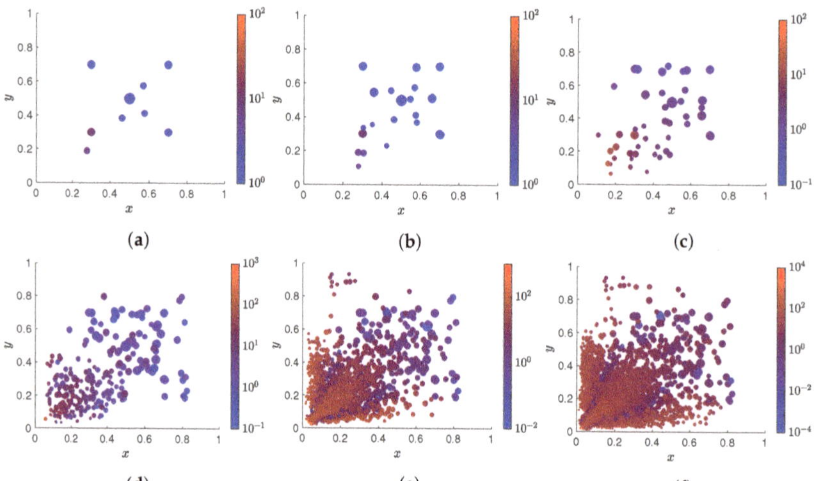

Figure 4. *Strategy #1*: Patches used to train the MF-VPINN with $C_M = 4$. Each dot represents a patch P_i, its position is the center $\mathbf{c}_{P_i}$ of the patch, its size is proportional to the patch size h_i^2, and its color is associated with the quantity η_i^γ. **(a)** Representation of $\mathcal{P}_2$; **(b)** Representation of $\mathcal{P}_3$; **(c)** Representation of $\mathcal{P}_4$; **(d)** Representation of $\mathcal{P}_6$; **(e)** Representation of $\mathcal{P}_8$; **(f)** Representation of $\mathcal{P}_9$.

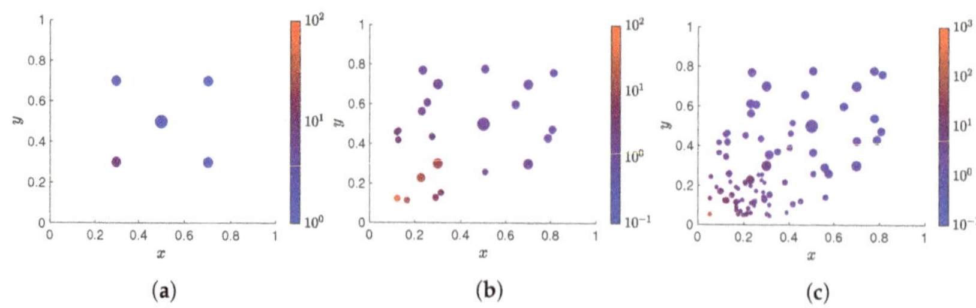

Figure 5. *Cont.*

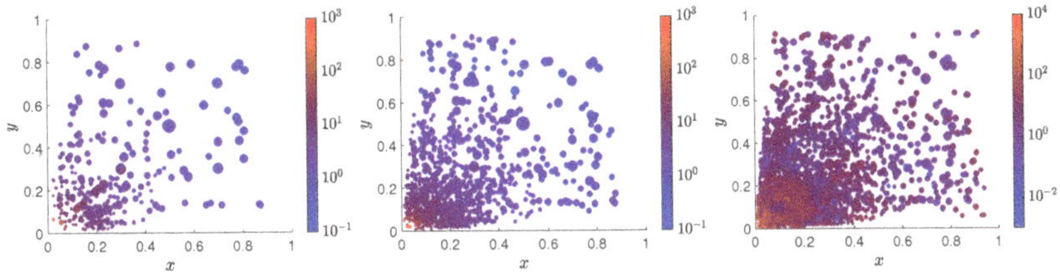

(d) (e) (f)

Figure 5. *Strategy #1*: Patches used to train the MF-VPINN with $C_M = 9$. Each dot represents a patch P_i, its position is the center $\boldsymbol{c}_{P_i}$ of the patch, its size is proportional to the patch size h_i^2, and its color is associated with the quantity η_i^{γ}. (**a**) Representation of $\mathcal{P}_1$; (**b**) Representation of $\mathcal{P}_2$; (**c**) Representation of $\mathcal{P}_3$; (**d**) Representation of $\mathcal{P}_4$; (**e**) Representation of $\mathcal{P}_5$; (**f**) Representation of $\mathcal{P}_6$.

Training an MF-VPINN with such a strategy leads to more accurate results. The error decays are shown in Figure 6, whereas a comparison with the previous one will be presented in Section 3.3. The patch distributions, for $C_M = 4$ and $C_M = 9$, are shown in Figures 7 and 8, respectively. Analyzing such distributions, it can be noted that the patches still accumulate near the origin as expected. However, it is possible to observe that there are regions that are only covered by the largest patches. This phenomenon is more evident when $C_M = 4$. To avoid such a phenomenon, we aim at inserting more patches far from the origin in order to train the MF-VPINN in the entire domain with a more balanced set of patches.

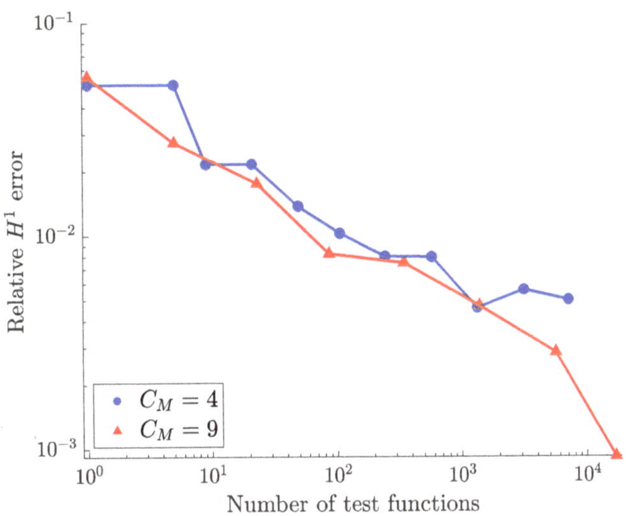

Figure 6. *Strategy #2*: Relative H^1 errors obtained at the end of each training iteration for $C_M = 4$ (blue circles) and $C_M = 9$ (red triangles).

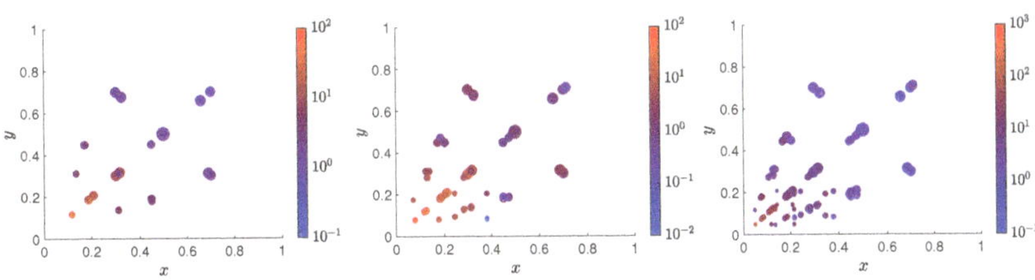

(a) (b) (c)

Figure 7. *Cont.*

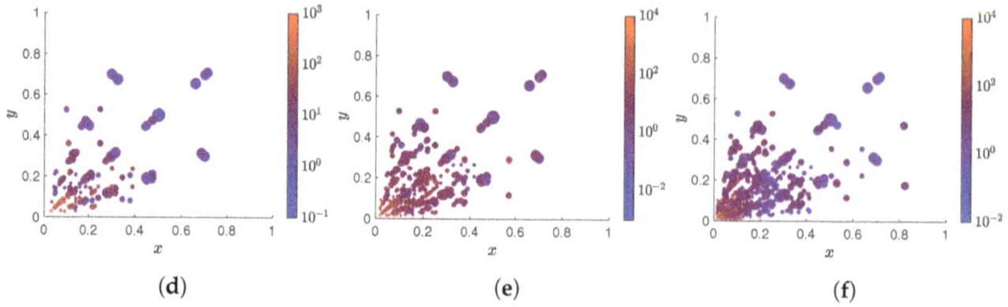

Figure 7. *Strategy #2*: Patches used to train the MF-VPINN with $C_M = 4$. Each dot represents a patch P_i, its position is the center $\boldsymbol{c}_{P_i}$ of the patch, its size is proportional to the patch size h_i^2, and its color is associated with the quantity η_i^γ. (**a**) Representation of $\mathcal{P}_3$; (**b**) Representation of $\mathcal{P}_4$; (**c**) Representation of $\mathcal{P}_5$; (**d**) Representation of $\mathcal{P}_6$; (**e**) Representation of $\mathcal{P}_7$; (**f**) Representation of $\mathcal{P}_8$.

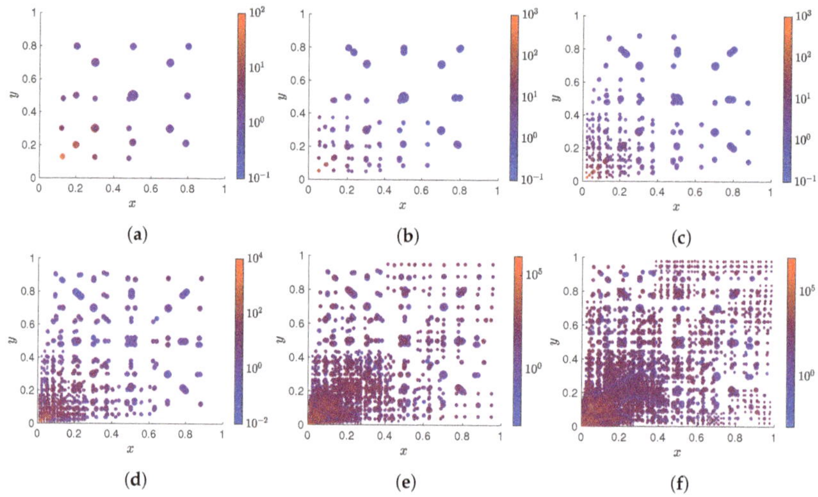

Figure 8. *Strategy #2*: Patches used to train the MF-VPINN with $C_M = 9$. Each dot represents a patch P_i, its position is the center $\boldsymbol{c}_{P_i}$ of the patch, its size is proportional to the patch size h_i^2, and its color is associated with the quantity η_i^γ. (**a**) Representation of $\mathcal{P}_2$; (**b**) Representation of $\mathcal{P}_3$; (**c**) Representation of $\mathcal{P}_4$; (**d**) Representation of $\mathcal{P}_5$; (**e**) Representation of $\mathcal{P}_6$; (**f**) Representation of $\mathcal{P}_7$.

Strategy #3: Fixed patch centers and small level gap strategy

In order to ensure better patch distributions, let us consider a new criterion to choose the position and the size of the new patches. We name this strategy the *small-level gap strategy* because it penalizes patch distributions with large differences between the levels of the smallest patches and the ones of the largest patches.

We denote by k-th level patch any patch P_i such that $P_i \in \mathcal{P}_k$ and $P_i \notin \mathcal{P}_{k'}$ for any $k' < k$. With this notation, it is possible to group all the patches according to their level. To do so, we denote by L_ℓ the set of k-th level patches with $k \leq \ell$. Let us consider the m-th training iteration. We define $\boldsymbol{\eta}^\ell_{\text{sort}}$ as the array containing the elements η_i^γ of $\boldsymbol{\eta}_{\text{sort}}$ (maintaining the same ordering) such that $P_i \in L_\ell$. We also denote by $\overline{\boldsymbol{\eta}}_{m,\ell}$ the array containing the first $\tau_m^\ell = \min\{\tau_m, \dim(L_\ell)\}$ elements of $\boldsymbol{\eta}^\ell_{\text{sort}}$. Note that $\overline{\boldsymbol{\eta}}_{m,\ell}$ is the equivalent of $\overline{\boldsymbol{\eta}}_m$ for patches in L_ℓ.

In order to generate the new patches in $\mathcal{P}_{m+1} \backslash \mathcal{P}_m$, let us add C_M new patches in any patch P_i such that $\eta_i^\gamma \in \overline{\boldsymbol{\eta}}_m \cup \overline{\boldsymbol{\eta}}_{m,\ell}$. The centers and sizes of the new patches are chosen as in *Strategy #2*. This allows us to exploit the fact that $P_i \subsetneq \cup_{k=1}^{C_M} P_i^k$ to remove the patches P_i such that $\eta_i^\gamma \in \overline{\boldsymbol{\eta}}_m \cup \overline{\boldsymbol{\eta}}_{m,\ell}$ from the new set of patches $\mathcal{P}_{m+1}$. We remark that such patches cannot be removed when the centers are randomly chosen as in *Strategy #1* because, in that case, $\mathcal{P}_{m+1}$ would not be a cover of Ω anymore.

We also highlight that, removing the patches P_i such that $\eta_i^\gamma \in \overline{\boldsymbol{\eta}}_m \cup \overline{\boldsymbol{\eta}}_{m,\ell}$ and choosing $A_{\text{ratio}} = 1$, it is possible to satisfy the inequality

$$\sum_{P_i \in \mathcal{P}_{m+1}} |P_i| \leq C|\Omega|,$$

for any $m \in \mathbb{N}$ and with $C > 0$ independent of m. Such a bound on the sum of the area of the patches is useful to ensure that there exists a number $N_{\text{patch\_per\_point}}$ such that any point inside Ω belongs to at most $N_{\text{patch\_per\_point}}$ patches. This property is useful to derive global error indicators. We choose to maintain $A_{\text{ratio}} = 1.25$ to compare the numerical results with the ones obtained using the previous strategies and to consider overlapping patches.

We train an MF-VPINN with $C_M = 4$ and $C_M = 9$ as in the previous tests. The corresponding error decays are shown in Figure 9. It can be observed that the error decreases in a smoother way and that, as in the previous tests, choosing $C_M = 4$ or $C_M = 9$ does not lead to significant differences in the error behavior. The patches used during the training are represented in Figures 10 and 11. We highlight that, when compared with the patch distributions in *Strategy #2*, there exist much more patches far from the origin, and, most importantly, the closer the center of a patch to the origin, the smaller its size. Even though the error decays with $C_M = 4$ and $C_M = 9$ are qualitatively similar, it should be noted that the patch distribution with $C_M = 9$ is more skewed. In fact. its patches can be clustered into two subgroups: the first one containing larger patches and covering most of the domain the second one containing only small patches with centers very close to the origin. A similar distribution is obtained with $C_M = 4$, even though it is characterized by a smoother transition between large and small patches.

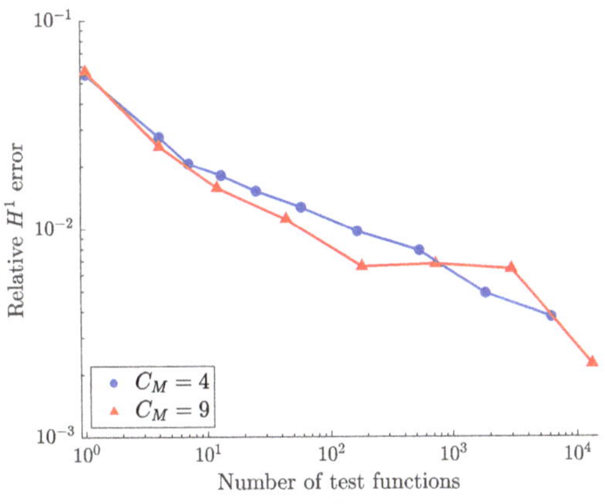

Figure 9. *Strategy #3*: Relative H^1 errors obtained at the end of each training iteration for $C_M = 4$ (blue circles) and $C_M = 9$ (red triangles).

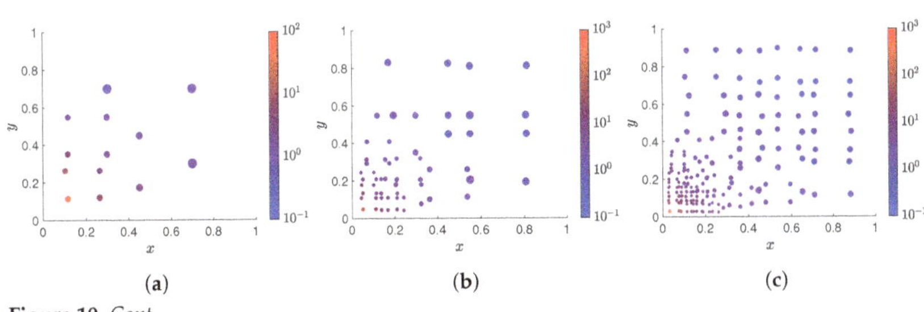

Figure 10. *Cont.*

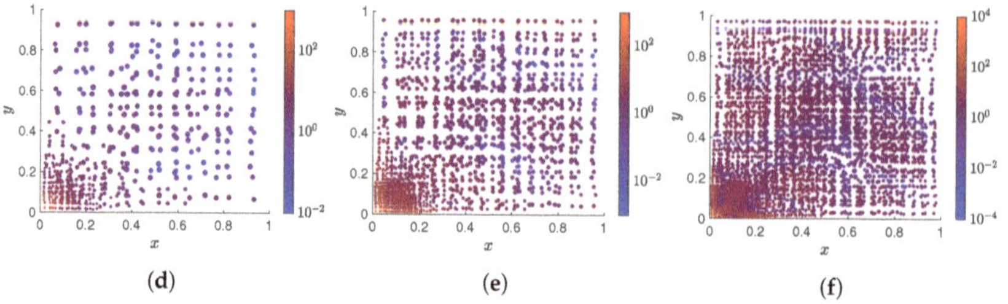

Figure 10. *Strategy #3*: Patches used to train the MF-VPINN with $C_M = 4$. Each dot represents a patch P_i, its position is the center $\boldsymbol{c}_{P_i}$ of the patch, its size is proportional to the patch size h_i^2, and its color is associated with the quantity η_i^γ. (**a**) Representation of $\mathcal{P}_3$; (**b**) Representation of $\mathcal{P}_5$; (**c**) Representation of $\mathcal{P}_6$; (**d**) Representation of $\mathcal{P}_7$; (**e**) Representation of $\mathcal{P}_8$; (**f**) Representation of $\mathcal{P}_9$.

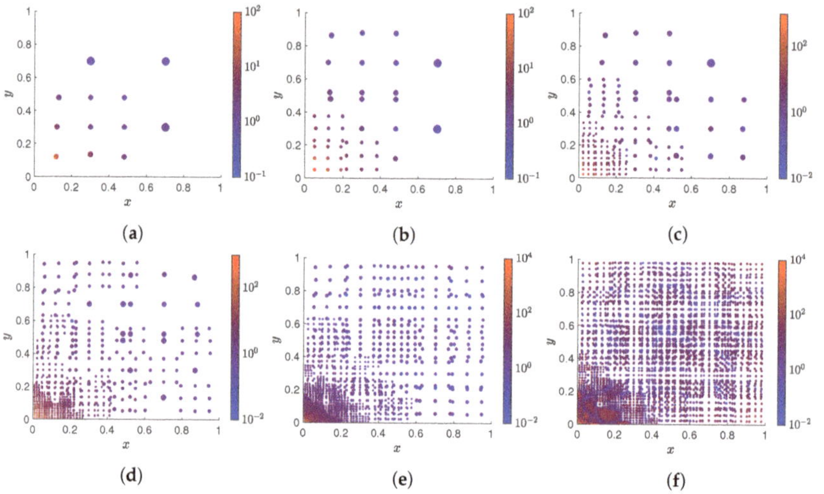

Figure 11. *Strategy #3*: Patches used to train the MF-VPINN with $C_M = 9$. Each dot represents a patch P_i, its position is the center $\boldsymbol{c}_{P_i}$ of the patch, its size is proportional to the patch size h_i^2, and its color is associated with the quantity η_i^γ. (**a**) Representation of $\mathcal{P}_2$; (**b**) Representation of $\mathcal{P}_3$; (**c**) Representation of $\mathcal{P}_4$; (**d**) Representation of $\mathcal{P}_5$; (**e**) Representation of $\mathcal{P}_6$; (**f**) Representation of $\mathcal{P}_7$.

In both cases, it can be observed that there are no large patches very close to small ones. This is in contrast with the distributions obtained in *Strategy #2* and leads to more stable solvers. Indeed, even though the test functions are not related to a global triangulation on the entire domain Ω, the current loss function is very similar to the one used in a standard VPINN with a good-quality mesh, i.e., a mesh in which neighboring elements are similar in size and shape. On the other hand, in *Strategy #2*, there exist large patches that are very close to small ones; this is equivalent to training a VPINN on a very poor-quality mesh. Such meshes, in the context of FEM, are strictly related to convergence and accuracy issues.

3.3. The Importance of the Error Indicator

As discussed in the previous sections, we use the error indicator described in Section 2.3 to interrupt the training and to decide where the new patches have to be inserted to maximize the accuracy. In this section, the advantages of such a choice are described.

Since each set $\mathcal{P}_m$ is a cover of Ω, the quantity $ES_m = \sum_{i=1}^{\dim(\mathcal{P}_m)} \eta_i$ is an indicator of the global H^1 error $\|u - u^{\mathcal{NN}}\|_1$ on the entire domain Ω. Therefore, tracking its behavior during the training is equivalent to tracking that of the unknown H^1 error. Such information is used to implement an early stopping strategy to reduce the computational cost of the iterative training. At the beginning of the m-th training iteration, all the vectors and sparse matrices required to compute ES_m are computed in a preprocessing phase. When such data structures are available, the error indicator can be assembled suitably combining basic algebraic operations.

We assemble ES_m every N_{check} epochs and store the best value obtained during the training, together with the corresponding neural network trainable parameters. Then, if no improvements are obtained in $p \cdot N_{\text{check}}$ epochs, the training is interrupted and the neural network parameters associated with the best value of ES_m are restored. Here, p is a tunable parameter named *patience*. The first N_{negl}^m epochs are neglected because they are often characterized by strong oscillations due to the optimizer initialization and the different loss functions. In the numerical experiment, we use $N_{\text{check}} = 10$, $p = 10$, $N_{\text{negl}}^m = 100(m+1)$.

Two typical scenarios are shown in Figure 12. In the top row, the behaviors of ES_m and of $c\|u - u^{\mathcal{NN}}\|_1$ are shown. Here, c is a scaling parameter used for visualization purposes, chosen such that ES_m and $c\|u - u^{\mathcal{NN}}\|_1$ coincide at the beginning of the training. Indeed, $\|u - u^{\mathcal{NN}}\|_1$ is about two orders of magnitude smaller than ES_m. Nevertheless, it can be noted that these two quantities display very similar behaviors during the training. In the bottom row, instead, we represent the corresponding loss function decay. The left column is associated with the training performed using the patches in $\mathcal{P}_6$ shown in Figure 5f and the right column with the one performed using the patches in $\mathcal{P}_2$ in Figure 11a. We remark that the loss function, ES_m and $c\|u - u^{\mathcal{NN}}\|$ are evaluated in the same epochs and that, in real applications, it is not possible to explicitly compute $c\|u - u^{\mathcal{NN}}\|$ since u is not known. Moreover, since we use the L-BFGS optimizer, the neural network is evaluated multiple times on the entire training set in each epoch. Therefore, on the x-axis of Figure 12 we show the number of neural network evaluations instead of the number of epochs.

It can be noted that the behavior of the quantities shown in the left column is qualitatively different from the ones in the right column. In fact, when the MF-VPINN is trained with $\mathcal{P}_6$ of Figure 5f, the error, the error indicator, and the loss function decrease in similar ways. Therefore, there is no need to interrupt the training early since the accuracy is improving, minimizing the loss function. On the other hand, when the MF-VPINN is trained with the $\mathcal{P}_2$ of Figure 11a, the loss decreases even when the error and the error indicator increase or remain constant. In this case, it is convenient to interrupt the training, since minimizing the loss function further would lead to more severe overfitting phenomena and a loss in accuracy and efficiency. At the end of the training, the neural network's trainable parameters corresponding to the best value of ES_2 are restored. We highlight that such a phenomenon, observed in [46] too, highlights the fact that the minimization of the loss function generates spurious oscillations that cannot be controlled and ruin the model accuracy. The issue can be partially alleviated with the adopted regularization or completely removed using inf-sup stable models as in [47].

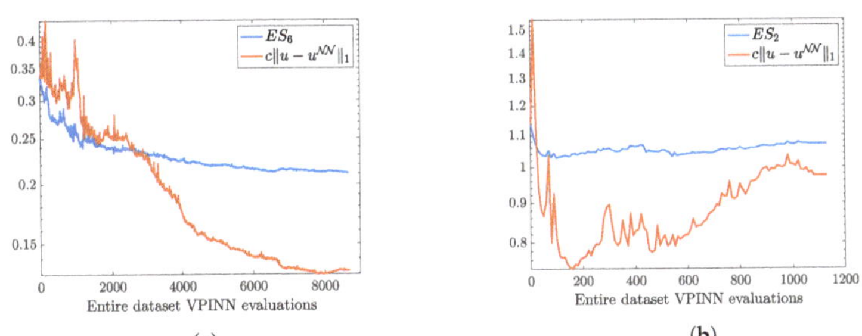

Figure 12. *Cont.*

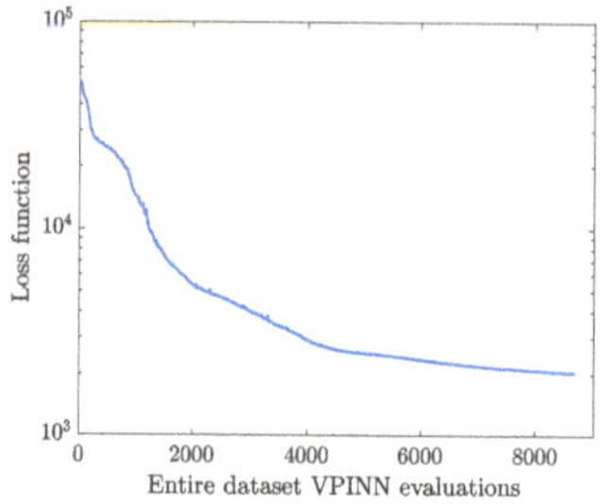

(c)

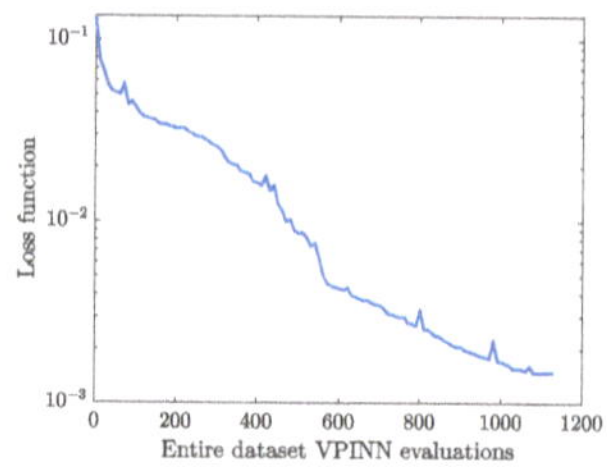

(d)

Figure 12. Top row: error indicator ES_m and rescaled H^1 error $c\|u - u^{\mathcal{NN}}\|$. Bottom row: loss function. Left column: curves for the training with patches in $\mathcal{P}_6$ shown in Figure 5f. Right column: curves for the training with patches in $\mathcal{P}_2$ in Figure 11a. (**a**) ES_6 and $c\|u - u^{\mathcal{NN}}\|$ for patches in Figure 5f; (**b**) ES_2 and $c\|u - u^{\mathcal{NN}}\|$ for patches in Figure 11a; (**c**) Loss function for patches in Figure 5f; (**d**) Loss function for patches in Figure 11a.

Strategy #4: Adaptive strategy without the error indicator

Let us now analyze the consequences of choosing the position of the new patches without using the error indicator. To do so, we consider *Strategy #1* but, instead of considering the new centers inside the patches P_i with the highest values of η_i^γ, we add them inside the patches with the highest values of $r^2_{h,i}(u^{\mathcal{NN}})$. Using the equation residuals is a common choice in PINN adaptivity because the residuals describe how accurately the neural network satisfies the PDE at that point. The obtained error decay is shown in Figure 13. It can be seen that the accuracy is worse than the ones obtained with the other strategies and that the convergence rate with respect to the number of patches is lower. In such a figure, we also compare the MF-VPINN with a standard VPINN trained with test functions defined on Delaunay meshes. Note that, when *Strategy #2* or *Strategy #3* is adopted, the MF-VPINN is more accurate than a simple VPINN, even though its main advantage resides in being a meshfree method.

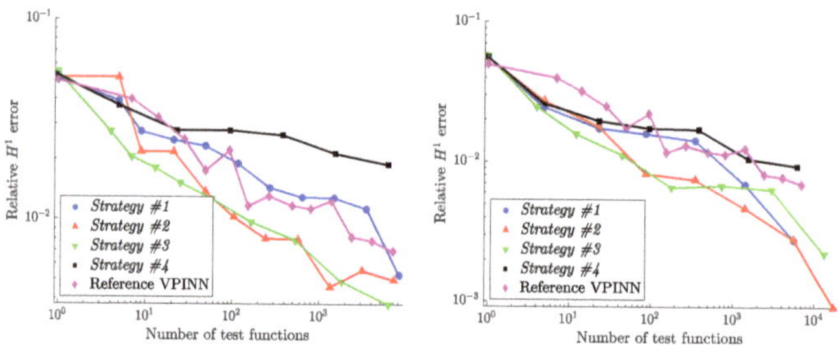

(a) (b)

Figure 13. Comparison between the relative H^1 errors obtained at the end of each training iteration with different strategies to choose the position of the new patches. (**a**) $C_M = 4$; (**b**) $C_M = 9$.

We highlight that, due to the low regularity of the solution, the expected convergence rate with respect to the number of test functions of an FEM solution computed on uniform refinements is $-1/3$. Note that the convergence rate of the proposed MF-VPINN method is still close to $-1/3$, even though it is a meshfree method (see Table 1). For completeness, we also remark that, if an adaptive FEM is used, the rate of convergence depends on the FEM order.

Table 1. Rates of convergence with respect to the number of test functions.

C_M	*Strategy #1*	*Strategy #2*	*Strategy #3*	*Strategy #4*	**Reference VPINN**
4	−0.213	−0.295	−0.283	−0.105	−0.232
9	−0.294	−0.376	−0.287	−0.182	−0.232

Coherently with Figure 13, the best strategies are *Strategy #2* and *Strategy #3*, whereas the worst one is *Strategy #4*, which does not exploit the error indicator. The poor performance of *Strategy #4* can also be explained by analyzing the corresponding patch distribution. Such distribution is shown in Figure 14 for $C_M = 4$ and in Figure 15 for $C_M = 9$. These plots highlight that the patches do not accumulate near the origin because the residuals of the patches closer to it are not significantly higher than the other ones. For example, note the different colors in Figures 4 and 14, since in both cases, we randomly choose the position of $C_M = 4$ centers inside the selected patches. Such a property is explained by the fact that, in order to minimize the loss function, the optimizer does not focus on specific regions of the domain. Therefore, the orders of magnitude of all the residuals with similar sizes are very close to each other regardless of the position of the corresponding patches. As discussed regarding Figure 12, we can conclude that the value of the residuals is not a good indicator of the actual error.

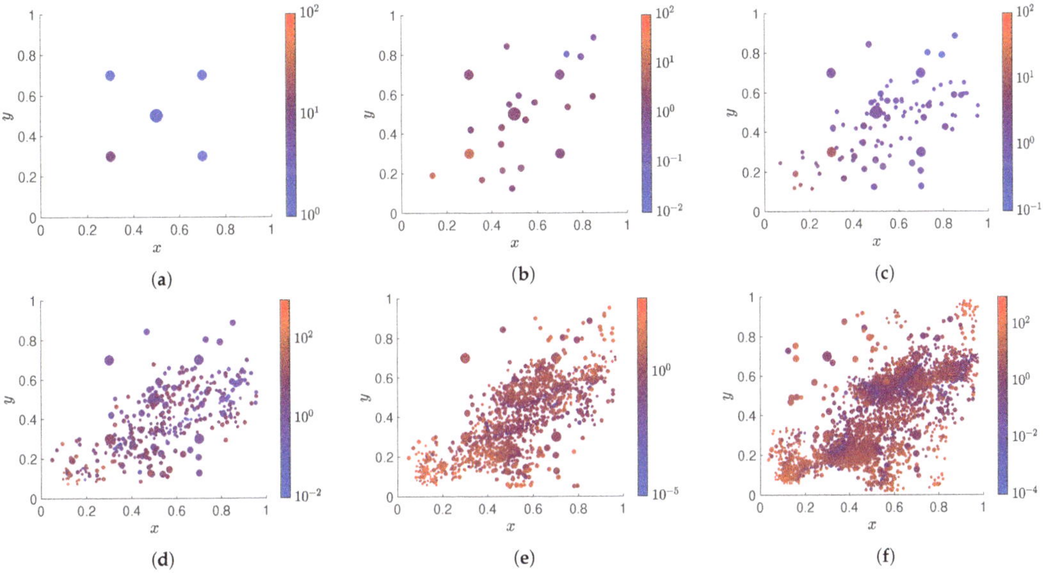

Figure 14. *Strategy #4*: Patches used to train the MF-VPINN with $C_M = 4$. Each dot represents a patch P_i, its position is the center $\boldsymbol{c}_{P_i}$ of the patch, its size is proportional to the patch size h_i^2, and its color is associated with the quantity $r_{h,i}^2(u^{\mathcal{NN}})$. (**a**) Representation of $\mathcal{P}_1$; (**b**) Representation of $\mathcal{P}_2$; (**c**) Representation of $\mathcal{P}_3$; (**d**) Representation of $\mathcal{P}_4$; (**e**) Representation of $\mathcal{P}_5$; (**f**) Representation of $\mathcal{P}_6$.

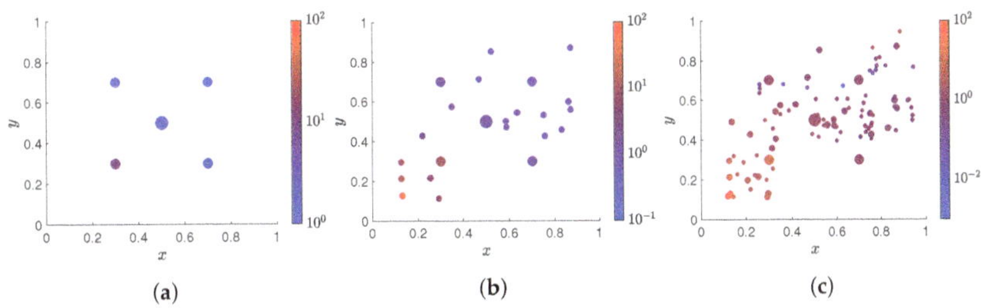

Figure 15. *Cont.*

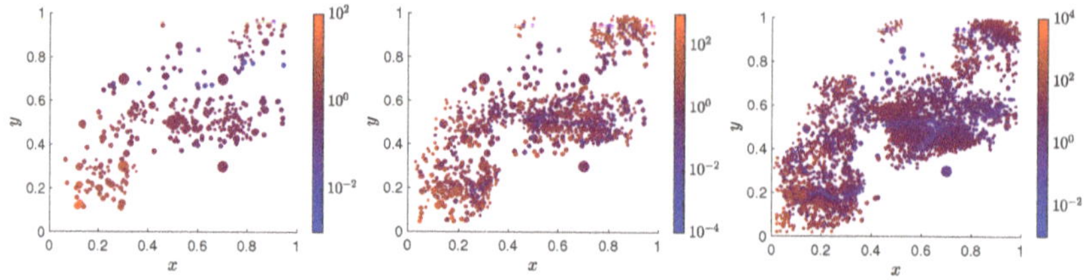

(d) (e) (f)

Figure 15. *Strategy #4*: Patches used to train the MF-VPINN with $C_M = 9$. Each dot represents a patch P_i, its position is the center c_{P_i} of the patch, its size is proportional to the patch size h_i^2, and its color is associated with the quantity $r_{h,i}^2(u^{\mathcal{NN}})$. (**a**) Representation of $\mathcal{P}_1$; (**b**) Representation of $\mathcal{P}_2$; (**c**) Representation of $\mathcal{P}_3$; (**d**) Representation of $\mathcal{P}_4$; (**e**) Representation of $\mathcal{P}_5$; (**f**) Representation of $\mathcal{P}_6$.

3.4. Extension to More a Complex Domain

In this section, we present some ideas that can be used to apply the method to more complex domains.

Let us consider a domain Ω_2 with some internal holes and boundary $\partial\Omega_2 = \Gamma_2$. In particular, $\Omega_2 = (0,1)^2 \setminus (\cup_{i=1}^4 H_i)$, where H_i, $i = 1,2,3,4$ are rectangular holes with centers $\mathbf{c}_{H_i}$ defined as

$$\mathbf{c}_{H_1} = \left(\frac{9}{26}, \frac{9}{34}\right), \qquad \mathbf{c}_{H_2} = \left(\frac{17}{26}, \frac{9}{34}\right),$$
$$\mathbf{c}_{H_3} = \left(\frac{9}{26}, \frac{25}{34}\right), \qquad \mathbf{c}_{H_4} = \left(\frac{17}{26}, \frac{25}{34}\right),$$

and basis and height equal $\frac{1}{26}$ and $\frac{1}{34}$, respectively.

In this domain, we consider the Poisson problem:

$$\begin{cases} -\Delta u = f & \text{in } \Omega_2, \\ u = g & \text{on } \Gamma_2, \end{cases} \tag{35}$$

with f and g such that the exact solution is

$$u(x,y) = \frac{1}{C_u}\left[x(x-1)\left(x-\frac{4}{13}\right)\left(x-\frac{5}{13}\right)\left(x-\frac{8}{13}\right)\left(x-\frac{9}{13}\right)\cdot \right. \\ \left. y(y-1)\left(y-\frac{4}{17}\right)\left(y-\frac{5}{17}\right)\left(y-\frac{12}{17}\right)\left(y-\frac{13}{17}\right)\right], \tag{36}$$

normalized through the constant $\frac{1}{C_u}$ to assume value 1 in $\left(\frac{2}{13}, \frac{2}{17}\right)$. This function is represented in Figure 16.

We extend the approaches proposed in Section 3.2 by adding a cutting procedure after the generation of the new patches. Note that, in particular, all the patches are already completely inside the square $[0,1]^2$ when we apply the cutting procedure, and we can thus focus only on the holes. When a patch intersects more than one hole, we recursively remove it from $\mathcal{P}_m$, we subdivide the corresponding region in 4 overlapping patches, and we add them to $\mathcal{P}_m$ until all the generated patches intersect at most one hole. Moreover, we observe that the region $P_i \setminus H_j$ inside the patch $P_i \in \mathcal{P}_m$ and outside the hole H_j, $j = 1,2,3,4$, can always be covered by the union of at most four rectangles. When a generated patch intersects a hole, we thus remove the patch and generate the minimum number of patches (at most four) that are as large possible and whose union is the region $P_i \setminus H_j$.

To avoid numerical instabilities, when this cutting procedure generates a patch with an aspect ratio larger than 100 or with an area more than 100 times smaller than the original uncutted patch, the new patches are removed from $\mathcal{P}_m$. This implies that it is not possible to remove the patches associated with the highest error indicators as in *Strategy #3* because otherwise, the union of all the patches would not cover the entire domain. We thus present numerical results only for *Strategy #1* and *Strategy #2*.

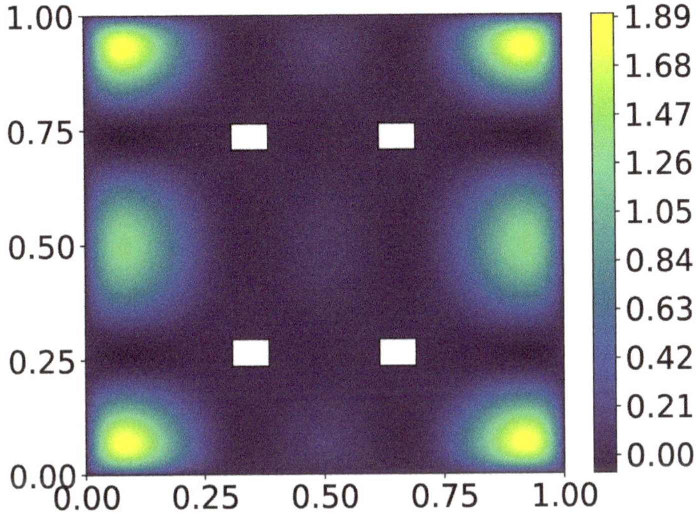

Figure 16. Graphical representation of the solution u in (36).

The obtained error decays are shown in Figure 17 for *Strategy #1* and *Strategy #2* with $C_M = 4$ and $C_M = 9$. The first and second errors are computed with the patches generated by cutting the patches in $\mathcal{P}_0$ and $\mathcal{P}_1$, respectively, whereas the third and fourth errors are obtained by refining the previous patches with the error indicator as previously described. Note that the first and second errors are very close for all the curves since the strategy and the value of C_M does not influence the training and that both strategies converge better with $C_M = 4$. The final patch distributions are displayed in Figure 18. Here, we can see that the inner part of the domain is covered by a few large patches, whereas the distribution is denser closer to the external boundary of Ω_2, where the solution is more oscillating.

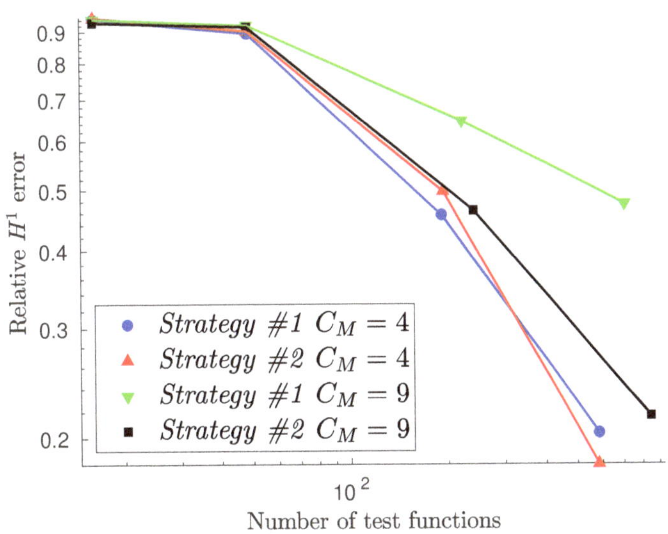

Figure 17. Relative H^1 errors obtained by solving problem (35).

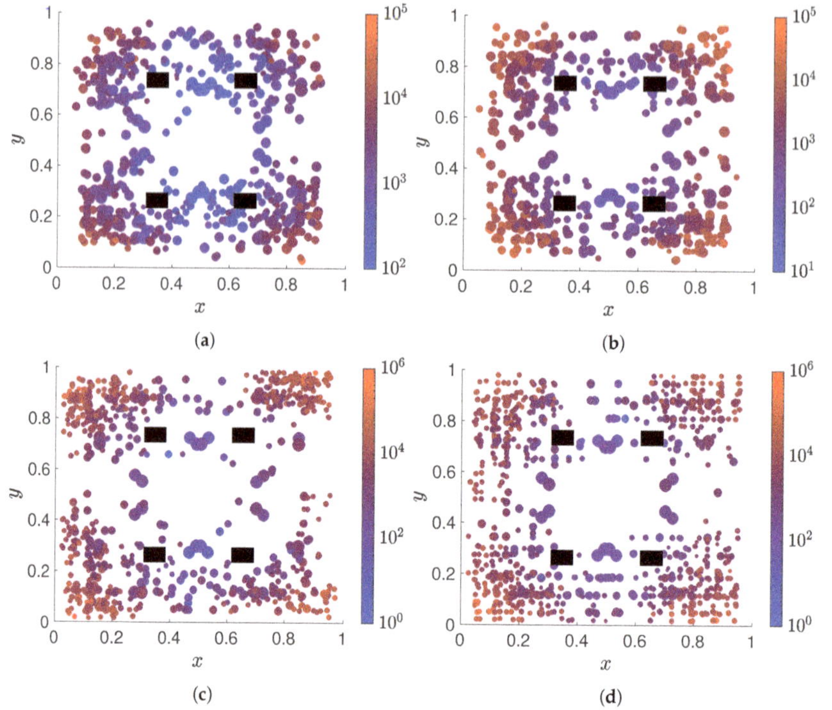

Figure 18. Problem (35): Representation of the last set of patches obtained with the different strategies. Each dot represents a patch P_i, its position is the center c_{P_i} of the patch, its size is proportional to the patch size h_i^2, and its color is associated with the quantity $r_{h,i}^2(u^{\mathcal{NN}})$. The black rectangles represent the holes H_i, $i = 1,2,3,4$. (**a**) *Strategy #1*, $C_M = 4$; (**b**) *Strategy #2*, $C_M = 4$; (**c**) *Strategy #1*, $C_M = 9$; (**d**) *Strategy #2*, $C_M = 9$.

4. Conclusions and Discussion

In this work, we presented a Meshfree Variational-Physics-Informed Neural Network (MF-VPINN). It is a PINN trained using the PDE variational formulation that does not require the generation of a global triangulation of the entire domain. In order to generate the test functions involved in the loss computation, we use an a posteriori error estimator based on the one discussed in [46]. Using such an error estimator, it is possible to add test functions only in regions in which the error is higher, thus increasing the efficiency of the method.

We highlight that the main advantages of the method are that it is meshfree, as it requires only a covering of the domain with patches that can be of different shapes and that it automatically improves the solution with the application of local patches without requiring a global mesh manipulation. It can be therefore used in domains where it is expensive or impossible to generate a mesh. On the other hand, if a mesh suitable to describe the solution can be generated, a standard VPINN is preferable since the implementation is simpler and the convergence rate with respect to the number of test functions is higher.

We discuss several strategies to generate the set of test functions. We observe that adding a few test functions inside the patches associated with higher errors while ensuring a smooth transition between regions with large patches and regions with small patches is the best way to obtain accurate solutions. We also show that, if the a posteriori error indicator is not used, the model's accuracy decreases and the training is slower.

In this paper, we only focus on second-order elliptic problem even though VPINNs can be used to solve more complex problems. In a forthcoming paper, we will adapt the a posteriori error estimator and analyze the MF-VPINN performance on other PDEs. Moreover, we are interested in the analysis of the approach in more complex domains (in which the patches have to be suitably deformed) and in high-dimensional problems, where using a standard VPINN is not practical.

Author Contributions: Conceptualization, S.B. and M.P.; methodology, S.B. and M.P.; software, M.P.; validation, M.P.; formal analysis, S.B. and M.P.; investigation, S.B. and M.P.; resources, S.B.; data curation, M.P.; writing—original draft preparation, M.P.; visualization, M.P.; supervision, S.B.; project administration, S.B.; funding acquisition, S.B. All authors have read and agreed to the published version of the manuscript.

Funding: The author S.B. kindly acknowledges partial financial support provided by PRIN project "Advanced polyhedral discretisations of heterogeneous PDEs for multiphysics problems" (No. 20204LN5N5_003) and by PNRR M4C2 project of CN00000013 National center for HPC, Big Data and Quantum Computing (HPC) (CUP: E13C22000990001). The author M.P. kindly acknowledges the financial support provided by the Politecnico di Torino where the research was carried out.

Data Availability Statement: The data are available upon reasonable request.

Conflicts of Interest: The authors declare no conflicts of interest.

References

1. Lagaris, I.; Likas, A.; Fotiadis, D. Artificial neural network methods in quantum mechanics. *Comput. Phys. Commun.* **1997**, *104*, 1–14. [CrossRef]
2. Lagaris, I.; Likas, A.; Fotiadis, D. Artificial neural networks for solving ordinary and partial differential equations. *IEEE Trans. Neural Netw.* **1998**, *9*, 987–1000. [CrossRef] [PubMed]
3. Lagaris, I.; Likas, A.; Papageorgiou, D. Neural-network methods for boundary value problems with irregular boundaries. *IEEE Trans. Neural Netw.* **2000**, *11*, 1041–1049. [CrossRef] [PubMed]
4. Abadi, M.; Agarwal, A.; Barham, P.; Brevdo, E.; Chen, Z.; Citro, C.; Corrado, G.S.; Davis, A.; Dean, J.; Devin, M .; et al. TensorFlow: Large-Scale Machine Learning on Heterogeneous Systems. 2015. Available online : https://www.tensorflow.org (accessed on 15 September 2024).
5. Paszke, A.; Gross, S.; Massa, F.; Lerer, A.; Bradbury, J.; Chanan, G.; Killeen, T.; Lin, Z.; Gimelshein, N.; Antiga, L.; et al. PyTorch: An Imperative Style, High-Performance Deep Learning Library. In *Advances in Neural Information Processing Systems 32*; Curran Associates, Inc.: Red Hook, NY, USA, 2019; pp. 8024–8035.
6. Bradbury, J.; Frostig, R.; Hawkins, P.; Johnson, M.J.; Leary, C.; Maclaurin, D.; Necula, G.; Paszke, A.; VanderPlas, J.; Wanderman-Milne, S.; et al. JAX: Composable Transformations of Python+NumPy Programs. 2018. Available online: http://github.com/google/jax (accessed on 15 September 2024).
7. Raissi, M.; Perdikaris, P.; Karniadakis, G. Physics informed deep learning (part i): Data-driven solutions of nonlinear partial differential equations. *arXiv* **2017**, arXiv:1711.10561.
8. Raissi, M.; Perdikaris, P.; Karniadakis, G. Physics informed deep learning (part ii): Data-driven solutions of nonlinear partial differential equations. *arXiv* **2017**, arXiv:1711.10566.
9. Raissi, M.; Perdikaris, P.; Karniadakis, G. Physics-informed neural networks: A deep learning framework for solving forward and inverse problems involving nonlinear partial differential equations. *J. Comput. Phys.* **2019**, *378*, 686–707. [CrossRef]
10. Pu, J.; Li, J.; Chen, Y. Solving localized wave solutions of the derivative nonlinear Schrödinger equation using an improved PINN method. *Nonlinear Dyn.* **2021**, *105*, 1723–1739. [CrossRef]
11. Yuan, L.; Ni, Y.; Deng, X.; Hao, S. A-PINN: Auxiliary physics informed neural networks for forward and inverse problems of nonlinear integro-differential equations. *J. Comput. Phys.* **2022**, *462*, 111260. [CrossRef]
12. Guo, Q.; Zhao, Y.; Lu, C.; Luo, J. High-dimensional inverse modeling of hydraulic tomography by physics informed neural network (HT-PINN). *J. Hydrol.* **2023**, *616*, 128828. [CrossRef]
13. Demo, N.; Strazzullo, M.; Rozza, G. An extended physics informed neural network for preliminary analysis of parametric optimal control problems. *Comput. Math. Appl.* **2023**, *143*, 383–396. [CrossRef]
14. Gao, H.; Sun, L.; Wang, J. PhyGeoNet: Physics-informed geometry-adaptive convolutional neural networks for solving parameterized steady-state PDEs on irregular domain. *J. Comput. Phys.* **2021**, *428*, 110079. [CrossRef]
15. Yuyao, C.; Lu, L.; Karniadakis, G.; Dal Negro, L. Physics-informed neural networks for inverse problems in nano-optics and metamaterials. *Opt. Express* **2020**, *28*, 11618–11633. [CrossRef]
16. Tartakovsky, A.; Marrero, C.; Perdikaris, P.; Tartakovsky, G.; Barajas-Solano, D. Learning parameters and constitutive relationships with physics informed deep neural networks. *arXiv* **2018**, arXiv:1808.03398.
17. Chen, Z.; Liu, Y.; Sun, H. Physics-informed learning of governing equations from scarce data. *Nat. Commun.* **2021**, *12*, 6136. [CrossRef] [PubMed]

18. Weinan, E.; Yu, B. The Deep Ritz method: A deep learning-based numerical algorithm for solving variational problems. *Commun. Math. Stat.* **2018**, *6*, 1–12.
19. Müller, J.; Zeinhofer, M. Error estimates for the deep Ritz method with boundary penalty. In Proceedings of the Mathematical and Scientific Machine Learning, PMLR, Beijing, China, 15–17 August 2022; pp. 215–230.
20. Lu, Y.; Lu, J.; Wang, M. A priori generalization analysis of the deep Ritz method for solving high dimensional elliptic partial differential equations. In Proceedings of the Conference on Learning Theory. PMLR, Boulder, CO, USA, 15–19 August 2021; pp. 3196–3241.
21. Sirignano, J.; Spiliopoulos, K. DGM: A deep learning algorithm for solving partial differential equations. *J. Comput. Phys.* **2018**, *375*, 1339–1364. [CrossRef]
22. Al-Aradi, A.; Correia, A.; Jardim, G.; de Freitas Naiff, D.; Saporito, Y. Extensions of the deep Galerkin method. *Appl. Math. Comput.* **2022**, *430*, 127287. [CrossRef]
23. Li, J.; Zhang, W.; Yue, J. A deep learning Galerkin method for the second-order linear elliptic equations. *Int. J. Numer. Anal. Model.* **2021**, *18*, 427–441.
24. Smith, B.F. Domain decomposition methods for partial differential equations. In *Parallel Numerical Algorithms*; Springer: Berlin/Heidelberg, Germany, 1997; pp. 225–243.
25. Toselli, A.; Widlund, O. *Domain Decomposition Methods-Algorithms and Theory*; Springer Science & Business Media: Berlin/Heidelberg, Germany, 2006; Volume 34.
26. Jagtap, A.; Kharazmi, E.; Karniadakis, G. Conservative physics-informed neural networks on discrete domains for conservation laws: Applications to forward and inverse problems. *Comput. Methods Appl. Mech. Eng.* **2020**, *365*, 113028. [CrossRef]
27. Shukla, K.; Jagtap, A.D.; Karniadakis, G.E. Parallel physics-informed neural networks via domain decomposition. *J. Comput. Phys.* **2021**, *447*, 110683. [CrossRef]
28. Jagtap, A.; Karniadakis, G. Extended physics-informed neural networks (XPINNs): A generalized space-time domain decomposition based deep learning framework for nonlinear partial differential equations. *Commun. Comput. Phys.* **2020**, *28*, 2002–2041.
29. Moseley, B.; Markham, A.; Nissen-Meyer, T. Finite Basis Physics-Informed Neural Networks (FBPINNs): A scalable domain decomposition approach for solving differential equations. *Adv. Comput. Math.* **2023**, *49*, 62. [CrossRef]
30. Viana, F.; Nascimento, R.; Dourado, A.; Yucesan, Y. Estimating model inadequacy in ordinary differential equations with physics-informed neural networks. *Comput. Struct.* **2021**, *245*, 106458. . [CrossRef]
31. Yang, L.; Meng, X.; Karniadakis, G. B-PINNs: Bayesian physics-informed neural networks for forward and inverse PDE problems with noisy data. *J. Comput. Phys.* **2021**, *425*, 109913 . [CrossRef]
32. Yang, L.; Zhang, D.; Karniadakis, G. Physics-Informed Generative Adversarial Networks for Stochastic Differential Equations. *SIAM J. Sci. Comput.* **2020**, *42*, A292–A317. [CrossRef]
33. Yucesan, Y.; Viana, F. Hybrid physics-informed neural networks for main bearing fatigue prognosis with visual grease inspection. *Comput. Ind.* **2021**, *125*, 103386. . [CrossRef]
34. Zhu, Y.; Zabaras, N.; Koutsourelakis, P.; Perdikaris, P. Physics-constrained deep learning for high-dimensional surrogate modeling and uncertainty quantification without labeled data. *J. Comput. Phys.* **2019**, *394*, 56–81. [CrossRef]
35. Pang, G.; Lu, L.; Karniadakis, G.E. fPINNs: Fractional physics-informed neural networks. *SIAM J. Sci. Comput.* **2019**, *41*, A2603–A2626. [CrossRef]
36. Liu, Z.; Wang, Y.; Vaidya, S.; Ruehle, F.; Halverson, J.; Soljačić, M.; Hou, T.Y.; Tegmark, M. Kan: Kolmogorov-arnold networks. *arXiv* **2024**, arXiv:2404.19756.
37. Koenig, B.C.; Kim, S.; Deng, S. KAN-ODEs: Kolmogorov-Arnold Network Ordinary Differential Equations for Learning Dynamical Systems and Hidden Physics. *arXiv* **2024**, arXiv:2407.04192.
38. Qian, K.; Kheir, M. Investigating KAN-Based Physics-Informed Neural Networks for EMI/EMC Simulations. *arXiv* **2024**, arXiv:2405.11383.
39. Kumar, V.; Gleyzer, L.; Kahana, A.; Shukla, K.; Karniadakis, G.E. Mycrunchgpt: A llm assisted framework for scientific machine learning. *J. Mach. Learn. Model. Comput.* **2023**, *4*, 41–72. [CrossRef]
40. Beck, C.; Hutzenthaler, M.; Jentzen, A.; Kuckuck, B. An overview on deep learning-based approximation methods for partial differential equations. *Discret. Contin. Dyn. Syst. B* **2022**, *28*, 3697–3746. [CrossRef]
41. Cuomo, S.; Di Cola, V.S.; Giampaolo, F.; Rozza, G.; Raissi, M.; Piccialli, F. Scientific Machine Learning Through Physics-Informed Neural Networks: Where we are and What's Next. *J. Sci. Comput.* **2022**, *92*, 88. [CrossRef]
42. Lawal, Z.; Yassin, H.; Lai, D.; Che Idris, A. Physics-Informed Neural Network (PINN) Evolution and Beyond: A Systematic Literature Review and Bibliometric Analysis. *Big Data Cogn. Comput.* **2022**, *6*, 140. [CrossRef]
43. Viana, F.A.; Subramaniyan, A.K. A survey of Bayesian calibration and physics-informed neural networks in scientific modeling. *Arch. Comput. Methods Eng.* **2021**, *28*, 3801–3830. [CrossRef]
44. Kharazmi, E.; Zhang, Z.; Karniadakis, G. VPINNs: Variational physics-informed neural networks for solving partial differential equations. *arXiv* **2019**, arXiv:1912.00873.
45. Kharazmi, E.; Zhang, Z.; Karniadakis, G. *hp*-VPINNs: Variational physics-informed neural networks with domain decomposition. *Comput. Methods Appl. Mech. Eng.* **2021**, *374*, 113547. [CrossRef]

46. Berrone, S.; Canuto, C.; Pintore, M. Solving PDEs by variational physics-informed neural networks: An a posteriori error analysis. *Ann. Univ. Ferrara* **2022**, *68*, 575–595. [CrossRef]
47. Berrone, S.; Canuto, C.; Pintore, M. Variational-Physics-Informed Neural Networks: The role of quadratures and test functions. *J. Sci. Comput.* **2022**, *92*, 100. [CrossRef]
48. Berrone, S.; Pieraccini, S.; Scialò, S. Towards effective flow simulations in realistic discrete fracture networks. *J. Comput. Phys.* **2016**, *310*, 181–201. [CrossRef]
49. Sukumar, N.; Srivastava, A. Exact imposition of boundary conditions with distance functions in physics-informed deep neural networks. *Comput. Methods Appl. Mech. Eng.* **2022**, *389*, 114333. [CrossRef]
50. Berrone, S.; Canuto, C.; Pintore, M.; Sukumar, N. Enforcing Dirichlet boundary conditions in physics-informed neural networks and variational physics-informed neural networks. *Heliyon* **2023**, *9*, e18820. [CrossRef] [PubMed]
51. Kingma, D.; Ba, J. Adam: A method for stochastic optimization. *arXiv* **2014**, arXiv:1412.6980 .
52. Wright, S.; Nocedal, J. *Numerical Optimization*; Springer: Berlin/Heidelberg, Germany, 1999; Volume 35, p. 7.
53. Baydin, A.; Pearlmutter, B.; Radul, A.; Siskind, J. Automatic differentiation in machine learning: A survey. *J. Mach. Learn. Res.* **2018**, *18*, 5595–5637.
54. Prechelt, L. Early stopping-but when? In *Neural Networks: Tricks of the Trade*; Springer: Berlin/Heidelberg, Germany, 1998; pp. 55–69.

algorithms

MDPI

Article

Cubic q-Bézier Triangular Patch for Scattered Data Interpolation and Its Algorithm

Owen Tamin [1] and Samsul Ariffin Abdul Karim [2,*]

[1] Faculty of Science and Natural Resources, Universiti Malaysia Sabah, Jalan UMS, Kota Kinabalu 88400, Malaysia; owentamin1996@yahoo.com

[2] School of Quantitative Sciences, UUM College of Arts and Sciences, Universiti Utara Malaysia, Sintok 06010, Malaysia

* Correspondence: samsul.ariffin@uum.edu.my

Abstract: This paper presents an approach to scattered data interpolation using q-Bézier triangular patches via an efficient algorithm. While existing studies have formed q-Bézier triangular patches through convex combination, their application to scattered data interpolation has not been previously explored. Therefore, this study aims to extend the use of q-Bézier triangular patches to scattered data interpolation by achieving C^1 continuity throughout the data points. We test the proposed scheme using both established data points and real-life engineering problems. We compared the performance of the proposed interpolation scheme with a well-known existing scheme by varying the q parameter. The comparison was based on visualization and error analysis. Numerical and graphical results were generated using MATLAB. The findings indicate that the proposed scheme outperforms the existing scheme, demonstrating a higher coefficient of determination (R^2), smaller root mean square error (RMSE), and faster central processing unit (CPU) time. These results highlight the potential of the proposed q-Bézier triangular patches scheme for more accurate and reliable scattered data interpolation via the proposed algorithm.

Keywords: patches; scattered data interpolation; continuity; visualization; surface reconstruction

1. Introduction

Scattered data interpolation is crucial in computational science, providing a method to estimate values at arbitrary points within datasets characterized by irregularly distributed data points. Unlike structured datasets, which are arranged on grids or meshes, scattered data often arise from real-world observations or simulations where data points are irregularly positioned in multidimensional space. Interpolating these data involves constructing a continuous function that approximates the dataset's underlying behavior, allowing for the inference of values at unobserved locations within the domain [1].

There are two primary approaches to scattered data interpolation: triangulation-based methods and meshless techniques. Triangulation-based approaches involve creating a mesh from the scattered data points. However, not all interpolation methods require such a mesh. For example, some variants of the Shepard method do not require creating a mesh; instead, barycentric coordinates are used to obtain the explicit expression of the polynomial interpolant [2]. Common basis functions used in these methods include Bézier, B-spline, and radial basis functions (RBFs). To mitigate noise in the scattered data, least squares approximation is often employed.

Liu [3] developed a local multi-level scattered data interpolation method by introducing layered datasets and adjusting compactly supported radial basis functions. Borne and Wende [4,5] explored meshless methods using specific RBFs for interpolation, employing domain decomposition methods to generate symmetric saddle point systems. Skala et al. [5] presented an innovative variant of the normalized RBFs known as the Squared Normalized RBFs (SN-RBFs), specifically designed for spatio-temporal data. This new formulation involves the normalization of the rows of the RBF matrix, a process that significantly improves the conditioning of the RBF matrix. As a result, the SN-RBF method enhances the robustness and accuracy of RBF interpolation, making it more suitable for a wide range of applications.

Joldes et al. [6] enhanced the moving least squares (MLS) technique by incorporating polynomial bases to improve scattered data interpolation. Brodlie et al. [7] examined constrained surface interpolation using the Shepard interpolant, addressing the problem with optimization techniques. Lai and Meile [8] reviewed the use of non-negative bivariate triangular splines for maintaining the shape of scattered data. Schumaker and Speleers [9] investigated preserving non-negativity using macro-element spline spaces with Clough–Tocher macro-elements.

Karim et al. [10,11] explored spatial interpolation for rainfall data and terrain data using cubic Bézier and cubic Ball triangular patches. These studies demonstrated that the scattered data interpolation approach by Said and Rahmat [12] does not exhibit C^1 continuity, prompting the establishment of new criteria for C^1 continuity. Their method yielded a differentiable surface with a seamless appearance.

Feng and Zhang [13] introduced segmented bivariate Hermite interpolation using triangulation for precise surface reconstruction from extensive scattered datasets. Sun et al. [14] developed a bivariate rational interpolation on a triangular domain but noted increased computation time with rational splines. Bozzini et al. [15] suggested using polyharmonic splines to estimate noisy scattered data.

Goodman and Said [16] developed a C^1 triangular interpolant using a convex combination scheme, differing from Foley and Opitz's C^1 cubic triangular convex combination technique [17]. Said and Rahmat's method using cubic Ball triangular patches yielded results equivalent to cubic Bézier triangular patches with reduced computation [12]. Karim and Saaban's study confirmed the non-continuous differentiability of Said and Rahmat's function [18].

Hussain et al. [19] introduced rational cubic Bézier triangular methods for positivity-preserving interpolation, asserting continuous differentiability. Chan and Ong [20] examined range-restricted interpolation using cubic Bézier triangles with partial derivatives calculated using Goodman's approach.

Research has also explored quartic Bézier triangular patches for scattered data interpolation. Saaban et al. [21] and Piah et al. [22] developed C^1 (or G^1) methods using quartic Bézier triangles, optimizing inner Bézier points to maintain range and positivity. Hussain et al. [23] expanded this to convexity-preserving interpolation with rational quartic Bézier triangles, addressing optimization challenges for C^1 continuity. Comprehensive surveys on scattered data interpolation are available in sources like [1,24–27].

Among the various generalizations of Bézier curves, q-Bézier curves introduce greater versatility to the classical Bézier curves through the use of the parameter q. This parameter allows for more flexible and dynamic curve modeling, accommodating a wider range of shapes and applications [28–33]. Lewanowicz et al. [30] expanded the q-Bernstein basis of univariate polynomials to a triangular domain, providing a novel framework for handling scattered data.

However, the evaluation algorithm they proposed does not generally involve linear convex combinations, which limits its practical utility in some applications [30]. Foley [17] addressed this limitation by extending the evaluation algorithm to incorporate convex combinations, enhancing the algorithm's applicability and robustness. Inspired by these advancements, we aim to further explore and apply these extended q-Bézier curve techniques to the challenging problem of scattered data interpolation, leveraging their flexibility to achieve more accurate and efficient interpolation results.

2. q-Bézier Triangular Patch

Definition 1. *Given a non-negative integer n and three indices $i, j, k \geq 0$ such that $i + j + k = n$, the triangular q-Bernstein from [34] is defined as*

$$B^n_{i,j,k}(u,v) = \left[\begin{array}{c} n \\ k \end{array} \right] \left(\begin{array}{c} i+j \\ i \end{array} \right) u^i v^j \prod_{s=0}^{k-1} (1 - q^s u - q^s v), \quad i,j,k \geq 0, \quad i+j+k=n \tag{1}$$
$$\text{where } q \in [0,1]$$

In order to construct a q-Bézier triangular patch for scattered data interpolation, the existing triangular Bernstein polynomials $B^n_{i,j,k}(u,v)$ are extended to $B^n_{i,j,k}(u,v,w)$, which is defined in Proposition 1.

Proposition 1. *The triangular q-Bernstein polynomials from Equation (1) are extended to $B^n_{i,j,k}(u,v,w)$, which are given by*

$$B^n_{i,j,k}(u,v,w) = \left[\begin{array}{c} n \\ k \end{array} \right] \left(\begin{array}{c} i+j \\ i \end{array} \right) u^i v^j \prod_{s=0}^{k-1} (1 - q^s + q^s w), \quad i,j,k \geq 0, \quad i+j+k=n \tag{2}$$
$$\text{where } q \in [0,1]$$

Proof. Given a non-negative integer n and three indices $i,j,k \geq 0$ such that $i+j+k=n$,

$$B^n_{i,j,k}(u,v) = \left[\begin{array}{c} n \\ k \end{array} \right] \left(\begin{array}{c} i+j \\ i \end{array} \right) u^i v^j \prod_{s=0}^{k-1} (1 - q^s u - q^s v) \tag{3}$$

$$= \left[\begin{array}{c} n \\ k \end{array} \right] \left(\begin{array}{c} i+j \\ i \end{array} \right) u^i v^j \prod_{s=0}^{k-1} (1 - q^s u - q^s v) + q^s - q^s \tag{4}$$

By factorizing Equation (4), we have

$$B^n_{i,j,k}(u,v) = \left[\begin{array}{c} n \\ k \end{array} \right] \left(\begin{array}{c} i+j \\ i \end{array} \right) u^i v^j \prod_{s=0}^{k-1} (1 - q^s + q^s(1 - u - v)) \tag{5}$$

Taking into account that $w = 1 - u - v$, we can therefore rewrite $B^n_{i,j,k}(u,v)$ as

$$B^n_{i,j,k}(u,v,w) = \left[\begin{array}{c} n \\ k \end{array} \right] \left(\begin{array}{c} i+j \\ i \end{array} \right) u^i v^j \prod_{s=0}^{k-1} (1 - q^s + q^s w) \tag{6}$$

□

Definition 2. *Let $D := \{(u,v,w) \mid u+v+w=1, u>0, v>0, w>0\}$. The following ten functions are a cubic q-Bézier basis function over the triangular domain D shown in Figure 1.*

Let $\alpha = 1 + q + q^2$, $\beta = 1 - q + qw$, and $\gamma = 1 - q^2 + q^2 w$.

$$\begin{aligned}
B^3_{3,0,0}(u,v,w) &= u^3, \\
B^3_{0,3,0}(u,v,w) &= v^3, \\
B^3_{0,0,3}(u,v,w) &= \beta\gamma w, \\
B^3_{2,1,0}(u,v,w) &= 3u^2 v, \\
B^3_{2,0,1}(u,v,w) &= \alpha u^2 w, \\
B^3_{1,2,0}(u,v,w) &= 3uv^2, \\
B^3_{0,2,1}(u,v,w) &= \alpha v^2 w, \\
B^3_{1,0,2}(u,v,w) &= \alpha\beta uw, \\
B^3_{0,1,2}(u,v,w) &= \alpha\beta vw, \\
B^3_{1,1,1}(u,v,w) &= 1 - \sum_{i+j+k=3, i,j,k \neq 1} B^3_{i,j,k}(u,v,w)
\end{aligned}$$

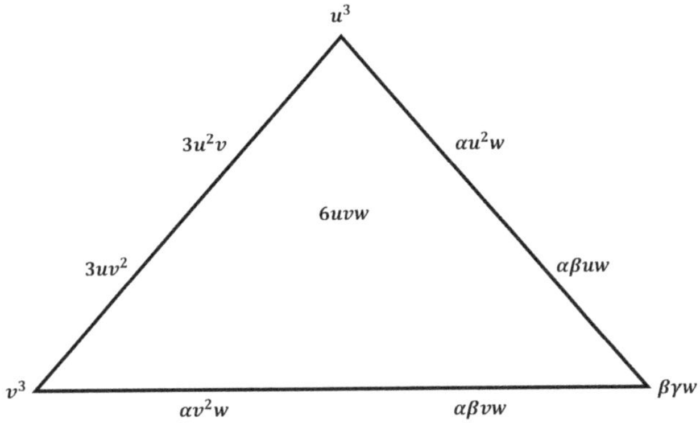

Figure 1. Bases for q-Bézier triangular.

Theorem 1. *The basis function defined over a triangular domain has the following properties:*

(a) *Partition of unity:* $\sum_{i+j+k=3} B^3_{i,j,k}(u,v,w) = 1$.

(b) *Non-negativity:* $B^3_{i,j,k}(u,v,w) \geq 0,\ i,j,k \in \mathbb{N},\ i+j+k=3$.

(c) *Property at the corner points: for* $i,j,k \in \mathbb{N},\ i+j+k=3$, *we have*

$$B^3_{i,j,k}(1,0,0) = \begin{cases} 1, & i=3, \\ 0, & i \neq 3, \end{cases}$$

$$B^3_{i,j,k}(0,1,0) = \begin{cases} 1, & j=3, \\ 0, & j \neq 3, \end{cases}$$

$$B^3_{i,j,k}(0,0,1) = \begin{cases} 1, & k=3, \\ 0, & k \neq 3, \end{cases}$$

(d) *When* $q=1$, *the cubic q-Bézier triangular is reduced to the standard cubic Bézier.*

Definition 3. *A cubic q-Bézier triangular patch is defined as*

$$R(u,v,w) = \sum_{i+j+k=3} B^3_{i,j,k}(u,v,w) b_{i,j,k} \tag{7}$$

where $B^3_{i,j,k}(u,v,w)$ *is defined by Equation (6) for all* $q \in [0,1]$.

Figure 2 presents a comparison of the interpolation surfaces for one cubic q-Bézier patch, with varying values of q where the values are 0.25, 0.5, 0.75, and 1.0. As the value of q increases, the surface gradually shifts closer to the control polygon, reflecting the influence of q on the overall interpolation. When $q = 1.0$, the surface conforms more closely to the shape of the control polygon, while for smaller q values (e.g., 0.25), the surface tends to be smoother and less influenced by the control points. This demonstrates the flexibility of the q-Bézier interpolation scheme in controlling the tension and shape of the interpolating surface.

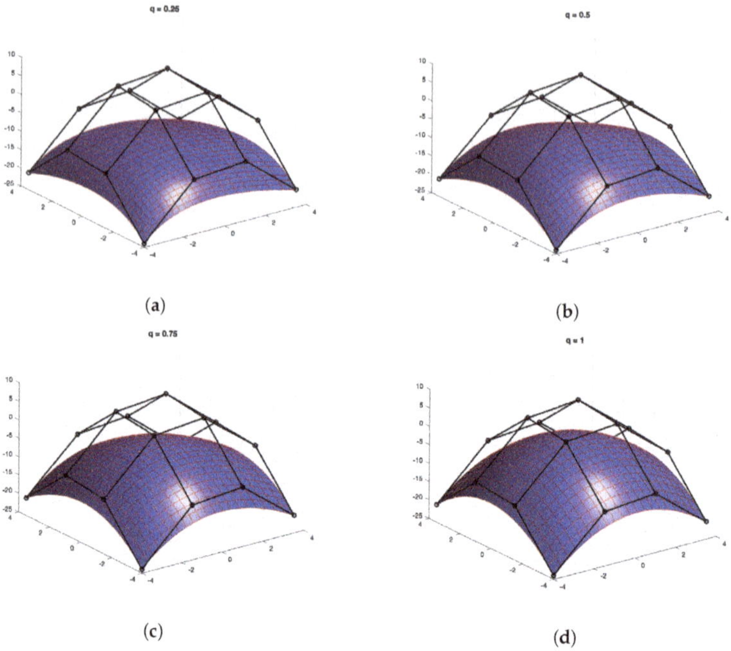

Figure 2. Cubic q-Bézier triangular with different values of q.

3. Scattered Data Interpolation Using q-Bézier Triangular Patch

In this section, we will extend the cubic q-Bézier triangular patch discussed in Section 2 to the application of scattered data interpolation. Figure 3 shows the arrangement of control points on one patch of a cubic q-Bézier triangular patch.

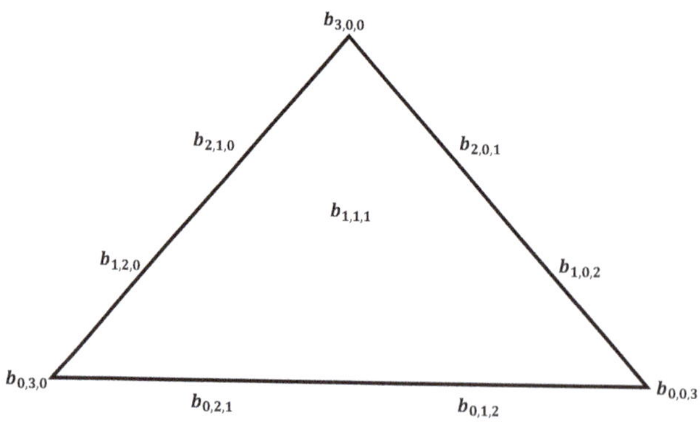

Figure 3. Control points for cubic q-Bézier triangular patch.

3.1. *Derivative Estimation*

The partial derivatives $f_x(V_i)$ and $f_y(V_i)$ at the vertices $(x_i, y_i), i = 1, 2, 3, \ldots, n$ of th triangle are estimated using the method proposed in [16]. These partial derivatives are used to find derivatives in the direction of ε, which is an edge of the triangle that is shown in Figure 4.

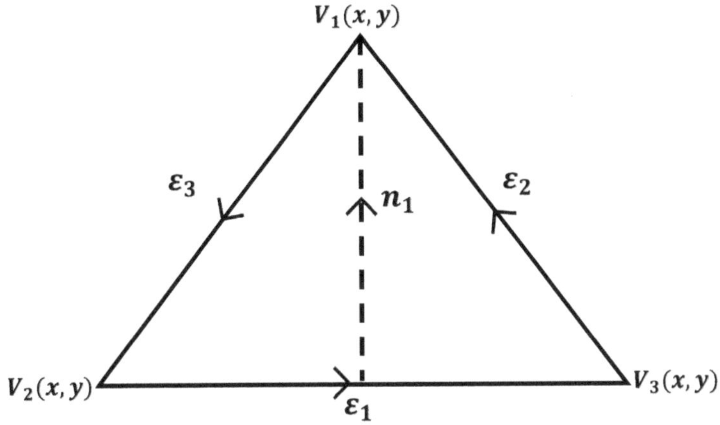

Figure 4. Directionals ε_1, ε_2, and ε_3.

The derivative along the side ε_i is given by

$$\frac{\partial f}{\partial \varepsilon_i} = (x_{i-1} - x_{i+1})\frac{\partial f}{\partial x} + (y_{i-1} - y_{i+1})\frac{\partial f}{\partial y} \tag{8}$$

3.2. Boundary Bézier Ordinates

Boundary Bézier ordinates are calculated using the C^1 continuity condition at the vertices of the triangle. At each vertex there are two directional derivatives in the direction of both the edges incident at the vertex.

The q-Bézier triangular patch is C^1 at the vertices of the triangle if the boundary Bézier ordinates are

$$\begin{aligned}
b_{3,0,0} &= F(V_1) \\
b_{2,0,1} &= \frac{(q^3-1)b_{1,0,2} + (-q^3+q^2+q-1)F(V_3) + 3F(V_1)}{q^2+q+1} - \frac{1}{q^2+q+1}\frac{\partial F(V_1)}{\partial e_2} \\
b_{2,1,0} &= F(V_1) + \frac{1}{3}\frac{\partial F(V_1)}{\partial e_3} \\
b_{0,3,0} &= F(V_2) \\
b_{0,2,1} &= \frac{(q^3-1)b_{0,1,2} + (-q^3+q^2+q-1)F(V_3) + 3F(V_2)}{q^2+q+1} + \frac{1}{q^2+q+1}\frac{\partial F(V_2)}{\partial e_1} \\
b_{1,2,0} &= F(V_2) - \frac{1}{3}\frac{\partial F(V_2)}{\partial e_3} \\
b_{0,0,3} &= F(V_3) \\
b_{0,1,2} &= F(V_3) - \frac{1}{q^2+q+1}\frac{\partial F(V_3)}{\partial e_1} \\
b_{1,0,2} &= F(V_3) + \frac{1}{q^2+q+1}\frac{\partial F(V_3)}{\partial e_2}
\end{aligned}$$

3.3. Inner Bézier Ordinates

The remaining $b^i_{1,1,1}, i = 1,2,3$ is obtained using the cubic precision method of Foley and Opitz [17], as shown in Figure 5. For complete derivation, the reader can refer to [17].

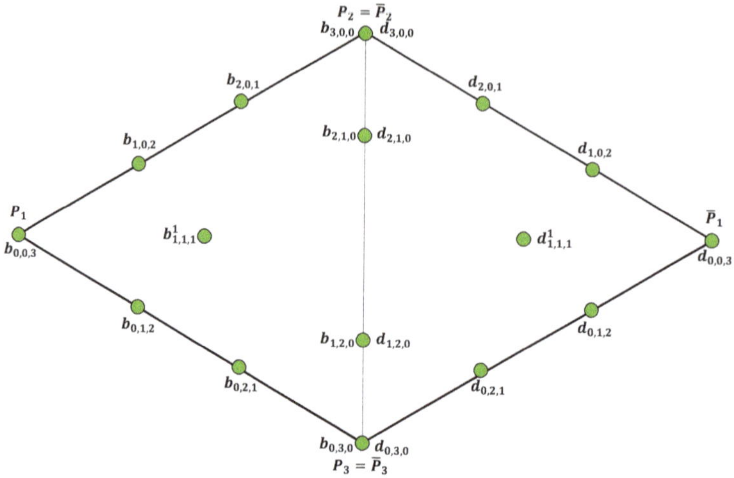

Figure 5. Two adjacent q-Bézier triangular patches.

Let (r, s, t) be the barycentric coordinates of $\bar{P}_1$ with respect to the triangle $P_1P_2P_3$. In order to achieve C^1 continuity along all edges, the following equations must be satisfied:

$$d_{0,1,2} = r^2 b_{0,3,0} + 2rsb_{1,2,0} + 2rtb_{0,2,1} + s^2 b_{2,1,0} + 2stb^1_{1,1,1} + t^2 b_{0,1,2} \tag{9}$$

$$d_{1,0,2} = r^2 b_{1,2,0} + 2rsb_{2,1,0} + 2stb_{2,0,1} + s^2 b_{3,0,0} + 2rtb^1_{1,1,1} + t^2 b_{1,0,2} \tag{10}$$

In reversing the triangles where (u, v, w) are the barycentric coordinates of P_1 with respect to the triangle $\bar{P}_1\bar{P}_2\bar{P}_3$, the following equation is obtained:

$$b_{0,1,2} = u^2 d_{0,3,0} + 2uvd_{1,2,0} + 2uwd_{0,2,1} + v^2 d_{2,1,0} + 2vwd^1_{1,1,1} + w^2 d_{0,1,2} \tag{11}$$

$$b_{1,0,2} = u^2 d_{1,2,0} + 2uvd_{2,1,0} + 2vwd_{2,0,1} + v^2 d_{3,0,0} + 2uwd^1_{1,1,1} + w^2 d_{1,0,2} \tag{12}$$

The value of d^1_{111} is obtained by adding Equations (10) and (11) together. Thus, we obtain

$$\begin{aligned} d^1_{1,1,1} = \frac{1}{2w(u+v)}\Big(&-u^2 d_{0,3,0} - u^2 d_{1,2,0} - 2uvd_{1,2,0} - 2uvd_{2,1,0} - 2uwd_{0,2,1} \\ &- v^2 d_{2,1,0} - v^2 d_{3,0,0} - 2vwd_{2,0,1} - w^2 d_{0,1,2} - w^2 d_{1,0,2}\Big) \end{aligned} \tag{13}$$

Similarly, by adding Equations (8) and (9), we obtain

$$\begin{aligned} b^1_{1,1,1} = \frac{1}{2t(r+s)}(&-r^2 b_{0,3,0} - r^2 b_{1,2,0} - 2rsb_{1,2,0} - 2rsb_{2,1,0} - 2rtb_{0,2,1} \\ &- s^2 b_{2,1,0} - s^2 b_{3,0,0} - 2stb_{2,0,1} - t^2 b_{0,1,2} - t^2 b_{1,0,2}) \end{aligned} \tag{14}$$

The final scattered data interpolation scheme using the cubic q-Bézier triangular patch can be expressed as follows:

$$R(u, v, w) = \sum_{i+j+k=3} B^3_{i,j,k}(u, v, w) b_{i,j,k} + 6uvw(\tau_1 b^1_{1,1,1} + \tau_2 b^2_{1,1,1} + \tau_3 b^3_{1,1,1}) \tag{15}$$

where

$$\tau_1 = \frac{vw}{uv + vw + wu}, \quad \tau_2 = \frac{wu}{uv + vw + wu}, \quad \tau_3 = \frac{uv}{uv + vw + wu}$$

The computational complexity of our proposed scattered data interpolation scheme for a single triangular patch involves 38 multiplications, 13 additions, and 1 division. This complexity is comparable to that of classical methods, such as those in [16], which require a total of 52 operations.

The overall algorithm used in this study with descriptions is below.

4. Results and Discussion

In this study, we implemented the proposed scheme using MATLAB 2023a to conduct the experiment. The MATLAB coding was developed based on Algorithm 1, and the processor used was 1.6 GHz Dual-Core Intel Core i5 (Intel, Santa Clara, CA, USA). The error norms were computed using a 33×33 uniform rectangular grid of evaluation points in the unit square. The error measurements used were the (a) coefficient of determination, R^2; (b) root mean square error (RMSE); and (c) central processing unit (CPU) time.

Algorithm 1 q-Bézier Triangular Patches for Scattered Data Interpolation

Input: Scattered data points $\{(x_i, y_i, z_i)\}_{i=1}^{n}$

Output: Interpolated surface, R^2, RMSE, and CPU time

Step 1: Estimate Partial Derivatives

- Compute the partial derivatives $\frac{\partial f}{\partial \varepsilon_1}, \frac{\partial f}{\partial \varepsilon_2}$ and $\frac{\partial f}{\partial \varepsilon_3}$ at each data point using Equation (8).
- MATLAB functions: `diff`, `gradient`

Step 2: Perform Delaunay Triangulation

- Use a Delaunay triangulation algorithm to create a mesh from the scattered data points.
- MATLAB functions: `delaunay`, `triplot`

Step 3: Calculate Boundary Control Points

- Determine the boundary control points for each triangle in the triangulated mesh.
- MATLAB functions: `convhull`, `boundary`

Step 4: Calculate Inner Control Points

- For each triangle, compute the inner control points for the local scheme $b^i_{1,1,1}, i = 1, 2, 3$.
- MATLAB function: custom implementation based on Foley and Opitz [17]

Step 5: Construct the Interpolated Surface

- Use the convex combination method to combine the three local schemes defined by Equation (15).
- MATLAB functions: `sum`, `minus`, `divide`

Step 6: Evaluate Performance Metrics

- Calculate R^2, RMSE, and CPU time.
- MATLAB functions: `sum`, `abs`, `sqrt`, `mean`, `tic`, `toc`

The error bound for the proposed scattered data interpolation scheme was adopted from [35]. Let the proposed scheme be denoted as p and Φ be the convex hull of points $\{(x_i, y_i)\}, i = 1, 2, \ldots, N$. Let Δ be a triangulation of Φ.

Theorem 2. *For every $f \epsilon C^4(\Phi)$. Its C^1 q-Bézier interpolant p satisfies*

$$\|f - p\| \leq K|\Delta|^4 |f|_4$$

where the constant K depends on the smallest angle in Δ. Furthermore, $|\Delta|$ is the size of the triangulation, i.e., the Δ largest length of edges of Δ.

The proposed scheme was tested using three well-known test functions $F_1(x, y)$, $F_2(x, y)$, and $F_3(x, y)$ for surface interpolation (Karim et al. [36]):

1. Sphere function:

$$F_1(x, y) = 0.75e^{\left(-\frac{(9x-2)^2+(9y-2)^2}{4}\right)} + 0.75e^{\left(-\frac{(9x+1)^2}{49} - \frac{(9y+1)^2}{10}\right)} + 0.5e^{\left(-\frac{(9x-7)^2+(9y-3)^2}{4}\right)} - 0.2e^{\left(-(9x-4)^2-(9y-7)^2\right)}$$

2. Cliff function:

$$F_2(x, y) = \frac{(1.25 + \cos(5.4y))}{6 + 6(3x - 1)^2}$$

3. Gentle function:

$$F_3(x, y) = \frac{1}{3}e^{-\frac{81}{16}((x-0.5)^2-(y-0.5)^2)}$$

Figure 6 shows the Delaunay triangulation of 36, 65, and 100 data points on the rectangular domain of [0, 1] × [0, 1] with uniform distribution. The Delaunay triangulation presented in Figure 6 is constrained, ensuring that specific segments are included as edges. This approach is particularly useful for maintaining specific boundary conditions and incorporating predefined features into the mesh structure [37].

Figures 7–14 display the surface interpolation and corresponding contour plots with various q parameters and data points derived from the $F_1(x, y)$ function. The visual inspection reveals that surface interpolation and its contour plots exhibit noticeable smoothness with 100 data points compared to 36 and 65 data points. Despite the change in the q parameter, differentiating the resulting surfaces proves visually challenging.

Similarly, Figures 15–22 illustrate surface interpolation and contour plots based on the $F_2(x, y)$ function under a varying q parameter and data points. Consistent with the observations from the $F_1(x, y)$ function, the surface interpolation and its contour plots are visually smoother with 100 data points compared to 36 and 65 data points. However, like before, distinguishing the effects of different q values remains difficult.

Figures 23–30 present the surface interpolation and contour plots for the $F_3(x, y)$ function, again with different q values and data points. The visual smoothness of surface interpolation for 100 data points, as opposed to 36 and 65 data points, mirrors the trends observed with the $F_1(x, y)$ and $F_2(x, y)$ functions. Across all three test functions, the interpolation results generated with 100 data points are accurate, suggesting that an increased number of data points enhances the interpolation accuracy relative to the true surface. However, the similarity in results across different q parameters indicates that visual inspection alone is insufficient to determine the optimal q parameter value.

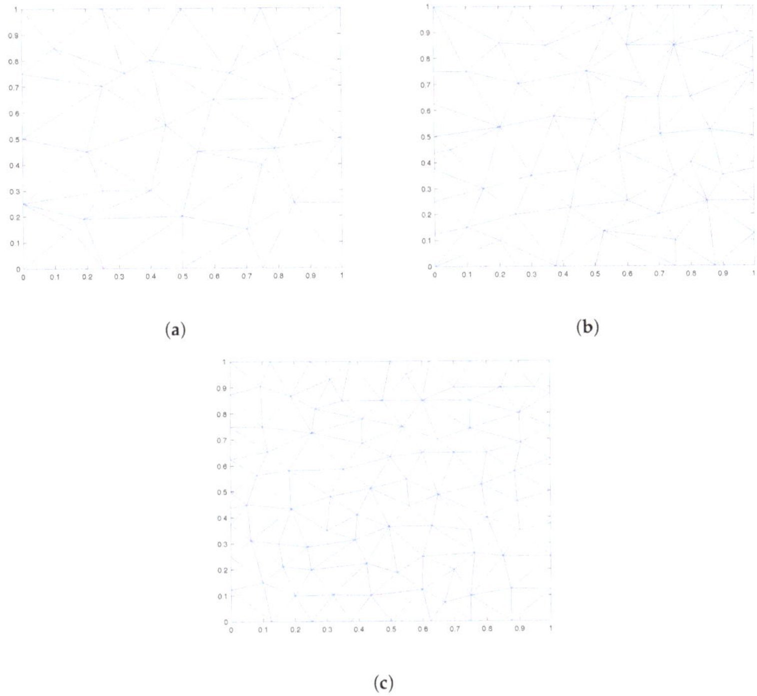

Figure 6. Triangulation domain for (**a**) 36 data points, (**b**) 65 data points, and (**c**) 100 data points.

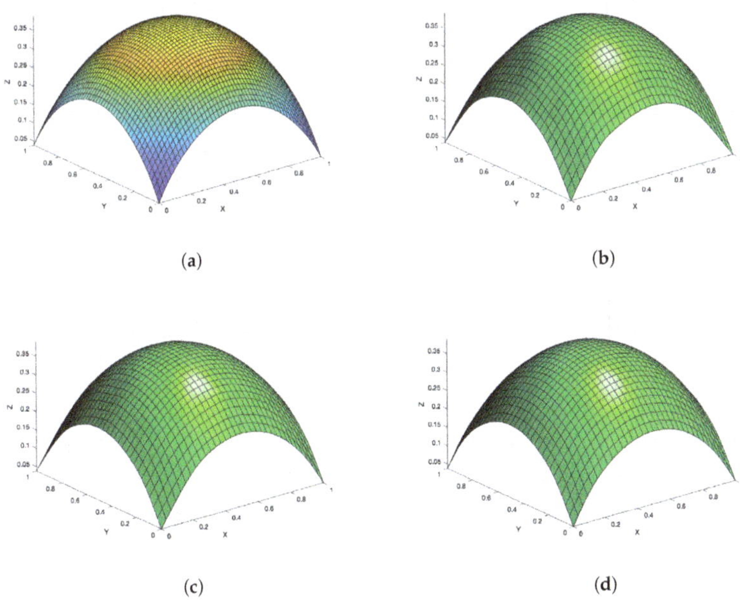

Figure 7. Surface interpolation based on F_1 function with 36, 65, and 100 data points when $q = 1$: (**a**) true surface; (**b**) 36 data points; (**c**) 65 data points; (**d**) 100 data points.

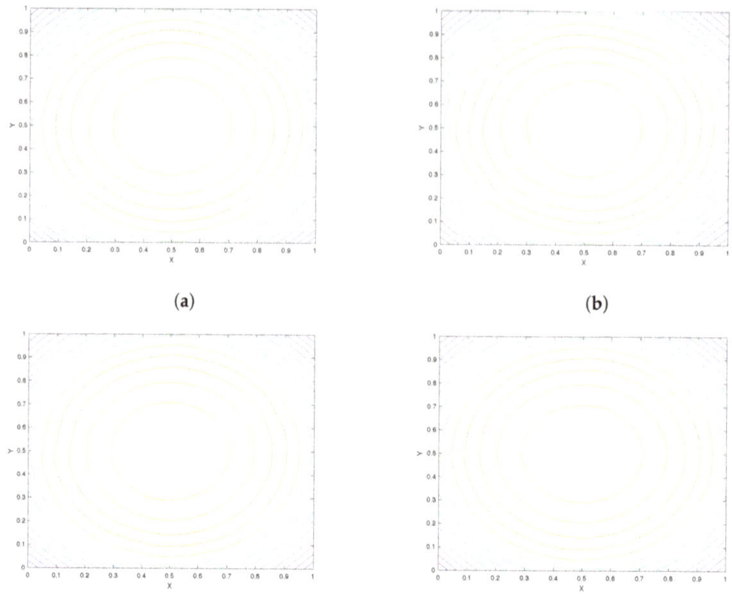

(a) (b) (c) (d)

Figure 8. Contour plots based on F_1 function with 36, 65, and 100 data points when $q = 1$: (**a**) true surface; (**b**) 36 data points; (**c**) 65 data points; (**d**) 100 data points.

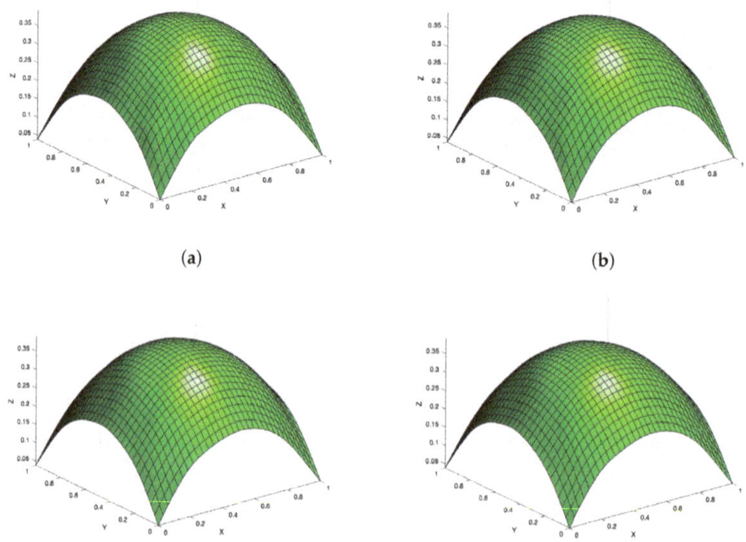

(a) (b) (c) (d)

Figure 9. Surface interpolation for 36 data points based on F_1 function. (**a**) $q = 0.25$. (**b**) $q = 0.5$. (**c**) $q = 0.75$. (**d**) $q = 1$.

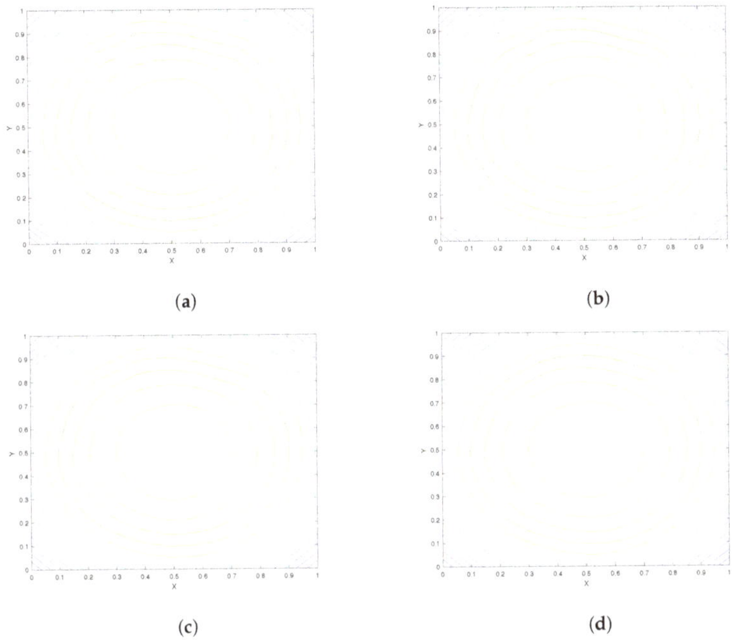

Figure 10. Contour plots for 36 data points based on F_1 function. (**a**) $q = 0.25$. (**b**) $q = 0.5$. (**c**) $q = 0.75$. (**d**) $q = 1$.

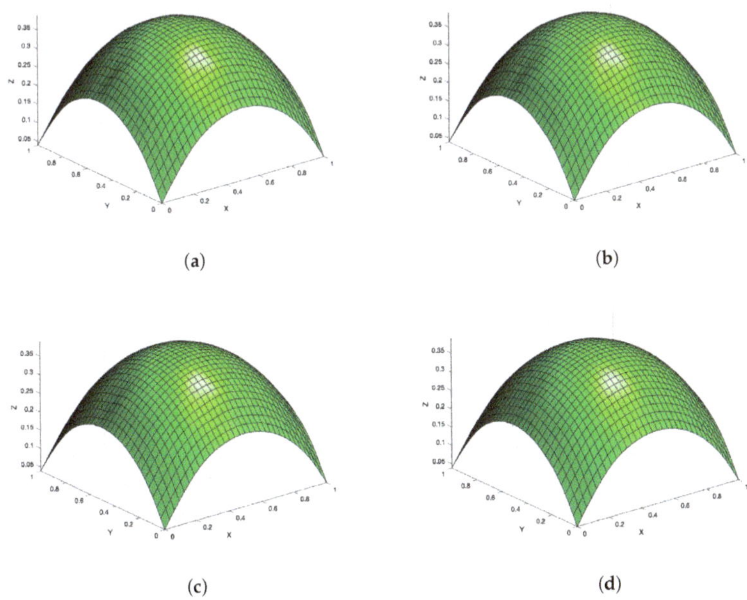

Figure 11. Surface interpolation for 65 data points based on F_1 function. (**a**) $q = 0.25$. (**b**) $q = 0.5$. (**c**) $q = 0.75$. (**d**) $q = 1$.

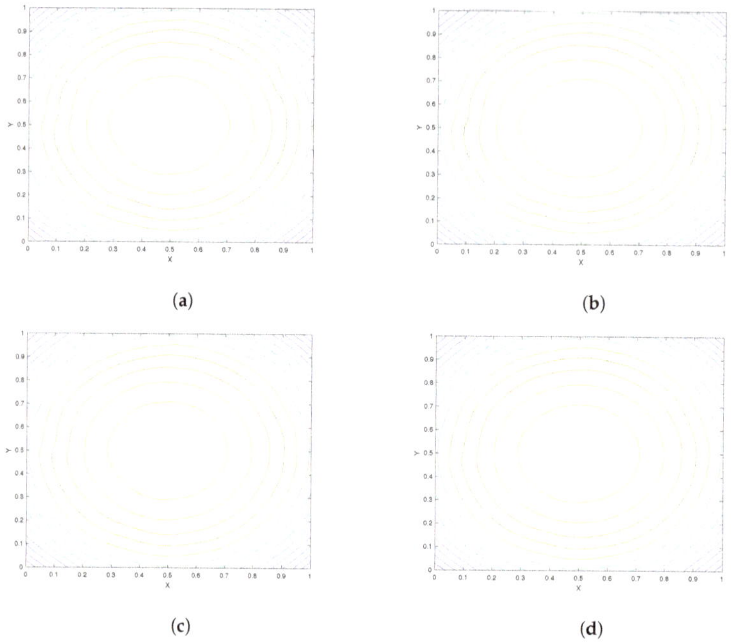

Figure 12. Contour plots for 65 data points based on F_1 function. (**a**) $q = 0.25$. (**b**) $q = 0.5$. (**c**) $q = 0.75$. (**d**) $q = 1$.

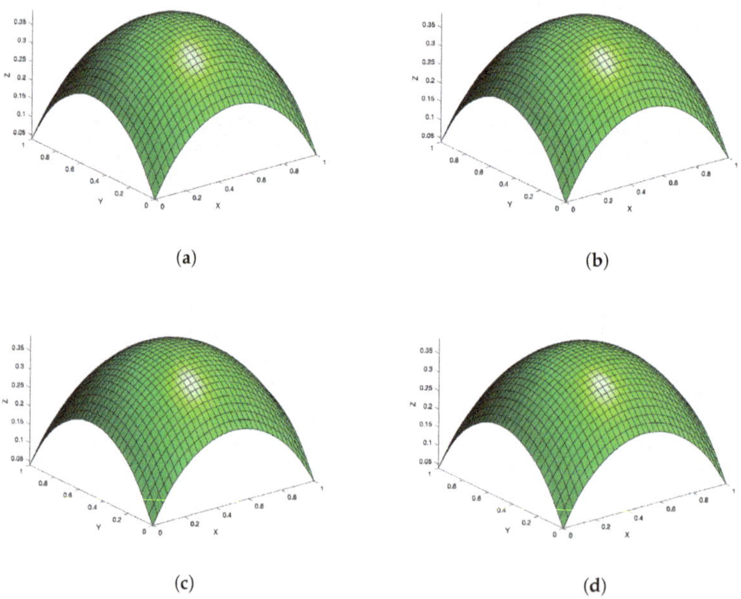

Figure 13. Surface interpolation for 100 data points based on F_1 function. (**a**) $q = 0.25$. (**b**) $q = 0.5$. (**c**) $q = 0.75$. (**d**) $q = 1$.

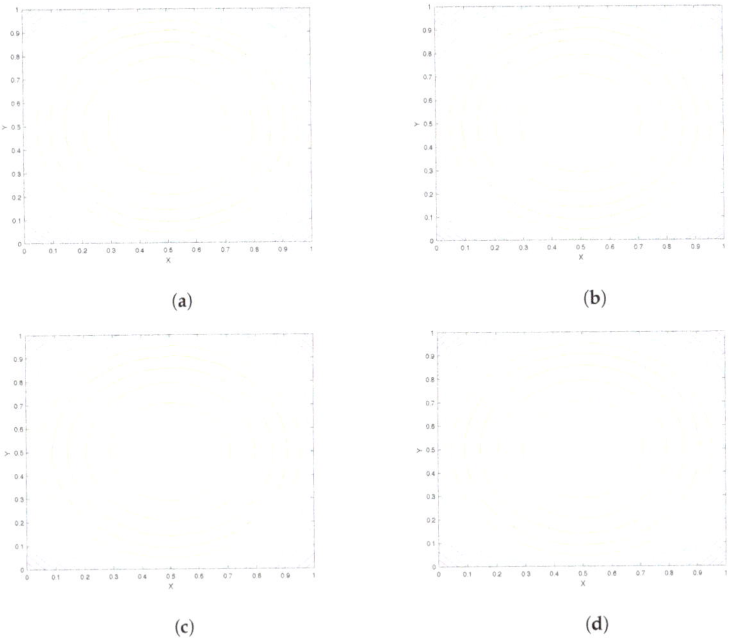

Figure 14. Contour plots for 100 data points based on F_1 function. (**a**) q = 0.25. (**b**) q = 0.5. (**c**) q = 0.75. (**d**) q = 1.

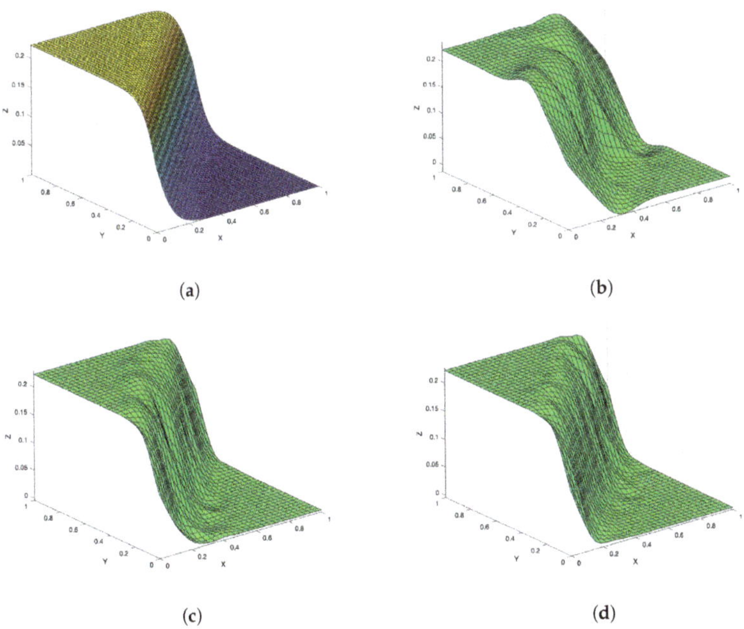

Figure 15. Surface interpolation based on F_2 function with 36, 65, and 100 data points when $q = 1$: (**a**) true surface; (**b**) 36 data points; (**c**) 65 data points; (**d**) 100 data points.

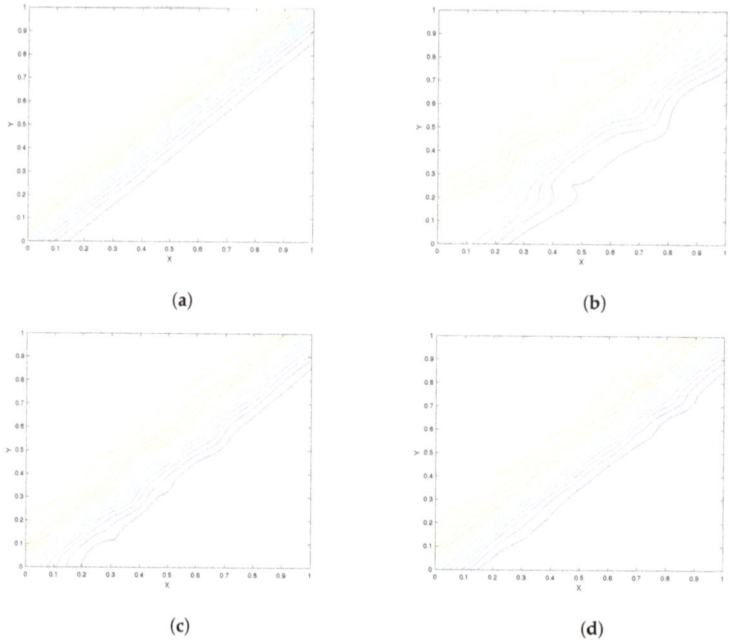

Figure 16. Contour plots based on F_2 function with 36, 65, and 100 data points when $q = 1$: (**a**) true surface; (**b**) 36 data points; (**c**) 65 data points; (**d**) 100 data points.

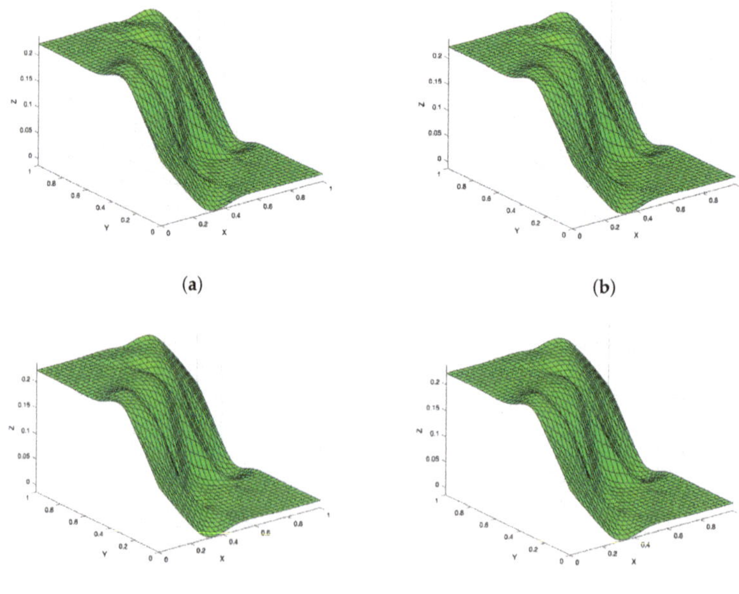

Figure 17. Surface interpolation for 36 data points based on F_2 function. (**a**) $q = 0.25$. (**b**) $q = 0.5$. (**c**) $q = 0.75$. (**d**) $q = 1$.

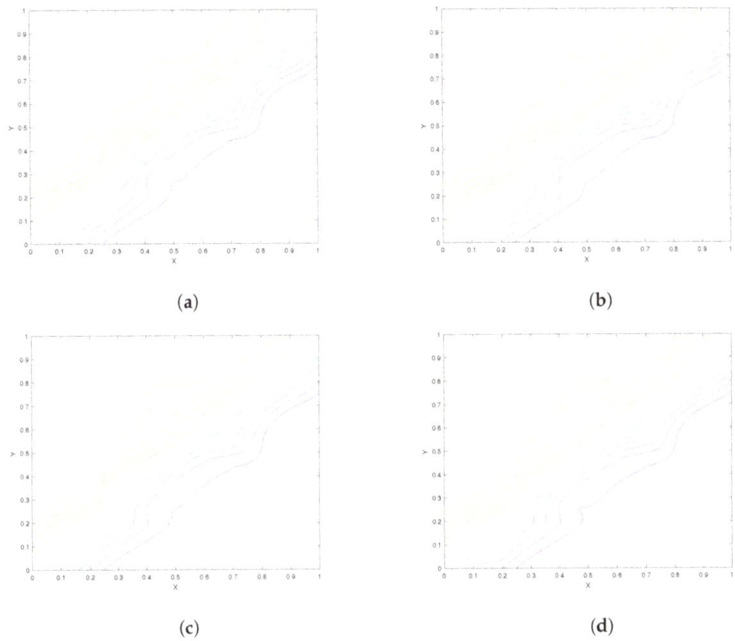

Figure 18. Contour plots for 36 data points based on F_2 function. (**a**) $q = 0.25$. (**b**) $q = 0.5$. (**c**) $q = 0.75$. (**d**) $q = 1$.

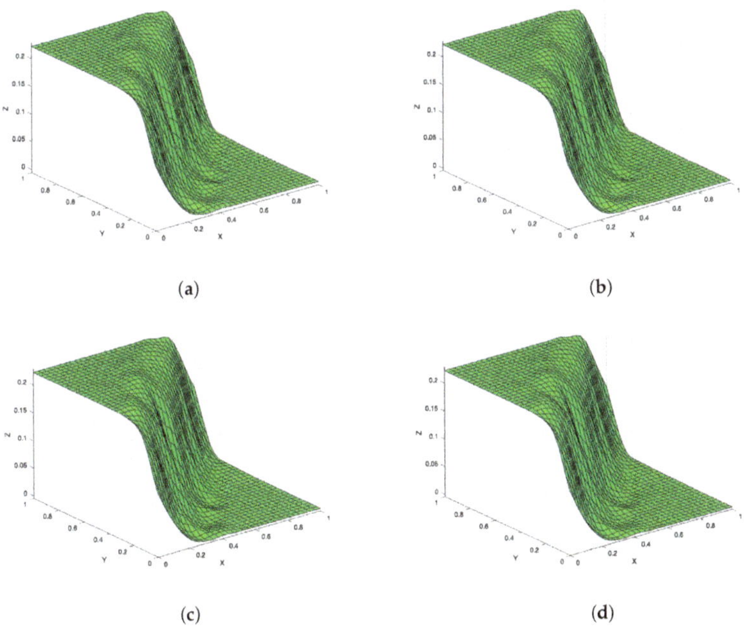

Figure 19. Surface interpolation for 65 data points based on F_2 function. (**a**) $q = 0.25$. (**b**) $q = 0.5$. (**c**) $q = 0.75$. (**d**) $q = 1$.

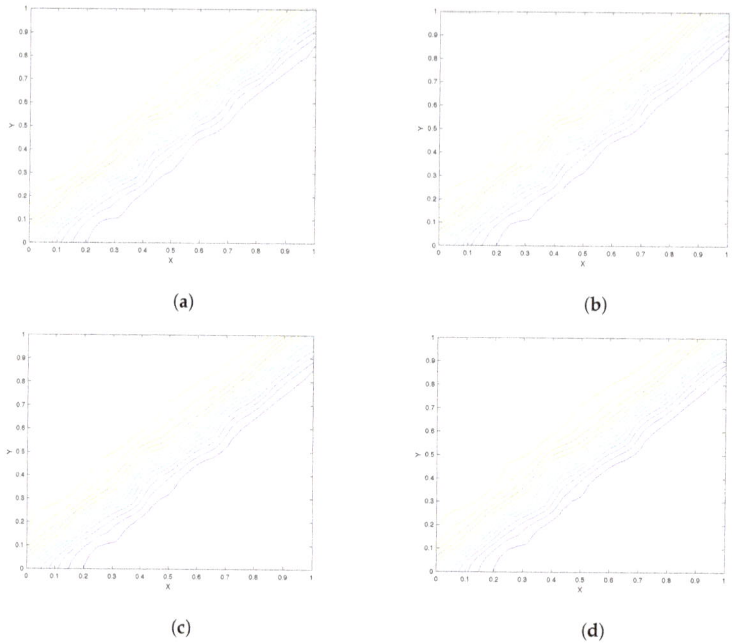

Figure 20. Contour plots for 65 data points based on F_2 function. (**a**) $q = 0.25$. (**b**) $q = 0.5$. (**c**) $q = 0.75$. (**d**) $q = 1$.

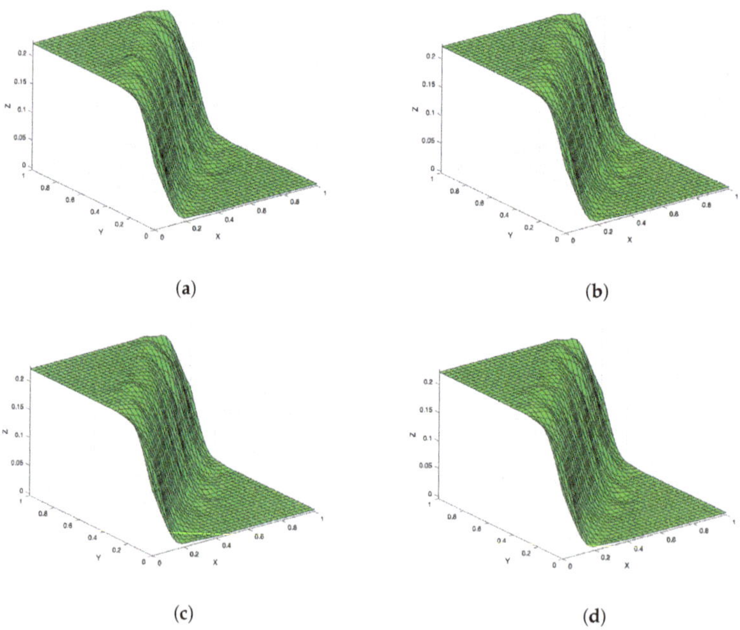

Figure 21. Surface interpolation for 100 data points based on F_2 function. (**a**) $q = 0.25$. (**b**) $q = 0.5$. (**c**) $q = 0.75$. (**d**) $q = 1$.

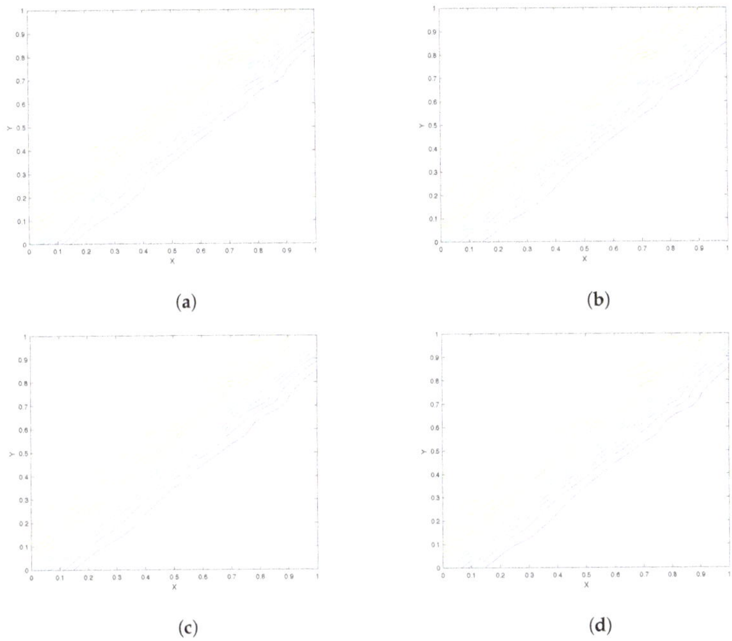

Figure 22. Contour plots for 100 data points based on F_2 function. (a) $q = 0.25$. (b) $q = 0.5$. (c) $q = 0.75$. (d) $q = 1$.

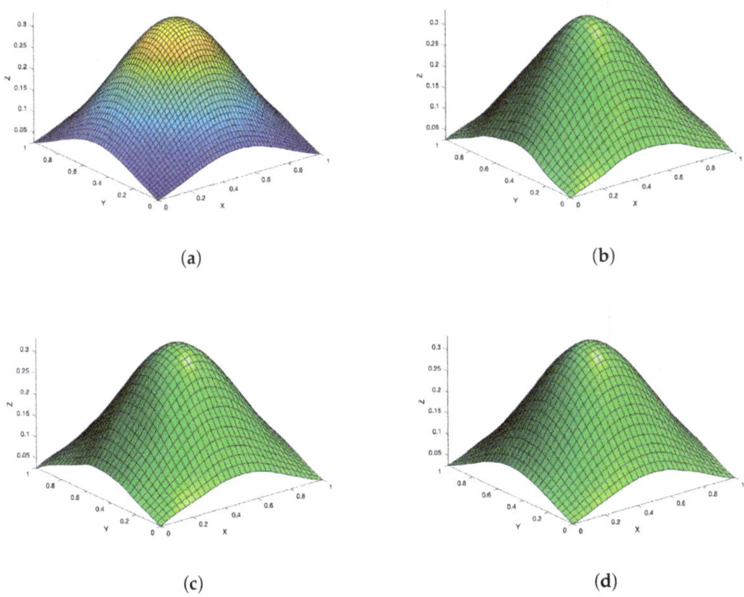

Figure 23. Surface interpolation based on F_3 function with 36, 65, and 100 data points when $q = 1$: (a) true surface; (b) 36 data points; (c) 65 data points; (d) 100 data points.

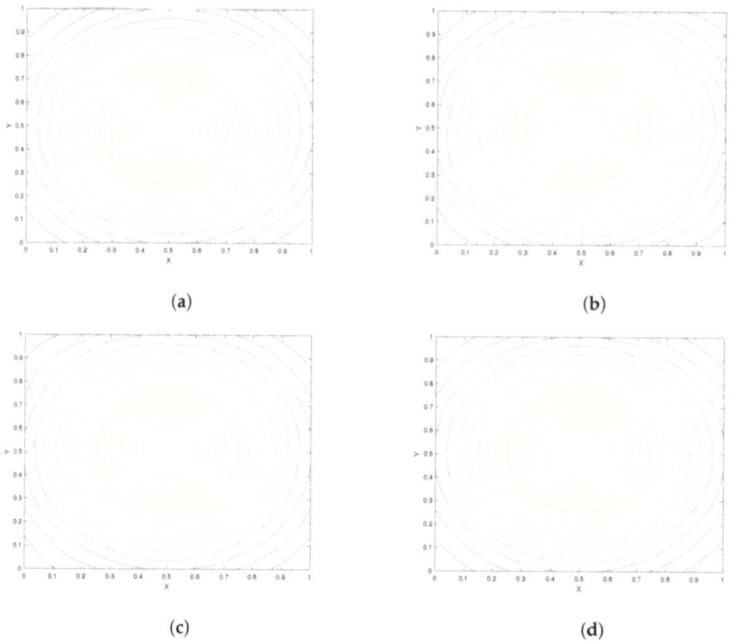

Figure 24. Contour plots based on F_3 function with 36, 65, and 100 data points when $q = 1$: (**a**) true surface; (**b**) 36 data points; (**c**) 65 data points; (**d**) 100 data points.

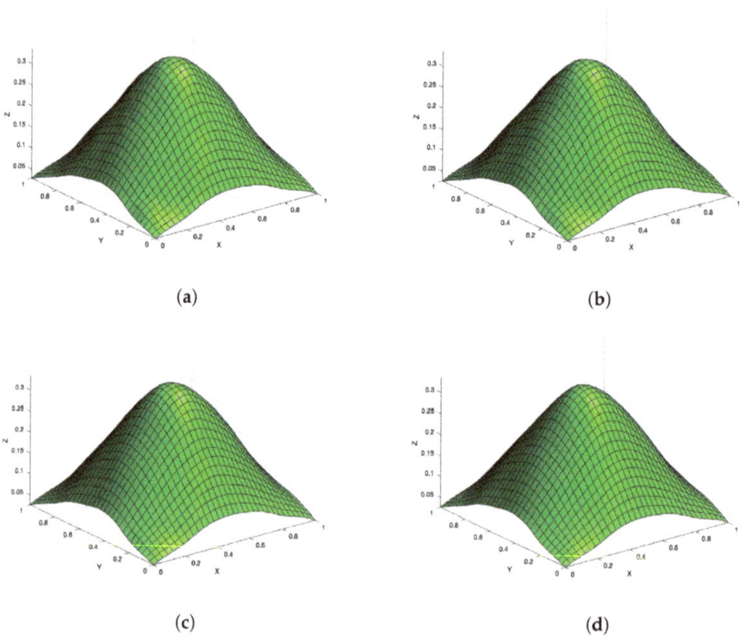

Figure 25. Surface interpolation for 36 data points based on F_3 function. (**a**) $q = 0.25$. (**b**) $q = 0.5$. (**c**) $q = 0.75$. (**d**) $q = 1$.

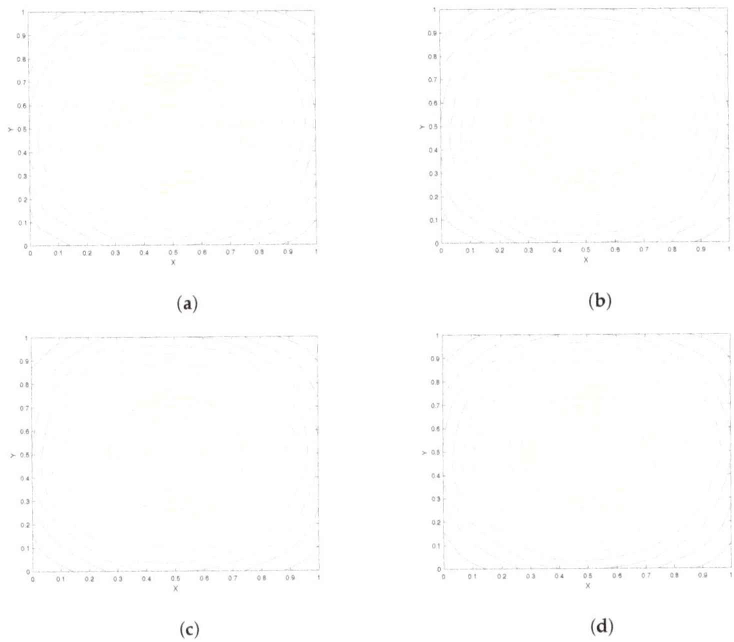

Figure 26. Contour plots for 36 data points based on F_3 function. (**a**) $q = 0.25$. (**b**) $q = 0.5$. (**c**) $q = 0.75$. (**d**) $q = 1$.

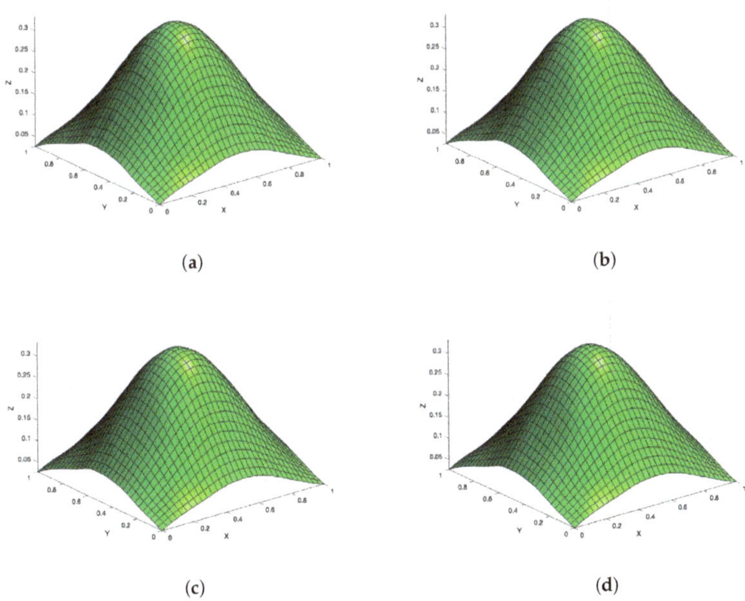

Figure 27. Surface interpolation for 65 data points based on F_3 function. (**a**) $q = 0.25$. (**b**) $q = 0.5$. (**c**) $q = 0.75$. (**d**) $q = 1$.

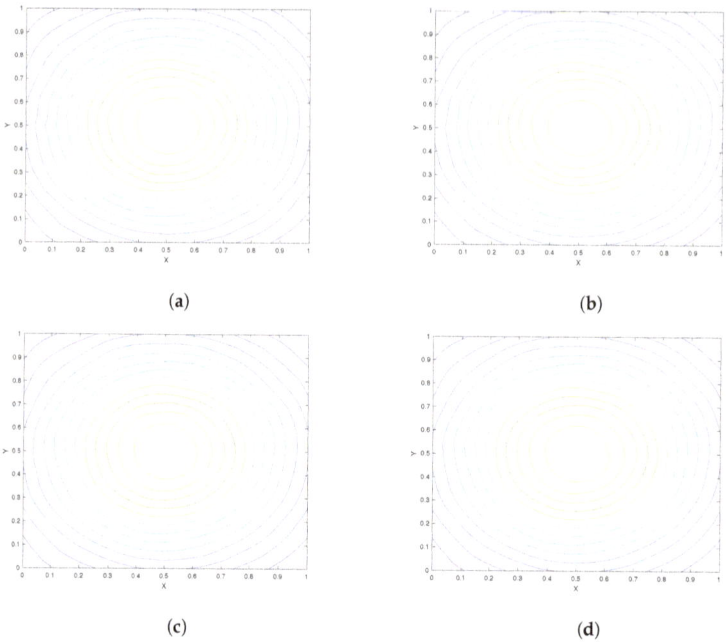

Figure 28. Contour plots for 65 data points based on F_3 function. (**a**) q = 0.25. (**b**) q = 0.5. (**c**) q = 0.75. (**d**) q = 1.

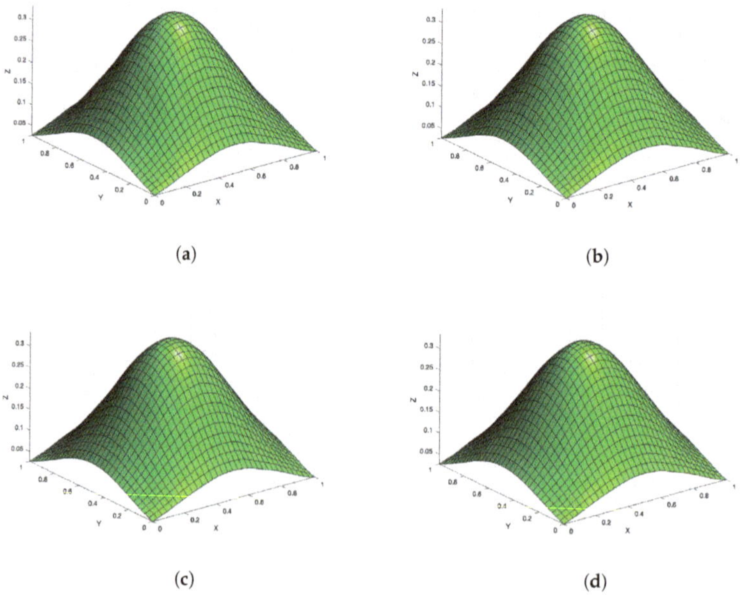

Figure 29. Surface interpolation for 100 data points based on F_3 function. (**a**) q = 0.25. (**b**) q = 0.5. (**c**) q = 0.75. (**d**) q = 1.

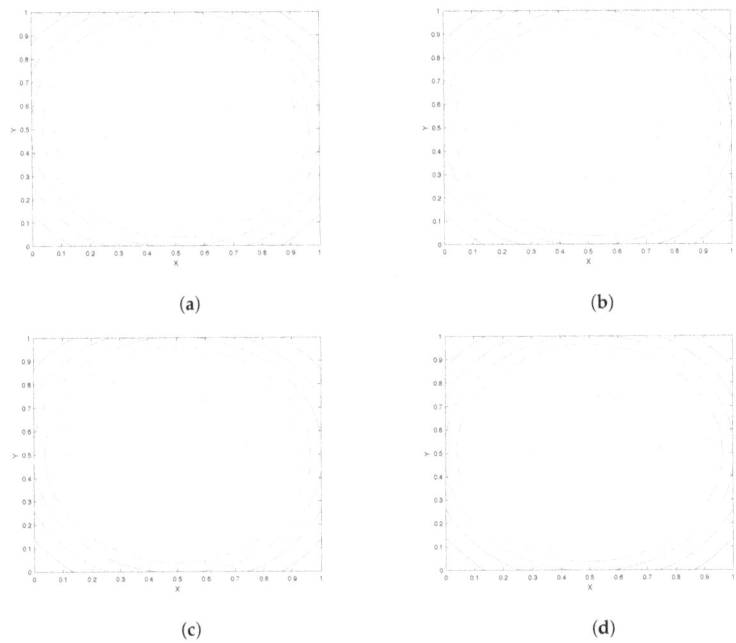

Figure 30. Contour plots for 100 data points based on F_3 function. (**a**) q = 0.25. (**b**) q = 0.5. (**c**) q = 0.75. (**d**) q = 1.

Since visualization alone is not sufficient to determine which q parameter yields the best interpolation results, error analysis was employed to provide a more robust comparison. Tables 1 and 2 present the overall measurements for 36, 65, and 100 data points with different q values for the proposed q-Bézier interpolation scheme. Based on these tables, it is evident that as the q parameter increases, there is an improvement in the interpolation quality, reflected by higher R^2 values, a reduced RMSE, and faster CPU time. Specifically, the q values of 0.75 and 1.00 show significant enhancements, indicating a trend toward the properties of the classical Bézier curve when q parameter equals 1.

Table 1. Overall error measurements for 36, 65, and 100 data points for q-Bézier with values of 0.25 and 0.50.

Function	Data Points	R^2		RMSE		CPU Time (s)	
		Q [1]	Q [2]	Q [1]	Q [2]	Q [1]	Q [2]
F_1	36	0.9974	0.9974	0.0038	0.0037	0.0981	0.1734
	65	0.9990	0.9991	0.0023	0.0023	0.2621	0.2743
	100	0.9998	0.9998	0.0011	0.0010	0.4495	0.4800
F_2	36	0.9832	0.9832	0.0129	0.0129	0.1554	0.1819
	65	0.9970	0.9971	0.0055	0.0053	0.2714	0.3210
	100	0.9986	0.9987	0.0037	0.0036	0.5213	0.4021
F_3	36	0.9973	0.9975	0.0042	0.0040	0.1336	0.1562
	65	0.9993	0.9994	0.0021	0.0020	0.2817	0.2079
	100	0.9998	0.9999	0.0011	0.0010	0.3653	0.3178

[1] $q = 0.25$; [2] $q = 0.50$.

The error analysis further supports the visual observations: the results for 100 data points consistently outperform those for 36 and 65 data points. This improvement is expected, as a higher number of data points provides a more detailed representation of the underlying function, thus enhancing the interpolation accuracy. Additionally, the CPU time analysis reveals that higher q values lead to faster computations. The combination of a higher R^2, lower RMSE, and reduced CPU time demonstrates the effectiveness of using a higher q parameter. The subsequent subsection will compare the proposed method with another state-of-the-art interpolation technique, focusing on the two best-performing q parameter values.

Table 2. Overall error measurements for 36, 65, and 100 data points for q-Bézier with values of 0.75 and 1.00.

Function	Data Points	R^2		RMSE		CPU Time (s)	
		Q [3]	Q [4]	Q [3]	Q [4]	Q [3]	Q [4]
F_1	36	0.9976	0.9977	0.0036	0.0035	0.1966	0.1893
	65	0.9991	0.9992	0.0022	0.0020	0.2424	0.2545
	100	0.9999	0.9999	0.0009	0.0009	0.3431	0.2557
F_2	36	0.9831	0.9829	0.0129	0.0130	0.2106	0.1758
	65	0.9973	0.9975	0.0051	0.0049	0.2550	0.1910
	100	0.9987	0.9988	0.0035	0.0035	0.2885	0.2807
F_3	36	0.9977	0.9978	0.0038	0.0037	0.2145	0.1850
	65	0.9995	0.9995	0.0019	0.0018	0.2301	0.2583
	100	0.9999	0.9999	0.0008	0.0007	0.3059	0.3024

[3] $q = 0.75$; [4] $q = 1.00$.

4.1. Comparison with Goodman and Said's Method

Based on the results from Tables 1 and 2, the optimal q values for the proposed interpolation method are between 0.75 and 1.00. To further validate the effectiveness of the proposed method, we compare it with the cross-derivative method of Goodman and Said [16]. Tables 3 and 4 present the overall error measurements between the proposed method and the cross derivative method for q values of 0.75 and 1.00 across three test functions.

The result indicate that the proposed method outperforms the cross-derivative method, particularly when the q value is 1.00. Specifically, the proposed method achieves higher R^2 values and lower RMSEs for all three test functions. For instance, with a q value of 1.00, the R^2 values are consistently higher, and the RMSE values are lower, indicating a more accurate fit to the data.

Although the differences in R^2 and RMSE between the proposed method and the cross-derivative method are not significant, the CPU time required for the proposed method is significantly shorter. The proposed method only takes 0.2557 s, 0.2807 s, and 0.2078 s to construct the surface interpolation for the $F_1(x, y)$, $F_2(x, y)$, and $F_3(x, y)$ functions, respectively. In contrast, the cross-derivative method requires substantially more time for similar tasks.

These findings suggest that the proposed method not only provides a high degree of accuracy but also offers significant computational advantages. Given these promising results, the next subsection will explore the application of the proposed method on real-life data.

Table 3. Overall error measurements for 36, 65, and 100 data points between the proposed method and Goodman and Said's method for q-Bézier with value of 0.75.

Function	Data Points	R^2		RMSE		CPU Time (s)	
		PS [1]	GS [2]	PS [1]	GS [2]	PS [1]	GS [2]
F_1	36	0.9976	0.9976	0.0036	0.0036	0.1966	0.1670
	65	0.9991	0.9992	0.0022	0.0021	0.2424	0.2045
	100	0.9999	0.9999	0.0009	0.0008	0.3431	0.3071

Table 3. *Cont.*

Function	Data Points	R^2		RMSE		CPU Time (s)	
		PS [1]	GS [2]	PS [1]	GS [2]	PS [1]	GS [2]
F_2	36	0.9831	0.9824	0.0129	0.0132	0.2106	0.1640
	65	0.9973	0.9973	0.0051	0.0052	0.2550	0.1768
	100	0.9987	0.9987	0.0035	0.0036	0.2885	0.2968
F_3	36	0.9977	0.9977	0.0038	0.0038	0.2145	0.1746
	65	0.9995	0.9994	0.0019	0.0019	0.2301	0.1817
	100	0.9999	0.9999	0.0008	0.0007	0.3059	0.2988

[1] Proposed method; [2] Goodman and Said's method.

Table 4. Overall error measurements for 36, 65, and 100 data points between the proposed method and Goodman and Said's method for q-Bézier with value of 1.00.

Function	Data Points	R^2		RMSE		CPU Time (s)	
		PS [1]	GS [2]	PS [1]	GS [2]	PS [1]	GS [2]
F_1	36	0.9977	0.9976	0.0035	0.0036	0.1893	0.1986
	65	0.9992	0.9992	0.0020	0.0021	0.2166	0.2374
	100	0.9999	0.9999	0.0009	0.0008	0.2557	0.3051
F_2	36	0.9829	0.9824	0.0130	0.0132	0.1758	0.1769
	65	0.9975	0.9973	0.0049	0.0052	0.1910	0.2638
	100	0.9988	0.9987	0.0035	0.0036	0.2807	0.3589
F_3	36	0.9978	0.9977	0.0037	0.0038	0.1373	0.2090
	65	0.9995	0.9994	0.0018	0.0019	0.1570	0.2496
	100	0.9999	0.9999	0.0007	0.0007	0.2078	0.4441

[1] Proposed method; [2] Goodman and Said's method.

4.2. Testing on Seamount Real Data Points

In this subsection, we apply the proposed method to reconstruct real-life geologic data from the Seamount dataset. This dataset, obtained from MATLAB, represents the underwater surface of a seamount located on the Louisville Ridge in the South Pacific, mapped in 1984. Figure 31 illustrates the Delaunay triangulation of the Seamount data, comprising 294 data points, along with the corresponding 3D interpolation. In Figure 32, the surface interpolation of the Seamount data points is presented for various q values. Furthermore, Figure 33 showcases the 3D interpolation derived from Figure 32.

Table 5 provides a comparative analysis of CPU times for different q values based on the average of three experimental runs. Notably, the q value of 1.00 consistently results in the shortest average time of 157.99 s. This observation aligns with previous findings, reinforcing that a q value of 1.00 not only ensures computational efficiency but also maintains interpolation accuracy.

Upon examining the visualizations, it is evident that the interpolations with a q value of 1.00 yield a smoother and more accurate representation of the Seamount surface compared to other q values. The reduced computational time further emphasizes the practicality of the proposed method for large and complex datasets. The following subsections will present a more detailed comparative analysis applied to engineering-related problems.

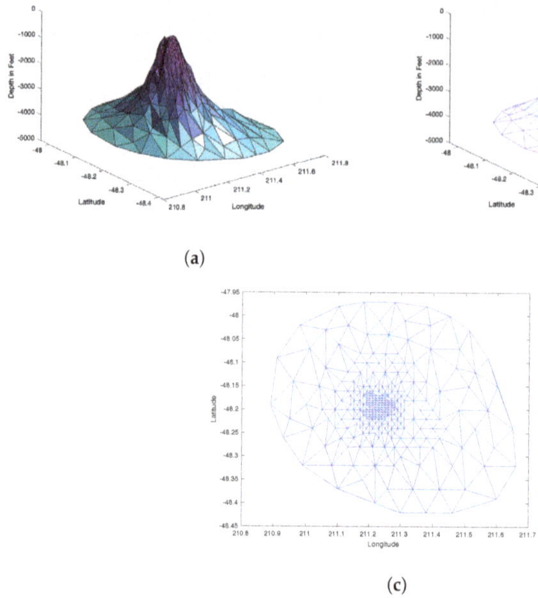

Figure 31. Example for (**a**) true surface, (**b**) 3D interpolation, and (**c**) Delaunay triangulation of Seamount data points.

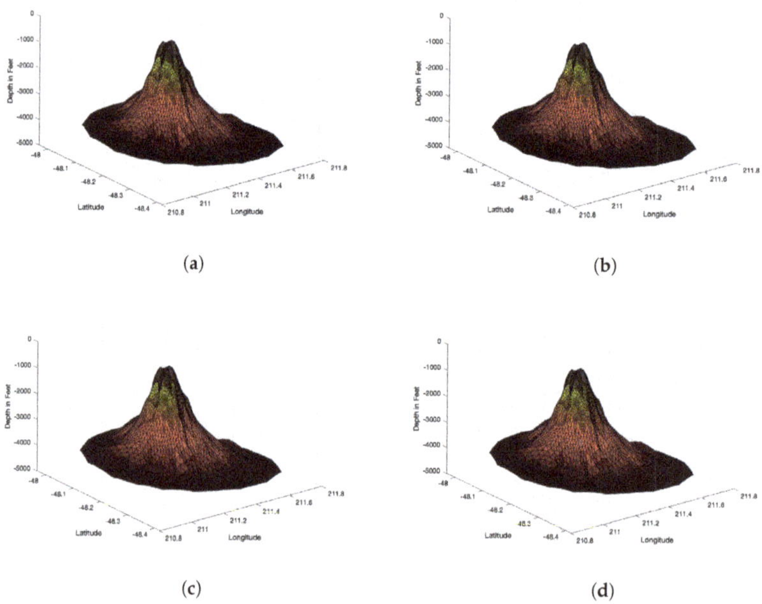

Figure 32. Surface interpolation with different q values based on Seamount data points. (**a**) $q = 0.25$. (**b**) $q = 0.5$. (**c**) $q = 0.75$. (**d**) $q = 1$.

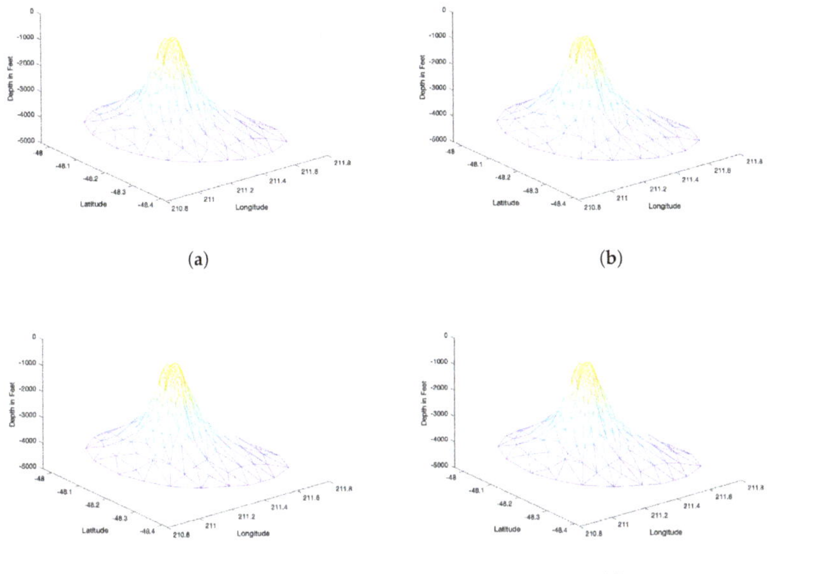

Figure 33. Three-dimensional interpolation with different q values based on Seamount data points. (**a**) $q = 0.25$. (**b**) $q = 0.5$. (**c**) $q = 0.75$. (**d**) $q = 1$.

Table 5. CPU time comparison between different q values.

q Values	CPU Time (s)			
	First Run	**Second Run**	**Third Run**	**Average**
0.25	163.25	159.80	158.05	160.37
0.50	160.45	158.11	157.78	158.78
0.75	157.74	157.76	159.17	158.22
1.00	157.88	158.37	157.73	157.99

4.3. Application to the Electric Potential of Two Point Charges

The scattered data interpolation method obtained in the previous subsection can be applied directly to the problem of the electric potential of two point charges proposed in [38].

The equation can be written as

$$V = \frac{1}{4\pi\kappa_0}\frac{g}{h} \tag{16}$$

where $\kappa_0 = 8.8541878 \times 10^{-12}\frac{C}{Nm^2}$ represents the permittivity constant. g is the magnitude of the charge in coulombs, whereas h is the distance from the particle in meters. Normally, the electric field of two or more particles is calculated using superposition. Based on Figure 34, the electric potential at a point due to two particles is given by

$$V = \frac{1}{4\pi\kappa_0}(\frac{g_1}{h_1} + \frac{g_2}{h_2}) \tag{17}$$

where g_1, g_2, h_1, and h_2 are the charges of the particles and the distance from the point to the corresponding particle, respectively. In taking $g_1 = 2 \times 10^{-10}C$ and $g_2 = 3 \times 10^{-10}C$, which are positioned in the $x\,y$ plane at points $(0.25, 0, 0)$ and $(-0.25, 0, 0)$, the scattered data interpolation that uses the q-Bézier triangular patch can solve the proposed problem. The existing Delaunay triangulation for the proposed problem is illustrated in Figure 35 as presented by Ali et al. [39]. Figure 36 showcases the surface interpolation and its corresponding contour plot for both the true function and the existing scheme of Ali et al. [39]. However, our observation indicates that their Delaunay triangulation could be enhanced for better interpolation accuracy by creating a more scattered data distribution. Another critical observation from Figure 36 is that the method of Ali et al. does not approximate the problem accurately, as the results do not closely resemble the actual contour plot.

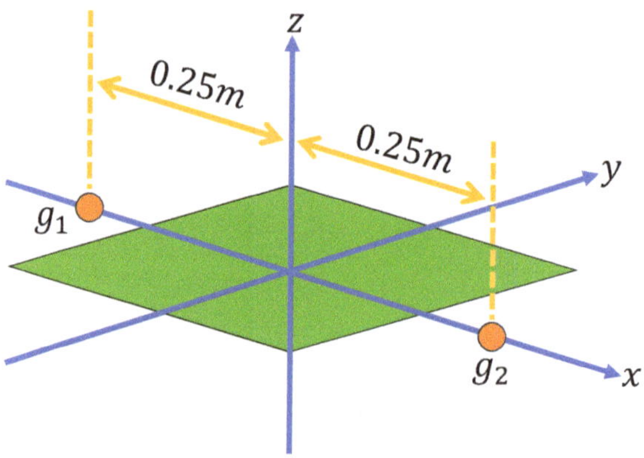

Figure 34. Example of electric potential of two point charges (Adopted from [38]).

To address these issues, we manually constructed our own Delaunay triangulation by carefully plotting scattered data with an uneven distribution, not only for 25 data points but also for 36, 65, and 100 data points, as shown in Figure 37. This extension aims to provide a more comprehensive comparative analysis of the proposed scheme. Figures 38 and 39 illustrate the surface interpolation and its contour plot when the q parameter is set to 1. Based on the figures, using 100 data points results in a smoother and more refined surface interpolation and contour plot that closely resembles the true function of the proposed problem. This pattern is consistent with previous sections of this study, where an increased number of data points provided more detailed information and features, better representing the proposed solution.

Tables 6 and 7 present the error analysis comparing Ali et al.'s method [39] and our proposed method across various q parameter values. Our proposed method achieves the best performance with 100 data points, showing an R^2 value of 0.9908 and an RMSE of 0.5753 when the q parameter is 0.25. This is particularly interesting because, in contrast to the previous sections with different functions, our proposed method performed best when the q parameter was equal to 1. This discrepancy suggests the potential for further exploration of optimal q parameter values to enhance scattered data interpolation schemes. This observation also highlights that different q values may yield the best results depending on the specific function and the nature of the real-life engineering problem being addressed.

Another noteworthy observation is that different arrangements of points, even with the same number of points, yield similar Delaunay triangulation results, as illustrated in Tables 6 and 7. The performance metrics for these different point arrangements do not show significant variation, indicating that the spatial positioning of points, given a fixed number of points, is not a crucial factor in enhancing the performance of the proposed scheme. This observation suggests that the robustness of the Delaunay triangulation method is maintained irrespective of point arrangement.

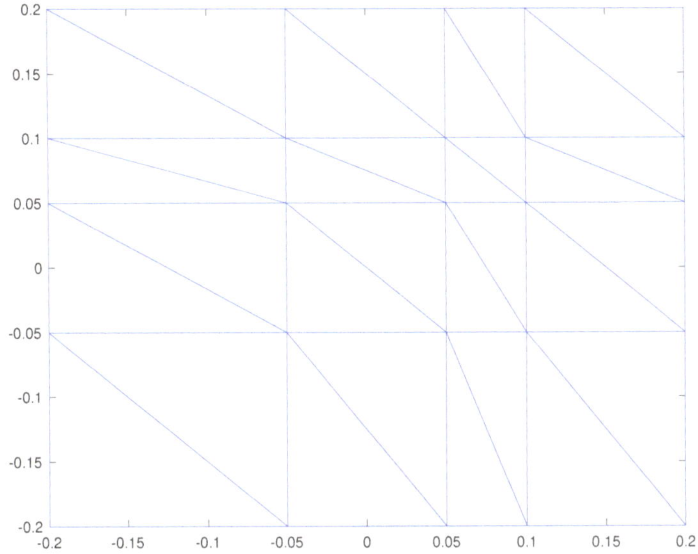

Figure 35. Delaunay triangulation of 25 data points (adopted from [39]).

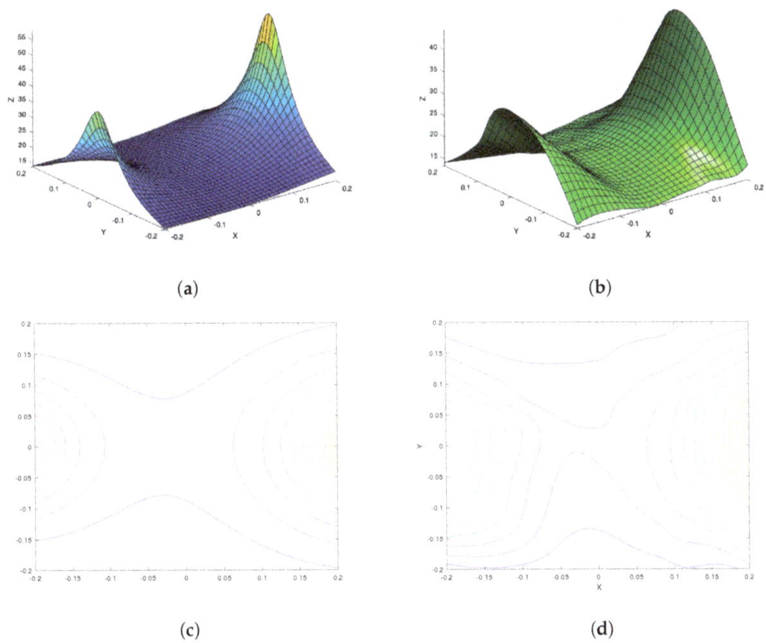

Figure 36. Surface and contour plot for (**a**) true function and (**b**) Ali et al. [39].

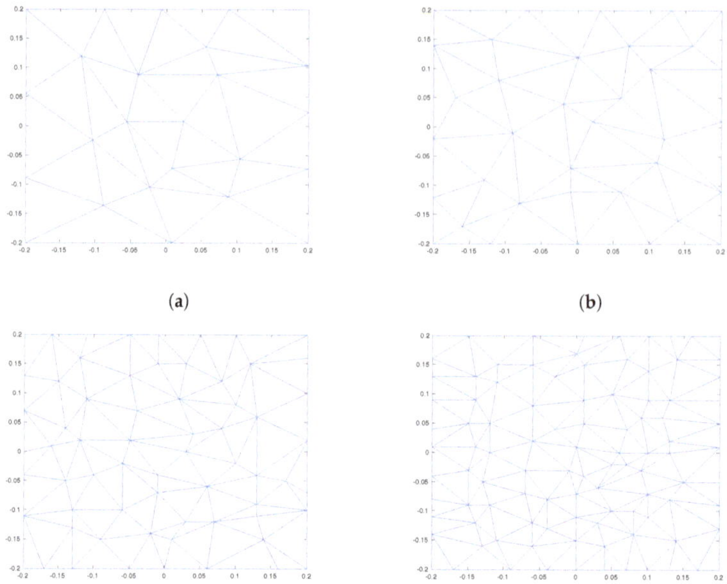

(a) (b) (c) (d)

Figure 37. Delaunay triangulation of 25, 35, 65, and 100 data points: (**a**) 25 data points; (**b**) 36 data points; (**c**) 65 data points; (**d**) 100 data points.

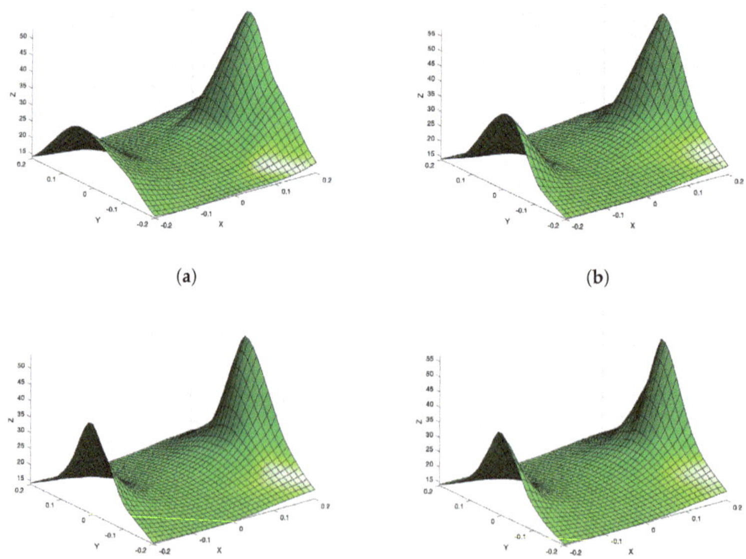

(a) (b) (c) (d)

Figure 38. Surface interpolation with 25, 36, 65, and 100 data points for q value of 1: (**a**) 25 data points; (**b**) 36 data points; (**c**) 65 data points; (**d**) 100 data points.

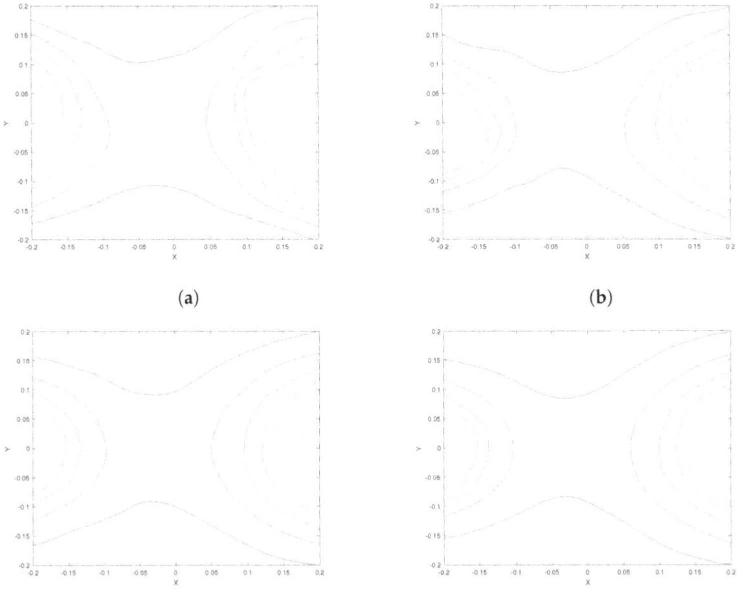

Figure 39. Contour plot with 25, 36, 65, and 100 data points for q value of 1: (a) 25 data points; (b) 36 data points; (c) 65 data points; (d) 100 data points.

Table 6. Overall error measurements for 25, 36, 65, and 100 data points for q-Bézier.

Method	Data Points	R^2		RMSE		CPU Time (s)	
		Q [1]	Q [2]	Q [1]	Q [2]	Q [1]	Q [2]
Fatin	25	0.8958	0.8934	1.9374	1.9592	0.1244	0.0989
Proposed	25	0.8868	0.8871	2.0195	2.0166	0.0743	0.1184
Proposed	36	0.9356	0.9356	1.5233	1.5225	0.1463	0.1746
Proposed	65	0.9791	0.9789	0.8674	0.8709	0.1954	0.2669
Proposed	100	0.9908	0.9907	0.5753	0.5777	0.2696	0.3106

[1] $q = 0.25$; [2] $q = 0.50$.

Table 7. Overall error measurements for 25, 36, 65, and 100 data points for q-Bézier.

Method	Data Points	R^2		RMSE		CPU Time (s)	
		Q [3]	Q [4]	Q [3]	Q [4]	Q [3]	Q [4]
Fatin	25	0.8901	0.8852	1.9898	2.0335	0.1654	0.1502
Proposed	25	0.8881	0.8893	2.0079	1.9971	0.0715	0.0995
Proposed	36	0.9361	0.9364	1.5173	1.5136	0.1160	0.1495
Proposed	65	0.9786	0.9780	0.8776	0.8898	0.1889	0.1958
Proposed	100	0.9905	0.9902	0.5836	0.5956	0.3125	0.2644

[3] $q = 0.75$; [4] $q = 1.00$.

4.4. Optimal Value of q

This section describes the optimal value of the q parameter. Previous results from visualization and error analysis indicated that the proposed scheme performs best when the q parameter is set to 1. To explore the potential for other values close to 1, we conducted additional simulations starting from a q value of 0.9 to 1.00 with a step size of 0.01. Table 8 presents the error measurements for 36, 65, and 100 data points across three test functions, ranging from $q = 0.90$ to $q = 1.00$ in increments of 0.01.

The results indicate that the R^2 and RMSE values are nearly identical for q values of 0.95, 0.96, 0.97, and 1. To facilitate visualization, we compare these four best q values using error analysis metrics in Figures 40–42. The bar chart visualization reveals no significant differences in the R^2 and RMSE values among these selected q values. Notably, the R^2 value is lowest for 36 data points based on the F_2 function, corresponding to the highest RMSE value observed for the same data points.

However, a q value of 0.96 outperforms the other three q values in terms of CPU time. Therefore, considering all error analysis metrics, a q value of 0.96 is optimal. This finding is intriguing, suggesting deviation from the classical Bézier curve typically used in state-of-the-art methods, despite $q = 0.96$ being close to 1. This presents opportunities for future research to explore optimal q values for enhancing scattered data interpolation methods.

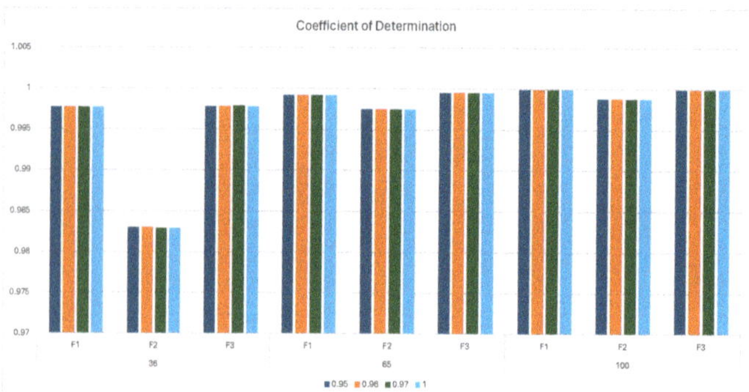

Figure 40. R^2 values for 36, 65, and 100 data points across three test functions, evaluated at four selected q values.

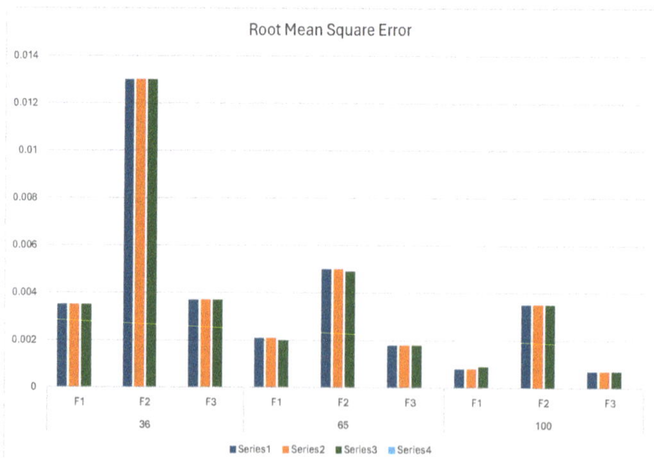

Figure 41. RMSE values for 36, 65, and 100 data points across three test functions, evaluated at four selected q values.

Table 8. Overall error measurements for 36, 65, and 100 data points for q-Bézier with values of from 0.90 to 1.00 with step size of 0.01.

q Value	Data Points								
	36			65			100		
	F_1	F_2	F_3	F_1	F_2	F_3	F_1	F_2	F_3
0.90	0.9977	0.9830	0.9978	0.9992	0.9974	0.9995	0.9999	0.9988	0.9999
	0.0036	0.0130	0.0037	0.0021	0.0050	0.0018	0.0008	0.0035	0.0007
	0.1449	0.1419	0.1487	0.1752	0.2267	0.2210	0.2256	0.3326	0.2341
0.91	0.9977	0.9830	0.9978	0.9992	0.9975	0.9995	0.9999	0.9988	0.9999
	0.0036	0.0130	0.0037	0.0021	0.0050	0.0018	0.0008	0.0035	0.0007
	0.1758	0.1598	0.1827	0.2049	0.2051	0.1908	0.3662	0.3654	0.3049
0.92	0.9977	0.9830	0.9978	0.9992	0.9975	0.9995	0.9999	0.9988	0.9999
	0.0036	0.0130	0.0037	0.0021	0.0050	0.0018	0.0008	0.0035	0.0007
	0.1895	0.1722	0.1731	0.2196	0.2047	0.1917	0.3483	0.3253	0.3222
0.93	0.9977	0.9830	0.9978	0.9992	0.9975	0.9995	0.9999	0.9988	0.9999
	0.0036	0.0130	0.0037	0.0021	0.0050	0.0018	0.0008	0.0035	0.0007
	0.1509	0.1966	0.2206	0.2154	0.2106	0.2609	0.3322	0.3873	0.3712
0.94	0.9977	0.9830	0.9978	0.9992	0.9975	0.9995	0.9999	0.9988	0.9999
	0.0036	0.0130	0.0037	0.0021	0.0050	0.0018	0.0008	0.0035	0.0007
	0.1748	0.1850	0.1643	0.2268	0.2256	0.1848	0.2829	0.3162	0.3039
0.95	0.9977	0.9830	0.9978	0.9992	0.9975	0.9995	0.9999	0.9988	0.9999
	0.0035	0.0130	0.0037	0.0021	0.0050	0.0018	0.0008	0.0035	0.0007
	0.1530	0.1695	0.1820	0.2018	0.2054	0.2528	0.3995	0.3927	0.3929
0.96	0.9977	0.9830	0.9978	0.9992	0.9975	0.9995	0.9999	0.9988	0.9999
	0.0035	0.0130	0.0037	0.0021	0.0050	0.0018	0.0008	0.0035	0.0007
	0.1670	0.1601	0.1737	0.1840	0.2091	0.2310	0.2172	0.3576	0.2682
0.97	0.9977	0.9829	0.9979	0.9992	0.9975	0.9995	0.9999	0.9988	0.9999
	0.0035	0.0130	0.0037	0.0021	0.0050	0.0018	0.0008	0.0035	0.0007
	0.1738	0.1344	0.1882	0.1901	0.2289	0.2058	0.2811	0.5423	0.3327
0.98	0.9977	0.9829	0.9978	0.9992	0.9975	0.9995	0.9999	0.9988	0.9999
	0.0035	0.0130	0.0037	0.0021	0.0050	0.0018	0.0009	0.0035	0.0007
	0.1689	0.0850	0.1773	0.2088	0.2134	0.2185	0.3095	0.2941	0.6070
0.99	0.9977	0.9829	0.9978	0.9992	0.9975	0.9995	0.9999	0.9988	0.9999
	0.0035	0.0130	0.0037	0.0021	0.0050	0.0018	0.0009	0.0035	0.0007
	0.1689	0.0852	0.1309	0.1898	0.1944	0.1814	0.2792	0.2571	0.2659
1.00	0.9977	0.9829	0.9978	0.9992	0.9975	0.9995	0.9999	0.9988	0.9999
	0.0035	0.0130	0.0037	0.0020	0.0049	0.0018	0.0009	0.0035	0.0007
	0.1893	0.1758	0.1850	0.2545	0.1910	0.2583	0.2557	0.2807	0.3024

Note: For each q value, the three corresponding sub-rows represent: (i) R^2, (ii) RMSE, (iii) CPU time (in seconds).

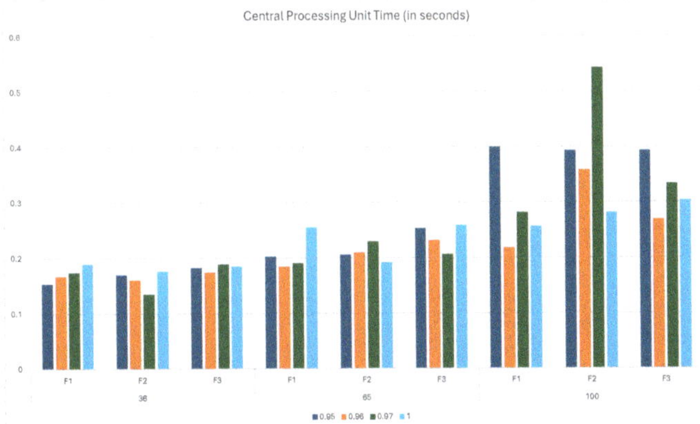

Figure 42. CPU time (in seconds) for 36, 65, and 100 data points across three test functions, evaluated at four selected q values.

5. Conclusions

In conclusion, this study successfully extends the application of q-Bézier curves to a scattered data interpolation scheme on well-known mathematical functions as well as real-life data. The proposed interpolation scheme was constructed using three local schemes that are blended through a convex combination via an efficient algorithm. The performance of the scheme was rigorously tested with 36, 65, and 100 uniformly distributed data points. Our results indicate that the proposed scheme achieves optimal visualization and error analysis outcomes when the q parameter is set between 0.75 and 1. This finding suggests that the careful tuning of the q parameter is crucial for enhancing interpolation accuracy.

The method presented in this paper is novel and still under exploration, particularly in finding the optimal q parameter. Our current approach is heuristic, and we recognize the need for a more systematic method to determine this parameter. One promising direction is to employ deep learning techniques, particularly Convolutional Neural Networks (CNNs) to optimize the q parameter. This approach involves training a CNN on a comprehensive dataset of scattered data points and their Delaunay triangulations to predict the optimal q parameter.

Additionally, our study is limited by the relatively small number of data points. Increasing the dataset size could provide more comprehensive validation of the proposed scheme and enhance its applicability to a wider range of problems.

Future research will also explore the application of the proposed scheme to image interpolation, particularly for RGB images. This extension could provide significant improvements in image processing tasks, where accurate interpolation is critical for enhancing image quality and resolution.

Author Contributions: Conceptualization, O.T.; data curation, O.T.; formal analysis, O.T. and S.A.A.K.; funding acquisition, S.A.A.K.; investigation, O.T.; methodology, O.T. and S.A.A.K.; project administration, S.A.A.K.; software, O.T. and S.A.A.K.; supervision, S.A.A.K.; validation, O.T. and S.A.A.K.; visualization, O.T. and S.A.A.K.; writing—original draft, O.T.; writing—review and editing, O.T. and S.A.A.K. All authors have read and agreed to the published version of the manuscript.

Funding: This research was fully supported by the Ministry of Higher Education (MOHE) of Malaysia through Fundamental Research Grant Scheme [FRGS/1/2023/ICT06/ UMS/02/1] (New Scattered Data Interpolation Scheme Using Quasi Cubic Triangular Patches for RGB Image Interpolation) and Universiti Malaysia Sabah.

Institutional Review Board Statement: Not applicable.

Informed Consent Statement: Not applicable.

Data Availability Statement: The data are shared in this paper.

Acknowledgments: The authors give special thanks to the Faculty of Computing and Informatics, Universiti Malaysia Sabah and School of Quantitative Sciences, College of Arts and Sciences, Universiti Utara Malaysia for the tremendous computing facility support.

Conflicts of Interest: The authors declare no conflicts of interest.

Abbreviations

The following abbreviations are used in this manuscript:

RBFs	Radial Basis Functions
SNRBFs	Square Normalized Radial Basis Functions
MLS	Moving Least Square
RMSE	Root Mean Square Error
CPU	Central Processing Unit

References

1. Franke, R.; Nielson, G. Scattered data interpolation of large sets of scattered data. *Int. J. Numer. Methods Eng.* **1991**, *15*, 1–691.
2. Cavoretto, R.; De Rossi, A.; Dell'Accio, F.; Di Tommaso, F. Fast computation of triangular Shepard interpolants. *J. Comput. Appl. Math.* **2019**, *354*, 457–470. [CrossRef]
3. Liu, Z. Local multilevel scattered data interpolation. *Eng. Anal. Bound. Elem.* **2018**, *92*, 101–107. [CrossRef]

4. Le Borne, S.; Wende, M. Domain decomposition methods in scattered data interpolation with conditionally positive definite radial basis functions. *Comput. Math. Appl.* **2019**, *77*, 1178–1196. [CrossRef]
5. Skala, V.; Mourycova, E. Meshfree Interpolation of Multidimensional Time-Varying Scattered Data. *Computers* **2023**, *12*, 243. [CrossRef]
6. Joldes, G.R.; Chowdhury, H.A.; Wittek, A.; Doyle, B.; Miller, K. Modified moving least squares with polynomial bases for scattered data approximation. *Appl. Math. Comput.* **2015**, *266*, 893–902. [CrossRef]
7. Brodlie, K.W.; Asim, M.R.; Unsworth, K. Constrained visualization using the Shepard interpolation family. *Comput. Graph. Forum* **2005**, *24*, 809–820. [CrossRef]
8. Lai, M.J.; Meile, C. Scattered data interpolation with nonnegative preservation using bivariate splines and its application. *Comput. Aided Geom. Des.* **2015**, *34*, 37–49. [CrossRef]
9. Schumaker, L.L.; Speleers, H. Nonnegativity preserving macro-element interpolation of scattered data. *Comput. Aided Geom. Des.* **2010**, *27*, 245–261. [CrossRef]
10. Karim, S.A.; Saaban, A.; Hasan, M.K.; Sulaiman, J.; Hashim, I. Interpolation using cubic Bézier triangular patches. *Int. J. Adv. Sci. Eng. Inf. Technol.* **2018**, *8*, 1746–1752. [CrossRef]
11. Karim, S.A.A.; Saaban, A.; Skala, V.; Ghaffar, A.; Nisar, K.S.; Baleanu, D. Construction of new cubic Bézier-like triangular patches with application in scattered data interpolation. *Adv. Differ. Equ.* **2020**, *2020*, 151. [CrossRef]
12. Said, H.; Wirza, R. A Cubic Ball Triangular Patch for the Scattered Data Interpolation. Master's Thesis, Pulau Pinang, Malaysia, 1994.
13. Feng, R.; Zhang, Y. Piecewise Bivariate Hermite Interpolations for large sets of scattered data. *J. Appl. Math.* **2013**, *2013*, 239703 . [CrossRef]
14. Sun, Q.; Bao, F.; Zhang, Y.; Duan, Q. A bivariate rational interpolation based on scattered data on parallel lines. *J. Vis. Commun. Image Represent.* **2013**, *24*, 75–80. [CrossRef]
15. Bozzini, M.; Lenarduzzi, L.; Rossini, M. Polyharmonic splines: An approximation method for noisy scattered data of extra-large size. *Appl. Math. Comput.* **2010**, *216*, 317–331. [CrossRef]
16. Goodman, T.; Said, H. A C^1 triangular interpolant suitable for scattered data interpolation. *Commun. Appl. Numer. Methods* **1991**, *7*, 479–485. [CrossRef]
17. Foley, T.A.; Opitz, K. Hybrid cubic Bézier triangle patches. In *Mathematical Methods in Computer Aided Geometric Design II*; Elsevier: Amsterdam, The Netherlands, 1992; pp. 275–286. [CrossRef]
18. Karim, S.A.B.A.; Saaban, A. Visualization terrain data using cubic Ball triangular patches. *MATEC Web Conf.* **2018**, *225*, 06023. [CrossRef]
19. Hussain, M.Z.; Hussain, M. C1 positivity preserving scattered data interpolation using rational Bernstein-Bézier triangular patch. *J. Appl. Math. Comput.* **2011**, *35*, 281–293. [CrossRef]
20. Chan, E.; Ong, B. Range restricted scattered data interpolation using convex combination of cubic Bézier triangles. *J. Comput. Appl. Math.* **2001**, *136*, 135–147. [CrossRef]
21. Azizan, S.; Piah, A.R.M.; Ahmad, A.M. G1 scattered data interpolation with minimized sum of squares of principal curvatures. In Proceedings of the International Conference on Computer Graphics, Imaging and Visualization (CGIV'05), Beijing, China, 26–29 July 2005.
22. Piah, A.R.M.; Saaban, A.; Majid, A.A. Range restricted positivity-preserving scattered data interpolation. *Malays. J. Fundam. Appl. Sci.* **2006**, *2*, 63–75. [CrossRef]
23. Hussain, M.; Hussain, M.Z.; Buttar, M. C^1 positive Bernstein-Bézier rational quartic interpolation. *Triangle* **2014**, *1*, 3.
24. Farin, G.E. *Curves and Surfaces for CAGD: A Practical Guide*, 5th ed.; Morgan Kaufmann: Burlington, MA, USA, 2002; pp. 1–499. [CrossRef]
25. Franke, R. Scattered data interpolation: Tests of some methods. *Math. Comput.* **1982**, *38*, 181–200. [CrossRef]
26. Franke, R.; Nielson, G.M. Scattered data interpolation and applications: A tutorial and survey. In *Geometric Modeling: Methods and Applications*; Springer: Berlin/Heidelberg, Germany, 1991; pp. 131–160.
27. Lodha, S.K.; Franke, R. Scattered data techniques for surfaces. In Proceedings of the Scientific Visualization Conference (Dagstuhl'97), Dagstuhl, Germany, 9–13 June 1997; IEEE: Piscataway, NJ, USA, 1997; p. 181.
28. Goldman, R.; Simeonov, P. Two essential properties of (q, h)-Bernstein–Bézier curves. *Appl. Numer. Math.* **2015**, *96*, 82–93. [CrossRef]
29. Goldman, R.; Simeonov, P.; Simsek, Y. Generating functions for the q-Bernstein bases. *SIAM J. Discret. Math.* **2014**, *28*, 1009–1025. [CrossRef]
30. Lewanowicz, S.; Woźny, P.; Area, I.; Godoy, E. Multivariate generalized Bernstein polynomials: Identities for orthogonal polynomials of two variables. *Numer. Algorithms* **2008**, *49*, 199–220. [CrossRef]
31. Oruç, H.; Phillips, G.M. q-Bernstein polynomials and Bézier curves. *J. Comput. Appl. Math.* **2003**, *151*, 1–12. [CrossRef]
32. Phillips, G.M. A de Casteljau algorithm for generalized Bernstein polynomials. *BIT Numer. Math.* **1997**, *37*, 232–236. [CrossRef]
33. Simeonov, P.; Zafiris, V.; Goldman, R. q-Blossoming: A new approach to algorithms and identities for q-Bernstein bases and q-Bézier curves. *J. Approx. Theory* **2012**, *164*, 77–104. [CrossRef]
34. Delgado, J.; Orera, H.; Peña, J.M. An evaluation algorithm for q-Bézier triangular patches formed by convex combinations. *J. Comput. Appl. Math.* **2023**, *428*, 115184. [CrossRef]

35. Schumaker, L.L. *Spline Functions: Computational Methods*; SIAM: Philadelphia, PA, USA, 2015.
36. Karim, S.A.A.; Saaban, A.; Nguyen, V.T. Scattered data interpolation using quartic triangular patch for shape-preserving interpolation and comparison with mesh-free methods. *Symmetry* **2020**, *12*, 1071. [CrossRef]
37. Fortune, S. A sweepline algorithm for Voronoi diagrams. In Proceedings of the Second Annual Symposium on Computational Geometry, Yorktown Heights, NY, USA, 2–4 June 1986; pp. 313–322. [CrossRef]
38. Gilat, A. *MATLAB: An Introduction with Applications*; John Wiley & Sons: Hoboken, NJ, USA, 2004.
39. Ali, F.A.M.; Karim, S.A.A.; Dass, S.C.; Skala, V.; Hasan, M.K.; Hashim, I. Efficient visualization of scattered energy distribution data by using cubic timmer triangular patches. In *Energy Efficiency in Mobility Systems*; Springer: Singapore, 2020; pp. 145–180. [CrossRef]

algorithms

MDPI

Article

Analytical Solution for the Problem of Point Location in Arbitrary Planar Domains

Vitor Santos

Department of Mechanical Engineering, Institute for Electronics Engineering and Informatics of Aveiro, University of Aveiro, 3810-193 Aveiro, Portugal; vitor@ua.pt

Abstract: This paper presents a general analytical solution for the problem of locating points in planar regions with an arbitrary geometry at the boundary. The proposed methodology overcomes the traditional solutions used for polygonal regions. The method originated from the explicit evaluation of the contour integral using the Residue and Cauchy theorems, which then evolved toward a technique very similar to the winding number and, finally, simplified into a variant of ray-crossing approach slightly more informed and more universal than the classic approach, which had been used for decades. The very close relation of both techniques also emerges during the derivation of the solution. The resulting algorithm becomes simpler and potentially faster than the current state of the art for point locations in arbitrary polygons because it uses fewer operations. For polygonal regions, it is also applicable without further processing for special cases of degeneracy, and it is possible to use in fully integer arithmetic; it can also be vectorized for parallel computation. The major novelty, however, is the extension of the technique to virtually any shape or segment delimiting a planar domain, be it linear, a circular arc, or a higher order curve.

Keywords: cauchy theorem; residue theorem; Jordan curve theorem; generalized polygons; complex calculus; parametric curves; Bézier segments; winding number; negative real axis intersection

1. Introduction

Testing whether a point is enclosed by a polygon is a problem that many programmers in numerous fields, such as in computer graphics, geographic information systems, machine vision, or robotics, among others, have certainly come across more than once, as many authors have been asserting for a long time [1]. Indeed, the issue is very important and is the object of several concerns in Computational Geometry. It has already been studied and solved for practically all cases of polygonal regions, despite the fact that some earlier algorithms occasionally needed iterative computations, were ambiguous in special cases of on-boundary point locations, or show numerical instability in situations of extreme proximity to borders and vertices. Nonetheless, these limitations are often not very relevant in terms of practical implementation, and the algorithms are actually used on many applications in the software industry. What these algorithms are not aimed at, though, is the analytical evaluation of whether a point lies inside a more complex region bounded by mixed linear and curved segments. Actually, there is no published solution known to the author that indicates how it is possible, with the very same base algorithm and in a straightforward and scalable procedure, to determine analytically, i.e., without linearizing the contour or performing geometric approximations, whether a point is inside or outside regions such as those illustrated in Figure 1.

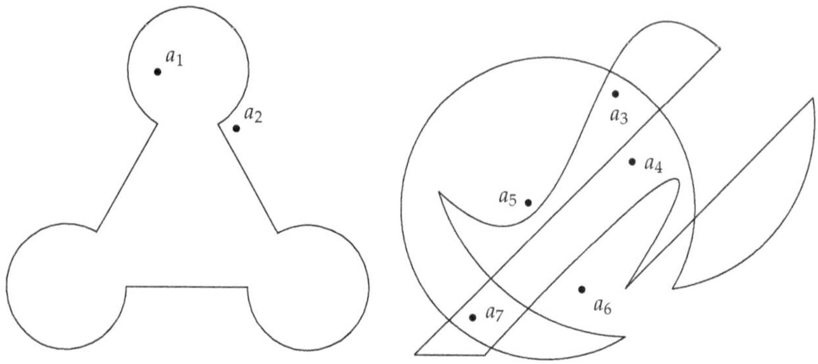

Figure 1. Examples of shapes bounded by linear, circular, or Bézier segments, including twisting and self-intersection, and for which lacks a procedure to determine with the same algorithmic process the relative location of the points illustrated. The case on the right is much more complex than the one on the left, and that can be challenging, even for humans.

2. Related Work

The first known reference to an algorithm for the point location problem dates back to 1962 by M. Shimrat [2]. That early algorithm had some limitations pointed out soon after by R. Hacker [3], and later by W. Randolph Franklin, who also states to have delivered, in 1970, a code written in FORTRAN for this same problem [4]. The books by Preparata and Shamos, published in 1985 [5], and by Sedgewick, published in 1990 [6], also covered the point in the polygon problem and have been useful references for many other authors and programmers who have developed specific implementations with adaptations for their own set-ups. Still, nowadays, it is easy to find papers in conferences and journals—in very diverse contexts—where this problem is addressed, even if the novelties have stabilized a long time ago around the two main principles, as described further below.

Eventually, what has become one of the most well-known references in the literature was the publication *Point in Polygon Strategies*, in 1994, by Eric Haines [7], and also some years later, the work of Hormann and Agathos, in 2001 [8], which is the algorithm being used in Matlab for point location in arbitrary polygons.

A clarification is recommended here to better focus the scope of the paper. Actually, there are two kinds of challenges in this problem: the "point in region" problem, and the "point location in subdivisions of the plane". Although related, there are some formal differences in these problems.

In the first problem, there is only one region, and we are testing whether a point lies inside or outside of it. In the second problem, there is a group of regions (typically a subdivision of the plane), and we need to identify which specific region the point belongs to. The first problem is to be solved using containment tests specific to the region (e.g., point-in-polygon), and it has constant-time complexity relative to the number of regions but may depend on the complexity of the region's boundary. The second problem typically involves building a data structure for efficient spatial querying across many regions, with logarithmic or sublinear complexity relative to the number of regions in the subdivision. The second problem usually involves triangulated regions (using Delaunay triangulation), and several works stand out in that front, ranging from the fundamentals from Voronoi diagrams [9] and Delaunay triangulation, including later optimizations for better data structure management [10], up to actual point location algorithms in such triangular subdivisions of the plane [11,12].

This paper focuses specifically on the first problem and proposes an alternative solution by including additionally the ability to extend the regions to general curve-shaped boundaries. So, in this paper, the expression "point location" refers to assessing whether some point lies inside or outside a given planar region.

2.1. Ray-Crossing Approaches

For a long time, one of the most popular algorithms was the one based on the idea of defining an infinite line starting at the point being analyzed: the number of intersections with the polygon boundary gives indications for the solution, as described, for example, by Preparata and Shamos [5]. Briefly stated, a point belongs to a polygon if an infinite straight line starting on it intersects the polygon an odd number of times in one direction (Figure 2). This algorithm is also known as the ray-crossing [13], the crossing number [14] algorithm, and the even-odd rule algorithm [15].

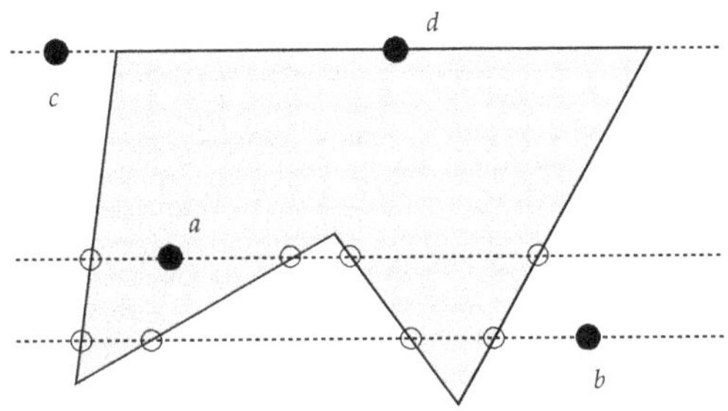

Figure 2. The traditional ray-crossing algorithm is based on the parity of intersection points. An odd number of intersections of a horizontal line starting on the point being tested along one of its two sides indicates that the point is inside the polygon, and an even number of intersections (also 0) indicates that the point is outside.

One traditional issue with the ray-crossing algorithm was that it may require further tests to account for the cases where the intersections of the line coincide with one or more vertices or even with an entire edge of the polygon itself. Implementations of this algorithm can be found in many places, and the earlier most well-known implementations occur in the works of Sedgewick [6], Haines [7], and O'Rourke [16].

The ray-crossing algorithm is simple and of limited computational cost, $\{\mathcal{O}(n)\}$, but, for more functionality, it may require additional steps, hence reducing its elegance. Additionally, applying its technique to generally shaped areas other than polygons, although possible in theory, could pose huge computational and practical difficulties since systems of simultaneous equations might need to be managed and solved. Still, and although it surely has not been intended for that, its applicability for more general self-intersecting (non-simple) domains may fail, as illustrated in Figure 3, where the point e would be evaluated as being in the "outside", which contradicts both the mathematical concept based on the winding number [17], and the human perception of "insideness" on a folded (and/or twisted) polygon.

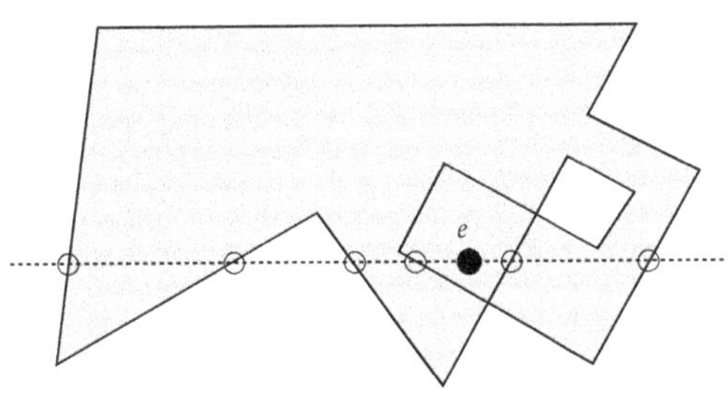

Figure 3. The classic ray-crossing algorithm is not meant to deal with non-simple or self-intersecting polygons. Point e would be detected as "outside", but it is "inside".

2.2. Winding Number Based Approaches

The problem of non-simple polygons (self-intersecting) has nonetheless been practically solved with algorithms based on the winding number [7], also named nonzero-rule algorithms, which have been refined and even made as efficient as the ray-crossing approach [14]. Its formulation, based on the sum of all the oriented angles with a center in the test point and delimited by each pair of successive vertices of the polygon, allows to obtain the winding number around a point (the number of times a closed curve winds around a given point in the positive sense), and then conclude whether the point is inside or outside the region. If $\alpha_1, \alpha_2, \ldots, \alpha_M$ are the M oriented angles defined by the testing point and the M segments defined with the M vertices of a polygon, the winding number is calculated by $W_N = \frac{1}{2\pi}\sum_{i=1}^{M}\alpha_i$. If W_N is equal to zero, then the number of winds of the polygon around the point is zero, meaning that the point is "outside" the polygon. Any other value for W_N will indicate that the point is "inside" the polygon.

Figure 4 illustrates the geometric principle of the procedure for a simple polygon.

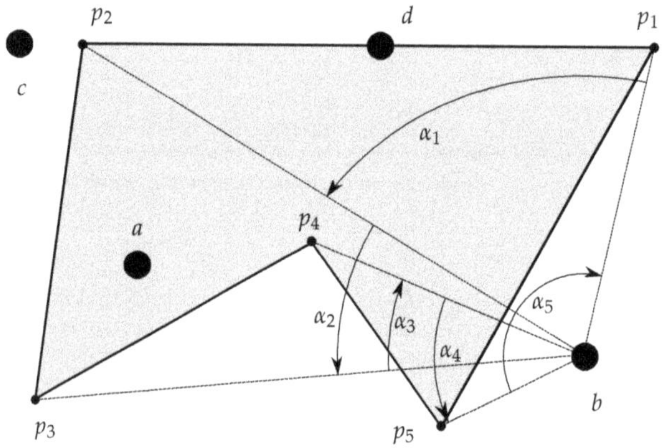

Figure 4. Algorithm based on the winding number. If $\sum_i \alpha_i = 0$, then the testing point (b in this case) is outside the polygon.

However, the traditional winding number-based algorithms may be ambiguous if the point lies on the boundary and nearly all implementations show some degree of false positives and missed detections, as demonstrated by S. Shirra in [18]. The technique described generically in Figure 4 is often seen in the literature as the calculation of the winding number contribution of each linear segment ($\overline{p_1 p_2}$) relative to a point a, where all of the contributions are added up. Mathematically, the procedure is simple, as presented, for example in [19], and can be easily understood as the calculation of the winding number contribution of that segment relative to the given test point a:

$$\begin{cases} \vec{u} = p_1 - a \\ \vec{v} = p_2 - a \\ \alpha = \mathrm{atan2}(\vec{u} \times \vec{v}, \vec{u} \cdot \vec{v}) \end{cases} . \tag{1}$$

Demonstration of (1) is straightforward because $\vec{u} \times \vec{v} = |\vec{u}||\vec{v}|\sin(\alpha)$ and $\vec{u} \cdot \vec{v} = |\vec{u}||\vec{v}|\cos(\alpha)$. In this context $\vec{u} \times \vec{v}$ is actually supposed to mean the third component of the actual cross product $(u_x, u_y, 0) \times (v_x, v_y, 0)$. Possible alternative notations would be to replace $\vec{u} \times \vec{v}$ in (1) by $\det\begin{bmatrix}\vec{u}^\top & \vec{v}^\top & \mathbb{1}^\top\end{bmatrix}$, where $\mathbb{1} = \begin{bmatrix}1 & 1 & 1\end{bmatrix}$, or $(\vec{u} \times \vec{v})_3$ or, of course, $(u_x v_y - u_y v_x)$.

When a point falls over the boundary of a region, more than one interpretation may arise. An example of that occurs in Figure 5, where two adjacent regions R_1 and R_2 (delimited by different shades and without the border lines being drawn to visually enhance the effect) seem to share a point a because it falls precisely on the boundaries of the regions. This raises the question of whether point a belongs to both regions, to only one of them, or neither of them!

Figure 5. Example of two adjacent regions R_1 and R_2 where point a lies precisely on the border, raising the question to which region it belongs.

The possibility of belonging to neither of the regions must be discarded since a lies in a area covered by both regions altogether, seeming absurd to affirm that a does not belong to neither of the regions and at the same time belonging to their union. So, there remains the possibility of belonging to only R_1 or R_2, or both at the same time. This is where the variation of interpretation can take place. If regions are analyzed independently, and using the winding number approach, any conclusion (belonging or not to R_1 or R_2) can be taken depending on the circulation sense of the polygonal lines, as described next. If, on the other hand, the two regions are taken as a degenerate self-intersecting polygon where the common edge is actually a double edge, it will be demonstrated further that a belongs naturally to the compound region ($R_1 \cup R_2$), independently of the circulation sense.

As mentioned earlier, when the testing point is over the boundary, the winding number approach has a variable behavior, depending on the circulation sense, which can be illustrated with this simple example based on a triangle with vertices $p_1 = (-1,0)$, $p_2 = (0,-1)$, $p_3 = (0,1)$ and a test point $a = (0,0)$. For the triangle longer segment, we have $\vec{u} = p_1 - (0,0)$ and $\vec{v} = p_2 - (0,0)$. Also, $\vec{u} \times \vec{v} = 1 \times 1 \times \sin(\pi) = 0$ and $\vec{u} \cdot \vec{v} = 1 \times 1 \times \cos(\pi) = -1$, and consequently $\alpha = \text{atan2}(0,-1) = \pi$. If we swap p_1 with p_2, which is equivalent to reversing the circulation sense, the result is the same for this segment (because it still holds $\alpha_{12} = +\pi$), which indicates that the approach does not distinguish the sense of circulation (winding number contribution) along a segment when the point is over it. But, for the remainder two segments, the winding number contribution is detected accordingly in different values and, therefore, the overall winding number changes! In conclusion, there are cases where the point will be considered inside and others outside the polygon, depending on the circulation sense. Figure 6 shows the elements that provide a simple example demonstrating this situation, which is occasionally mentioned by some authors, but also usually dismissed as irrelevant or avoided with specific argumentation, such as in [4].

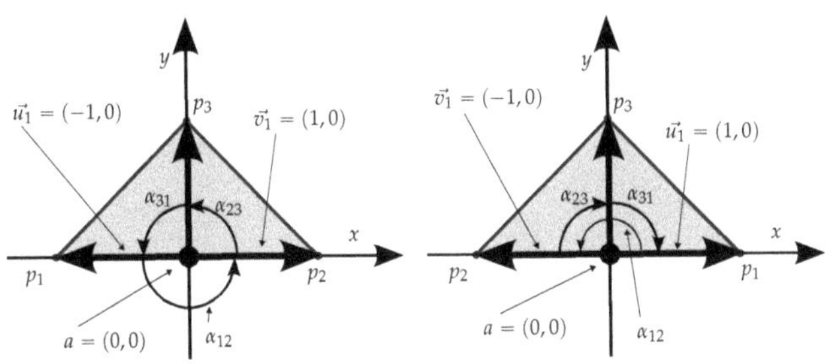

Figure 6. Example of the behavior of the winding number technique for a point on the border, $a = (0,0)$, of a triangle when the circulation sense is reversed (the order $p_1 \to p_2 \to p_3 \to p_1$ is always assumed). For the inclusion of a, the sum of the three angles is expected to be $\sum \alpha_{ij} = \pm 2\pi$, which does not occur on the right (clockwise circulation sense) because $\alpha_{12} = +\pi$, causing a net $\sum \alpha_{ij} = 0$. For the sake of clarity, auxiliary vectors $\vec{u}$ and $\vec{v}$ (from Equation (1)) are shown only for the first oriented segment ($p_1 \to p_2$).

Although apparently inconvenient and potentially unexplainable (the sense of circulation was not expected to determine the conclusion of the point inclusion test), this behavior is actually predicted by the Jordan curve theorem.

Indeed, an important concept related to the problem of point location is the well-known Jordan curve theorem (JCT) that, in a simple redaction, as is done in [20], states that "Every Jordan curve (a non-self-intersecting continuous loop in the plane) separates the plane into exactly two components". This means that any point is either in one region or in the other, and implicitly, there is no distinct third region, like, for example, "the border". So, formally, any results derived from the JCT, namely those based on the winding number, may find ambiguity in assessing which of the two regions is the point lying on the boundary, and this is an unavoidable fact that must be managed when using techniques based on circulation or winding number. Actually, for simple curves, when the observer is traveling along the curve counter-clockwise, the points on this curve (boundary), by convention, will belong to the region on the left hand side; so, the identification of which is the "inside" region depends on the traveling (circulation) sense of the observer, and this is where the ambiguity may take place.

2.3. Generalization to Generic Shapes

The generalization of the problem for domains beyond polygons (simple or non-simple) can be carried out by using the winding number approach. However, the literature has not shown a definitive viable implementation to do it, which includes the case of S. Gatilov in [19], who gives a methodology to calculate the winding angle associated with a circular arc. But Gatilov's approach counts on the winding number technique described in (1), with its ambiguity for points over segments, which in the case of arcs would be points over the chord of the arc. That is not viable since the point will not even be on the boundary, and the calculation may fail depending on the sense of the circulation! Let us consider the example in Figure 7 with $a = (0,0)$. The contribution for the winding number only by the arc $p_1 \frown p_2$ depends on the circulation sense, as happens with any segment, but in case of this arc, the value is not symmetric: it is π or 0! Since the contributions of the other two segments are symmetric depending on the circulation sense ($\pm\pi$), this all adds up to ambiguity in detecting whether a lies inside or outside the region, which is absurd because point a is clearly inside the region.

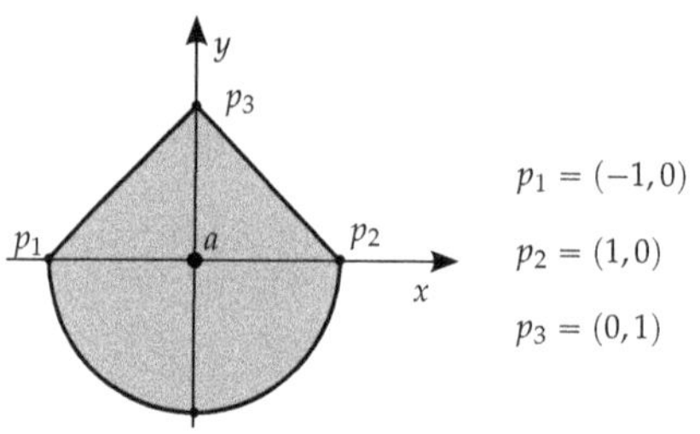

Figure 7. Case where point a is without doubt included in the delimited region, including a circular arc, independently of the circulation sense, but which could fail by the application of the traditional winding number technique proposed by [19].

Actually, the problem of point location in regions that may have curve segments was first mentioned, although not actually solved in practice, by Edelsbrunner and Maurer in 1981 [21], and the general problem is seen by Kirkpatrick in 1983 as a challenge that may require a completely new approach:

> [...] While the algorithm [proposed by Kirkpatrick] can be adapted to certain other situations (for example, when all internal regions are star shaped), the general problem of optimal search in subdivisions formed from arbitrary curve segments may require a totally new approach. ([22])

The winding number approach has been a well-accepted solution for the problem of point location, and the calculation of the winding number of a polygonal curve Γ around some point a is easy to calculate by determining the number of intersections of Γ with the real (horizontal) axis, as first pointed out in [23] and later implemented into a specific algorithm by [1]. However, in the literature, nothing specific is added concerning the winding number of other curves for the purpose of point inclusion, although algorithms exist to calculate the winding number contribution of Bézier curves [24].

2.4. Background and Scope of This Paper

This paper demonstrates and gives implementations of a unified approach that covers the traditional solutions based on ray-crossing or winding numbers, solves the cases where those algorithms may find limitations or ambiguities, and extends and demonstrates a practical, viable solution for generic contours that can be expressed in a parametric form, which is its major novelty.

In the early 1990's, information was not so easy to obtain or track, and faced with the need, the author proposed, during his Ph.D. Thesis [25], an approach for the problem of point location in polygons based on the Cauchy and Residue theorems, which is closely related to the winding number approach, although unfamiliar at the time.

The algorithm operates by explicitly evaluating contour integrals in the complex plane of a specially chosen function. The solution, found by accidental simplification (it was not fully demonstrated then), proved to work well with a very simple and elegant mathematical formulation. Only later, when formal full mathematical demonstrations were needed, was the current solution developed, as presented in this paper. Hence, it can be said that the solution came to be after the usage of the Cauchy and Residue theorems, but it later became as it it now in its present state. The next sections describe in detail the foundations of the technique, including algorithmic implementations, followed by illustrative results and conclusions.

Code for this research and demonstrations of the algorithms described in the remainder of this paper are publicly available in a GitHub repository (https://github.com/vitoruapt/PointInclusion (accessed on 2 October 2024)).

3. Base Approach and Related Theorems

The relative location of a point in a polygon, or actually any other form of closed contour in the plane, can be explored by the results provided by the theorem of Residue and the theorem of Cauchy, which are introduced next.

Theorem 1 (Residue Theorem). *If Γ is a closed curve of the plane, and $f(z)$ is an analytic function in the domain $R(\Gamma)$ enclosed by Γ (included), except in a finite number of points inside R, then:*

$$\oint_{\Gamma} f(z)\,\mathrm{d}z = 2\pi \mathrm{j} \sum_{k} r_k \tag{2}$$

being r_k the residues of $f(z)$ at the singular points inside R (being $\mathrm{j} = \sqrt{-1}$).

The residue r of a complex function $f(z)$ at a point a, which is a pole of order n of f, is calculated by $r = \frac{1}{(n-1)!} \lim\limits_{z \to a} \frac{\mathrm{d}^{n-1}}{\mathrm{d}z^{n-1}}[(z-a)^n f(z)]$. A complex function is analytic on a region R if it is complex differentiable at every point in R. A common alternative designation is "holomorphic function", usually the preferable variant by some mathematicians.

Another important fact is given by the Cauchy theorem (or Cauchy–Goursat theorem, since Edouard Goursat (1858–1936) proved this theorem without imposing the condition that $f'(z)$ should be continuous inside $R(\Gamma)$ and on the boundary, which was an additional condition left by Cauchy [26]) that states the following:

Theorem 2 (Cauchy–Goursat Theorem). *If $R(\Gamma)$ is a simply or multiply connected region whose boundary Γ is sectionally smooth, and if $f(z)$ is analytic in $R(\Gamma)$, then the following result applies:*

$$\oint_{\Gamma} f(z)\,\mathrm{d}z = 0. \tag{3}$$

The previous statement is actually the formalization of the common expression, which states that a line integral of any analytic function is independent of the path. So, the advantage here is to use non-analytic functions on the regions of interest. After selecting a function that is non-defined in only one point inside $R(\Gamma)$, and whose residue is non-null on that point, by using the Residue and Cauchy theorems, the following reasoning can be engaged:

$$f(z) \text{ is analytic inside } \Gamma \Rightarrow \oint_{\Gamma} f(z)\,\mathrm{d}z = 0 \tag{4}$$

or, conversely:

$$\oint_{\Gamma} f(z)\,\mathrm{d}z \neq 0 \Rightarrow f(z) \text{ is not analytic inside } \Gamma. \tag{5}$$

Asserting that $f(z)$ is not analytic inside Γ means that $f(z)$ has at least one discontinuity inside Γ. Consequently, and being aware that a is the unique pole (discontinuity) of $f(z)$ in $R(\Gamma)$, i.e., the unique point where it is not defined, a function such as $f(z) = \frac{1}{z-a}$ allows us to state this final and most valuable conclusion: if the contour integral of $f(z)$ along a given contour Γ is not null, then its pole lies in the region delimited by that contour, or, in mathematical terms:

$$\begin{cases} a \text{ is the pole of } f(z) \\ \oint_{\Gamma} f(z)\,\mathrm{d}z \neq 0 \end{cases} \Leftrightarrow a \in R(\Gamma). \tag{6}$$

It is now necessary to draw some considerations to select an appropriate function and the procedures to apply and take advantage of the previous result. Let a boundary Γ be decomposed in M parts Ω_k with $1 \leqslant k \leqslant M$. After reminding ourselves that a contour integral is no more than a line integral where the integration path is a closed curve, the following is clear:

$$\oint_{\Gamma} f(z)\,\mathrm{d}z = \sum_{i=1}^{M} \int_{\Omega_i} f(z)\,\mathrm{d}z. \tag{7}$$

Although other possibilities exist, we can adopt for $f(z)$ the one suggested earlier and presented in (8):

$$f(z) = \frac{1}{z-a}. \tag{8}$$

The residue of $f(z)$ at pole a has the value 1. The option for $f(z)$ given in (8) is due to the fact that the integrand function should be simple for concerns of computational cost, and that it must possess a non-null residue, which would otherwise lead to inconclusive results. For example, all functions of the type $f(z) = \frac{1}{(z-a)^n}$, with $n > 1$, or the type $f(z) = \frac{z}{(z-a)^n}$, with $n > 2$, have a residue of 0 on point a, making them useless for this purpose. On the other hand, there are many other possibilities, all with residue 1 at point a, such as the following: $f(z) = \frac{z}{(z-a)^2}$, $f(z) = \frac{1}{1-\mathrm{e}^{(z-a)}}$ or $f(z) = \frac{1}{\sin(z-a)}$. However, in all these (usable) alternatives, there would always be the need to calculate a complex logarithm for the anti-derivative. Therefore, the simplest and most obvious case found was precisely the function in expression (8), but, indeed, the principle is independent of the function, which is, however, useful only if it has poles and non-null residues.

4. Calculation of the Contour Integral

If $f(z)$ is analytic in any path between z_1 and z_2, then the fundamental theorem of calculus along curves applies:

$$\int_{\Gamma_{z_1 z_2}} f(z)\,\mathrm{d}z = F(z_2) - F(z_1). \tag{9}$$

Expression (9) states a very well-known and useful technique, where $F(z)$ is the anti-derivative of $f(z)$; however, it can be directly applied only if $F(z)$ is continuous along the integration path and $f(z)$ is defined over the integration path; this later condition may not be satisfied when the point being tested is over the contour. Nonetheless, and based on the definition of the Cauchy Principal Value (CPV), the contour integral can still be calculated even if the function being integrated has a pole on the contour. That can be performed by using expression (10) assuming that the pole is enclosed by a circle of radius δ, and the fraction of the path outside that circle is named $\Gamma(\bar{\delta})$, where the function $f(z)$ is always integrable independently of how small δ becomes [27]:

$$\int_{\Gamma} f(z)\,\mathrm{d}z = \lim_{\delta\to 0} \int_{\Gamma(\bar{\delta})} f(z)\,\mathrm{d}z\,. \tag{10}$$

Having solved the possible problem of the non-definition of $f(z)$ in a finite number of points, it only remains the issue of the continuity of $F(z)$, which is going to be managed with a methodology derived further.

4.1. The Complex Logarithm

The anti-derivative of the chosen function (8) is the logarithm of a complex number, which, formally, is a multi-valued result. By definition, the logarithm of a complex number z is the number w that satisfies the expression $z = \mathrm{e}^{w}$. After this definition, it is then clear that the logarithm of z, $\mathrm{Ln}(z)$, is multi-valued; if $z = r\mathrm{e}^{\mathrm{j}\theta}$, using $\mathrm{j} = \sqrt{-1}$, and being true that $r\mathrm{e}^{\mathrm{j}\theta} = r\mathrm{e}^{\mathrm{j}(\theta+2k\pi)} = \mathrm{e}^{\ln r+\mathrm{j}(\theta+2k\pi)}$, the following arises:

$$\mathrm{Ln}(z) = \ln r + \mathrm{j}(\theta + 2k\pi), k = 0, \pm 1, \pm 2, \ldots \tag{11}$$

It must be noted that the logarithm of a complex number is here graphed with a capital letter to stress its multi-valued nature. With multi-valued expressions, it is not possible to establish comparisons or other relational operations without precautions. The meaning of $\mathrm{Ln}(z)$ depends on the branch where the logarithm is being evaluated, that is, the value of k in (11). However, the logarithm can be restricted to a single branch, called the principal value, where the imaginary part is unique, hence with only one possible argument, called the principal argument (often designated by the mathematical arg() function). The principal value is obtained by making $k = 0$ in (11) resulting in:

$$\ln(z) = \ln r + \mathrm{j}\arg(z) = \ln r + \mathrm{j}\theta. \tag{12}$$

When restricted to the principal values, logarithms exhibit some particularities like, for example, $\ln\left(\mathrm{e}^{-\mathrm{j}\frac{5}{4}\pi}\right)$ equals $\frac{3}{4}\pi\mathrm{j}$ rather than $-\frac{5}{4}\pi\mathrm{j}$, and, as well, $\ln\left(\mathrm{e}^{-\mathrm{j}\pi}\right) = +\pi\mathrm{j}$ and not $-\pi\mathrm{j}$! As expected, the restriction of the function output to an interval will certainly cause discontinuity of that function and, indeed, $\ln(z)$ is not continuous for all points of the non-positive real axis because it is clear that:

$$\begin{cases} \lim\limits_{\theta\to 0^-} \ln \mathrm{e}^{\mathrm{j}(\pi+\theta)} = \lim\limits_{\theta\to 0^+} \ln \mathrm{e}^{\mathrm{j}(\pi-\theta)} = +\pi\mathrm{j} \\ \lim\limits_{\theta\to 0^+} \ln \mathrm{e}^{\mathrm{j}(\pi+\theta)} = -\pi\mathrm{j} \end{cases}. \tag{13}$$

These issues of the principal value and continuity also affect the logarithms of products (or quotients) when expanded to sums (or subtractions) of logarithms; subtraction or addition of complex logarithms may not always give a result with the imaginary part restricted to the principal value and that is why the following definitions are stated for the logarithm of a quotient, or equivalently, subtraction of logarithms:

$$\ln\frac{z_1}{z_2} = \begin{cases} \ln z_1 - \ln z_2 & \Leftarrow -\pi < \arg z_1 - \arg z_2 \leqslant +\pi \\ \ln z_1 - \ln z_2 - 2\pi \mathrm{j} & \Leftarrow \quad \arg z_1 - \arg z_2 > +\pi \\ \ln z_1 - \ln z_2 + 2\pi \mathrm{j} & \Leftarrow \quad \arg z_1 - \arg z_2 \leqslant -\pi \end{cases} \tag{14}$$

and, conversely,

$$\ln z_1 - \ln z_2 = \begin{cases} \ln \frac{z_1}{z_2} & \Leftarrow -\pi < \arg z_1 - \arg z_2 \leqslant +\pi \\ \ln \frac{z_1}{z_2} + 2\pi \mathrm{j} & \Leftarrow \quad \arg z_1 - \arg z_2 > +\pi\,. \\ \ln \frac{z_1}{z_2} - 2\pi \mathrm{j} & \Leftarrow \quad \arg z_1 - \arg z_2 \leqslant -\pi \end{cases} \tag{15}$$

4.2. Line Integral for Linear Segments

Resuming back to the genesis of the algorithm, the essence of the entire procedure is then to evaluate explicitly $\oint_\Gamma \frac{dz}{z-a}$ and verify whether its value is null or not. It will not be null if a lies within or over the contour Γ. By applying (9) to a segment of a path between points z_1 and z_2, the following result would potentially occur:

$$\int\limits_{[z_1 z_2]} \frac{dz}{z-a} = \ln(z_2 - a) - \ln(z_1 - a). \tag{16}$$

4.2.1. Concerns When Calculating the Line Integral

Despite the elegant solution given by (16), there remains the uncertainty of its applicability due to the possible non-continuity of $\ln(z)$ along the path $[z_1 z_2]$. That non-continuity will occur precisely when the path from z_1 to z_2 crosses the negative real axis (NRA), that is, in a neighborhood where z on the path passes from the 2nd quadrant (Q2) to the 3rd quadrant (Q3) or vice-versa. The following notation is adopted to define the situations of a path crossing the NRA:

$$\begin{array}{ll} \downarrow\circ & \text{path crosses the NRA from Q2 to Q3.} \\ \uparrow\circ & \text{path crosses the NRA from Q3 to Q2.} \\ \downarrow\uparrow\circ & \text{path crosses the NRA in any sense.} \\ \bot\circ & \text{path touches the NRA from Q2.} \\ \top\circ & \text{path touches the NRA from Q3.} \\ \nleftarrow\circ & \text{path does not cross the NRA.} \end{array} \tag{17}$$

So, the problem is then to calculate the line integral in such conditions of discontinuity, which is explained next. The discontinuity of $\ln(z)$ occurs when z crosses the negative real axis, but this effect may occur with other functions apart from the logarithm. So, let us assume that some generic function f can be parametrized on θ (to ease the practical calculation, knowing that $z = re^{j\theta}$), and its definite integral is to be evaluated from $(\pi - \alpha)$ to $(\pi + \beta)$, where α and β are positive angles smaller than π radians, i.e., $\alpha, \beta \in]0, \pi[$. Being δ an infinitesimal positive value, then the following can be written:

$$\int\limits_{\pi-\alpha}^{\pi+\beta} f(\theta)\, d\theta = \int\limits_{\pi-\alpha}^{\pi-\delta} f(\theta)\, d\theta + \int\limits_{\pi-\delta}^{\pi+\delta} f(\theta)\, d\theta + \int\limits_{\pi+\delta}^{\pi+\beta} f(\theta)\, d\theta\,. \tag{18}$$

From the three terms on the right side of (18), the first and the last are always defined for any value of δ (for the functions involved in this analysis). So, taking the limit when $\delta \to 0$ will also force the middle term to be zero because the following holds true for any function $f(\theta)$:

$$\lim_{\delta \to 0} \int_{\pi-\delta}^{\pi+\delta} f(\theta)\,\mathrm{d}\theta = \int_{\pi}^{\pi} f(\theta)\,\mathrm{d}\theta = 0. \tag{19}$$

Therefore, and being $F(\theta)$ the anti-derivative of $f(\theta)$, expression (18) can be developed into (20):

$$\begin{aligned} \int_{\pi-\alpha}^{\pi+\beta} f(\theta)\,\mathrm{d}\theta &= \lim_{\delta \to 0^+} \left[F(\theta)\Big|_{\pi-\alpha}^{\pi-\delta} + F(\theta)\Big|_{\pi+\delta}^{\pi+\beta} \right] \\ &= \lim_{\delta \to 0^+} [F(\pi-\delta) - F(\pi-\alpha) + F(\pi+\beta) - F(\pi+\delta)] \\ &= F(\pi+\beta) - F(\pi-\alpha) + \lim_{\delta \to 0^+} [F(\pi-\delta) - F(\pi+\delta)]. \end{aligned} \tag{20}$$

Expression (20) shows that, when the integration path crosses the negative horizontal axis, the line integral of a function f, has an additional term of $\lim_{\delta \to 0^+}[F(\pi-\delta) - F(\pi+\delta)]$, whose concrete value depends obviously on the continuity of $F(\theta)$. If $F(\theta)$ happens to be continuous on the value $\theta = \pi$, then that additional term is null, and we fall back to the fundamental theorem (9). Resuming to the case in discussion, using $F(\theta)=\ln\left(r\mathrm{e}^{\mathrm{j}\theta}\right)$, then, by taking into account (13), the following holds:

$$\lim_{\delta \to 0^+} \left[\ln\left(r\mathrm{e}^{\mathrm{j}(\pi-\delta)}\right) - \ln\left(r\mathrm{e}^{\mathrm{j}(\pi+\delta)}\right) \right] = [\ln(r) + \pi\mathrm{j}] - [\ln(r) - \pi\mathrm{j}] = +2\pi\mathrm{j}, \tag{21}$$

which, by recovering the original variable ($z = r\mathrm{e}^{\mathrm{j}\theta}$), yields:

$$\int_{[z_1 z_2]} \frac{\mathrm{d}z}{z-a} = \ln(z_2 - a) - \ln(z_1 - a) + 2\pi\mathrm{j}, \quad \text{for } \downarrow\!\!\!\multimap. \tag{22}$$

A similar reasoning is easy to replicate for the case where the sense of the path is reversed, e.g., swap the integral limits in (20), that is, when the crossing occurs from 3rd quadrant ($Q3$) to 2nd quadrant ($Q2$), and the result for that case is:

$$\int_{[z_1 z_2]} \frac{\mathrm{d}z}{z-a} = \ln(z_2 - a) - \ln(z_1 - a) - 2\pi\mathrm{j}, \quad \text{for } \uparrow\!\!\!\multimap. \tag{23}$$

In summary, it can be concluded that the line integral for a path between two points z_1 and z_2 of $\frac{1}{z-a}$ is given by:

$$\int_{[z_1 z_2]} \frac{\mathrm{d}z}{z-a} = \begin{cases} \ln(z_2 - a) - \ln(z_1 - a) + 2\pi\mathrm{j} & \Leftarrow \downarrow\!\!\!\multimap \\ \ln(z_2 - a) - \ln(z_1 - a) - 2\pi\mathrm{j} & \Leftarrow \uparrow\!\!\!\multimap. \\ \ln(z_2 - a) - \ln(z_1 - a) & \Leftarrow \times\!\!\!\multimap \end{cases} \tag{24}$$

Considering points z_1 and z_2 to be such that

$$\begin{cases} (z_1 - a) \in Q2 & \Rightarrow +\frac{\pi}{2} < \arg(z_1 - a) < \pi \\ (z_2 - a) \in Q3 & \Rightarrow -\pi < \arg(z_2 - a) < -\frac{\pi}{2} \end{cases}, \tag{25}$$

makes it simple to assert that, for this case, we have:

$$\lim\left(\max\{\arg(z_2 - a)\} - \min\{\arg(z_1 - a)\}\right) = -\pi, \tag{26}$$

and consequently, the following always holds true:

$$\arg(z_2 - a) - \arg(z_1 - a) < -\pi. \tag{27}$$

Similarly, if we have the reverse situation,

$$\begin{cases} (z_1 - a) \in Q3 \\ (z_2 - a) \in Q2 \end{cases}, \tag{28}$$

then, the following can also be demonstrated:

$$\arg(z_2 - a) - \arg(z_1 - a) > +\pi. \tag{29}$$

In summary, when NRA intersection occurs from $Q2$ to $Q3$ ($\downarrow\!\circ$), we have the first case in (24) which, by using the equality of case 3 from (15), certified by (27), allows to state the following:

$$\begin{aligned} \int\limits_{[z_1 z_2]} \frac{\mathrm{d}z}{z - a} &= \ln(z_2 - a) - \ln(z_1 - a) + 2\pi\mathrm{j} \\ &= \left(\ln\frac{z_2 - a}{z_1 - a} - 2\pi\mathrm{j}\right) + 2\pi\mathrm{j} = \ln\frac{z_2 - a}{z_1 - a} \end{aligned} \tag{30}$$

Similarly, when NRA intersection occurs from $Q3$ to $Q2$ ($\uparrow\!\circ$), we have the second case in (24), which allows us to state the following by using the equality of case 2 from (15), as confirmed by (29):

$$\begin{aligned} \int\limits_{[z_1 z_2]} \frac{\mathrm{d}z}{z - a} &= \ln(z_2 - a) - \ln(z_1 - a) - 2\pi\mathrm{j} \\ &= \left(\ln\frac{z_2 - a}{z_1 - a} + 2\pi\mathrm{j}\right) - 2\pi\mathrm{j} = \ln\frac{z_2 - a}{z_1 - a} \end{aligned} \tag{31}$$

In conclusion, and by comparing expressions (24) and (15), the following final result is obtained for any $a \neq z_1, z_2$:

$$\int\limits_{[z_1 z_2]} \frac{\mathrm{d}z}{z - a} = \ln\frac{z_2 - a}{z_1 - a}. \tag{32}$$

A non-null value for expression (32) as the indicator of point inclusion in polygons is precisely the methodology firstly used by the author in [25] (pp. 149–152), in the early 1990's, as mentioned before, and whose validity has just been formally demonstrated.

4.2.2. Simplifying the Calculation of the Contour Integral

The result expressed by (32) allows to assert the full methodology because it calculates, directly and in a straightforward mode, the circulation of a path made up of linear segments. In the particular case of Γ defined as M linear segments, and with $f(z) = \frac{1}{z-a}$, expression (7) with the results of expression (32) yields the following generic result:

$$\oint_{\Gamma} f(z)\mathrm{d}z = \ln\frac{z_2 - a}{z_1 - a} + \ln\frac{z_3 - a}{z_2 - a} + \cdots + \ln\frac{z_M - a}{z_{M-1} - a} + \ln\frac{z_1 - a}{z_M - a}$$
$$= \sum_{n=1}^{M} \ln\frac{z_{(n \bmod M)+1} - a}{z_n - a}. \tag{33}$$

Furthermore, to simplify the calculation and avoid using at all the complex logarithms, the previous calculations can be even more simplified since the possible variations of the contour calculations only occur in the imaginary part, as it is intuitive from previous statements, and easily demonstrable by expression (34). Indeed, being $z_k = r_k e^{j\theta_k}$ and $z_m = r_m e^{j\theta_m}$ and therefore $\Re\left\{\ln\frac{z_k}{z_m}\right\} = \Re\left\{\ln\frac{r_k}{r_m} + \mathrm{j}\arg\frac{z_k}{z_m}\right\} = \ln\frac{r_k}{r_m}$, the calculation of the real part of (33) for the polygon with M vertices is always null, as shown next:

$$\Re\left\{\oint_{\Gamma} f(z)\,\mathrm{d}z\right\} = \Re\left\{\ln\frac{z_2}{z_1} + \cdots + \ln\frac{z_M}{z_{M-1}} + \ln\frac{z_1}{z_M}\right\}$$
$$= \ln\frac{r_2}{r_1} + \cdots + \ln\frac{r_M}{r_{M-1}} + \ln\frac{r_1}{r_M} \tag{34}$$
$$= \ln\left(\frac{r_2}{r_1}\frac{r_3}{r_2}\cdots\frac{r_M}{r_{M-1}}\frac{r_1}{r_M}\right) = \ln(1) = 0.$$

Finally, it may then be stated that the circulation can be obtained by simply calculating arg() operations using (35):

$$\oint_{\Gamma} f(z)\,\mathrm{d}z = \sum_{n=1}^{M} \arg\frac{z_{(n \bmod M)+1} - a}{z_n - a}. \tag{35}$$

Reminding the definition in (14), and being $z_1 = r_1 e^{j\theta_1}$ and $z_2 = r_2 e^{j\theta_2}$, the definition of the arg() function for a quotient of two complex numbers is, of course, given by (36):

$$\arg\frac{z_1}{z_2} = \begin{cases} \theta_1 - \theta_2 & \Leftarrow -\pi < \theta_1 - \theta_2 \leqslant +\pi \\ \theta_1 - \theta_2 - 2\pi & \Leftarrow \quad \theta_1 - \theta_2 > +\pi \\ \theta_1 - \theta_2 + 2\pi & \Leftarrow \quad \theta_1 - \theta_2 \leqslant -\pi \end{cases}. \tag{36}$$

Resorting to the actual geometric problem, and denoting the testing point by $a = (x_a, y_a)$, and a generic point to delimit segments by $P_n = (x_n, y_n)$, it is immediate to state that

$$\theta_n = \arg(P_n - a) = \mathrm{atan2}(y_n - a_y, x_n - a_x) \tag{37}$$

and, hence, Algorithm 1 on the following page can be promptly established. Moreover, since the circulation of $f(z)$ always has a null real part (34), only the imaginary part is relevant and, therefore, the variant for the calculation given by (35) is preferred (simpler) than the calculation using expression (33).

In conclusion, in a closed contour, the neat value of (35) will be null if, along the path, there are no transitions between quadrants Q2 and Q3, but if they occur, additional terms of $\pm 2\pi\mathrm{j}$ must be added at each time, as expressed by (36). So, the following proposition can be asserted:

Theorem 3 (NRA intersection theorem). *In a region enclosing a singularity of a function $f(z)$, the value of the contour integral (hence, the point inclusion) depends exclusively on the number and sense of path transitions over the negative real axis (NRA).*

This section has demonstrated how the contour integral is indeed the translation of the winding number approach introduced earlier as one of the well-known techniques for point location in polygons. Moreover, as stated in the NRA intersection theorem, it is also clear that the issues of the winding number calculation (angle contributions) occur only when the path crosses the NRA, which resembles very much the ray crossing method.

Algorithm 1: Polygon inclusion test using real arctan()

Require: $a = (a_x, a_y)$, $P_n = (x_n, y_n)$ for $n = 1, 2, \ldots, M$

if $a \in \{P_1, P_2, \ldots, P_n\}$ **then**
    return true //point a is one of the vertices
end if
for $n = 1$ **to** M **do**
    $\theta_n \leftarrow \text{atan2}(y_n - a_y, x_n - a_x)$
end for
$S \leftarrow 0$
for $n = 1$ **to** M **do**
    $\theta \leftarrow \theta_{(n \bmod M)+1} - \theta_n$
    if $\theta > \pi$ **then**
        $\theta \leftarrow \theta - 2\pi$
    else if $\theta \leqslant -\pi$ **then**
        $\theta \leftarrow \theta + 2\pi$
    end if
    $S \leftarrow S + \theta$
end for
if $S \neq 0$ **then**
    return true //point a is included
else
    return false //point a is not included
end if

4.2.3. Points on the Boundary of Polygons with Undefined Orientation

Algorithm 1 is valid for all combinations of point a and polygon $\{P_1, P_2, \ldots, P_M\}$, and requires no post-processing; however, there can be ambiguity in a special situation described next, which is nevertheless easily solvable with a more general approach.

The special situation just mentioned concerns the points over the boundary when the orientation of the polygon (coarsely, the order in which the vertices are browsed) is not the direct orientation, which may give ambiguous results, as expected by the Jordan curve theorem, and mentioned at the beginning of this paper. Indeed, the contour integral $\oint_\Gamma \frac{dz}{z-a}$, when a is over the path Γ, can result either in 0 or $2\pi j$ or, equivalently, a winding number of 0 or 1. Additionally, if the polygon is not simple (i.e., is self-intersecting), the polygon orientation is not adjustable (by reversing the order of vertices, for example) because the orientation changes locally after each self-intersection of the boundary line! For those cases, the contour integral is nonetheless always defined, and Figure 8 shows several examples of points near a self-intersecting polygon.

In Figure 8, several observations immediately arise: the first is that clear "outside" and "inside" points are always well determined (winding number of zero or non-zero, respectively). A second observation is that points on the boundary may yield a winding number of 0 or 1, depending on the circulation sense.

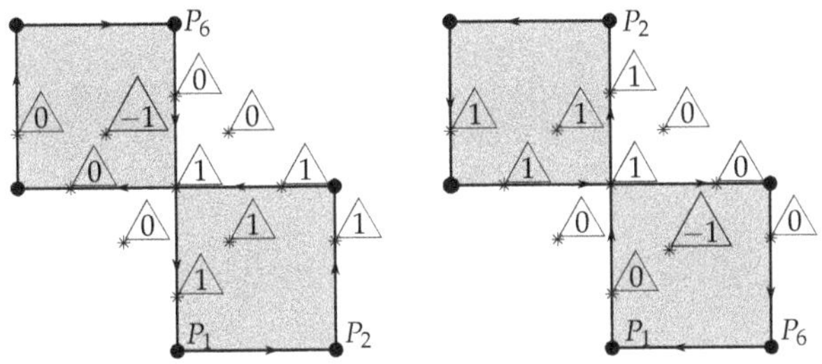

Figure 8. Several examples of winding numbers for testing points (* in the image) in different situations on a non-simple (self-intersecting) polygon. On the left, the direct sense of circulation was used when starting on P_1 and moving toward P_2, and on the right the reverse sense was used (also starting on P_1 but toward a new P_2). Some points over the boundary result in an ambiguous winding number that depends on the circulation sense.

There is also a particular case of a point being twice on the boundary (in the intersections), which is always considered "inside", no matter the sense of circulation. The mentioned ambiguity is however not unsolvable since it is detectable: simply, in case of doubt about the sense of the polygon orientation, to cover even for points over the boundary, a double calculation can be performed: one with the given vertex point order, and the other with its reverse sense; in case one of the calculations yields zero and the other non-zero, then it is the case of a point over the boundary, and further decision can be taken (normally to be considered included in the polygon). This avoids the need of special analysis to detect points on the boundary, which could be absolutely not recommend on general shaped contours, as managed ahead in this paper.

The challenge of points over the boundary is recurrent in all algorithms in the literature. Even one of the most common approaches in the state of the art (the already cited work of Hormann and Agathos [8] used in Matlab) has a special treatment for the points on the boundary that the authors call *boundary version* and is included in Algorithm 7 in their paper, and deals efficiently with points on boundaries of polygons.

In conclusion, if there can be points anywhere on the boundary that need to be precisely assessed, and there is no knowledge whether a sequence of points $\{P_1, P_2, \ldots, P_M\}$ of a simple polygon is not given in the positive circulation sense, or whether the polygon is not simple, then Algorithm 1 must be called twice, being the second time done with the polygon vertices in the reverse order. If the two calls of the algorithm yield different results, then we are in the presence of a point over the boundary, and it is (usually) to be considered inside the polygon (Algorithm 2 on the following page).

Algorithm 2 on the next page requires double the computation resources since it calls the actual calculation algorithm twice. This is required only for the absolutely general case of totally random polygons and testing points, although total randomness will hardly generate situations of points exactly over the border. Anyway, alternatives will be presented ahead.

Despite being effective and elegant, in practice, Algorithm 1 is computationally demanding because one trigonometric function (arctan) is used per each side of the polygon, making it less efficient computationally than other alternatives. That, too, is a reason for a change in the paradigm of the calculation, which will align with the existing state of the art algorithms for polygons but also go beyond their performance, not only by outperforming the computational cost but also going where they do not go actually, that is, extension for shapes other than polygons with linear segments. That paradigm is the parametric definition of the integration path.

Algorithm 2: Unambiguous test for generalized polygons

Require: $a = (a_x, a_y)$, $P_n = (x_n, y_n)$ for $n = 1, 2, \ldots, M$
Create $R_m = (x_m, y_m)$ for $m = M, M-1, \ldots, 1$
$V_D \leftarrow$ **call** Algorithm1(P_n, a)
$V_R \leftarrow$ **call** Algorithm1(R_m, a)
if $V_D \neq V_R$ **then**
  return true //point a is over the boundary
else
  return V_D //True/false if a is inside/outside
end if

4.3. Parametric Definition of the Integration Path

An observation that can be made about the technique described in the previous sections is that calculating the line integral for a function with a non-continuous anti-derivative requires care because of the sense of the integration path crossing the NRA, which results in different contributions, as seen. Anyway, for linear segments, the method just described circumvents that issue; however, for non-straight paths to be discussed further, that may not be so clear. Hence, let us adopt an alternative approach to obtain (32) by using parametric integration paths because that explicitly defines a path from a start to an end, along with the variation of the parameter that describes the curve. Let us use a parameter t that varies from 0 to 1 to cover the full path segment, and let us start with the case of linear segments. Being z_1 and z_2 the extremities of a linear path, any point of the path from z_1 to z_2 is given by:

$$z = z_1 + t(z_2 - z_1)\,, \; 0 \leqslant t \leqslant 1. \tag{38}$$

After adjusting the integration variable, $\mathrm{d}z = (z_2 - z_1)\mathrm{d}t$, the line integral can now be calculated as follows:

$$\int\limits_{[z_1 z_2]} \frac{\mathrm{d}z}{z - a} = \int\limits_0^1 \frac{(z_2 - z_1)\,\mathrm{d}t}{z_1 + t(z_2 - z_1) - a} \tag{39}$$

that results in:

$$\Big[\ln(z_1 + t(z_2 - z_1) - a) \Big]_0^1 = \ln(z_2 - a) - \ln(z_1 - a), \tag{40}$$

which is the same as (16), but shares the same risks of discontinuity because the anti-derivative is still a logarithm, despite the fact that the integration variable is now t. Therefore, the solution is exactly the same as before, given in (22) and (23). In the case of NRA crossing, and depending on the sense ($\downarrow\circ$ or $\uparrow\circ$), there will be a contribution of $\pm j2\pi$ to the contour integral, or ± 1 in the NRA crossings counter.

Although it could have been stated and adopted earlier, working for a segment $\overline{z_1 z_2}$ in respect to a point a is actually equivalent to working with segment $\overline{(z_1 - a)(z_2 - a)}$ in respect to $a = 0$. So, from now onwards, when z points are used, it is assumed that they result from the original points subtracted from a, i.e., $(z - a)$, and that the reference point in analysis is the system origin $(0, 0)$.

5. NRA Crossing in Parametric Paths

We can now settle the procedures to calculate the contour integral along a closed path or, equivalently, the inclusion of a point by a region based solely on the intersection of that path with the NRA.

5.1. NRA Crossings and Their Sense for Linear Segments

What we seek to detect is whether there is an intersection and what is its sense. And this is straightforward to calculate after (38), which can be expanded into:

$$\begin{cases} \Re\{z\} = x = x_1 + t(x_2 - x_1) & \Leftarrow 0 \leqslant t \leqslant 1 \\ \Im\{z\} = y = y_1 + t(y_2 - y_1) & \Leftarrow 0 \leqslant t \leqslant 1 \end{cases}. \tag{41}$$

A NRA intersection occurs at some point z_c when $\Im\{z_c\} = 0$ and $\Re\{z_c\} \leqslant 0$ in $z_c = z_1 + t_c(z_2 - z_1)$, $0 \leqslant t_c \leqslant 1$, i.e., intersection occurs for the value t_c that verifies:

$$\begin{cases} y_1 + t_c(y_2 - y_1) = 0 \\ x_1 + t_c(x_2 - x_1) \leqslant 0 \end{cases} \Leftrightarrow \begin{cases} t_c = \frac{y_1}{y_1 - y_2} \\ t_c(x_1 - x_2) \geqslant x_1 \end{cases}. \tag{42}$$

Notice that the case $y_1 = y_2$ represents a horizontal segment, and "intersection" only occurs for $y_1 = y_2 = 0$. However, that is not actually an intersection of the NRA, as explained later, and this special situation is to be discarded early in the algorithms (also to avoid the division by 0). Apart from that case, if a valid t_c is found, then it is necessary to determine the sense of the NRA crossing, and that is easily verifiable by the derivative of the imaginary component $\Im\{z\}$ in relation to t. If the derivative is positive for that value of t_c, it means that the $y = \Im\{z\}$ component is increasing with t, therefore the path segment is passing from $Q3$ to $Q2$, or ↑∘, that is, there is a contribution of -1 to the intersection counting (I_C), or $+1$ for the reverse case ($Q2$ to $Q3$, or ↓∘). This can be formalized as:

$$I_C = -\operatorname{sgn}\left(\left.\frac{\mathrm{d}y}{\mathrm{d}t}\right|_{t=t_c}\right). \tag{43}$$

For linear segments, the previous expression is constant for any point of the segment where the NRA crossing might occur:

$$I_C = -\operatorname{sgn}(y_2 - y_1) = \operatorname{sgn}(y_1 - y_2). \tag{44}$$

If, by chance, the testing point happens to be an intersection point of the polygon boundary with the NRA, either along the path segment ($t_c \neq 0$ and $t_c \neq 1$) or strictly on a vertex (t_c is 0 or 1), that can be promptly detected because its real coordinate would be zero: $x_1 + t_c(x_2 - x_1) = 0$. This indicates that the point is over the boundary, and a decision can be made immediately without further calculations. Formally: if $x_1 = t_c(x_1 - x_2) = t_c \Delta x$, then the point is on the boundary, and can be dispatched earlier in the algorithm.

There are also the particular cases of $t_c = 0$ or $t_c = 1$; these represent the cases where the segment starts or ends exactly on the NRA (┬∘ or ┴∘). In those cases, this means that another segment starts or ends there too! To leave the contributions separate, but allowing them to add up or cancel, for those cases, the intersection contribution calculated above should be halved, that is, a ± 0.5 contribution for segments that originate or end on the NRA is to be applied. In other words, each crossing of the NRA can be accounted with ± 1, but reaching or leaving it can be accounted as a half-crossing, or ± 0.5. Formally, this can be stated as in (45):

$$t_c = 0 \vee t_c = 1 \Rightarrow I_C = -0.5 \operatorname{sgn}\left(\left.\frac{\mathrm{d}y}{\mathrm{d}t}\right|_{t=t_c}\right). \tag{45}$$

Figure 9 illustrates the situation of I_C values for several testing points ($a_1, a_2, \ldots, a_9$) in a polygonal self-intersecting region; the associated table details the partial and total values of NRA intersection counting. At each case (point a_i), the imaginary axis would be a vertical line that crosses the enclosed domain exactly at each of the a_i points.

In summary, if a path does not intersect the NRA (↛∘), the calculation of the line integral is simple (actually, not necessary in practice) because there are no discontinuities in the process and, in the end, all terms of the closed path will cancel altogether; only the intersections of the path with the NRA will affect contour evaluation or the winding number. Also, notice that horizontal segments ($y_1 = y_2$) never intersect NRA; they may even lie over it but never actually cross it, so their contribution to the I_C is null, as can be seen in the segment that contains the a_2 point in Figure 9.

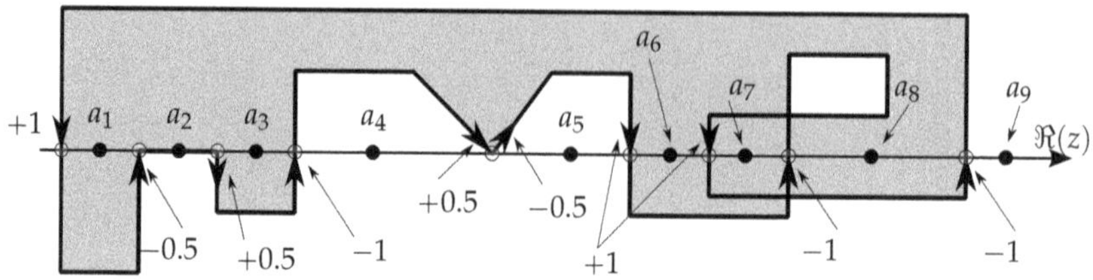

	a_1	a_2	a_3	a_4	a_5	a_6	a_7	a_8	a_9
$\downarrow\circ$	+1	+1	+1.5	+1.5	+2	+3	+4	+4	+4
$\uparrow\circ$	0	−0.5	−0.5	−1.5	−2	−2	−2	−3	−4
Total I_C	+1	+0.5	+1	0	0	+1	+2	+1	0
Is a_i inside?	yes	yes	yes	no	no	yes	yes	yes	no

Figure 9. Contributions of NRA intersections/touchings on the left side of nine example test points. The sense of NRA reaching/crossing is relevant for the final sum. Since the imaginary axis would lie over each a_i point, the numbers in the table indicate the I_C partial and total values, i.e., the counting of intersections (crossings) of the NRA. For example, the total down I_C for a_6 is given by summing all downward NRA intersections at its left: $+3 = +1 + 0.5 + 0.5 + 1$ and, similarly, for the total up I_C, the count is $-2 = -0.5 - 1 - 0.5$. The net total is $I_C = +1$, hence a_6 is inside. Curiously, the result for a_7 is $I_C = +2$, that is, we could consider that a_7 is inside "twice", which is graspable from the figure!

The major novelties in this approach for polygonal regions are that the existence of NRA crossing can be determined by detecting the value of a single parameter t and the sense of crossing can be determined by a simple comparison operation.

5.2. The Case of Points on the Border, Again!

In Figure 9 specific situations arise if the testing point is exactly a NRA crossing point or, which is to say, the testing point is over the border. Once again, this situation forbids the possibility of seamlessly applying a unique algorithm for points on the border. For example, still in Figure 9, if the testing point occurs on the first crossing point (the one on the left of a_1), since it has no other crossing on its left, this implies being outside the region; so, its own intersection contribution (I_C) should be counted to consider it inside. But, if this reasoning is applied to the 4th intersection (between a_3 and a_4), then the overall contribution at that point ($+1 - 0.5 + 0.5 - 1$) would result in 0 and that point would be considered outside the region, contrarily to the intention of considering it as inside. The conclusion is that there is no unique solution to distinguish with the same operation whether a point on the border is always considered inside or outside the region: this is unavoidable, as supported by the Jordan curve theorem presented earlier, where the sense of crossing (circulation) affects the winding contribution, and therefore the conclusion about inclusion. The solution, as already mentioned, is to check those situations before further calculation.

5.3. Universal Algorithm for Arbitrary Polygons

All of the previous procedures are integrated in Algorithm 3 on the following page, which shows all the steps to test polygonal inclusion of any point a on any polygon with M vertices (for any $M > 0$, but only $M > 2$ is really meaningful), normal or self-intersecting.

The computational cost of Algorithm 3 is $\mathcal{O}(n)$, and the arithmetic operations involved are only sums (or subtractions) and at most M floating point divisions per polygon, although optimizable, as described further.

Algorithm 3: Universal inclusion test in arbitrary polygons

Require: $a = (a_x, a_y)$, $P_n = (x_n, y_n)$ for $n = 1, 2, \ldots, M$
1: $P_n \leftarrow P_n - a$ for $n = 1, 2, \ldots, M$
2: $I_C \leftarrow 0$ //NRA Intersection counter
3: **for** $n = 1$ **to** M **do**
4: $Y_1 \leftarrow y_n$; $Y_2 \leftarrow y_{(n \bmod M)+1}$
5: $X_1 \leftarrow x_n$; $X_2 \leftarrow x_{(n \bmod M)+1}$
6: **if** $\text{sgn}(Y_1) == \text{sgn}(Y_2)$ **then**
7: continue //no NRA crossing: horizontal segment or segment entirely in one quadrant. Jump to next segment.
8: **end if**
9: $\Delta Y \leftarrow Y_1 - Y_2$
10: $\Delta X \leftarrow X_1 - X_2$
11: $t_c \leftarrow \frac{Y_1}{\Delta Y}$
12: **if** $t_c \Delta X == X_1$ **then**
13: **return true** //point *a* is on the boundary (included)
14: **end if**
15: **if** $t_c \geqslant 0$ **and** $t_c \leqslant 1$ **and** $t_c \Delta X \geqslant X_1$ **then**
16: $I_P \leftarrow \text{sgn}(\Delta Y)$
17: **if** $t_c == 0$ **or** $t_c == 1$ **then**
18: $I_P \leftarrow I_P \times 0.5$ //touch NRA only
19: **end if**
20: $I_C \leftarrow I_C + I_P$
21: **end if**
22: **end for**
23: **if** $I_C \neq 0$ **then**
24: **return true** //point *a* is included
25: **else**
26: **return false** //point *a* is not included
27: **end if**

In Algorithm 3, the points on the border, including vertices, are considered inside the polygon, but, if wanted otherwise, this can be easily modified to exclude them. It can be done either with an optional initial checking for *a* being one of the vertices or during the normal flow of the algorithm after test from line 12 that results from the NRA crossing conditions in expression (42). The decision on line 13 of Algorithm 3 could be set to **false** if the point on the boundary is defined as being outside.

Notice also that the apparent risk of division by zero in line 11 of the algorithm never occurs because of early analysis in line 6 of the algorithm; indeed, if $Y_1 == Y_2$, we also have $\text{sgn}(Y_1) == \text{sgn}(Y_2)$ and the algorithm skips to the next segment since there is no contribution of the current linear segment to the value of I_C. Illustration of the functionality and results of Algorithm 3 can be found in the GitHub repository indicated earlier.

5.4. Multi-Ring Polygons and Multi-Polygons

There is a category of planar domains commonly used in Geographical Information Systems (GIS), namely to represent contours of countries and other geographical entities, that accounts both for multiple separate polygons (multi-polygons) or for polygons that include "holes" defined by one (or more) separate closed lines fully included in the outer polygon; each of these separate contours is named a ring. A simple polygon has one ring, and a polygon with two "holes" has three rings [28]. There are also strict cases of multi-ring multi-polygons where rings nest inside each other. All these situations can be handled with the same concepts described earlier using the sense or circulation of a contour. Figure 10 illustrates three situations of multi-ring polygons.

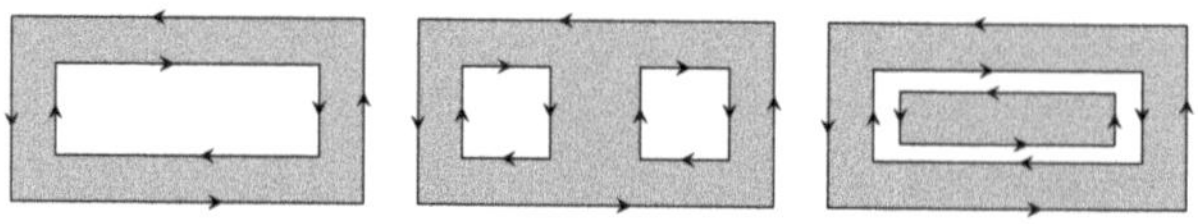

Figure 10. Multi-ring polygons as used in GIS contexts. The orientations are indicative, but appropriate relative senses must be ensured to correctly represent, for example, a polygon with a "hole".

The figure also illustrates the need for defining correctly the sequence of the polygon vertices in order to apply the same concepts of winding sense and, therefore, obtain the number of NRA intersections. Extension of the algorithm to this type of region is achieved by simply applying the procedure on each of the 2, 3, or more contours (rings) and sum up the separate results of NRA intersections.

5.5. *Optimizing Algorithm 3*

Algorithm 3 on the preceding page can be computationally alleviated by avoiding unnecessary divisions in case the NRA intersection does not occur. Indeed, as required by expression (42), for NRA intersection to occur we must have the following: $0 \leqslant \frac{y_1}{y_1 - y_2} \leqslant 1$. In other words, if either $\frac{y_1}{y_1 - y_2} > 1$ or $\frac{y_1}{y_1 - y_2} < 0$, then the NRA intersection does not occur, and no more calculations need to be performed. So, a few comparisons (3 at most) can be made before asserting the need to perform the division to find the actual t_c of NRA intersection. Since $y_1 - y_2 = \Delta y$, consider the following:

$$\frac{y_1}{\Delta y} < 0 \Rightarrow (\Delta y > 0 \wedge y_1 < 0) \vee (\Delta y < 0 \wedge y_1 > 0) \tag{46}$$

$$\frac{y_1}{\Delta y} > 1 \Rightarrow (\Delta y > 0 \wedge y_2 > 0) \vee (\Delta y < 0 \wedge y_2 < 0), \tag{47}$$

which can be combined. We are allowed to state that there is no NRA intersection if the following occurs:

$$\begin{cases} \Delta y > 0 \wedge (y_1 < 0 \vee y_2 > 0) \\ \Delta y < 0 \wedge (y_2 < 0 \vee y_1 > 0) \end{cases} \Rightarrow \nrightarrow o. \tag{48}$$

Hence, just before line 11 of Algorithm 3 a few lines could be added to optimize the procedure, demonstrating also that at most 3 comparisons are needed to conclude whether there is NRA intersection or not before calculating $t_c = Y_1/\Delta Y$, as proposed next:

```
    ⋮
    if ΔY > 0 then
        if Y1 < 0 or Y2 > 0 then
            continue //no NRA intersection
        end if
    else
        if Y2 < 0 or Y1 > 0 then
            continue //no NRA intersection
        end if
    end if
11: tc ← Y1/ΔY
    ⋮
```

Previous expressions can be further optimized because when they are reached, it is already known that $\text{sgn}(Y_1) \neq \text{sgn}(Y_2)$. Therefore, $Y_1 < 0 \vee Y_2 > 0$ is equivalent to only one of them, e.g., $Y_1 < 0$. A similar reasoning applies to the "else". This further reduces the number of comparisons from 3 to 2.

In case there is intersection (i.e., the previous tests concluded that there may be a NRA intersection because $0 \leqslant t_c \leqslant 1$), it is then necessary to calculate its precise value to test the second requirement of expression (42), that is, check the condition $t_c(x_1 - x_2) \geqslant x_1$, which at first glance seems unavoidable to calculate through the division. But, actually, that division operation can be avoided and replaced by two multiplications and a comparison, as is shown next: being $x_1 - x_2 = \Delta x$ and $t_c\Delta x \geqslant x_1 \Leftrightarrow \frac{y_1}{\Delta y}\Delta x \geqslant x_1$, if $\Delta y > 0$ then, for NRA crossing, we need to have: $y_1\Delta x \geqslant x_1\Delta y$, otherwise, we need to have the following: $y_1\Delta x \leqslant x_1\Delta y$.

The conclusion is summarized by expression (49):

$$\begin{cases} \Delta y > 0 \wedge y_1\Delta x \geqslant x_1\Delta y \\ \Delta y \leqslant 0 \wedge y_1\Delta x < x_1\Delta y \end{cases} \Rightarrow \downarrow\!\!\!\!\uparrow\!\circ. \tag{49}$$

This last simplification not only avoids the mathematical division butalso allows the algorithm to fully operate in integer arithmetic representation, which would improve its usage in an approach based purely on the integer representation of points/vertices.

Yet, in line 17, Algorithm 3 still requires to check whether t_c is equal to 0 or to 1 to conclude if the "crossing" is actually a "touching", which would imply an adjustment on the intersection counter. But that checking turns out redundant because those cases correspond to $y_1 = 0$ or $y_2 = 0$, that is, there is no need to calculate t_c for those cases either.

In summary, the algorithm in its full extension and covering all these optimizations, is detailed as Algorithm 4 on the next page. It can be noted, in line 32, that a multiplication by 0.5 is present, which, although being a statistically rare situation, could be modified for further optimization; that would be to count each full crossing as an integer of value ± 2, by, for example, changing line 30 into $I_P \leftarrow \text{sgn}(\Delta Y) + \text{sgn}(\Delta Y)$, and a half crossing being half of that value (± 1) with the final conclusions still holding, and integer computations could be present throughout the entire algorithm.

In conclusion, besides the simplicity of the formulation, the algorithm has also the virtue of a straightforward operation with degeneracy situations, such as point over segments or over vertices, or even as null length segments, which could happen in situations of rounding the point representation to integers.

5.6. Vectorization of Algorithm 4

Although not necessarily a breakthrough in reducing computational costs, Algorithm 4 on the following page can be easily vectorized for parallel computing, at least for the test of the inclusion of one point. Follows in Figure 11 a fully operational excerpt in Matlab code (listing) that accepts a matrix P with the polygon vertices (with the last vertex replicated from the first) and a point A. The procedure is vector based and operates with an entire polygon "at once".

Despite the possibility of some finer tuning, the vectorized approach, nevertheless, forces all the operations to be performed through the entire chain of tests. Some tests could be dismissed early in a sequential approach, but the vector approach forces all the operations for all cases. Hence, it may not be suited for better performance in all computational setups.

Algorithm 4: An optimized version of Algorithm 3

```
Require: a = (a_x, a_y), P_n = (x_n, y_n) for n = 1, 2, ..., M
P_n ← P_n − a for n = 1, 2, ..., M
I_C ← 0 //initialize NRA Intersection counter
for n = 1 to M do
  Y_1 ← y_n ; Y_2 ← y_{(n mod M)+1}
  X_1 ← x_n ; X_2 ← x_{(n mod M)+1}
  if sgn(Y_1) == sgn(Y_2) then
    continue //no NRA crossing: horizontal line, etc.
  end if
  ΔY ← Y_1 − Y_2, ΔX ← X_1 − X_2
  if ΔY > 0 then
    if Y_1 < 0 then
      //or Y_2 > 0 redundant
      continue //no NRA intersection
    end if
  else
    if Y_2 < 0 then
      //or Y_1 > 0 redundant
      continue //no NRA intersection
    end if
  end if
  if ΔY ⩾ 0 then
    if Y_1 × ΔX < X_1 × ΔY then
      continue //no NRA intersection
    end if
  else
    if Y_1 × ΔX ⩾ X_1 × ΔY then
      continue //no NRA intersection
    end if
  end if
  I_P ← sgn(ΔY) //There is NRA intersection
  if Y_1 == 0 or Y_2 == 0 then
    I_P ← I_P × 0.5 //touch NRA only
  end if
  I_C ← I_C + I_P //Update NRA crossing counter
end for
if I_C ≠ 0 then
  return true //point a is included
else
  return false //point a is not included
end if
```

```
Q=P-repmat(A,1,size(P,2));%translate to origin
Y1=Q(2,1:end-1);
Y2=Q(2,2:end);
DY=Y1-Y2;
X1=Q(1,1:end-1);
X2=Q(1,2:end);
DX=X1-X2;
Y1DX=Y1.*DX;
X1DY=X1.*DY;
noCrossSgn=(sign(Y1)==sign(Y2));
noCrossYA=(DY >0 & Y1<0);
noCrossYB=(DY<=0 & Y2<0);
noCrossXA=(sign(DY)> 0 & Y1DX < X1DY);
noCrossXB=(sign(DY)< 0 & Y1DX > X1DY);
Ip=~(noCrossSgn | noCrossYA | noCrossYB | noCrossXA | noCrossXB );
Ip=Ip.*sign(DY).*(1-0.5*(Y1==0 | Y2==0));
IC=sum(Ip);
```

Figure 11. Example of vectorization of Algorithm 4 in Matlab code.

5.7. Comparison to the State-of-the-Art Algorithm

For several years, Matlab has been using a version based on the algorithm from Hormann and Agathos [8], valid for point location in arbitrary polygons. According to the original paper, that algorithm is based on three steps, which, on the optimized variant, includes the following operations:

- Evaluation of the determinant: performs 2 multiplications and 1 subtraction per segment;
- Quadrant classification: uses 6 comparisons per vertex;
- Determination of winding number: performs 2 subtractions/sums and up to 4 + 1 comparisons per vertex.

Algorithm 4 has more steps (though simpler) than Algorithm 3 and requires:

- Initial test: 1 sign comparison per vertex;
- Coordinate differences: 2 per vertex;
- Y coordinate conditions: up to 3 comparisons per vertex;
- Test NRA: 2 multiplications and 2 comparisons per vertex;
- Test special cases end points: 2 comparisons with zero;
- Crossing counter: 1 addition per NRA crossing.

The comparison is given side-by-side in Table 1, where it can be seen that Algorithm 4 has fewer comparisons (1 + 2 + 2 + 2 = 7 vs. 10 + 1 = 11) and potentially much less if tests dismiss early; there is a similar number of additions/sums: 2 + 1 vs. 3; the multiplications are in equal number but the advantage is that in Algorithm 4 they are needed only when x-axis intersection occurs. In the previous analysis, $2M$ subtractions were omitted, reporting to the relocation of the polygon around point $(0,0)$, but which is a common operation to all.

Table 1. Comparison of estimated number of operations per vertex/segment in the algorithms.

	Algorithm 3	Algorithm 4	Algorithm from [8]
Comparisons	1 + 3 + 2	1 + 2 + 2 + 2	10 + 1
Sums/subtractions	2 + 1	2 + 1	3
Multiplications	1	2	2
Divisions	1	0	0

An illustrative example of simple benchmarking is shown in Figure 12 with a polygon with more than 2200 sides and thousands of testing points in two situations: on the left, 5000 completely random points, and on the right more than 2200 points near the border. Results are shown for Algorithm 3 and the state-of-the-art algorithm in Matlab [8]. The result shown is the average of 100 runs and had the same outcome in both algorithms. In both cases, Algorithm 3 proposed in this paper outperforms by a factor of about 5.

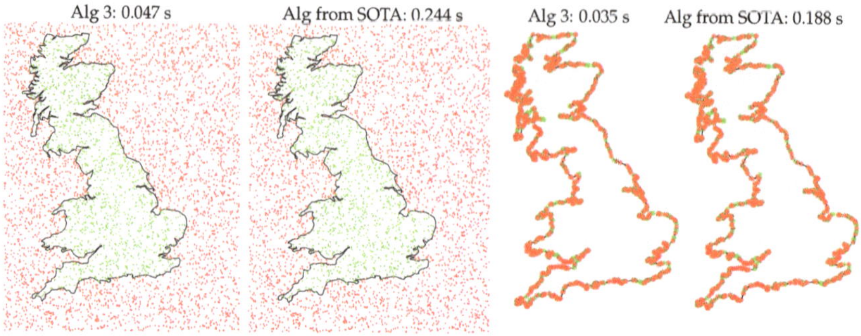

Figure 12. Example of points tested in a 2216-side polygon: 5000 random on the left and more than 2200 random points very close to the borders on the right. As indicated at the top of each image, Algorithm 3 ran about 5 times faster than the state-of-the-art (SOTA) Matlab's native algorithm [8]. Tested in a HP ZBook Fury 15 with Intel Core i7-10750H CPU 2.60 GHz with 32 GBytes of RAM in Ubuntu Linux 22.04.3.

6. NRA Crossing and Sense for Circular Arcs

For circular arcs, a few changes occur relative to linear segments. What is the most apparent is the fact that circular arcs can cross the NRA more than once or may even be tangent (Figure 13). Moreover, the parametric definition of arcs is a little more elaborate than for straight lines, as described next.

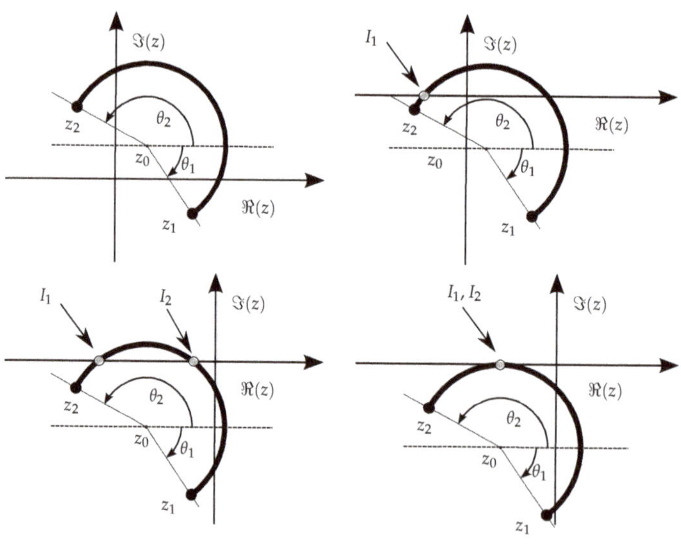

Figure 13. Examples of arc segments with center in z_0 starting on point z_1 and ending at point z_2. Depending on z_0, the very same arc can intersect the negative real axis (NRA), zero times, once (I_1), twice (I_1 and I_2), or be tangent ($I_1 \equiv I_2$), which, for the purposes of path integral evaluation, is equivalent to intersect the NRA twice, in opposite senses.

6.1. Parametric Expression for Circular Arcs

Any point of a circular arc with center at z_0, and evolving in the positive sense (counter-clockwise—CCW), between point $z_1 = z_0 + re^{j\theta_1}$ and point $z_2 = z_0 + re^{j\theta_2}$ is expressed by $z = z_0 + re^{j\theta}$, where θ covers the interval defined by θ_1 and θ_2, and $r = |z_0 - z_1| = |z_0 - z_2|$.

To simplify the parametrization and unify the procedures used for the linear segment, the following parametric representation is used:

$$z = z_0 + re^{j[\theta_1 + t(\theta_2 - \theta_1)]}, 0 \leqslant t \leqslant 1. \tag{50}$$

However, to ensure an unambiguous interpretation of the path described for the positive sense, it must be $\theta_1 \leqslant \theta_2$ where it is expected that the angles are obtained by the following operations: $\theta_1 = \arg(z_1 - z_0)$ and $\theta_2 = \arg(z_2 - z_0)$. But if that is not the case, then one of the angles must be converted by modulus 2π, that is, force θ_1 and θ_2 to fall in the interval $]-2\pi, +2\pi]$, by applying either $\theta_2 \leftarrow \theta_2 + 2\pi$ or $\theta_1 \leftarrow \theta_1 - 2\pi$, depending on the necessary case.

All the situations in Figure 13 respect the proper condition, $\theta_1 \leqslant \theta \leqslant \theta_2$, but different situations are illustrated in Figure 14 where θ encounters a discontinuity if restricted to the interval $]-\pi, +\pi]$. As the figure explains, in one case the solution for a proper parametrization is to convert θ_2 into $\theta_2 + 2\pi$ and the other to convert θ_1 into $\theta_1 - 2\pi$.

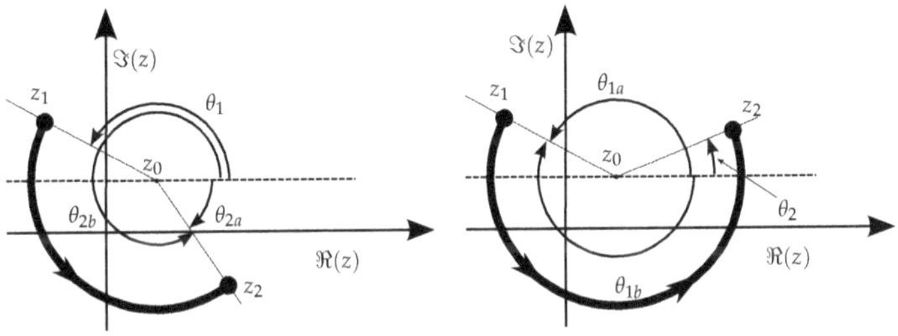

Figure 14. Example of arc segments where the *CCW* evolution of θ from θ_1 to θ_2 would not be monotonic when $\theta_i \in]-\pi, +\pi]$, but that can be made monotonic by taking the equivalent angles: on the left, $\theta_2 \leftarrow \theta_2 + 2\pi$, that is, use θ_{2b} instead of θ_{2a}, and on the right $\theta_1 \leftarrow \theta_1 - 2\pi$, that is, use θ_{1b} instead of θ_{1a}.

On the other hand, if a negative (or clockwise, *CW*) sense is intended, an equivalent operation must be taken into account, as illustrated in Figure 15.

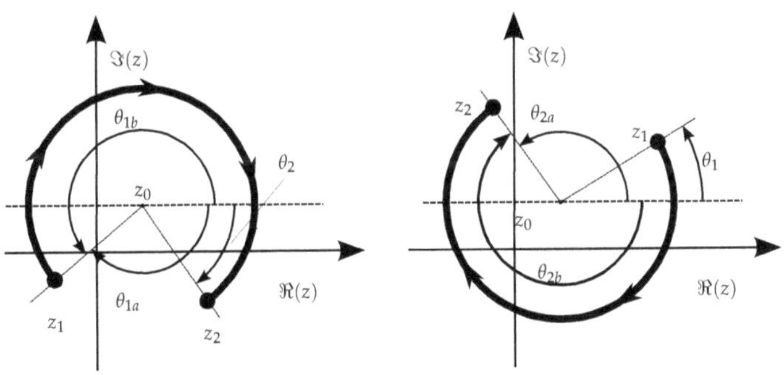

Figure 15. Example of arc segments where the *CW* evolution of θ from θ_1 to θ_2 would not be monotonic when $\theta_i \in]-\pi, +\pi]$, but that can be made monotonic by taking the equivalent angles: on the left, $\theta_1 \leftarrow \theta_1 + 2\pi$, that is, use θ_{1b} instead of θ_{1a}, and on the right $\theta_2 \leftarrow \theta_2 - 2\pi$, that is, use θ_{2b} instead of θ_{2a}.

In summary, to ensure the proper and unambiguous parametrization of a circular segment, the following steps have to be carried out when defining a circular boundary segment:

- Indicate z_1, z_2, and the center z_0 (or some other means to obtain z_0, like a third point z_p, or a radius and the relative placement of the center);
- Obtain $\theta_1 = \arg(z_1 - z_0)$ and $\theta_2 = \arg(z_2 - z_0)$;
- Define circulation sense (*CCW* or *CW*);
- Check the potential angle adjustments given by (51):

$$\begin{cases} \theta_2 \leqslant \theta_1 \wedge CCW & \begin{cases} \theta_2 < 0 & \Rightarrow \theta_2 \leftarrow \theta_2 + 2\pi \\ \theta_2 \geqslant 0 & \Rightarrow \theta_1 \leftarrow \theta_1 - 2\pi \end{cases} \\ \theta_1 \leqslant \theta_2 \wedge CW & \begin{cases} \theta_1 < 0 & \Rightarrow \theta_1 \leftarrow \theta_1 + 2\pi \\ \theta_1 \geqslant 0 & \Rightarrow \theta_2 \leftarrow \theta_2 - 2\pi \end{cases} \end{cases}. \tag{51}$$

In expression (51), the verification of the equality between θ_1 and θ_2 is also considered to include the situation of complete circles (where θ_1 and θ_2 would coincide, but actually one of them should be 2π larger than the other to ensure that the possibility of a full circle is covered).

Having the circular path segments duly defined and parametrized, the calculation of the NRA intersections is as straightforward as for the linear segment. Hence, to adapt for the circular arc, there must exist one (or two) values for t that generate an NRA crossing, and let us call them t_c:

$$x_c + jy_c = (x_0 + jy_0) + re^{j[\theta_1 + t_c(\theta_2 - \theta_1)]}, 0 \leqslant t_c \leqslant 1 \tag{52}$$

or in expanded view:

$$\begin{cases} x_c = x_0 + r\cos[\theta_1 + t_c(\theta_2 - \theta_1)] & 0 \leqslant t_c \leqslant 1 \\ y_c = y_0 + r\sin[\theta_1 + t_c(\theta_2 - \theta_1)] & 0 \leqslant t_c \leqslant 1 \end{cases}. \tag{53}$$

For the same conditions as before, $x_c \leqslant 0$ and $y_c = 0$ as seen in Section 5, we then have:

$$\begin{cases} x_0 \leqslant -r\cos[\theta_1 + t_c(\theta_2 - \theta_1)] & 0 \leqslant t_c \leqslant 1 \\ y_0 = -r\sin[\theta_1 + t_c(\theta_2 - \theta_1)] & 0 \leqslant t_c \leqslant 1 \end{cases}. \tag{54}$$

However, since the circle can cross the NRA twice, two solutions would be expected for t_c in the second equation of (54); indeed, the total number of solutions in θ of an equation of the type $x = \sin(\theta)$ is given by the following:

$$\theta = \begin{cases} \arcsin x \pm 2k\pi & k = 0,1,2,\ldots \\ \pi - \arcsin x \pm 2k\pi & k = 0,1,2,\ldots \end{cases} \tag{55}$$

that, when restricted to the $[-2\pi, +2\pi]$ interval, simplifies into, at most, the following 5 solutions:

$$\theta = \begin{cases} \arcsin x \\ \arcsin x \pm 2\pi \\ -\arcsin x \pm \pi \end{cases}. \tag{56}$$

Resuming to the variables from (54), we have

$$\begin{cases} t_{c_1} = \dfrac{\theta_1 + \arcsin \frac{y_0}{r}}{\theta_1 - \theta_2} \\ t_{c_{2,3}} = \dfrac{\theta_1 + \arcsin \frac{y_0}{r} \pm 2\pi}{\theta_1 - \theta_2} \\ t_{c_{4,5}} = \dfrac{\theta_1 - \arcsin \frac{y_0}{r} \pm \pi}{\theta_1 - \theta_2} \end{cases} \tag{57}$$

where all those potential solutions, besides needing to verify the condition $0 \leqslant t_c \leqslant 1$, must also verify the first condition of (54), that is, $x_0 \leqslant -r\cos[\theta_1 + t_c(\theta_2 - \theta_1)]$.

Naturally, at most two of the solutions from (57) are valid (two intersections of the x axis), but there can be only one, or even no, solutions at all. As just mentioned, both potential solutions (let us name them t_{c_A} and t_{c_B}) must be tested whether they verify expression (54). Nonetheless, there are obvious cases that can be tested beforehand to accelerate the numeric algorithms: one is $|y_0| > r$, resulting in a clear impossibility, and the other case is when $|y_0| = r$, implying that the arc is tangent to the x axis.

For the case of circular arcs, the situation of tangency may be interpreted as either a non-intersection or a double touching of the NRA (entering and leaving the NRA at the same point). The net additional contribution for the calculation of the line integral is, therefore, zero. Hence, if $|y_0| \geqslant r$, then the circular segment does not affect the intersection counter I_C. The exception occurs in case the tangency point ($|y_0| = r$) happens for t_c being 0 or 1, which is a "half intersection", as described next.

6.2. Sense of NRA Crossing for Circular Arcs

As before, in case there is an NRA intersection, its sense must be determined, and expression (43) can be used and adapted for the case of arcs:

$$I_C = -\operatorname{sgn}\{(\theta_2 - \theta_1) r \cos[\theta_1 + t_c(\theta_2 - \theta_1)]\}. \tag{58}$$

Expression (58) can be further simplified avoiding the need to calculate the trigonometric function. Knowing that $r > 0$ and being $\theta(t_c) = (\theta_1 + t_c(\theta_2 - \theta_1)) \bmod 2\pi$, we have:

$$I_C = \begin{cases} \operatorname{sgn}(\theta_1 - \theta_2) & \Leftarrow -\frac{\pi}{2} < \theta(t_c) < \frac{\pi}{2} \\ \operatorname{sgn}(\theta_2 - \theta_1) & \Leftarrow \text{Otherwise} \end{cases}. \tag{59}$$

Also, as occurred for the linear segments, the particular cases of $t_c = 0$ or $t_c = 1$ correspond to "half intersections" (one extremity of the arc is over the NRA) and, therefore, the I_C calculated with (59) must be adjusted the same way, as given by (45), or formally stated: $t_c = 0 \vee t_c = 1 \Rightarrow I_C \leftarrow 0.5 \times I_C$.

7. Application to Bézier Curves

Most planar domains can be defined by closed splines or other arbitrary curve segments, delimited by two extreme (anchor) points and some control points in a parametric way, which is most frequently of a polynomial nature.Each segment can however be treated the same way as was demonstrated earlier for linear and circular segments. For example, Bézier curves of order n in the complex plane can be defined by:

$$B_n(t) = \sum_{k=0}^{n} t^k \left(\frac{n!}{(n-k)!} \sum_{i=0}^{k} \frac{(-1)^{i+k} z_i}{i!(k-i)!} \right) \tag{60}$$

where z_i are the control points, starting at z_0 and ending at z_n (the anchor points), and with $0 \leqslant t \leqslant 1$.

Segments are described by polynomials, making it easy both to differentiate and find roots, except for higher orders (like 5, 6, or more), and in that case, it can represent a challenge for analytical approaches, although numerical approaches are always possible. Bézier curves can also be expressed in the more compact notation:

$$B_n(t) = \sum_{i=0}^{n} \binom{n}{i} (1-t)^{n-i} t^i z_i, 0 \leqslant t \leqslant 1 \tag{61}$$

from which follows an example of a Bézier segment of the 3rd order, which is one of the most commonly used, thus, representative of many challenges in computational geometry:

$$B_3(t) = (1-t)^3 z_0 + 3(1-t)^2 t z_1 + 3(1-t) t^2 z_2 + t^3 z_3 \tag{62}$$

or, in full expansion:

$$B_3(t) = (3z_1 - z_0 - 3z_2 + z_3)t^3 + 3(z_0 - 2z_1 + z_2)t^2 + 3(z_1 - z_0)t + z_0. \tag{63}$$

As described earlier, detecting the NRA crossing of such a segment requires the solution of $B_3(t) = 0$, which would yield up to 3 results for t in the $[0, +1]$ interval. If less than 3 NRA intersections occur, this means that some solutions for t are complex, and are to be discarded. To determine the sense of growing at the intersection points with the NRA, the derivative of $B_3(t)$ in order to t is required, and that too is a direct operation easily allowed by the polynomial representation of the curve segment:

$$\frac{\mathrm{d}B_3(t)}{\mathrm{d}t} = 3(3z_1 - z_0 - 3z_2 + z_3)t^2 + 6(z_0 - 2z_1 + z_2)t + 3(z_1 - z_0). \tag{64}$$

Figure 16 shows a region defined by a cubic Bézier curve and two linear segments. In this case, the Bézier anchor points are $z_0 = (0,0)$ and $z_3 = (2,1)$ and the two control points are $z_2 = (0,-1)$ and $z_3 = (2,1)$. The two linear segments are $\overline{z_3z_4}$ and $\overline{z_4z_0}$. Points $a_1 = (0.9, 0.6)$ and $a_2 = (1.1, 0.4)$ will be tested but only a_1 is included in the region.

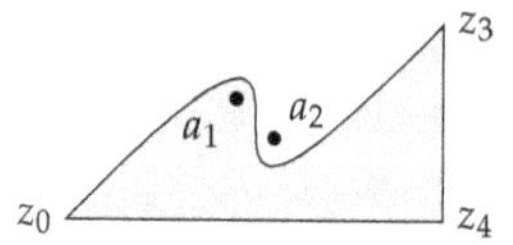

Point	Coordinate
z_0	$(0,0)$
z_1	$(2,2)$
z_2	$(0,-1)$
z_3	$(2,1)$
z_4	$(2,0)$

Figure 16. Region delimited by a cubic Bézier and two linear segments. Points z_1 and z_2 (not plotted in the figure) are the Bézier control points.

To test points a_1 and a_2, the procedure seems to require the adjustment of the "vertices" values reported to each testing point and calculate the new resulting polynomials. Actually, that turns out unnecessary in the parametric form because the new polynomial only changes by an offset, and the generic expression (63) simply becomes $B_3(t) \leftarrow B_3(t) - a_i$, where a_i is the point being tested.

An additional relevant aspect for computational purposes is that the expression for the Bézier polynomial derivative (64) remains the same whatever the testing point a_i, which saves computation resources when testing the inclusion of many different points for the same region. The generic equations to test the NRA crossing and contributions (I_C) for a given $a_i = (a_{x_i}, a_{y_i})$ are:

$$\begin{cases} x_{B_3} = 8t^3 - 12t^2 + 6t - a_{x_i} \\ y_{B_3} = 10t^3 - 15t^2 + 6t - a_{y_i} \\ \dfrac{\mathrm{d}y_{B_3}}{\mathrm{d}t} = 30t^2 - 30t + 6 \end{cases}.$$

To solve the problem for this particular situation, the procedure is implemented as follows:

1. Find the roots of $y_{B_3} = 10t^3 - 15t^2 + 6t - a_{y_i}$;
2. Test whether the found roots satisfy the condition of NRA intersection, that is, $x_{B_3} = 8t^3 - 12t^2 + 6t - a_{x_i} \leqslant 0$;
3. For each root that satisfies the previous expression there is a NRA intersection whose sense needs to be assessed by the sign of $\frac{\mathrm{d}y_{B_3}}{\mathrm{d}t} = 30t^2 - 30t + 6$.

Table 2 shows some numerical results that illustrate the functionality of the algorithm for a cubic Bézier segment for two test points a_1 and a_2.

Figures 17 and 18 illustrate the plots concerning NRA intersections by the cubic Bézier segment for the cases of point a_1 and a_2, respectively.

The intersection contributions I_C (detailed in Table 2) clearly translate the situations of inclusion of the tested points. For a_1, total I_C is -1 and for a_2 total I_C is $+1 - 1 = 0$. Actually, the I_C contributions from the two linear segments were not analyzed, but it is simple to confirm that they do not contribute at all because they do not cross the NRA: the horizontal segment will never contribute, and the vertical segment does not cross the NRA for the points being tested.

Table 2. Detailed analysis of operations concerned with NRA crossings by a cubic Bézier segment for two testing points in the case illustrated in Figures 16–18.

Testing Points	t_i	$x_{B_3}(t_i)$	$\updownarrow\!\circ$	$\frac{dy_{B_3}}{dt}(t_i)$	I_C	$\sum I_C$
	$t_1 = 0.1517$	−0.2379	yes	2.139	−1	
a_1	$t_2 = 0.4312$	0.0974	no	−1.358	0	−1
	$t_3 = 0.9171$	0.6805	no	3.719	0	
	$t_1 = 0.0829$	−0.6805	yes	3.719	−1	
a_2	$t_2 = 0.5688$	−0.0974	yes	−1.358	+1	0
	$t_3 = 0.8483$	0.2379	no	2.139	0	

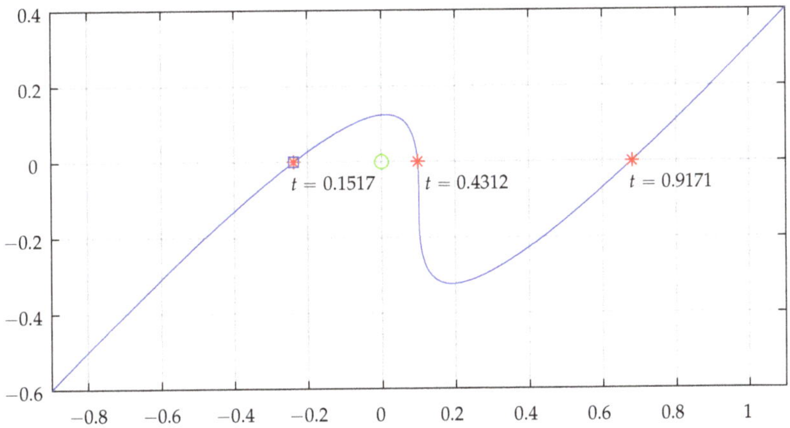

Figure 17. Bézier real axis intersections for testing point a_1. The "*" are the points where the curve intersects the horizontal axis, the green circle represents the point being tested when shifted to the system origin, and the square indicates a NRA crossing that effectively contributes to the intersection counting. Check Table 2 for more details.

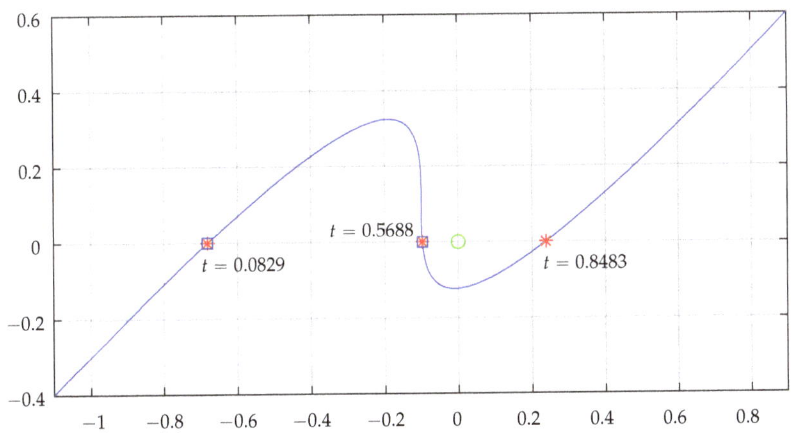

Figure 18. Bézier real axis intersections for testing point a_2. The "*" are the points where the curve intersects the horizontal axis, the green circle represents the point being tested when shifted to the system origin, and the squares indicate a NRA crossing that effectively contributes to the intersection counting. Check Table 2 for more details.

With higher order curves, such as these 3rd order Bézier, it may happen that the intersection point is tangent to the NRA and expression (43) would result in 0 (inconclusive). This is an easy situation to detect (one further comparison) and concerns a case where all three roots t_c are equal. There are two solutions for that very special case to allow the detection of the intersection sense: one would be to calculate expression (43) in a "small" neighborhood of t_c, for example, $t_c - \delta$ instead of t_c (where the value of δ can be on the range of the precision involved—e.g., 10^{-6}); the other solution, formally more elegant, is simply to check the quadrants of z_1 and z_2, i.e., use the value of $\text{sgn}(y_1 - y_2) = \text{sgn}(\Delta Y_b)$.

Application to the Challenges Given

By using the previous techniques for linear, circular, and Bézier segments in an integrated and complementary form, it is possible to test situations as those illustrated in Figure 1 for a set of random points, which results as shown in Figure 19.

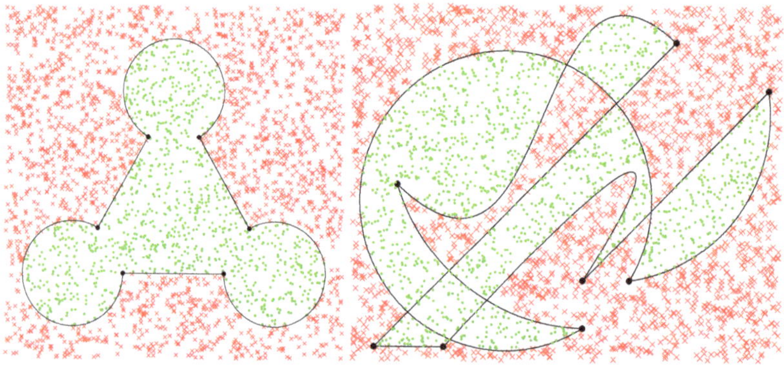

Figure 19. Results of algorithms applied to the regions of Figure 1. Due to limited resolution in the display above, some points may seem to be on the border, but actually, none of them are.

From the point of view of programming, three functions were created and used: `NRAintLin()`, `NRAintArc()` and `NRAintBez()`, whose source code is also available at the GitHub repository indicated earlier. These functions return the NRA counting for each respective segment when invoked accordingly to the type of segment: 3 linear and 3 circular arcs on the region on the left, and 2 linear, 3 circular arcs and 2 Bézier segments on the region on the right. These three functions are essential to establish the general algorithm for regions delimited by linear, circular, or Bézier segments, which is the subject of the next section.

8. Algorithm for Arbitrary Planar Domains

After the statements, demonstrations, and examples described in the previous sections, it is now straightforward to establish the basis of a general algorithm for arbitrary planar domains that can be described by parametric equations, which have associated a sense of circulation along the boundary. The procedure can be stated as follows:

1. Set up the closed region by segments defined parametrically in such a way that each segment is defined by two extremes z_1 and z_2, a parameter t, being $0 \leqslant t \leqslant 1$, and optional points or parameters depending on the type of the segments (linear, circular arcs, Bézier curves, etc.)
    - Linear segments do not require any further parameters or points;
    - circular arcs require a third point (center), or else a radius along with the definition of its orientation;
    - other curves (of second or higher order) require additional elements to fully define them (3rd order Bézier curves require two more points).
2. Normalize the domain delimiting points (vertices) relative positions by subtracting the test point a to all vertices.

3. Check the intersections of this newly relocated region with the negative half part of the horizontal axis (x). Intersections occur from $Q2$ to $Q3$ or vice-versa.
   Full intersections are accounted for an integer value ($+1$ or -1, depending on the sense, $Q2$ to $Q3$ (↓∘) or $Q3$ to $Q2$ (↑∘), respectively), or a partial value ($+0.5$ or -0.5) if the segment starts or ends on the very horizontal axis (⊤∘, ⊥∘), and the positive or negative contribution is determined once again if the segment goes or comes to/from quadrants $Q2$ or $Q3$.
4. The verification of the sense of the evolution of the path must be asserted in these intersection/touching points: it is straightforward for linear segments, but for other types of segments, the descendant gradient of y with respect to t must be used at that point to determine the sense of cross/approach to the horizontal axis.

   Algorithm 5 formalizes the required steps.

Algorithm 5: Universal inclusion test in arbitrary regions

Require: $a = (a_x, a_y)$, $P_n = (x_n, y_n)$ for $n = 1, 2, \ldots, M$
Require: $L_n \in \{0, 1, 2, 3, \ldots\}$ //type of segment/number of control points
Require: $C_{n_k} = \{(xx_k, yy_k)\}_n$ for $k = 0, 1, \ldots, L_n$ //Control points //It is admitted that $P_{M+1} = P_1$.

if $a \in \{P_1, P_2, \ldots, P_M\}$ **then**
    return true //point a is one of the vertices
end if
$P_n \leftarrow P_n - a$ for $n = 1, 2, \ldots, M$
$C_n \leftarrow C_n - a$ for $n = 1, 2, \ldots, M$
$I_C \leftarrow 0$
for $n = 1$ **to** M **do**
    $[X(t), Y(t)] = f(P_n, P_{n+1}, \{C_n\}, L_n)$
    $T \leftarrow$ solutions of $Y(t) = 0$
    for all t_c in T **do**
        if $X(t_c) == 0$ **then**
            return true //point a is on border (included).
        end if
        if $t_c \geqslant 0$ **and** $t_c \leqslant 1$ **and** $X(t_c) \leqslant 0$ **then**
            $I_P \leftarrow -\operatorname{sgn}\left(\frac{dY(t)}{dt}\Big|_{t=t_c}\right)$ //intersects NRA
            if $I_P == 0$ **and** L_n is a cubic Bézier **then**
                $I_P \leftarrow \operatorname{sgn} \Delta Y_b$ //Cf. text for definition of ΔY_b
            end if
        end if
        if $t_c == 0$ **or** $t_c == 1$ **then**
            $I_P \leftarrow I_P \times 0.5$ //touch NRA only
        end if
        $I_C \leftarrow I_C + I_P$
    end for
end for
if $I_C \neq 0$ **then**
    return true //point a is included
else
    return false //point a is not included
end if

A most relevant point when comparing this general algorithm to the algorithms presented in the preceding sections is the input of data: besides the testing point and the list of vertices that require no changes at all, there are now more parameters to be given to distinguish and define the types of segments.

The algorithm then requires the following data as inputs:

- The testing point a (same as before)
- The vertices (extreme points) of the region segments (same as before)
- A set of numbers indicating the type of segment or number of control points (e.g., 0 for linear, 1 for circular, 2 for 3rd order Bézier segments, etc.)
- A set of control points for each segment

The steps of the algorithm that vary with the type of segment occur mainly in lines 8 and 9, where the parametric equations of the segment must be obtained and solved, and also in line 15 to calculate a derivative and obtain the sense of NRA crossing.

As mentioned before, the GitHub repository made available also includes the code and illustration of the application of this general algorithm with several examples besides the ones in Figure 19. No benchmarking is performed with this general algorithm because, in the literature, there are no other algorithms for this same purpose to compare with.

9. Discussion and Conclusions

This paper presents a methodology to solve, in analytical form, the problem of point inclusion in planar domains for regions of virtually unlimited complexity, namely with boundary curves of order greater than linear segments. Specifically, for the cases for regions bounded arbitrarily by polygonal, circular arcs and Bézier lines have been derived and implemented, and the methodology for regions with other shapes has been described and is ready to be applied to the general algorithm.

The approaches described are handled in analytical form, so they are independent of any numerical approximations, hence independent of scale. Nonetheless, when performing actual calculations in floating point operations, the limitation of the numeric representation of the computational system naturally occurs. However, due to the nature of the operations (sums and multiplications in the Algorithms 3 and 4 for polygonal regions only), the propagation of uncertainty is minimal, and operations can even be performed in integer representation, which is suitable for some embedded systems. On the other hand, this is not the case for the general approach of non-linear segments (covered in Algorithm 5) because boundary segments are non-linear, and non-linear calculations are required, including trigonometric or root operations. Since the accumulated chain of non-linear operations is short, uncertainty propagation is contained, and results have shown that very good accuracy (in 12 or more decimal places) is kept in the final outcome. This can be checked, for example, by "zooming" into the plots extensively and verifying the correct result of point inclusion. The mentioned GitHub repository allows these tests to be performed by other users.

The paper also demonstrates why a simple winding number based technique for detecting point inclusion will always be ambiguous if the testing point is on the border. Nonetheless, the proposed algorithms detect those situations early, and the technique relies on the need of a single pass operation, i.e., it requires no iterative procedure or further verifications.

Besides this advantage of a single operation, the possibility of extending the algorithm to regions of more complex shapes opens wider directions in the field of simulation and computational geometry. The usage for curves of high orders needs to deal with negative real axis intersection, which is equivalent to finding the roots of polynomials or other transcendental curves. This implies unavoidable additional computational costs when compared to straight polygonal shapes. Nevertheless, the pure analytical formulation of the method may remain an advantage against the alternatives of linearization or approximation of curves to polygonal lines.

Along these lines, a final remark is worth mentioning, which concerns a potential alternative to this approach for general shapes. We could think of the possibility of triangulating a region such as the ones shown in Figure 19 and apply point location techniques cited in Section 2. Although the technique is usually applied for a problem not addressed in this paper, a region could be triangulated and the test of inclusion of a point in the overall region could be performed by identifying which subregion, if any, includes that point. Besides the variable computational costs of this extreme solution, two main limitations immediately arise. The first is concerning self-intersecting regions and regions with holes (like the ones in Figure 10) that require additional care and pre-processing to be triangulated. Secondly, and even more relevant, if the region boundaries are curved, they must be linearized (sampled) to perform the triangulation. This has two drawbacks: the computational demands of the preprocessing and an additional computational cost that depends on the scale. The dependency on scale is particularly relevant near the boundaries: if very high accuracy is required, the samples over the curved boundaries must be more dense; hence, a larger number of triangles is generated in the mesh. This dependency on scale (with consequences on the accuracy in areas closer to the borders) does not occur in the proposed approach because it is analytical.

The next list of topics explains the advantages of the proposed approach in this paper, the reasons for better performance than existing solutions, and the reason for that to happen.

- It involves fewer operations: besides the common sums and subtractions, at most only one floating point division is required per segment (or two multiplications instead) and also fewer comparisons, for the case of polygonal shaped regions. This makes it faster than state of the art solutions (about 5 times as fast was demonstrated for polygons with many sides and a large number of data points for the algorithm implemented in Matlab [8]).
- It has no restrictions on which points define the polygons (or the generically shaped region) nor on the testing points that can be over the border, or being themselves the very vertices, and no particular post processing or operations are required.
- It accepts the geometry of boundary segments to be other than straight lines and is applicable virtually to any type of curve that can be expressed parametrically.
- The parametric formulation of the boundary makes it straightforward to define the curve segments and their extremes (anchor points).
- The parametric definition is intrinsically associated with the concept of sense or circulation, which allows the detection of the sense of axis intersection, and not only the fact that there are intersections; and the sense of intersection is crucial to obtaining the global winding number.
- The parametric description allows a separation of coordinates, making algebraic operations independent, and allows us to overcome limitations in degenerate situations that would occur in two coordinate point representations but do not occur in separate coordinate representations.

Although the formal and analytic procedures for point location in generic planar domains have been fully established and demonstrated in this paper, the future may hold improvements in the computational component of the algorithms for specific curves. This can be true, namely for some curves of very high complexity, as well as for the detection of specific situations of point location relative to those more complex boundaries, in order to accelerate the calculation operations.

Funding: An early part of this work was developed under the Ph.D. Grant BD-1657/91_IA from JNICT, Programa CIENCIA, Portugal.

Data Availability Statement: Algorithms and data to support the results of this paper are available at a GitHub repository: https://github.com/vitoruapt/PointInclusion.

Acknowledgments: The author acknowledges the EC Joint Research Center of Ispra, Italy, for hosting him as a Ph.D. student in the early 1990s' where the basic idea that lead to this research started.

Conflicts of Interest: The author declares no conflicts of interest.

References

1. Alciatori, D.; Miranda, R. *A Winding Number and Point-in-Polygon Algorithm*; Technical Report; Department of Mechanical Engineering, Colorado State University: Fort Collins, CO, USA, 1995.
2. Shimrat, M. Algorithm 112: Position of Point Relative to Polygon. *Commun. ACM* **1962**, *5*, 434. [CrossRef]
3. Hacker, R. Certification of Algorithm 112: Position of Point Relative to Polygon. *Commun. ACM* **1962**, *5*, 606. [CrossRef]
4. Franklin, W.R. PNPOLY-Point Inclusion in Polygon Test. 1994–2006. Available online: https://wrfranklin.org/Research/Short_Notes/pnpoly.html (accessed on 4 October 2024).
5. Preparata, F.P.; Shamos, M.I. *Computational Geometry: An Introduction*; Springer: New York, NY, USA, 1985.
6. Sedgewick, R. *Algorithms in C*; Addison-Wesley Longman Publishing Co., Inc.: Boston, MA, USA, 1990.
7. Haines, E. Point in Polygon Strategies. In *Graphics Gems IV*; Heckbert, P.S., Ed.; Academic Press Professional, Inc.: San Diego, CA, USA, 1994; pp. 24–46.
8. Hormann, K.; Agathos, A. The point in polygon problem for arbitrary polygons. *Comput. Geom. Theory Appl.* **2001**, *20*, 131–144. [CrossRef]
9. Aurenhammer, F. Voronoi diagrams—A survey of a fundamental geometric data structure. *ACM Comput. Surv.* **1991**, *23*, 345–405. [CrossRef]
10. Devillers, O. The Delaunay Hierarchy. *Int. J. Found. Comput. Sci.* **2002**, *13*, 163–180. [CrossRef]
11. Devroye, L.; Mücke, E.P.; Zhu, B. A Note on Point Location in Delaunay Triangulations of Random Points. *Algorithmica* **1998**, *22*, 477–482. [CrossRef]
12. Mücke, E.P.; Saias, I.; Zhu, B. Fast randomized point location without preprocessing in two- and three-dimensional Delaunay triangulations. *Comput. Geom.* **1999**, *12*, 63–83. [CrossRef]
13. O'Rourke, J. How Do I Find If a Point Lies within a Polygon. 2003. Available online: http://www.faqs.org/faqs/graphics/algorithms-faq/ (accessed on 4 October 2024).
14. Sunday, D. Inclusion of a Point in a Polygon. 2012. Available online: https://web.archive.org/web/20130126163405/http://geomalgorithms.com/a03-_inclusion.html (accessed on 4 October 2024).
15. Foley, J.D.; van Dam, A.; Feiner, S.K.; Hughes, J.F. *Computer Graphics: Principles and Practice*, 2nd ed.; Addison-Wesley Longman Publishing Co., Inc.: Boston, MA, USA, 1990.
16. O'Rourke, J. *Computational Geometry in C*; Cambridge University Press: Cambridge, CA, USA, 1998.
17. Needham, T. *Visual Complex Analysis*; Oxford University Press: Oxford, UK, 1998.
18. Schirra, S. How Reliable Are Practical Point-In-Polygon Strategies? In Proceedings of the 16th Annual European Symposium on Algorithms (ESA '08), Karlsruhe Germany, 15–17 September 2008; pp. 744–755. [CrossRef]
19. Gatilov, S. Efficient Angle Summation Algorithm for Point Inclusion Test and Its Robustness. *J. Reliab. Comput.* **2013**, *19*, 1–25.
20. Ross, F.; Ross, W.T. The Jordan curve theorem is non-trivial. *J. Math. Arts* **2011**, *5*, 213–219. [CrossRef]
21. Edelsbrunner, H.; Maurer, H.A. A space-optimal solution of general region location. *Theor. Comput. Sci.* **1981**, *16*, 329–336. [CrossRef]
22. Kirkpatrick, D. Optimal search in planar subdivisions. *SIAM J. Comput.* **1983**, *12*, 28–35. [CrossRef]
23. Guibas, L.; Ramshaw, L.; Stolfi, J. A kinetic framework for computational geometry. In Proceedings of the 24th Annual Symposium on Foundations of Computer Science (1983), Washington, DC, USA, 7–9 November 1983; pp. 100–111. [CrossRef]
24. Jackowski, B. Computing the area and winding number for a Bézier curve. In *TUGboat*; Tex Users Group: Portland, OR, USA, 2012; Volume 33. Available online: https://tug.org/TUGboat/tb33-1/tb103jackowski.pdf (accessed on 4 October 2024).
25. Santos, V. Robot Autonomous Navigation: Sensorial Data Interpretation and Local Navigation. Ph.D. Thesis, Universidade de Aveiro, Aveiro, Portugal, 1995. Available online: http://hdl.handle.net/10773/17931 (accessed on 4 October 2024).
26. Wylie, C.; Barrett, L. *Advanced Engineering Mathematics*; McGraw-Hill: New York, NY, USA, 1982.
27. Kanwal, R. *Linear Integral Equations*; Birkhäuser Boston: Boston, MA, USA, 1996.
28. OGC. *OpenGIS Simple Features Specification for SQL, Revision 1.1*; Technical Report; Open Geospatial Consortium, Inc.: Arlington, VA, USA, 1999.

algorithms

MDPI

Article

Performance Analysis and Parallel Scalability of Numerical Methods for Fractional-in-Space Diffusion Problems with Adaptive Time Stepping

Svetozar Margenov * and Dimitar Slavchev

Institute of Information and Communication Technologies, Bulgarian Academy of Sciences, 1113 Sofia, Bulgaria; dimitargslavchev@parallel.bas.bg

* Correspondence: margenov@parallel.bas.bg

Abstract: We study numerical methods and algorithms for time-dependent fractional-in-space diffusion problems. The considered anomalous diffusion is modelled by the fractional Laplacian $(-\Delta)^\alpha$, $0 < \alpha < 1$, following the integral definition. Fractional diffusion is non-local, and the finite element method (FEM) discretization in space leads to a dense stiffness matrix. It is well known that numerically solving such non-local boundary value problems is expensive. Difficulties increase significantly when the problem is time-dependent. The aim of the article is to develop computationally efficient methods and algorithms. There are two main features of our approach. Hierarchical semi-separable (HSS) compression is applied for an approximate solution of the arising linear systems. For time discretization, we use an adaptive forward–backward Euler scheme. The properties of the composite algorithm thus obtained are investigated. In particular, the block representation of HSS compression allowed us to upgrade the HSS solver to efficiently handle varying diagonally perturbed transition matrices corresponding to changing time steps. The contribution of the paper is threefold. The methods are completely constructive, which allows for a clearly structured description of the algorithms. A theoretical estimate of the computational complexity is presented. It shows the advantages of the adaptive time stepping in combination with the HSS solver. Theoretical results are supported by representative numerical experiments. Both sequential and parallel scalability and efficiency are analyzed. The presented results provide convincing proof of the concept of the proposed methods and algorithms.

Keywords: fractional parabolic problems; fractional Laplacian; adaptive time stepping

1. Introduction

Anomalous diffusion has many applications in science and engineering. It is observed in flows in strongly heterogeneous porous media, superconductivity, diffusion of polymers in supercold media [1]; the electrodiffusion of ions into nerve cells [2] and photon diffusion diagnostics [3]; image processing and machine learning [4]; and the spread of viral diseases, computer viruses, and crime [5], to name a few. A more exhaustive review of its applications can be found in [6]. Anomalous diffusion is mathematically modelled with the fractional Laplace operator $(-\Delta)^\alpha$, where α is the fractional power. There are different approaches regarding the definition of fractional Laplacian. For example, the Balakrishnan formula, a formula involving semi-group, and the Dynkins definition based on probabilistic considerations are discussed in [7,8]. It is shown that they are all equivalent in the whole space $\mathbb{R}^d$ but differ significantly when homogeneous Dirichlet boundary conditions in a bounded domain are enforced. The work of comparing the properties of relevant models is an ongoing process in which various aspects of nonlocal phenomena are analyzed as well as how they describe the available experimental data. In recent years, mathematicians have been particularly active in developing numerical methods for multidimensional boundary value problems in computational domains of general geometry, applying and comparing results for different definitions of fractional Laplacian. In [9], a comparison between the spectral and integral definitions can be found. In this work, we will concentrate on the following integral definition of $(-\Delta)^{\alpha}$:

$$(-\Delta)^{\alpha} u(x) = C(d,\alpha)\ \text{P.V.} \int_{\mathbb{R}^d} \frac{u(x)-u(y)}{|x-y|^{d+2\alpha}} dy, \quad \alpha \in (0,1).$$

Here P.V. stands for principal value, d is the number of spatial dimensions, α is the fractional power, and $C(d,\alpha)$ is the following constant

$$C(d,\alpha) = \frac{2^{2\alpha}\alpha\Gamma\left(\alpha+\frac{d}{2}\right)}{\pi^{d/2}\Gamma(1-\alpha)},$$

where Γ is the gamma function. The fractional Laplacian is a non-local operator, and after applying a suitable discretization in space, the associated boundary value problems are reduced to solving systems of linear algebraic equations with dense matrices. For real-life problems, solving such systems can be computationally very expensive. If a standard LU or Cholesky decomposition is applied, the computational complexity to solve such a system is $O(N^3)$, where N is the number of degrees of freedom in space. For the considered parabolic problem, we need to solve a series of such linear systems, and the complexity becomes $O(mN^3)$, where m is the number of time steps. The aim of this research is to develop a more efficient numerical method with significantly less computational complexity.

In this work, we do not impose any special constraints on the geometry of the space computational domain Ω. Also, in the finite element method (FEM) discretization, the mesh is unstructured by default. This formulation of the problem excludes the possibility of using more specialized linear algebra methods, which are based on a special structure of the stiffness matrix.

In computational mathematics, numerical methods for evolution problems are naturally based on the best available solvers for the corresponding elliptic boundary value problems. In the case of standard (local) diffusion, the stiffness matrix is sparse, and the developed near-optimal implicit methods for parabolic problems use fast preconditioned iterative solution methods (such as algebraic multigrid) for the emergent linear systems. The situation changes substantially when fractional diffusion is involved. The matrices are dense, and the concept of iterative solvers is not applicable in the general case. Furthermore, it turns out that there are specific difficulties depending on the definition of the fractional Laplacian used.

The history of the development of efficient (near-optimal) numerical methods for space-fractional diffusion problems in multidimensional domains of general geometry is less than ten years. And despite the considerable number of publications in this field, there are few results of this type for parabolic problems. Without pretending to be exhaustive, we will briefly touch on some of them. The case of spectral fractional diffusion is considered in [10–14] . The common feature of these results is that they can all be interpreted in terms of certain rational approximations. A further step in the development of this approach, significantly expanding the scope of the investigated problems, can be found in [15–18]. These works mainly focus on convergence analysis, with less discussion of algorithmic implementation and computational complexity. The same applies to [19], where finite element approximations for integral-fractional evolution problems are analysed. This paper is thematically closest to the parabolic equation we are considering.

Here, we use a rather different pure algebraic approach. Hierarchical semi-separable (HSS) compression is applied to approximately solve the arising linear systems. Earlier results based on HHS compression have been published in [20], where the case of a uniform time step is experimentally analysed.

In this paper, we implement an adaptive forward–backward Euler scheme for discretization in time. The aim is to automatically adapt the method to the evolution of the regularity of the solution. The obtained composite algorithm is analysed. The block representation of HSS compression is used to upgrade the HSS solver. This allows us to efficiently handle different diagonally perturbed transition matrices corresponding to changing time steps. The main contributions of the paper are as follows. The method is fully constructive, leading to a well-structured algorithmic implementation. The theoretical estimate of the computational complexity clearly shows the efficiency of the adaptive time stepping in combination with the HSS solver. The presented numerical experiments support the theoretical results. Both sequential and parallel scalability and efficiency are analyzed. The conclusion is that a convincing proof of concept of the proposed method and algorithm is presented.

In short, the fundamental novelties of the results that will be presented are the complex solutions obtained as a result of the developed composite algorithm. The time discretization is optimized, where the variable time step size adapts to the dynamics in the behavior of the solution. The hidden structure of the dense fractional diffusion stiffness matrix is shown to fit well with the stochastic HSS compression approach, thereby allowing attractive parallel scalability to be reached.

The rest of the paper is organized as follows. The fractional-in-space parabolic problem is presented in the next section, which concludes with the adaptive forward–backward Euler scheme for time discretization. Section 3 is devoted to hierarchical semi-separable compression. A key aspect here is the construction of a modified algorithm for diagonally perturbed matrices. The most important from a theoretical point of view is Section 4, where the computational complexity of the composite algorithm is analysed. The test problem for numerical experiments is then discussed in the next section. An analysis of the results of the conducted sequential and parallel numerical tests is presented in Sections 6 and 7, respectively. In the next section, the relative error behaviour of the HSS solver is analyzed. The paper ends with a brief discussion of the results obtained.

2. Fractional-in-Space Parabolic Equation

2.1. Continuous Problem

The following parabolic problem with unknown function $u(x,t)$, $(x,t) \in \Omega \times [0,T]$ is considered

$$\left|\begin{array}{rcllll} \dfrac{\partial u(x,t)}{\partial t} + (-\Delta)^\alpha u(x,t) & = & f(x,t), & x \in \Omega, & t \in (0,T) \\ u(x,t) & = & 0, & x \in \Omega^\complement, & t \in (0,T) \\ u(x,0) & = & u^0(x), & x \in \Omega. & \end{array}\right. \tag{1}$$

Here, Ω is the open space domain, $\Omega^\complement = \mathbb{R}^d \setminus \Omega$ is the complement of Ω, and the open time domain is denoted with $t \in (0,T)$.

2.2. Finite Element Approximation in Space

For the discretization in space of the $(-\Delta)^\alpha$ operator, we use a finite element scheme as outlined by Acosta in [21]. Let us consider an admissible triangulation on Ω with $N_\mathcal{T}$ as number of linear conforming elements. We will denote with $\{\varphi_1, \ldots, \varphi_N\} \subset \mathbb{V}_h$ the nodal basis corresponding to the internal nodes $\{x_1, \ldots, x_N\}$. The basis functions have the form $\varphi_i(h_j) = \delta_i^j, i,j \in [1,N]$, where δ is the Kronecker delta. Thus, the $(-\Delta)^\alpha$ operator (1) is approximated by the fractional stiffness matrix $K = (K_{ij}) \in \mathbb{R}^{N\times N}$ such that

$$K_{ij} = \frac{C(d,\alpha)}{2} \langle \varphi_i, \varphi_j \rangle_{H^\alpha(\Omega)}.$$

where $H^\alpha(\Omega)$ is the following Sobolev space

$$H^\alpha(\Omega) = \{v \in L^2(\Omega) : |v|_{H^\alpha(\Omega)} < \infty\}.$$

It is defined with the Aronszajn–Slobodeckij seminorm

$$|v|_{H^{\alpha}(\Omega)} = \left(\iint_{\Omega\times\Omega} \frac{(v(x)-v(y))^2}{|x-y|^{d+2\alpha}} dxdy \right)^{\frac{1}{2}}.$$

The integrals that must be calculated in order to compute K_{ij} are carried out over the unbounded domain $\mathbb{R}^d$. A ball-shaped domain B is introduced around Ω to facilitate their calculations. The distance between the boundary $\partial\Omega$ and the compliment $B^{\complement}$ is discussed in [21]. An auxiliary admissible triangulation $\mathcal{T}_A$ is introduced on $B \setminus \Omega$ with n_A elements. We will denote with $n_{\tilde{\mathcal{T}}} = n + n_A$ the number of elements on the combined triangulation $\tilde{\mathcal{T}} = \mathcal{T} \cup \mathcal{T}_A$. Thus, the coefficients of K can be written as

$$K_{ij} = \frac{C(d,\alpha)}{2} \sum_{l=1}^{n_{\tilde{\mathcal{T}}}} \left(\sum_{k=1}^{n_{\tilde{\mathcal{T}}}} I_{l,k}^{i,j} + 2J_l^{i,j} \right), \quad l,k \in [1, n_{\tilde{\mathcal{T}}}].$$

The integrals I and J are

$$\begin{aligned} I_{l,k}^{i,j} &= \int_{T_l}\int_{T_m} \frac{\left(\varphi_i(x)-\varphi_i(y)\right)\left(\varphi_j(x)-\varphi_j(y)\right)}{|x-y|^{d+2\alpha}} dxdy \\ J_l^{i,j} &= \int_{T_l}\int_{B^{\complement}} \frac{\varphi_i(x)\varphi_j(x)}{|x-y|^{d+2\alpha}} dydx. \end{aligned} \tag{2}$$

Computing $I_{l,k}^{i,j}$ and $J_k^{i,j}$ is a numerically complex task. For this purpose, specialized methods and algorithms are utilized. Specifically, singular integrands are involved in the computation of I, while J is calculated over an unbounded domain. The derivation of the I and J integrals as well as the theoretical analysis of the described finite element scheme is given in [21].

2.3. Backward Euler Discretization in Time

The unconditionally stable backward Euler scheme is implemented, in which the lumped mass matrix M^L is used. M^L is obtained by summing all off-diagonal elements of the mass matrix M into the diagonal. In this way,

$$M_{ii}^L = \sum_{j=0}^{N} M_{ij}, \quad i = 1, \ldots N,$$

and thus, the time-stepping method is written in the form

$$M^L \frac{\mathbf{u}^{j+1} - \mathbf{u}^j}{\tau_j} + K\mathbf{u}^{j+1} = M^L \frac{\mathbf{f}^{j+1} + \mathbf{f}^j}{2}, \quad j = 0, \ldots, m-1. \tag{3}$$

Here, $\omega = \{t_0 = 0, t_1, \ldots, t_m = T\}$ is the finite difference mesh in the interval $[0, T]$, m is the number of time steps and τ_j the size of the jth step. The superscript j means that the value of the mesh function is taken at the node t_j. The implementation of the scheme (3) reduces to solving m systems of linear algebraic equations

$$\tilde{K} u^{j+1} = \tilde{b}^{j+1}, \quad j = 0, \ldots m-1, \tag{4}$$

where

$$\tilde{K}_j = K + \frac{M^L}{\tau_j}, \quad \tilde{b}^j = M^L \left(\frac{\mathbf{f}^{j+1} + \mathbf{f}^j}{2} + \frac{\mathbf{u}^j}{\tau_j} \right).$$

Let us assume that the mesh of the time stepping scheme is uniform; that is, $\tau_j = \tau$. Then, the matrix $\tilde{K}_j = \tilde{K}$ does not depend on j, and if, for example, LU factorization is applied to solve the linear systems in (4), only one factorization of $\tilde{K}$ will be needed. The computational complexity of the factorization is $O(N^3)$ and the complexity of solving the system with the factorized matrix is $O(N^2)$. Thus, the computational complexity of the backward Euler scheme under consideration will be $O(N^3 + mN^2)$. The parallel performance of LU- and HSS-based solvers for the case of a uniform time stepping is studied in [20].

The parameters τ_j control the time discretization error. When the regularity of the problem varies significantly with time, it may be advantageous to apply variable time steps, thus balancing the error. From a computational point of view, the motivation for using variable time steps is to significantly reduce their number compared to the uniform time step case while maintaining the required accuracy of the numerical solution. Such are, for example, the cases of an initial condition with jumps or jumps in time of the right-hand side. Let us assume that in a (by default) small number of subintervals of the time domain, we uniformly refine the step size. Then, the computational complexity of the LU-based solver method becomes $O(qN^3 + mN^2)$, where q denotes the number of times to perform the LU factorization, which corresponds to the number of times, when $\tau_j \neq \tau_{j+1}$, including the first step τ_1. The performance of this method for LU- and HSS-based solvers is examined in [22].

2.4. Adaptive Forward–Backward Euler Discretization in Time

The local refinement discussed above implies a more specific regularity of the solution in time that is a priori known. We do not need any such assumptions in this paper. The developed adaptive time stepping algorithm estimates the current error and automatically adjusts the next step size based on it. The procedure is as follows. The approximate solution is calculated using the stable backward Euler scheme (3), while a forward Euler scheme is used to estimate the size of the next time step. This type of methods are widely used in standard (local) diffusion problems. For example, adaptive time stepping for coupled multiscale problems is studied in [23,24], where the solution dynamics arise from nonlinear transitions of material properties. A parabolic equation with a jumping right-hand side is considered in [25], which is closest to the test problem we used in our numerical experiments. The time step adaptation algorithm explored in this paper is largely in the spirit of [25]. The great challenge and novelty of the results presented here are due to the non-locality of fractional diffusion.

Let us consider the semi-discrete Cauchy problem

$$\frac{\partial \mathbf{U}}{\partial t} + K\mathbf{U} = \mathbf{F} \tag{5}$$

corresponding to the FEM discretization in space of (1), where the vectors $\mathbf{U}$ and $\mathbf{F}$ represent the nodal functions of the unknown solution $u(x,t)$ and the right-hand side $f(x,t)$ on $\mathcal{T}$. The error $\mathbf{z}^{i+1}$ of the backward Euler scheme for (5) can be estimated by

$$||\mathbf{z}^{i+1}|| \leq \sum_{i=0}^{j} \tau_{i+1}||\Psi_{i+1}||, \tag{6}$$

where the truncation error Ψ_{i+1} has the form

$$\Psi_{i+1} = \mathbf{F}^{i+1} - \frac{\mathbf{U}^{\mathrm{i+1}} - \mathbf{U}^{\mathrm{i}}}{\tau_{i+1}} - K\mathbf{U}^{i+1}.$$

Following the notations introduced in the fully discrete case, the superscript j means that the values are taken in the node $t_j \in \omega$. Now, let $\delta > 0$ be the parameter defining the desired accuracy. We are looking for such a time step τ_{i+1} that the inequality

$$\Psi_{i+1} \leq \tau_{i+1}\delta \tag{7}$$

holds. If (7) is satisfied, from (6), we conclude that the error accumulates linearly and is majorized by

$$||\mathbf{z}^{i+1}|| \leq \delta t_{i+1}.$$

Unfortunately, it is not possible to apply (6) and (7) directly. This is because, to calculate the truncation error, we need the exact solution $\mathbf{U} = \mathbf{U}(t)$, which is of course unknown. In the algorithm for the time step selection presented below, we use an approximate truncation error. For this purpose, an auxiliary forward Euler step is performed.

Thus, the numerical solution of the parabolic problem is calculated by applying as the main discretization tool the unconditionally stable forward Euler scheme. Importantly, the additionally included inverse-free forward scheme is significantly cheaper from a computational point of view. It should be noted that this scheme is unstable, which is not a problem, however, because we only apply it locally to estimate τ_{i+1}.

Algorithm for Time Step Selection

The adaptive size of τ_{i+1} is computed in the following three steps:

1. Prognostic solution. Calculation of $\tilde{u}^{i+1}$ in $\tilde{t}_{i+1} = t_i + \tilde{\tau}_{i+1}$ using the forward Euler scheme (8).
2. Prognostic error. The prognostic solution $\tilde{u}^{i+1}$ is used to compute the prognostic truncation error $\tilde{\Psi}_{i+1}$ given by (9).
3. Step size selection. The time step τ_{i+1} is selected based on the prognostic error and the parameter δ that defines the desired accuracy.

When choosing a predictive time step, we always aim to increase the previous one by multiplying it by a predetermined factor $\gamma > 1$ as follows:

$$\tilde{\tau}_{i+1} = \gamma \tau_i.$$

The prognostic solution $\tilde{\mathbf{u}}^{i+1}$ is then calculated by the forward Euler scheme

$$M^L \frac{\tilde{\mathbf{u}}^{i+1} - \mathbf{u}^i}{\tilde{\tau}_i + 1} + K\mathbf{u}^i = \mathbf{f}^i. \tag{8}$$

Here, the approximate truncation error at the prognostic time step is computed in the form

$$\tilde{\Psi}_{i+1} = \tilde{\mathbf{f}}^{i+1} - M^L \frac{\tilde{\mathbf{u}}^{i+1} - \mathbf{u}^i}{\tilde{\tau}_{i+1}} - K\mathbf{u}^i \tag{9}$$

where $\tilde{\mathbf{f}}^{i+1}$ is the right-hand side at the prognostic time $\tilde{t}_{i+1} = t_i + \gamma \tau_i$. That is how we obtain

$$\overline{\tau}_{i+1} = \frac{\delta}{||\tilde{\Psi}_{i+1}||} \gamma \tau_i.$$

Finally, we apply the constraints that the time step τ_{i+1} cannot be less than τ_0 and cannot exceed the estimated time step, and so we obtain

$$\tau_{i+1} = \begin{cases} \tau_0, \text{ if } \overline{\tau}_{i+1} < \tau_0, \\ \tilde{\tau}_{i+1}, \text{ if } \overline{\tau}_{i+1} > \tilde{\tau}_{i+1}, \\ \overline{\tau}_{i+1}, \text{ otherwise.} \end{cases}$$

We also introduce the threshold parameter θ and impose the following step change constraint:

$$\text{if } \min\left\{ \left| \frac{\tau_{i+1} - \tau_i}{\tau_i} \right|, \left| \frac{\tau_i - \tau_{i+1}}{\tau_{i+1}} \right| \right\} < \theta \text{ then } \tau_{i+1} := \tau_i.$$

This rule allows us to avoid too small relative changes in the time steps.

One time step of the forward Euler scheme involves matrix-vector multiplication with computational complexity of $O(N^2)$, while the backward Euler scheme requires additionally solving a dense system of linear algebraic equations that has computational complexity of $O(N^3)$ in the general case when using LU factorization (or any other Gaussian elimination-type method). Thus, the total computational complexity of the discussed adaptive time-stepping method, if a LU factorization solver is used in the implementation, is $O(mN^3)$. The rest of the paper is aimed at improving the performance of the method by using HSS compression instead of LU factorization.

In conclusion, we note that, in general, the number of time steps m will be much smaller with the adaptive stepping scheme than if we use a uniform scheme with $\tau = \tau_0$.

3. Hierarchical Semi-Separable Compression

Hierarchical compression methods were introduced by Hackbusch in [26] with the $\mathcal{H}$-matrix (see also [27]). The main hypothesis here is that a data-sparse matrix A can be approximated by a matrix in compressed form H. Here, by data-sparse, we mean a matrix where the off-diagonal blocks can be expressed as a product of smaller vectors and matrices. In this work, we use the hierarchically semi-separable compression (HSS) method introduced by Martinsson in [28] and for the numerical experiments, we use the implementation within the STRUctured Matrices PACKage (STRUMPACK) [29,30].

Hierarchical compression was first introduced to solve systems of linear algebraic equations arising from the application of the boundary element method. It is noted by Acosta in [21] that the integrals (2) involved in the computation K are similar to those produced by discretization with the boundary element method. This makes K likely to be a a suitable candidate for hierarchical compression. In [31], we have experimentally verified the suitability of HSS compression for elliptic fractional diffusion problems.

3.1. Basic Relations and Algorithmic Steps

A system of linear algebraic equations in the form $Ax = b$ is solved with HSS in three steps.

1. *Hierarchical semi-separable compression.* The matrix A is divided in four blocks, and the off-diagonal blocks are approximated by a product of three matrices (also called generators) as follows:
$$A \approx \begin{pmatrix} D_1 & U_1 B_{1,2} V_2 \\ U_2 B_{2,1} V1 & D_2 \end{pmatrix}$$
For a suitable matrix A (with low off-diagonal blocks), U generators have few columns, B are small and square or rectangular, and V have few rows. The diagonal blocks D can also be compressed in a similar way and so on recursively until a threshold for the smallest block to be compressed is reached. The maximum rank r of the off-diagonal blocks occurring in the HSS compression of A is calculated. The computational efficiency of the compression can be roughly measured by comparing r to the size of the matrix N. For suitable problems $r \ll N$. In general, HSS compression is approximate, where random sampling is applied to calculate the generators. The user must supply a relative threshold ε_{rel} and an absolute one ε_{abs}. The computational complexity of the HSS compression is $O(rN^2)$.
2. *ULV-like factorization.* A special form of LU factorization called ULV-like factorization can be applied to the thus compressed matrix H. The compression implemented within STRUMPACK uses the structure of the generators, unlike the original ULV factorization, which uses orthogonal transformations [30]. The computational complexity of ULV-like factorization is $O(r^2N)$.
3. *Solving a factorized matrix system.* A system of linear algebraic equations with matrix H can be solved by ULV-like factorization with computational complexity $O(rN)$.

The compressed form of the matrix allows faster performance of linear algebraic operations. Thus, multiplying a vector by H has a computational complexity of $O(rN)$. It should also be noted that HSS compression is sensitive to the order of the unknowns. In this work, we use the recursive bisection algorithm to obtain a more suitable off-diagonal structure. This choice is based on results from [31,32].

3.2. Modified Algorithm for Diagonally Perturbed Matrices

We will recall that for the backward Euler scheme, we solve a series of systems of linear algebraic Equation (3). The matrix of each system has the form

$$\tilde{K}_i = \frac{M^L}{\tau_i} + K, \quad i = 1, \ldots, m.$$

It should be noted that the lumped mass matrix is diagonal and the only part of $\tilde{K}_i$ that would differ between time steps is the main diagonal, and only when the time step changes $\tau_i \neq \tau_{i-1}$. On the other hand, we can notice that HSS compression does not change the diagonal values. Thus, we can compress the stiffness matrix K once. Then, only the diagonal is updated when needed. The developed modified Algorithm 1 is implemented using STRUMPACK functionalities.

Algorithm 1 Algorithm for diagonally perturbed matrices with HSS compression

Input: $K, M^L, t = [0, T], \mathbf{f}_i = f(t), \theta, \mathbf{u}_0 = u(0), \gamma, \delta$
Output: $\mathbf{u}_i, \tau_i, m$
$i = 0, \ t = 0$
H = compress(K) ▷ Apply HSS compression on the stiffness matrix
while $t \leq T$ **do**
  $\tilde{\tau}_{i+1} = \tau_i \gamma$ ▷ Increase prognostic step
  $\tilde{\mathbf{u}}_{i+1} : M^L \frac{\tilde{\mathbf{u}}^{i+1} - \mathbf{u}^i}{\tilde{\tau}_{i+1}} + K\mathbf{u}^i = \mathbf{f}^i$ ▷ Compute prognostic solution
  $\tilde{\Psi}_{i+1} = \tilde{\mathbf{f}}^{i+1} - M^L \frac{\tilde{\mathbf{u}}^{i+1} - \mathbf{u}^i}{\tilde{\tau}_{i+1}} - K\mathbf{u}^i$ ▷ Compute truncation error
  $\overline{\tau}_{i+1} = \frac{\delta}{||\tilde{\Psi}_{i+1}||} \gamma \tau_i$ ▷ Calculate time step

  if $\overline{\tau}_{i+1} < \tau_0$ **then** ▷ Apply size limitations to new time step
    $\tau_{i+1} = \tau_0$
  else if $\overline{\tau}_{i+1} > \tilde{\tau}_{i+1}$ **then**
    $\tau_{i+1} = \tilde{\tau}_{i+1}$
  else if $|\frac{\overline{\tau}_{i+1} - \tau_i}{\tau_i}| < \theta$ **or** $|\frac{\overline{\tau}_{i+1} - \tau_i}{\tau_{i+1}}| < \theta$ **then**
    $\tau_{i+1} = \tau_i$
  end if
  if $t + \tau_{i+1} > T$ **then**
    $\tau_{i+1} = T - t$
  end if

  if $\tau_{i+1} \neq \tau_i$ **then**
    $H = \frac{M^L}{\tau_{i+1}} + H$ ▷ Perturb the diagonal of H with M^L and τ_{i+1}
    ULV = factor(H) ▷ Calculate the ULV factorization of H
  end if
  $\tilde{b}_i = M^L \left(\frac{\mathbf{f}^{i+1} + \mathbf{f}^i}{2} + \frac{\mathbf{u}^i}{\tau_i} \right)$ ▷ Compute right hand side
  $\mathbf{u}_{i+1}$ = solve(ULV, $\tilde{b}_{i+1}$) ▷ Compute the solution at time $t + \tau_{i+1}$

  $i = i + 1, t = t + \tau_{i+1}$ ▷ Prepare parameters for next time step
  if $\tau_{i+1} \neq \tau_i$ **then**
    $H = H - \frac{M^L}{\tau_{i+1}}$ ▷ Restore original compressed matrix
  end if
end while
$m = i$

4. Computational Complexity of the Time Stepping Algorithms

Here, we briefly compare the computational complexity of the considered time-stepping algorithms.

LU factorization (or any direct Gaussian elimination method) for solving dense linear systems requires $O(N^3)$ arithmetic operations. More precisely, the computational complexity of the factorization (forward elimination step) is $O(N^3)$ while solving the system with the factorized matrix (backward substitution step) costs $O(N^2)$ operations.

Alternatively, as discussed in Section 3, the solver based on HSS compression requires $O(rN^2)$ arithmetic operations, where r is the maximum rank of the off-diagonal blocks generated in the compression process. The costs of the three consequential steps here are as follows: HSS compression—$O(rN^2)$; ULV-like factorization—$O(r^2N)$; solving system with a factorized matrix—$O(rN)$.

Let us first consider the case of the backward Euler scheme with a uniform time step τ. Then, the matrix $\tilde{K} = K + M^L/\tau$ does not change during the time-stepping procedure, that is, the factorization is performed only once. Thus, the following estimates of the computational complexity of the time-stepping algorithm hold:

$$\mathcal{N}_{\text{uniform LU}} = O(N^3) + O(mN^2),$$

$$\mathcal{N}_{\text{uniform HSS}} = O(rN^2) + O(mrN).$$

A numerical comparison between these two methods can be found in [22]. Then, in [20], a special case of different time steps is considered, where the time domain is divided into (by default) a small number of subdomains with piecewise constant time steps.

The performance analysis that will be presented in the rest of the paper only applies to the case of adaptive time stepping. In the forward–backward Euler adaptive scheme, the matrix $\tilde{K}_i$ can generally change at each time step. In practice, this does not allow us to reuse the factorization computed at the previous time step.

Let us write the number of time steps m in the form

$$m = m_1 + m_2,$$

where m_1 denotes the number of times the adaptive forward–backward Euler scheme has changed the step size. Then,

$$\mathcal{N}_{\text{adaptive LU}} = O(m_1N^3 + mN^2).$$

As shown in Section 3.2, the HSS compression is performed only once for the adaptive scheme, while the ULV-like factorization must be performed m_1 times. Therefore, the computational complexity $\mathcal{N}_{\text{adaptive HSS}}$ can be estimated by

$$\mathcal{N}_{\text{adaptive HSS}} = O(rN^2 + m_1r^2N + mrN).$$

We should note that the number of steps of the adaptive method m as well as the part m_1 will depend on the parameters set by the user and the behaviour of the right-hand side $f(x,t)$.

In conclusion, the computational complexity analysis shows that we can expect the adaptive scheme to outperform the uniform one if the number of adaptive time steps m is significantly smaller than the number of uniform steps required to cover the same time domain $m_{\text{uniform}} = T/\tau_0$ and also when $m < N$.

5. Test Problem for Numerical Experiments

The test problem is defined in terms of the parabolic Equation (1), such that $\Omega = (-1,1) \times (-1,1)$, $T = 0.1$, and an initial condition $u^0(x) = 0$. In all numerical experiments, $\alpha = 0.5$, which is considered as a representative case to prove the concept of the analysed methods and algorithms.

We denote by $\tilde{f}(x)$ the checker-board function on $\Omega \setminus \partial\Omega = (-1,1) \times (-1,1)$

$$\tilde{f}(x) = \begin{cases} 1 & \text{if } x_1 x_2 > 0 \\ -1 & \text{otherwise} \end{cases}. \tag{10}$$

Then the right-hand side of $f(x,t) \in L^2(\Omega \times (0,T))$ has the form

$$f(x,t) = \begin{cases} 100\tilde{f}(x) & \text{if } t \in (0, \tilde{t}_1] \\ 200\tilde{f}(x) & \text{if } t \in (\tilde{t}_2, \tilde{t}_3] \\ 0 & \text{otherwise} \end{cases}, \tag{11}$$

where $\tilde{t}_1 = 0.01$, $\tilde{t}_2 = 0.05$ and $\tilde{t}_3 = 0.06$. For the adaptive time stepping scheme, we set the parameters $\gamma = 1.5$, $\delta = 0.01$, and $\theta = 0.1$. We also apply successive refinement of FEM meshes, resulting in an increase in degrees of freedom in space with $N = 2131$, 4167, 8030, 12,805, 16,184, 24,892, 32,302.

In the case of adaptive time stepping, the accuracy in time is controlled by the initial time-step τ_0, which must be aligned with the FEM accuracy in space. Furthermore, in the context of our test problem, τ_j must agree on jump points of $f(t)$.

In the performance analysis, we use also results for the backward Euler algorithm with uniform step τ. For this purpose, the relative error in the norm l_2 is used

$$R_j = \frac{\|\mathbf{u}^j - \mathbf{u}^{j+1}\|_{l_2}}{\|\mathbf{u}^j\|_{l_2}} = \frac{\sqrt{\sum_{i=1}^{N}(u_i^{j+1} - u_i^j)^2}}{\sqrt{\sum_{i=1}^{N}(u_i^j)^2}},$$

where the solution $\mathbf{u}^j$ is obtained with $m = 200 \times 2^j, j = 0,1,2,\ldots$. In the considered example, there are several points where f has jumps. To fit them, we add additional checks that will not allow the time step to go beyond these two points. If $t_i < 0.01$ and $\tilde{t}_{i+1} > 0.05$, we lower $\tilde{\tau}_i$ such that $\tilde{t}_i + \tilde{\tau}_i = 0.05$. In a similar way, we treat the interval $[0.05, 0.06]$.

In Figure 1, we present the relative errors for $N = 2131$ with decreasing constant time step τ (the behaviour is quite similar for all seven variants of N). The results shown are for $t = 0.025$, 0.05, 0.075, 0.1. The relative error between two consecutive values of the time steps becomes smaller than 10^{-3} at $m = 6400$ and smaller than 10^{-4} at $m =$ 204,800. Based on this simple observation, in the numerical experiments for the adaptive scheme, we will consider the cases $m_{\max} = 6400$ and $m_{\max} = 204{,}800$, which correspond to $\tau_0 = 1.5625 \times 10^{-5}$ and $\tau_0 = 4.8828125 \times 10^{-7}$. As you will see in the next section, the two chosen options for τ_0 turn out to be very representative for the purposes of the presented performance analysis.

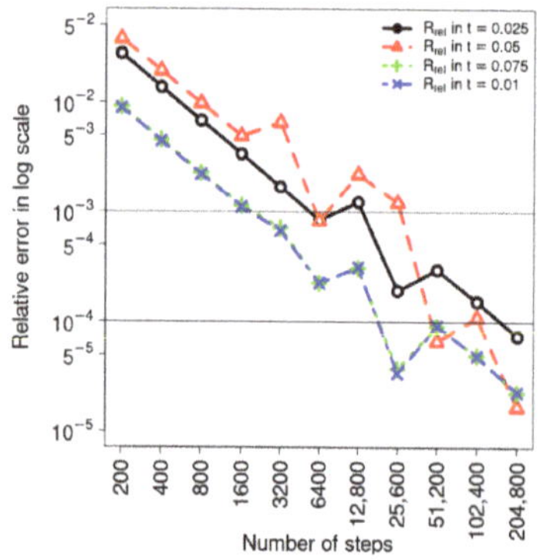

Figure 1. Backward Euler method with uniform time step: reduction in relative error R_j versus decrease in τ for $N = 2131$.

Finally, for the absolute and relative threshold used in the HSS compression implemented in STRUMPACK, we will use a fixed $\varepsilon_{\text{abs}} = 10^{-8}$ and vary $\varepsilon_{\text{rel}} = 10^{-2}, 10^{-4}, 10^{-6}$ and 10^{-8}.

6. Comparative Analysis of Sequential Performance

Here, we analyse the performance of the adaptive scheme compared to the backward Euler algorithm with uniform time step $\tau = \tau_0$. Figure 2 shows the computation times of the sequential algorithms with the LU factorization solver. We see that the adaptive time stepping scheme has a clear advantage in the case of a smaller time step $\tau_0 = 4.8828125 \times 10^{-7}$. This experimental result is fully consistent with the theoretical conclusion in Section 4 that the adaptive method has better computational complexity when $m < N$. The number of steps generated by the adaptive algorithm is presented in Figure 3. We observe that although the minimum step size is reduced by two orders of magnitude, the amount of adaptive time steps is increased by only a factor of approximately two. In other words, the discussed adaptive strategy leads to a logarithmic increase in the total number of time steps with respect to τ_0, that is, $m = O(|\log \tau_0|)$. This explains the much better computational times of the adaptive method when τ_0 is small enough.

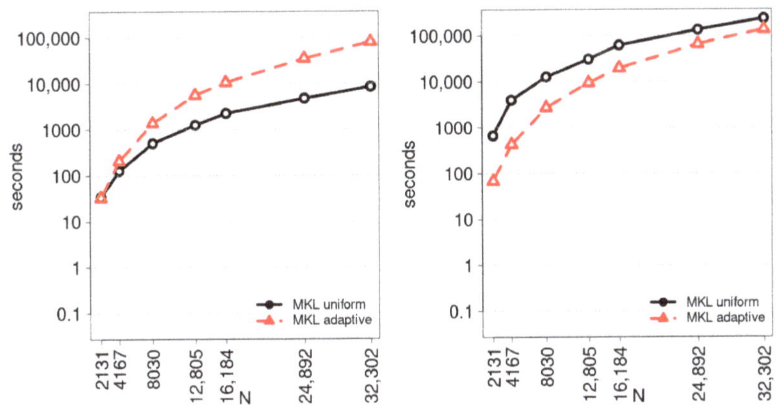

(**a**) $m_{\text{max}} = 6400$; $\tau_0 = 1.5625 \times 10^{-5}$ (**b**) $m_{\text{max}} = 204{,}800$; $\tau_0 = 4.8828125 \times 10^{-7}$

Figure 2. Sequential computational times of the uniform and adaptive schemes with LU factorization from the MKL package.

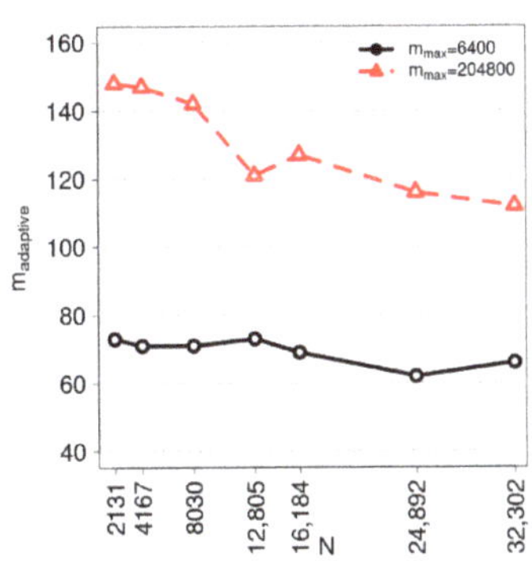

Figure 3. Number of adaptive steps used for both minimum step sizes.

Let us recall that the adaptive forward–backward Euler method developed is explicit. We do not need inner iterations to fit the error estimator. Avoiding inner iterations is an important sought-after property of the algorithm, taking into account the non-locality of the fractional diffusion and the fact that the stiffness matrix is dense.

The following Figure 4 plays a key role in the experimental proof-of-concept of the approach proposed in this paper. Here, we present a comparison of the computational times of the adaptive algorithm with LU-based and HSS-based solvers. For both variants of the minimum time step τ_0, hierarchical compression performs much better. It should be noted that there is little difference between the results with $\tau_0 = 1.5625 \times 10^{-5}$ and $\tau_0 = 4.8828125 \times 10^{-7}$. The reason for this is that most of the steps are relatively larger, except for the locally refined ones around the jump points of $f(x,t)$. Figure 5 illustrates this behaviour of the adaptive step sizes over the entire time interval. We also see the monotonically increasing time steps in the subintervals where the right-hand side does not change. This is consistent with the evolving smoothness of the solution.

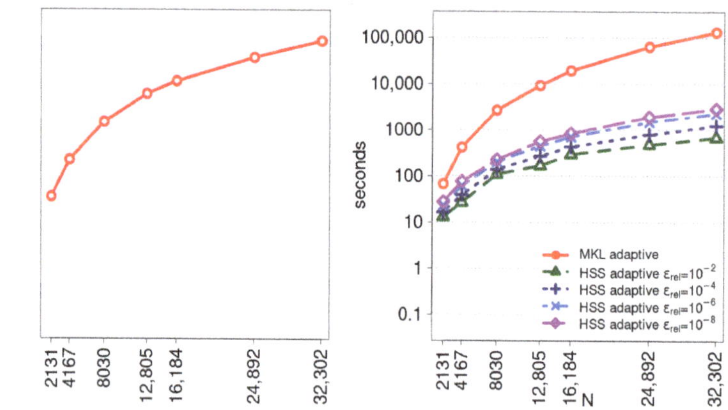

(**a**) $m_{\max} = 6400$; $\tau_0 = 1.5625 \times 10^{-5}$ (**b**) $m_{\max} = 204{,}800$; $\tau_0 = 4.8828125 \times 10^{-7}$

Figure 4. Sequential computational times of adaptive scheme with LU factorization and HSS compression.

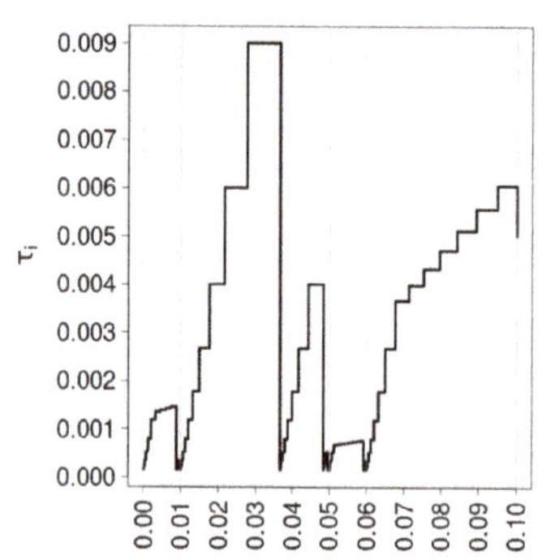

Figure 5. Adaptive time step size over the whole time interval for $N = 2131$ and $\tau_0 = 1.5625 \times 10^{-5}$ with LU solver.

7. Parallel Scalability

For large-scale problems, the efficient implementation of computing systems with parallel architecture becomes an increasingly important issue. Solving dense linear systems with block LU-based solvers is a commonly accepted standard for measuring the performance of supercomputers. This approach is used in the well-known LINPACK TOP 500 ranking [33]. And although numerical linear algebra is the backbone of computing in general, evaluating the overall performance of a real-life problem is always a more complex task than benchmarking a single routine. Ensuring a transparent comparison, we follow the steps from Section 6 here, now analyzing the parallel performance. Let us also recall that MKL and STRUMPACK were developed as parallel software packages.

In Figure 6, we present the parallel times for the backward Euler method with a uniform time step and the adaptive forward–backward Euler time stepping scheme. The block LU-based solver from MKL is used. The tests were run on a single server with 16 threads. Similar to the sequential tests, the adaptive scheme has better times for all N values at smaller τ_0; see the graph on the right-hand side. However, we also observe a relative improvement in parallel times in the case of $\tau_0 = 1.5625 \times 10^{-5}$. This is because the solution with the already factorized system has a lower parallel speed-up, which can be seen in Figure 7. With 16 threads, the adaptive method achieves up to $\sim$14 times better computation time than the sequential one, while the speed-up for the uniform time step method is only $\sim$5. Further details on the parallel speed-up of the LU factorization step and then of the solution step with the factorized matrix system can be found in [22].

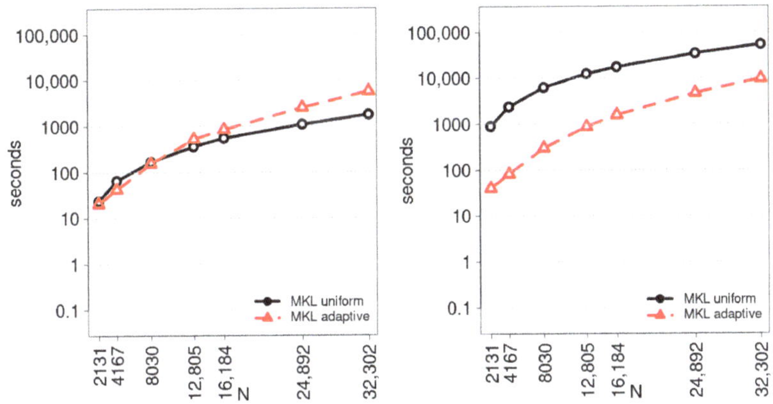

(a) $m_{\max} = 6400$; $\tau_0 = 1.5625 \times 10^{-5}$ (b) $m_{\max} = 204{,}800$; $\tau_0 = 4.8828125 \times 10^{-7}$

Figure 6. Parallel computational times of the uniform and adaptive LU solvers from MKL.

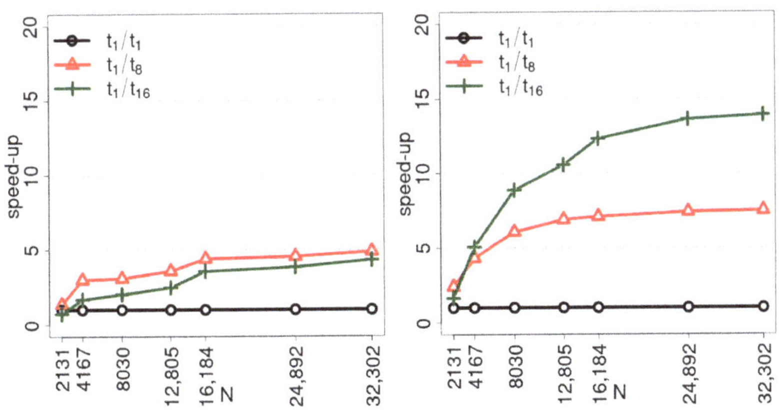

(a) Uniform with LU. (b) Adaptive with LU.

Figure 7. Parallel speed-up (parallel time divided by sequential time) of the uniform and adaptive methods with an LU factorization-based solver. The speed-up is calculated for $\tau_0 = 4.8828125 \times 10^{-7}$ experiments but is similar to the other value of τ_0.

In Figure 8, we present the parallel times with LU factorization and HSS-compression-based solvers for the adaptive method. For all experiments, HSS compression outperforms LU factorization. This means that the stiffness matrix obtained from the FEM discretization of the fractional Laplacian is suitable for hierarchical semi-separable compression. Overall, the adaptive algorithm with the HSS compression solver has a lower parallel speed-up of ∼6 (see Figure 9) compared to the adaptive method with LU factorization (see Figure 7), but its better performance is due to the substantially lower computational complexity.

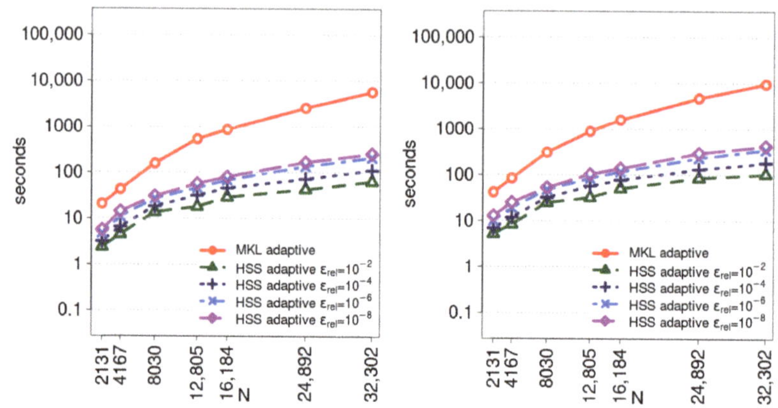

(a) $m_{\max} = 6400$; $\tau_0 = 1.5625 \times 10^{-5}$ **(b)** $m_{\max} = 204{,}800$; $\tau_0 = 4.8828125 \times 10^{-7}$

Figure 8. Parallel times of the uniform and adaptive schemes with HSS compression and ULV-like factorization from the STRUMPACK package.

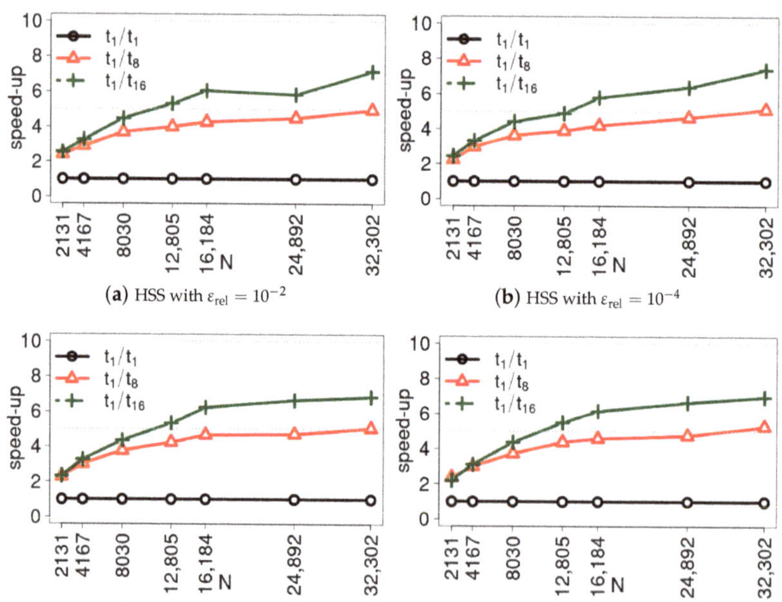

(a) HSS with $\varepsilon_{\text{rel}} = 10^{-2}$ **(b)** HSS with $\varepsilon_{\text{rel}} = 10^{-4}$

(c) HSS with $\varepsilon_{\text{rel}} = 10^{-6}$ **(d)** HSS with $\varepsilon_{\text{rel}} = 10^{-8}$

Figure 9. Parallel speed-up of the HSS compression and ULV-like factorization-based solver for the adaptive algorithm.

8. Analysis of Relative Errors of the HSS Compression Solver

The hierarchical semi-separable compressed representation of matrices is approximate. In the numerical experiments, we presented computational times and speed-up with four values of the relative threshold that is used in the compression algorithm implemented in the STRUMPACK package. In Figure 10, we show the relative errors with respect to the reference solution obtained by the uniform method with LU factorization. For the backward Euler method with a uniform time step, this relative error is of the order ε_{rel} (see [20]). Here, we show the calculated values of R_j at the time points $T/4$, $T/2$, $3T/4$, and T, recalling that $T = 0.1$. What we see is that for the adaptive method, decreasing the relative threshold after 10^{-4} does not improve the relative error. And this is exactly what we should expect from the numerical tests, where the desired accuracy parameter was set to $\delta = 0.01$.

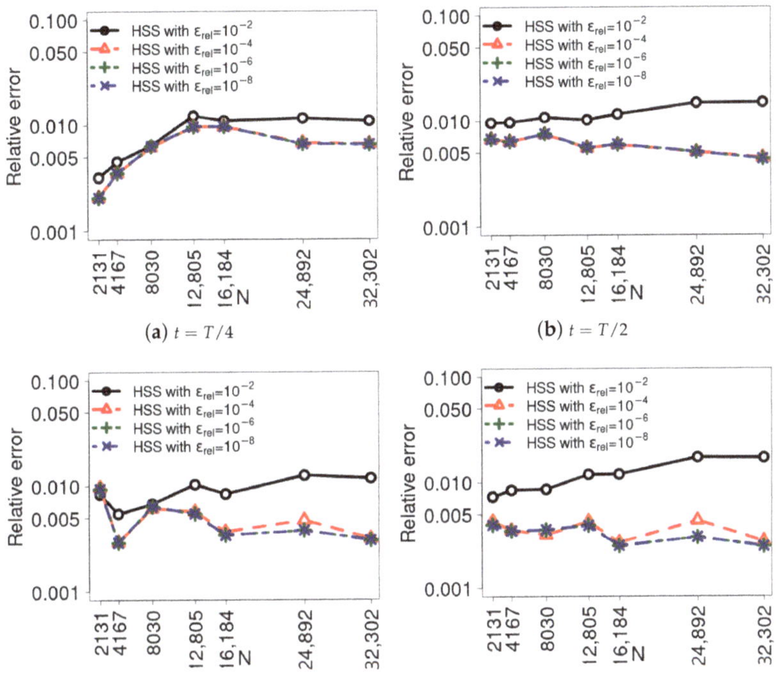

(**a**) $t = T/4$ (**b**) $t = T/2$ (**c**) $t = 3T/4$ (**d**) $t = T$

Figure 10. Relative errors of the solution with $\tau_0 = 4.8828125 \times 10^{-7}$ of the adaptive method with HSS compression in relation to the uniform method with LU factorization for $\varepsilon_{\text{rel}} = 10^{-2}, 10^{-4}, 10^{-6}, 10^{-8}$.

9. Discussion

In the case of standard (local) differential equations, there have been many years of successful work on the development of numerical methods for parabolic problems. In particular, the stiffness and mass matrices are sparse and positive definite if the FEM is applied to discretization in space. Thus, explicit methods have an optimal complexity with respect to N, with an analogous result obtained for implicit methods if fast (e.g., multigrid or multilevel) PCG iterative solvers are used. The methods and algorithms discussed here show how the challenges change substantially when moving from local to non-local problems.

Complex solutions are integrated to ensure the computational efficiency of the developed method. A successful combination of the near-optimal unconditionally stable adaptive backward–forward Euler scheme with the upgraded HSS compression-based solver for the diagonally perturbed stiffness matrices that appear in transient systems is achieved. In addition, although recursive by definition, HSS compression and its implementation in the STRUMPACK software package show promising parallel scalability within the discussed composite algorithm.

The numerical experiments show a very good agreement with the theoretical results, thus proving the concept of the applied research methodology. We note the following conclusions: the advantages of the adaptive scheme increase as the minimum time step decreases; the HSS compression-based solver outperforms the LU factorization; the overall accuracy is effectively controlled by the parameters of the adaptive scheme and the HSS compression.

10. Conclusions

In conclusion, the main contribution of the paper is determined by the development, analysis and implementation of a new highly efficient numerical method and algorithms for parabolic diffusion problems with a fractional Laplacian in space. The theoretical results, including, in particular, computational complexity estimates and scalability analyses, are clearly confirmed by the sequential and parallel numerical tests presented. The adaptive backward–forward Euler scheme optimizes computational complexity by avoiding the need for prior assumptions about the regularity of the solution. The algorithm essentially uses the fact that the time derivative is standard (local). For this reason, the concept of an adaptive time stepping will need extensive further development in the case of fractional (nonlocal) diffusion in time.

The progress made in this paper poses new challenges and creates new opportunities for future research in the development of numerical methods for time-dependent fractional diffusion problems in space. In this context, the following topics are on our short-term priority list: adaptive higher-order schemes; adaptive time steps for spectral fractional diffusion; and combining adaptive time stepping with local mesh refinement and hp-FEM in space.

The present investigations can naturally be extended to more general equations that include a fractional power of the diffusion operator (e.g., the Laplacian). This applies, for example, to the case of reaction and/or convection-type terms, where the monotonicity conditions of the schemes may be of particular interest along with stability issues.

The nonlinear case is a separate topic of great scientific value and practical importance. Here, there are very wide possibilities for implementing and upgrading the adaptive time-stepping methods and algorithms. Such an example is the time-dependent fractional in a space diffusion–reaction system of equations coupled by nonlinear reaction operators [34].

Author Contributions: Methodology, S.M.; investigation, S.M. and D.S.; writing—original draft preparation, D.S.; writing—review and editing, S.M.; visualization, D.S. All authors have read and agreed to the published version of the manuscript.

Funding: The work was partially supported by the Bulgarian National Science Fund under grant No. KP-06-N72/2. The second author was also supported by the Bulgarian Ministry of Education and Science under the National Research Programme Young scientists and postdoctoral students.

Data Availability Statement: Dataset available on request from the authors.

Acknowledgments: The access to the infrastructure of the Centre of Excellence in Informatics and ICT, Grant No BG05M2OP001-1.001-0003, financed by the Science and Education for Smart Growth Operational Program and co-financed by the European Union through the European Structural and Investment funds is acknowledged.

Conflicts of Interest: The authors declare no conflicts of interest.

References

1. Binder, K.; Bennemann, C.; Baschnagel, J.; Paul, W. Anomalous diffusion of polymers in supercooled melts near the glass transition. In *Anomalous Diffusion From Basics to Applications*; Pękalski, A., Sznajd-Weron, K., Eds.; Springer: Berlin/Heidelberg, Germany, 1999; pp. 124–139.

2. Langlands, T.; Henry, B.; Wearne, S. Fractional Cable Equation Models for Anomalous Electrodiffusion in Nerve Cells: Finite Domain Solutions. *SIAM J. Appl. Math.* **2011**, *71*, 1168–1203. [CrossRef]
3. Taitelbaum, H. Diagnosis using photon diffusion: From brain oxygenation to the fat of the atlantic salmon. In *Anomalous Diffusion from Basics to Applications*; Pękalski, A., Sznajd-Weron, K., Eds.; Springer: Berlin/Heidelberg, Germany, 1999; pp. 160–174.
4. Rosasco, L.; Belkin, M.; Vito, E.D. On Learning with Integral Operators. *J. Mach. Learn. Res.* **2010**, *11*, 905–934.
5. Chaturapruek, S.; Breslau, J.; Yazdi, D.; Kolokolnikov, T.; McCalla, S. Crime modeling with Lèvy flights. *SIAM J. Appl. Math.* **2013**, *73*, 1703–1720. [CrossRef]
6. Sun, H.; Zhang, Y.; Baleanu, D.; Chen, W.; Chen, Y. A new collection of real world applications of fractional calculus in science and engineering. *Commun. Nonlinear Sci. Numer. Simul.* **2018**, *64*, 213–231. [CrossRef]
7. Kwaśnicki, M. Ten equivalent definitions of the fractional laplace operator. *Fract. Calc. Appl. Anal.* **2017**, *20*, 7–51. [CrossRef]
8. Lischke, A.; Pang, G.; Gulian, M.; Song, F.; Glusa, C.; Zheng, X.; Mao, Z.; Cai, W.; Meerschaert, M.M.; Ainsworth, M.; et al. What is the fractional Laplacian? A comparative review with new results. *J. Comput. Phys.* **2020**, *404*, 109009. [CrossRef]
9. Harizanov, S.; Margenov, S.; Popivanov, N. Spectral Fractional Laplacian with Inhomogeneous Dirichlet Data: Questions, Problems, Solutions. In *Advanced Computing in Industrial Mathematics*; Georgiev, I., Kostadinov, H., Lilkova, E., Eds.; Springer: Cham, Switzerland, 2021; pp. 123–138. [CrossRef]
10. Bonito, A.; Lei, W.; Pasciak, J.E. The approximation of parabolic equations involving fractional powers of elliptic operators. *J. Comput. Appl. Math.* **2017**, *315*, 32–48. [CrossRef]
11. Markus Melenk, J.; Rieder, A. An exponentially convergent discretization for space–time fractional parabolic equations using hp-FEM. *IMA J. Numer. Anal.* **2022**, *43*, 2352–2376. [CrossRef]
12. Nochetto, R.H.; Otárola, E.; Salgado, A.J. A PDE Approach to Space-Time Fractional Parabolic Problems. *SIAM J. Numer. Anal.* **2016**, *54*, 848–873. [CrossRef]
13. Vabishchevich, P.N. Splitting schemes for non-stationary problems with a rational approximation for fractional powers of the operator. *Appl. Numer. Math.* **2021**, *165*, 414–430. [CrossRef]
14. Vabishchevich, P.N. Numerical Solution of Non-stationary Problems for a Space-Fractional Diffusion Equation. *Fract. Calc. Appl. Anal.* **2016**, *19*, 116–139. [CrossRef]
15. Čiegis, R.; Starikovičius, V.; Suboč, O.; Čiegis, R. On Construction of Partially Dimension-Reduced Approximations for Nonstationary Nonlocal Problems of a Parabolic Type. *Mathematics* **2023**, *11*, 1984. [CrossRef]
16. Danczul, T.; Hofreither, C.; Schöberl, J. A unified rational Krylov method for elliptic and parabolic fractional diffusion problems. *Numer. Linear Algebra Appl.* **2023**, *30*, e2488. [CrossRef]
17. Khristenko, U.; Wohlmuth, B. Solving time-fractional differential equations via rational approximation. *IMA J. Numer. Anal.* **2022**, *43*, 1263–1290. [CrossRef]
18. Yang, Y.; Huang, J. Double fast algorithm for solving time-space fractional diffusion problems with spectral fractional Laplacian. *Appl. Math. Comput.* **2024**, *475*, 128715. [CrossRef]
19. Acosta, G.; Bersetche, F.M.; Borthagaray, J.P. Finite Element Approximations for Fractional Evolution Problems. *Fract. Calc. Appl. Anal.* **2019**, *22*, 767–794. [CrossRef]
20. Slavchev, D.; Margenov, S. Performance Study of Hierarchical Semi-separable Compression Solver for Parabolic Problems with Space-Fractional Diffusion. In *Large-Scale Scientific Computing*; Lirkov, I., Margenov, S., Eds.; Springer: Cham, Switzerland, 2022; pp. 71–80.
21. Acosta, G.; Borthagaray, J. A Fractional Laplace Equation: Regularity of Solutions and Finite Element Approximations. *SIAM J. Numer. Anal.* **2017**, *55*, 472–495. [CrossRef]
22. Slavchev, D.; Margenov, S. On the Application of a Hierarchically Semi-separable Compression for Space-Fractional Parabolic Problems with Varying Time Steps. In *Numerical Methods and Applications*; Georgiev, I., Datcheva, M., Georgiev, K., Nikolov, G., Eds.; Springer: Cham, Switzerland, 2023; pp. 289–301.
23. Blaheta, R.; Byczanski, P.; Kohut, R.; Starý, J. Algorithms for Parallel Fem Modelling of Thermo-Mechanical Phenomena Arising from the Disposal of the Spent Nuclear Fuel. In *Elsevier Geo-Engineering Book Series*; Elsevier: Amsterdam, The Netherlands, 2004; Volume 2, pp. 395–400.
24. Georgiev, K.; Kosturski, N.; Margenov, S.; Starý, J. On adaptive time stepping for large-scale parabolic problems: Computer simulation of heat and mass transfer in vacuum freeze-drying. *J. Comput. Appl. Math.* **2009**, *226*, 268–274. [CrossRef]
25. Vabishchevich, P.N.; Vasil'ev, A.O. Time step selection for the numerical solution of boundary value problems for parabolic equations. *Comput. Math. Math. Phys.* **2017**, *57*, 843–853. [CrossRef]
26. Hackbusch, W. A Sparse Matrix Arithmetic Based on $\mathcal{H}$-Matrices. Part I: Introduction to $\mathcal{H}$-Matrices. *Computing* **1999**, *62*, 89–108. [CrossRef]
27. Hackbusch, W.; Grasedyck, L.; Börm, S. An introduction to hierarchical matrices. *Math. Bohem.* **2002**, *127*, 229–241. [CrossRef]
28. Martinsson, P.G. A Fast Randomized Algorithm for Computing a Hierarchically Semiseparable Representation of a Matrix. *SIAM J. Matrix Anal. Appl.* **2011**, *32*, 1251–1274. [CrossRef]
29. Xia, J.; Chandrasekaran, S.; Gu, M.; Li, X.S. Fast algorithms for hierarchically semiseparable matrices. *Numer. Lin. Alg. Appl.* **2010**, *17*, 953–976. [CrossRef]
30. Rouet, F.H.; Li, X.; Ghysels, P.; Napov, A. A Distributed-Memory Package for Dense Hierarchically Semi-Separable Matrix Computations Using Randomization. *ACM Trans. Math. Softw.* **2016**, *42*, 1–35. [CrossRef]

31. Slavchev, D.; Margenov, S.; Georgiev, I. On the application of recursive bisection and nested dissection reorderings for solving fractional diffusion problems using HSS compression. *AIP Conf. Proc.* **2020**, *2302*, 120008. [CrossRef]
32. Rebrova, E.; Chávez, G.; Liu, Y.; Ghysels, P.; Li, X.S. A Study of Clustering Techniques and Hierarchical Matrix Formats for Kernel Ridge Regression. In Proceedings of the 2018 IEEE International Parallel and Distributed Processing Symposium Workshops (IPDPSW), Vancouver, BC, Canada, 21–25 May 2018; pp. 883–892. [CrossRef]
33. Dongarra, J.J.; Luszczek, P.; Petitet, A. The LINPACK Benchmark: Past, present and future. *Concurr. Comput. Pract. Exp.* **2003**, *15*, 803–820. [CrossRef]
34. Georgiev, K.; Margenov, S. Numerical Methods for Fractional Diffusion-Reaction Problems Based on Operator Splitting and BURA. In *Advanced Computing in Industrial Mathematics*; Georgiev, I., Kostadinov, H., Lilkova, E., Eds.; Springer: Cham, Switzerland, 2023; pp. 73–84.

algorithms

MDPI

Article

The Nelder–Mead Simplex Algorithm Is Sixty Years Old: New Convergence Results and Open Questions

Aurél Galántai

Software Engineering Institute, John von Neumann Faculty of Informatics, Óbuda University, 1034 Budapest, Hungary; galantai.aurel@nik.uni-obuda.hu

Abstract: We investigate and compare two versions of the Nelder–Mead simplex algorithm for function minimization. Two types of convergence are studied: the convergence of function values at the simplex vertices and convergence of the simplex sequence. For the first type of convergence, we generalize the main result of Lagarias, Reeds, Wright and Wright (1998). For the second type of convergence, we also improve recent results which indicate that the Lagarias et al.'s version of the Nelder–Mead algorithm has better convergence properties than the original Nelder–Mead method. This paper concludes with some open questions.

Keywords: Nelder–Mead simplex methods; convergence; comparison

MSC: 65K10; 90C56

1. Introduction

The original Nelder–Mead (NM) simplex algorithm was constructed in 1965 by Nelder and Mead [1] as an improved version of the simplex method of Spendley, Hext and Himsworth [2]. During the almost six decades that have passed, the Nelder–Mead algorithm has became a truly popular and widely used method for the minimization problem

$$f(x) \to \min \quad (f : \mathbb{R}^n \to \mathbb{R})$$

in derivative-free optimization (see, e.g., [3–11]).

In spite of its widespread use, only a few convergence results are known. They are of essentially two types. The first type of results are related to the convergence of function values at the vertices of the simplex sequence. They say nothing about the vertices. The most famous results of this type are due to Lagarias, Reeds, Wright and Wright [12]. The other type of results are related to the convergence of simplex vertices or simplices. The first result of this type is due to Kelley [13], who proved that under a certain (Armijo-type) descent condition, the accumulation points of the simplex sequence are critical points of f. Recent results on the convergence of the simplex sequence to a common limit point are given in papers [14–16].

Wright [17,18] raised several questions concerning the Nelder–Mead method:

(a) Do the function values at all vertices necessarily converge to the same value?
(b) Do all vertices of the simplices converge to the same point?
(c) Why is it sometimes so effective (compared to other direct search methods) in obtaining a rapid improvement in f?
(d) One failure mode is known (McKinnon [19])—but are there other failure modes?
(e) Why, despite its apparent simplicity, should the Nelder–Mead method be difficult to analyze mathematically?

In fact, there are several recent examples that indicate different types of convergence behavior:

- The function values at the simplex vertices may converge to a common limit value, while the function f has no finite minimum and the simplex sequence is unbounded (Example 5 of [14]; Examples 1 and 2 of [15]; Example 4 of [20]).
- The simplex vertices may converge to a common limit point, but it is not a stationary point of f (McKinnon [19] and Example 3 of [15]).
- The simplex sequence may converge to a limit simplex of a positive diameter, resulting in different limit values of f at the vertices at the limit simplex (Theorem 1 of [21]; Example 4 of [14]; Examples 4 and 5 of [15]; Examples 1, 2 and 3 of [20]).
- The function values at the simplex vertices may converge to a common value, while the simplex sequence converges to a limit simplex with a positive diameter (Example of 5 [20]).

These examples are answering questions (a), (b) and (d) of Wright negatively.

There are several variants of the original Nelder–Mead method [1]. Here, we study two main versions of the Nelder–Mead algorithm: the 'original' (unordered) method of Nelder and Mead [1] (Algorithm 1) and the 'ordered' version of Lagarias, Reeds, Wright and Wright [12] (Algorithm 2). In Section 2, we introduce both algorithms, while Section 3 contains their matrix representations. In Section 4, we study the connection and spectral properties of these matrices. In Section 5, we prove convergence of the function values at the simplex vertices for both algorithms. These results are more general than those of Lagarias et al. [12], while their proof is quite simple.

In Section 6, we study the convergence of the generated simplex sequence. Here, we give necessary and sufficient conditions for the convergence of the simplex sequence (for both algorithms). This relates the convergence to the convergence of an infinite matrix product. We pay particular attention to the case when the simplex sequence converges to a common limit point. The results show why the convergence problem (questions (c) and (e) of Wright) is difficult. This paper closes with computational results, conclusions and open questions.

2. Two Versions of the Nelder–Mead Method

We first define the original Nelder–Mead simplex method in the following form (see [1,6,22]).

The simplex of iteration k is denoted by $S^{(k)} = \left[x_1^{(k)}, \ldots, x_{n+1}^{(k)}\right] \in \mathbb{R}^{n\times(n+1)}$, where $x_i^{(k)}$ denotes vertex i of iteration k. The function values at the vertices are denoted by $f_i^{(k)} = f\left(x_i^{(k)}\right)$ $(i = 1, \ldots, n+1)$. We define the index of the worst vertex (h_k), the best vertex (ℓ_k) and the second worst vertex (m_k) as follows:

$$h_k : f_{h_k}^{(k)} = \max_{1\le i\le n+1} f\left(x_i^{(k)}\right), \quad \ell_k : f_{\ell_k}^{(k)} = \min_{1\le i\le n+1} f\left(x_i^{(k)}\right),$$

$$m_k : f_{m_k}^{(k)} = \max_{1\le i\le n+1,\ i\ne h_k} f\left(x_i^{(k)}\right).$$

Note that $1 \le \ell_k, m_k, h_k \le n+1$, and they are not necessarily unique. Define the center $x_c^{(k)}$ and straight line $x^{(k)}(\alpha)$ as follows:

$$x_c^{(k)} = \frac{1}{n} \sum_{i=1,\ i\ne h_k}^{n+1} x_i^{(k)}, \quad x^{(k)}(\alpha) = (1+\alpha)x_c^{(k)} - \alpha x_{h_k}^{(k)},$$

The reflection, expansion, outside and inside contraction points of simplex $S^{(k)}$ are defined as

$$x_r^{(k)} = x^{(k)}(1), \quad x_e^{(k)} = x^{(k)}(2), \quad x_{oc}^{(k)} = x^{(k)}\left(\frac{1}{2}\right), \quad x_{ic}^{(k)} = x^{(k)}\left(-\frac{1}{2}\right).$$

The corresponding function values are denoted by $f_r^{(k)} = f\left(x_r^{(k)}\right)$, $f_e^{(k)} = f\left(x_e^{(k)}\right)$, $f_{oc}^{(k)} = f\left(x_{oc}^{(k)}\right)$ and $f_{ic}^{(k)} = f\left(x_{ic}^{(k)}\right)$, respectively.

The original Nelder–Mead simplex method is given by Algorithm 1.

Algorithm 1: Original Nelder–Mead algorithm.

```
Input: Set S^(0) ∈ R^{n×(n+1)} and k = 0.
while k ≥ 0, do
    Determine x_{h_k}^(k), x_{ℓ_k}^(k), x_{m_k}^(k), f_{h_k}^(k), f_{ℓ_k}^(k) and f_{m_k}^(k);      /* 1 */
    if f_{ℓ_k}^(k) ≤ f_r^(k) < f_{m_k}^(k) then                                  /* Reflect */
        x_{h_k}^(k) ← x_r^(k), k = k + 1 and go to 1.
    if f_r^(k) < f_{ℓ_k}^(k) and f_e^(k) < f_r^(k) then                          /* Expand */
        x_{h_k}^(k) ← x_e^(k), k = k + 1 and go to 1. ;
    else if f_r^(k) < f_{ℓ_k}^(k) and f_e^(k) ≥ f_r^(k) then                     /* Expand */
        x_{h_k}^(k) ← x_r^(k), k = k + 1 and go to 1.
    if f_{m_k}^(k) ≤ f_r^(k) < f_{h_k}^(k) and f_oc^(k) ≤ f_r^(k) then            /* Contract outside */
        x_{h_k}^(k) ← x_oc^(k), k = k + 1 and go to 1.
    if f_r^(k) ≥ f_{h_k}^(k) and f_ic^(k) < f_{h_k}^(k) then                      /* Contract inside */
        x_{h_k}^(k) ← x_ic^(k), k = k + 1 and go to 1.
    Evaluate f at the points                                                   /* Shrink */
        x_i^(k+1) := (x_i^(k) + x_{ℓ_k}^(k))/2, i = 1, 2, ..., n + 1,
    k = k + 1;
```

The "decision boundaries" are the same as in the ordered Nelder–Mead method of Lagarias et al. [12], which is given in Algorithm 2. Suppose that

$$f\left(x_1^{(0)}\right) \leq f\left(x_2^{(0)}\right) \leq \cdots \leq f\left(x_{n+1}^{(0)}\right) \tag{1}$$

holds for the initial simplex vertices and that this ordering is kept during the subsequent iterations. Here, $\ell_k = 1$, $m_k = n$ and $h_k = n+1$ for all k.

There are two rules of reindexing after each iteration. For a nonshrink step, $x_{n+1}^{(k)}$ is replaced by a new point $v \in \left\{x_r^{(k)}, x_e^{(k)}, x_{oc}^{(k)}, x_{ic}^{(k)}\right\}$. The following cases are possible: $f(v) < f_1^{(k)}$, $f_1^{(k)} \leq f(v) < f_n^{(k)}$ and $f_n^{(k)} \leq f(v) < f_{n+1}^{(k)}$. Define

$$j^{(k)} = \begin{cases} 1, & \text{if } f(v) < f_1^{(k)} \\ \left\{j : f_{j-1}^{(k)} \leq f(v) < f_j^{(k)}, 2 \leq j \leq n+1\right\}, & \text{otherwise} \end{cases}. \tag{2}$$

Then, the new simplex vertices are

$$x_i^{(k+1)} = \begin{cases} x_i^{(k)} & \left(1 \leq i \leq j^{(k)} - 1\right), \\ v & \left(i = j^{(k)}\right), \\ x_{i-1}^{(k)} & \left(i = j^{(k)} + 1, \ldots, n+1\right). \end{cases} \tag{3}$$

This 'insertion rule' puts v into the ordering with the highest possible index, guaranteeing that $f_i^{(k+1)} \le f_i^{(k)}$ $(i = 1, \ldots, n+1)$ and

$$f_1^{(k+1)} \le f_2^{(k+1)} \le \cdots \le f_{n+1}^{(k+1)} \quad (k \ge 0). \tag{4}$$

If shrinking occurs, then the new vertices must be ordered such that $x_1^{(k+1)} = x_1^{(k)}$ if $f(z_1) \le f(z_i)$ $(i = 2, \ldots, n+1)$.

For a given $S^{(0)}$ and f, the sequence $\left\{S^{(k)}\right\}$ is uniquely defined by Algorithm 2. Clearly, this is not true for Algorithm 1.

Algorithm 2: Ordered Nelder–Mead algorithm.

Input: Set $S^{(0)} \in \mathbb{R}^{n\times(n+1)}$ and $k = 0$.
while $k \ge 0$, **do**
  Order the vertices of $S^{(k)}$ so that $f_1^{(k)} \le f_2^{(k)} \le \cdots \le f_{n+1}^{(k)}$; /* Order */
  if $f_1^{(k)} \le f_r^{(k)} < f_n^{(k)}$ **then** /* Reflect */
    set $S^{(k+1)} = \left[x_1^{(k)}, x_2^{(k)}, \ldots, x_n^{(k)}, x_r^{(k)}\right]$;
  if $f_r^{(k)} < f_1^{(k)}$ *and* $f_e^{(k)} < f_r^{(k)}$ **then** /* Expand */
    set $S^{(k+1)} = \left[x_1^{(k)}, x_2^{(k)}, \ldots, x_n^{(k)}, x_e^{(k)}\right]$;
  else if $f_r^{(k)} < f_1^{(k)}$ *and* $f_r^{(k)} \le f_e^{(k)}$ **then** /* Expand */
    set $S^{(k+1)} = \left[x_1^{(k)}, x_2^{(k)}, \ldots, x_n^{(k)}, x_r^{(k)}\right]$;
  if $f_n^{(k)} \le f_r^{(k)} < f_{n+1}^{(k)}$ *and* $f_{oc}^{(k)} \le f_r^{(k)}$ **then** /* Contract outside */
    set $S^{(k+1)} = \left[x_1^{(k)}, x_2^{(k)}, \ldots, x_n^{(k)}, x_{oc}^{(k)}\right]$;
  if $f_r^{(k)} \ge f_{n+1}^{(k)}$ *and* $f_{ic}^{(k)} < f_{n+1}^{(k)}$ **then** /* Contract inside */
    set $S^{(k+1)} = \left[x_1^{(k)}, x_2^{(k)}, \ldots, x_n^{(k)}, x_{ic}^{(k)}\right]$;
  if $f_n^{(k)} \le f_r^{(k)} < f_{n+1}^{(k)} \wedge f_{oc}^{(k)} > f_r^{(k)}$ *or* $\min(f_r^{(k)}, f_{ic}^{(k)}) \ge f_{n+1}^{(k)}$ **then** /* Shrink */
    evaluate f at the points $z_i = \frac{1}{2}\left(x_1^{(k)} + x_i^{(k)}\right), i = 1, 2, \ldots, n+1,$
    and set $S^{(k+1)} = [z_1, z_2, \ldots, z_n, z_{n+1}]$;
  $k = k + 1$;

3. Matrix Representations for the Two Algorithms

The steps of both algorithms can be easily described by transformation matrices. We first consider Algorithm 1. If the incoming vertex $v = x^{(k)}(\alpha_k)$ $(\alpha_k \in \left\{\pm\frac{1}{2}, 1, 2\right\})$ belongs to $\left\{x_r^{(k)}, x_e^{(k)}, x_{oc}^{(k)}, x_{ic}^{(k)}\right\}$, then

$$S^{(k+1)} = \left[x_1^{(k)}, \ldots, x_{h_k-1}^{(k)}, v, x_{h_k+1}^{(k)}, \ldots, x_{n+1}^{(k)}\right],$$

where v replaces vertex $x_{h_k}^{(k)}$. For $j = 1, \ldots, n+1$, define the vector

$$c_j(\alpha) = [\frac{1+\alpha}{n}, \ldots, \frac{1+\alpha}{n}, -\alpha, \frac{1+\alpha}{n}, \ldots, \frac{1+\alpha}{n}]^T \in \mathbb{R}^{n+1} \tag{5}$$

with $-\alpha$ at the position j and the matrix

$$T_j(\alpha) = I + (c_j(\alpha) - e_j)e_j^T, \tag{6}$$

where e_j is the j-th unit vector. Then, $v = S^{(k)} c_{h_k}(\alpha_k)$ and

$$S^{(k+1)} = S^{(k)} T_{h_k}(\alpha_k) \quad \left(\alpha_k \in \left\{\pm\frac{1}{2}, 1, 2\right\}\right). \tag{7}$$

The shrinking iteration can be written as

$$S^{(k+1)} = S^{(k)} T_{\ell_k}^{shr} \quad \left(T_{\ell_k}^{shr} = \frac{1}{2} I + \frac{1}{2} e_{\ell_k} e^T\right), \tag{8}$$

where $e = [1, 1, \ldots, 1]^T$. It follows that

$$S^{(k+1)} = S^{(k)} T^{(k)} = S^{(0)} \prod_{i=0}^{k} T^{(i)} = S^{(0)} B_k \quad \left(T^{(i)} \in \widetilde{\mathcal{T}}\right), \tag{9}$$

where

$$\widetilde{\mathcal{T}} = \left\{T_\ell(\alpha) : 1 \le \ell \le n+1,\ \alpha = \pm\frac{1}{2}, 1, 2\right\} \cup \left\{T_j^{shr} : 1 \le j \le n+1\right\}. \tag{10}$$

The set $\widetilde{\mathcal{T}}$ has exactly $5(n+1)$ nonsingular matrices. Note that $e^T T_j(\alpha) = e^T$ and $e^T T_{\ell_k}^{shr} = e^T$.

For Algorithm 2, we have the following (see, e.g., [14,15]). Assume that simplex $S^{(k)}$ satisfies condition (4). Then, for nonshrinking iterations, the new simplex is given by

$$S^{(k+1)} = S^{(k)} T(\alpha) P_{j(k)} \quad (\alpha \in \left\{\pm\frac{1}{2}, 1, 2\right\}), \tag{11}$$

where

$$T(\alpha) = \begin{bmatrix} I_n & \frac{1+\alpha}{n} e \\ 0 & -\alpha \end{bmatrix} \in \mathbb{R}^{(n+1)\times(n+1)} \tag{12}$$

and

$$P_j = \left[e_1, \ldots, e_{j-1}, e_{n+1}, e_j, \ldots, e_n\right] \in \mathbb{R}^{(n+1)\times(n+1)} \quad (j = 1, \ldots, n+1) \tag{13}$$

is a permutation matrix. If shrinking occurs, the new simplex is

$$S^{(k+1)} = S^{(k)} T_{shr} P \quad \left(T_{shr} = \frac{1}{2} I_{n+1} + \frac{1}{2} e_1 e^T\right),$$

where the permutation matrix $P \in \mathcal{P}_{n+1}$ is defined by the ordering condition (4) and $\mathcal{P}_{n+1}$ is the set of all possible permutation matrices of order $n+1$.

Define the sets

$$\mathcal{T}_1 = \left\{T(\alpha) P_j : \alpha \in \left\{-\frac{1}{2}, \frac{1}{2}\right\},\ j = 1, \ldots, n+1\right\},$$

$$\mathcal{T}_2 = \{T(1) P_j : j = 1, \ldots, n\} \cup \{T(2) P_1\}, \quad \mathcal{T}_3 = \{T_{shr} P : P \in \mathcal{P}_{n+1}\}$$

and

$$\mathcal{T} = \mathcal{T}_1 \cup \mathcal{T}_2 \cup \mathcal{T}_3. \tag{14}$$

It follows that

$$S^{(k+1)} = S^{(k)} T_k P^{(k)} = S^{(0)} \prod_{i=0}^{k} T_i P^{(i)} = S^{(0)} B_k \quad \left(T_i P^{(i)} \in \mathcal{T}, k \ge 1\right), \tag{15}$$

where

$$B_k = \prod_{i=0}^{k} T_i P^{(i)} \quad \left(T_i P^{(i)} \in \mathcal{T}\right). \tag{16}$$

The set $\mathcal{T}$ consists of $3n+3+(n+1)!$ nonsingular matrices. For any $T_i P^{(i)} \in \mathcal{T}$, $e^T T_i P^{(i)} = e^T$.

Observe that both algorithms are nonstationary iterations. Note that the number of shrinking matrices for Algorithm 2 increased to $(n+1)!$ compared to the $n+1$ shrinking matrices of Algorithm 1.

4. Connections and Properties

Here, we investigate the connection between the transformation matrices of both algorithms, their consequences and the spectral properties of the matrices.

The corresponding transformation matrices of Algorithms 1 and 2 are connected by orthogonal similarities. For any $1 \leq s \leq n+1$,

$$P^T T_s(\alpha) P = T_{n+1}(\alpha) = T(\alpha) \tag{17}$$

with $P = [e_1, \ldots, e_{s-1}, e_{s+1}, \ldots, e_{n+1}, e_s]$ and

$$Q^T T_s^{shr} Q = \frac{1}{2} I + \frac{1}{2} e_1 e^T = T_{shr} \tag{18}$$

with $Q = [e_s, e_1, \ldots, e_{s-1}, e_{s+1}, \ldots, e_{n+1}]$.

With $\widetilde{S}^{(k)} = \left[\widetilde{x}_1^{(k)}, \ldots, \widetilde{x}_{n+1}^{(k)}\right]$ and $S^{(k)} = \left[x_1^{(k)}, \ldots, x_{n+1}^{(k)}\right]$, denote the kth simplices of Algorithms 1 and 2, respectively. If a permutation matrix $Q_k = \left[e_{r_1}, \ldots, e_{r_{n+1}}\right] \in \mathbb{R}^{(n+1)\times(n+1)}$ exists such that $\widetilde{S}^{(k)} Q_k = S^{(k)}$, $x_1^{(k)} = \widetilde{x}_{\ell_k}^{(k)}$ and $x_{n+1}^{(k)} = \widetilde{x}_{h_k}^{(k)}$, then we can ask if $\widetilde{S}^{(k+1)}$ and $S^{(k+1)}$ define the same simplex. The assumption implies that $e_{r_1} = e_{\ell_k}$ and $e_{r_{n+1}} = e_{h_k}$. Hence, $\widetilde{S}^{(k)} = S^{(k)} Q_k^T$ and

$$\widetilde{S}^{(k+1)} = \widetilde{S}^{(k)} T_{h_k}(\alpha_k) = S^{(k)}\left(Q_k^T + \left(Q_k^T c_{h_k}(\alpha_k) - Q_k^T e_{h_k}\right) e_{h_k}^T\right).$$

Since $e_{h_k} = Q_k e_{n+1}$, $Q_k^T e_{h_k} = e_{n+1}$ and $Q_k^T c_{h_k}(\alpha_k) = \left[e_{r_1}, \ldots, e_{r_{n+1}}\right]^T c_{h_k}(\alpha_k) = c_{n+1}(\alpha_k)$, we have

$$\begin{aligned}\widetilde{S}^{(k+1)} &= S^{(k)}\left(I + (c_{n+1}(\alpha_k) - e_{n+1}) e_{n+1}^T\right) Q_k^T = S^{(k)} T(\alpha_k) Q_k^T \\ &= S^{(k+1)} P_{j(k)}^T Q_k^T.\end{aligned}$$

In the case of shrinking,

$$\widetilde{S}^{(k+1)} = \widetilde{S}^{(k)} T_{\ell_k}^{shr} = S^{(k)}\left(\frac{1}{2} Q_k^T + \frac{1}{2} Q_k^T e_{\ell_k} e^T\right).$$

Note that $Q_k^T e_{\ell_k} = Q_k^T Q_k e_1 = e_1$. Hence,

$$\widetilde{S}^{(k+1)} = S^{(k)}\left(\frac{1}{2} Q_k^T + \frac{1}{2} e_1 e^T\right) = S^{(k)}\left(\frac{1}{2} I + \frac{1}{2} e_1 e^T Q_k\right) Q_k^T.$$

Since $e^T Q_k = e^T$, we have that

$$\widetilde{S}^{(k+1)} = S^{(k)} T_{shr} Q_k^T = S^{(k+1)} \left[P^{(k)}\right]^T Q_k^T.$$

Hence, the simplices of iteration $k+1$ are the same for the ordering of vertices. Since the continuation of the original Nelder–Mead method is not uniquely determined, unlike in the case of Algorithm 2, we can expect similar although not identical results for both algorithms.

The following results are obvious. For any $1 \le j \le n+1$,

$$T_j(\alpha)T_j(\beta) = T_j(-\alpha\beta).$$

Lemma 1. *The eigenvalues of $T_j(\alpha)$ are $\lambda_i(T_j(\alpha)) = 1$ $(i = 1, \ldots, n)$ and $\lambda_{n+1}(T_j(\alpha)) = -\alpha$ $(j = 1, \ldots, n+1)$. T_j^{shr} has the eigenvalues $\lambda_1\left(T_j^{shr}\right) = 1$ and $\lambda_i\left(T_j^{shr}\right) = \frac{1}{2}$ $(i = 2, \ldots, n+1)$ for $j = 1, \ldots, n+1$. Both $T_j(\alpha)$ and T_j^{shr} have a diagonal Jordan form.*

Corollary 1. *Sequences $\left\{T_j(\alpha)^k\right\}_{k=1}^{\infty}$ and $\left\{\left(T_j^{shr}\right)^k\right\}_{k=1}^{\infty}$ are uniformly bounded for $|\alpha| \le 1$ and convergent for $-1 \le \alpha < 1$ (see, e.g., [23–25]).*

Note that $\|T_j(\alpha)\|_2 = \|T(\alpha)\|_2 = \|T(\alpha)P_j\|_2 > 1$ and $\left\|T_j^{shr}\right\|_2 = \|T_{shr}\|_2 = \|T_{shr}P\|_2 >$ 1,
where

$$\|T(\alpha)\|_2 = \left(s_n + \sqrt{s_n^2 - \alpha^2}\right)^{\frac{1}{2}} \quad \left(s_n = \frac{1+\alpha^2}{2} + \frac{(1+\alpha)^2}{2n}\right) \tag{19}$$

and

$$\|T_{shr}\|_2 = \left(\frac{n+5}{8} + \sqrt{\frac{(n+5)^2}{64} - \frac{1}{4}}\right)^{\frac{1}{2}}. \tag{20}$$

The first expression follows from a result on the singular values of companion matrices (see, e.g., Barnett [26] or Kittaneh [27]), while the second follows from a result of Montano, Salas and Soto [28].

Matrices T_{shr}, $T_{shr}P$, T_j^{shr}, $T_j(\alpha)$, $T(\alpha)P_j$ $(-1 < \alpha < 0)$ are left-stochastic. Hence,

$$\|T_{shr}\|_1 = \|T_{shr}P\|_1 = \left\|T_j^{shr}\right\|_1 = \|T_j(\alpha)\|_1 = \|T(\alpha)P_j\|_1 = 1 \quad (-1 < \alpha < 0).$$

We note that $T_j(1)$ $(1 \le j \le n+1)$ is an involution. If $T(1)$ is multiplied by a permutation matrix P_j $(1 \le j \le n)$, this property is lost. However, for $n = 2$, $T(1)P_2$ is a 6-involutory matrix, which is exploited in proving Theorem 3 (for k-involutory matrices, see Trench [29]).

The multiplication of $T(\alpha)$ or T_{shr} by a permutation matrix P_j or P makes a significant change in the eigenstructure. The following results hold (see Theorem 3 and Corollary 1 of Galántai [15]).

Theorem 1. *(i) $T(\alpha)P_1$ has at least one eigenvalue $\lambda = 1$.*
(ii) $T(1)P_j$ $(1 \le j \le n+1)$ has at least two eigenvalues $\lambda = 1$.
(iii) If $\alpha > -1$ and $j \ge 2$, then $T(\alpha)P_j$ has exactly $j-1$ eigenvalues $\lambda = 1$.
(iv) For $|\alpha| < 1$, $T(\alpha)P_1$ has exactly one eigenvalue $\lambda_1 = 1$, while the remaining n eigenvalues are in the open unit disk.
(v) If $|\alpha| < 1$ and $2 \le j \le n+1$, then $T(\alpha)P_j$ has exactly $j-1$ eigenvalues $\lambda = 1$, while $n+2-j$ eigenvalues are in the open unit disk.
(vi) If $\alpha = 1$ and $1 \le j \le n+1$, then all the eigenvalues of $T(1)P_j$ are on the unit circle.
(vii) If $\alpha > 1$ and $j = 1$, then $T(\alpha)P_1$ has a second eigenvalue in the interval $\left(1, 1 + \frac{1+\alpha}{n}\right)$ and all its eigenvalues are in the annulus $1 \le |\lambda| \le 1 + \alpha$.

Corollary 2. *The eigenvalues of $T_{shr}P$ $(P \in \mathcal{P}_{n+1})$ are $\lambda_1(T_{shr}P) = 1$ and $|\lambda_i(T_{shr}P)| = \frac{1}{2}$ for $i = 2, \ldots, n+1$.*

The change in the eigenstructure of $T(\alpha)P_j$ is the most striking in contrast to that of $T_j(\alpha)$. It gives better convergence properties for Algorithm 2, as shown in Sections 6 and 7.

5. Convergence of the Function Values at the Vertices

Lagarias et al. [12] studied the convergence of function values at the simplex vertices for Algorithm 2. We cite their two most important results, before proving more general results for both Algorithms.

Lemma 2 (Lemma 3.3 of [12]). *If function f is bounded from below on $\mathbb{R}^n$ and only a finite number of shrink iterations occur, then each sequence $\left\{f_i^{(k)}\right\}_{k=0}^{\infty}$ (generated by Algorithm 2) converges to some limit f_i^* for $i = 0,1,\ldots,n+1$ and $f_1^* \leq f_2^* \leq \cdots \leq f_{n+1}^*$.*

Lemma 2 guarantees the convergence under rather general conditions. In general, the limit values $\{f_i^*\}_{i=1}^{n+1}$ can be different as shown by Examples 1 and 2 of [20] (also see [15,21]).

Theorem 2 (Theorem 5.1 of [12]). *Assume that f is a strictly convex function on $\mathbb{R}^2$ with bounded level sets. Assume that $S^{(0)}$ is non-degenerate. Then, the Nelder–Mead simplex method (Algorithm 2) converges in the sense that $f_1^* = f_2^* = f_3^* = f^*$.*

Theorem 2 is generally considered as the main result of [12]. In addition, it can be proved (see [20]) that if f is strictly convex on $\mathbb{R}^2$ with bounded level sets, $S^{(0)}$ is non-degenerate (affine invariant); thus, any accumulation point S' of the simplex sequence $\left\{S^{(k)}\right\}$ (generated by Algorithm 2) has the form $S' = [x', x', x']$.

We need the following concept and results.

Definition 1. *Let f be defined on a convex set $S \subset \mathbb{R}^n$. The function f is said to be quasiconvex on S if*

$$f(\lambda x_1 + (1-\lambda)x_2) \leq \max\{f(x_1), f(x_2)\}$$

for every $x_1, x_2 \in S$ and for every $\lambda \in [0,1]$.

Convexity implies quasiconvexity.

Lemma 3 (Lemma 6 of [20]). *If f is quasiconvex and bounded below on $\mathbb{R}^n$, then each sequence $\left\{f_i^{(k)}\right\}_{k=0}^{\infty}$ is monotone decreasing and converges to some limit f_i^* for $i = 0,1,\ldots,n+1$. Furthermore, $f_1^* \leq f_2^* \leq \cdots \leq f_{n+1}^*$.*

Corollary 3 (Corollary 2 of [20]). *If f is convex on $\mathbb{R}^n$ and an index $1 \leq j \leq n$ exists such that $f_j^{(k)} < f_{n+1}^{(k)}$, then $f_{ic}^{(k)} < f_{n+1}^{(k)}$ and no shrinking occurs in iteration k.*

Lemma 4 (Lemma 10 of [20]). *Assume that f is convex and bounded below on $\mathbb{R}^n$. If for some integer ℓ $(1 \leq \ell \leq n)$,*

$$f_\ell^* < f_{\ell+1}^*, \tag{21}$$

then there is an index $K > 0$ such that for all $k \geq K$, $j^{(k)} > \ell$, that is, the first ℓ vertices do not change.

Lemma 5 (Lemma 11 of [20]). *If f is convex and bounded below on $\mathbb{R}^n$, then $f_n^* = f_{n+1}^*$.*

The following generalization of Theorem 2 of Lagarias et al. was proved in [20]. Here, we present it with a new short proof.

Theorem 3. *Assume that f is a convex function on $\mathbb{R}^2$ and is bounded below. Then, the Nelder–Mead simplex method (Algorithm 2) converges in the sense that $f_1^* = f_2^* = f_3^*$.*

Proof. It follows (Lemmas 3 and 5) that $f_1^* \leq f_2^* = f_3^*$. Assume that $f_1^* < f_2^*$. It follows from Lemma 4 that there exists an index $K > 0$ such that for $k \geq K$, $x_1^{(k)} = x_1^{(K)}$. Hence, only $x_2^{(k)}$ and $x_3^{(k)}$ may change. The insertion rule and the impossibility of shrinking (Corollary 3) imply that only the following cases are possible:
(i) $f_1^{(k)} \leq f_r^{(k)} < f_2^{(k)}$;
(ii) $f_2^{(k)} \leq f_r^{(k)} < f_3^{(k)}$ and $f_1^{(k)} \leq f_{oc}^{(k)} \leq f_r^{(k)}$;
(iii) $f_3^{(k)} \leq f_r^{(k)}$ and $f_1^{(k)} \leq f_{ic}^{(k)} < f_3^{(k)}$.

We assume that $K > 0$ is big enough, so that for $k \geq K$, $f_1^* \leq f_1^{(k)} < f_1^* + \varepsilon$, $f_2^* \leq f_i^{(k)} < f_2^* + \varepsilon$ $(i = 2,3)$, where $\varepsilon > 0$ is such that $f_1^* + 4\varepsilon < f_2^*$. Depending on the selected case, $f_r^{(k)} \geq f_2^*$, $f_{oc}^{(k)} \geq f_2^*$ or $f_{ic}^{(k)} \geq f_2^*$ must hold. In case (i), $x_r^{(k)}$ replaces $x_2^{(k)}$ and $S^{(k+1)} = S^{(k)}T(1)P_2$. In case (ii), since $x_{oc}^{(k)} = \frac{1}{2}\left(x_c^{(k)} + x_r^{(k)}\right)$, we have

$$f_{oc}^{(k)} \leq \frac{1}{4}f_1^{(k)} + \frac{1}{4}f_2^{(k)} + \frac{1}{2}f_r^{(k)} \leq \frac{1}{4}(f_1^* + \varepsilon) + \frac{3}{4}(f_2^* + \varepsilon) < f_2^*.$$

Hence, case (ii) cannot occur. Similarly, in case (iii),

$$f_{ic}^{(k)} \leq \frac{1}{4}f_1^{(k)} + \frac{1}{4}f_2^{(k)} + \frac{1}{2}f_3^{(k)} \leq \frac{1}{4}(f_1^* + \varepsilon) + \frac{3}{4}(f_2^* + \varepsilon) < f_2^*$$

showing that case (iii) cannot occur. Hence, we can only have the simplex sequence

$$S^{(k+K)} = S^{(K)}[T(1)P_2]^k, \quad k \geq 0.$$

Since $T(1)P_2$ is a 6-involutory matrix ($[T(1)P_2]^6 = I_3$), $\left\{S^{(k+K)}\right\}$ is a periodic sequence and $S^{(6+K)} = S^{(K)}$, which is impossible by the insertion rule. This is a contradiction; thus, $f_1^* = f_2^* = f_3^*$. □

For Algorithm 1, we prove the following results. First, observe the following facts. For nonshrink steps, we have
(i) $x_i^{(k+1)} = x_i^{(k)}$ for $i \neq h_k$, $x_{h_k}^{(k)} = v$;
(ii) $f_i^{(k+1)} = f_i^{(k)}$ $(i \neq h_k)$ and $f_{h_k}^{(k+1)} < f_{h_k}^{(k)}$;
(iii) $f_{\ell_k}^{(k)} = \min_i f_i^{(k)} \leq f_i^{(k)} \leq \max_i f_i^{(k)} = f_{h_k}^{(k)}$;
(iv) $f_{\ell_{k+1}}^{(k+1)} \leq f_{\ell_k}^{(k)}$ and $f_{h_{k+1}}^{(k+1)} \leq f_{h_k}^{(k)}$.

Hence, $f_i^{(k+1)} \leq f_i^{(k)}$ $(i = 1, \ldots, n+1)$.

Lemma 6 (Algorithm 1). *If $f : \mathbb{R}^n \to \mathbb{R}$ is bounded below and only a finite number of shrink operations occur, then there exist finite limit values f_i^* such that $f_i^{(k)} \to f_i^*$ $(i = 1, 2, \ldots, n+1)$.*

The only difference to Lemma 2 is that the function values $f_i^{(k)}$ and their limits are not ordered.

Lemma 7 (Algorithm 1). *If $f : \mathbb{R}^n \to \mathbb{R}$ is quasiconvex on $\mathbb{R}^n$ and bounded below, then there exist finite limit values f_i^* such that $f_i^{(k)} \to f_i^*$ $(i = 1, 2, \ldots, n+1)$.*

Proof. If f is quasiconvex, then in the case of shrinking, the inequality

$$f_i^{(k+1)} = f\left(\frac{1}{2}\left(x_i^{(k)} + x_{\ell_k}^{(k)}\right)\right) \leq \max\left\{f_i^{(k)}, f_{\ell_k}^{(k)}\right\} = f_i^{(k)} \quad (i = 1, \ldots, n+1)$$

also holds. So does Lemma 6. □

Lemma 8 (Algorithm 1). *If $f : \mathbb{R}^n \to \mathbb{R}$ is strictly convex, then no shrinking occurs.*

Proof. Shrinking happens if $f_{m_k}^{(k)} \le f_r^{(k)} < f_{h_k}^{(k)} \wedge f_{oc}^{(k)} > f_r^{(k)}$ or $f_{ic}^{(k)} \ge f_{h_k}^{(k)} \wedge f_r^{(k)} \ge f_{h_k}^{(k)}$ holds. In general, $f_c^{(k)} < \frac{1}{n}\sum_{i=1,\, i\ne h_k}^{n+1} f_i^{(k)} \le f_{m_k}^{(k)}$. If $f_{m_k}^{(k)} \le f_r^{(k)} < f_{h_k}^{(k)}$, then $f_{oc}^{(k)} = f\left(\frac{1}{2}\left(x_r^{(k)} + x_c^{(k)}\right)\right) < \frac{1}{2}\left(f_r^{(k)} + f_{m_k}^{(k)}\right) \le f_r^{(k)}$. If $f_r^{(k)} \ge f_{h_k}^{(k)}$, then $f_{ic}^{(k)} = f\left(\frac{1}{2}\left(x_c^{(k)} + x_{h_k}^{(k)}\right)\right) < \frac{1}{2}\left(f_c^{(k)} + f_{h_k}^{(k)}\right) \le f_{h_k}^{(k)}$. □

Remark 1 (Algorithm 1). *Assume that $f : \mathbb{R}^n \to \mathbb{R}$ is convex. Then, $f_c^{(k)} \le f_{m_k}^{(k)}$. If $f_{m_k}^{(k)} \le f_r^{(k)} < f_{h_k}^{(k)}$, then $f_{oc}^{(k)} \le f_r^{(k)}$. If $f_r^{(k)} \ge f_{h_k}^{(k)}$, then $f_{ic}^{(k)} \le f_{h_k}^{(k)}$. However, if at least for one $j \ne h_k$, $f_j^{(k)} < f_{h_k}^{(k)}$ holds, then $f_c^{(k)} \le \frac{1}{n}\sum_{i=1,\, i\ne h_k}^{n+1} f_i^{(k)} < f_{h_k}^{(k)}$ and $f_{ic}^{(k)} < f_{h_k}^{(k)}$. In such a case, there is no shrinking.*

Assume that $f : \mathbb{R}^n \to \mathbb{R}$ is bounded below and only a finite number of shrink operations occur. Then, we have finite numbers f_i^* such that $f_i^{(k)} \to f_i^*$ as $k \to \infty$, and each $\left\{f_i^{(k)}\right\}$ is monotone decreasing ($i = 1, \dots, n+1$). The values f_i^* are unordered. However, there is permutation $\{i_1, i_2, \dots, i_{n+1}\}$ of the indices $\{1, 2, \dots, n+1\}$ such that

$$f_{i_1}^* \le f_{i_2}^* \le \dots \le f_{i_n}^* \le f_{i_{n+1}}^*. \tag{22}$$

Lemma 9 (Algorithm 1). *If f is convex and bounded below on $\mathbb{R}^n$, then $f_{i_n}^* = f_{i_{n+1}}^*$.*

Proof. Assume that $f_{i_n}^* < f_{i_{n+1}}^*$. Lemma 6 and Remark 1 imply that for $0 < \varepsilon < f_{i_{n+1}}^* - f_{i_n}^*$, there exists an index $K > 0$ such that for $k \ge K$,

$$f_{i_j}^* \le f_{i_j}^{(k)} \le f_{i_j}^* + \varepsilon \quad (j = 1, \dots, n+1).$$

Note that $f_{i_j}^* \le f_{i_j}^{(k)} \le f_{i_n}^* + \varepsilon < f_{i_{n+1}}^* \le f_{i_{n+1}}^{(k)}$ ($j = 1, \dots, n$). Hence, there cannot be shrinking, and only the worst vertex $x_{i_{n+1}}^{(k)}$ can change for $k \ge K$ ($f(v) \ge f_{i_{n+1}}^*$). Clearly, $i_{n+1} = h_k$ for $k \ge K$. If $\alpha = 1, 2$, then $f\left(x_r^{(k)}\right) < f_{m_k}^{(k)} \le f_{i_n}^*$ by definition. Hence, only the cases $\alpha = \pm\frac{1}{2}$ are possible. For simplicity, let $\eta = i_{n+1}$. Then,

$$S^{(k+1)} = S^{(k)} T_\eta(\alpha) \quad \left(\alpha \in \left\{-\frac{1}{2}, \frac{1}{2}\right\}\right)$$

and

$$S^{(K+j)} = S^{(K)} \prod_{i=1}^{j} T_\eta(\alpha_i) \quad \left(\alpha_i \in \left\{-\frac{1}{2}, \frac{1}{2}\right\}\right).$$

Since $T_j(\alpha)T_j(\beta) = T_j(-\alpha\beta)$ ($j = 1, \dots, n+1$), it follows that

$$\prod_{i=1}^{j} T_\eta(\alpha_i) = T_\eta\left((-1)^{j+1}\prod_{i=1}^{j}\alpha_i\right) \to T_\eta(0) = \begin{bmatrix} I_{\eta-1} & \frac{1}{n}e & 0 \\ 0 & 0 & 0 \\ 0 & \frac{1}{n}e & I_{n+1-\eta} \end{bmatrix},$$

$$S^{(K+j)} \to S^{(K)} \begin{bmatrix} I_{\eta-1} & \frac{1}{n}e & 0 \\ 0 & 0 & 0 \\ 0 & \frac{1}{n}e & I_{n+1-\eta} \end{bmatrix} = S^*$$

and

$$x_{i_{n+1}}^{(k)} \to x_{i_{n+1}}^* = \frac{1}{n}\sum_{j=1,\, j\ne\eta}^{n+1} x_j^{(K)} = \frac{1}{n}\sum_{j=1}^{n} x_{i_j}^{(K)}.$$

Since f is convex, we have

$$f^*_{i_{n+1}} = f\left(x^*_{i_{n+1}}\right) = f\left(\frac{1}{n}\sum_{j=1}^{n} x^{(K)}_{i_j}\right) \le \frac{1}{n}\sum_{i=1}^{n} f^{(K)}_{i_j} < f^*_{i_{n+1}},$$

which is a contradiction. Hence, $f^*_{i_n} = f^*_{i_{n+1}}$ holds under convexity. □

Theorem 4. *If f is convex and bounded below on $\mathbb{R}^2$, then Algorithm 1 converges in the sense that $f^*_1 = f^*_2 = f^*_3$.*

Proof. Lemmas 6 and 9 imply that $f^*_{i_1} \le f^*_{i_2} = f^*_{i_3}$ for some permutation i_1, i_2, i_3 of $\{1,2,3\}$. Assume that $f^*_{i_1} < f^*_{i_2}$. Then, there exists an index K such that for $k \ge K$, $f^*_{i_1} \le f^{(k)}_{i_1} < f^*_{i_1} + \varepsilon < f^*_{i_2} \le f^{(k)}_{m_k} \le f^{(k)}_{h_k} < f^*_{i_2} + \varepsilon$ ($f^*_{i_1} + 4\varepsilon < f^*_{i_2}$) holds with $m_k, h_k \in \{i_2, i_3\}$. Since $f^{(k)}_{i_1} < f^*_{i_2}$, only vertices $x^{(k)}_{i_2}$ and $x^{(k)}_{i_3}$ can change for $k \ge K$. Also, note that $f^{(k)}_{i_1} < f^*_{i_2}$ implies that no shrinking may occur. It also follows that only $f^{(k)}_r, f^{(k)}_{oc}, f^{(k)}_{ic} \ge f^*_{i_2}$ are possible. Consider now the following 'possible' cases:
(i) $f^*_{i_2} \le f^{(k)}_r < f^{(k)}_{m_k}$;
(ii) $f^{(k)}_{m_k} \le f^{(k)}_r < f^{(k)}_{h_k}$ and $f^{(k)}_{oc} \le f^{(k)}_r$;
(iii) $f^{(k)}_{h_k} \le f^{(k)}_r$ and $f^{(k)}_{ic} < f^{(k)}_{h_k}$.

In the first case, $x^{(k+1)}_{m_{k+1}} = x^{(k+1)}_{h_k} = x^{(k)}_r$ and $x^{(k+1)}_{h_{k+1}} = x^{(k)}_{m_k}$. Hence, the operation $S^{(k+1)} = S^{(k)} T_{h_k}(1)$ must be followed by $S^{(k+2)} = S^{(k+1)} T_{m_k}(\alpha)$ ($h_k \ne m_k$) for some $\alpha \in \left\{-\frac{1}{2}, \frac{1}{2}, 1\right\}$. In the second case,

$$f^{(k)}_{oc} \le \frac{1}{4} f^{(k)}_{i_1} + \frac{1}{4} f^{(k)}_{m_k} + \frac{1}{2} f^{(k)}_r \le \frac{1}{4}\left(f^*_{i_1} + \varepsilon\right) + \frac{3}{4}\left(f^*_{i_2} + \varepsilon\right) < f^*_2,$$

which makes this particular step impossible. Similarly, in the third case,

$$f^{(k)}_{ic} \le \frac{1}{4} f^{(k)}_{i_1} + \frac{1}{4} f^{(k)}_{m_k} + \frac{1}{2} f^{(k)}_{h_k} \le \frac{1}{4}\left(f^*_{i_1} + \varepsilon\right) + \frac{3}{4}\left(f^*_{i_2} + \varepsilon\right) < f^*_2,$$

this step is impossible. Hence, the only possible step is $S^{(k+1)} = S^{(k)} T_{h_k}(1)$ with $m_{k+1} = h_k$ and $h_{k+1} = m_k$. In fact, we have $S^{(k+2)} = S^{(k)} T_{h_k}(1) T_{m_k}(1)$ for a pair of m_k and h_k, and this is repeated. Consider the following possible cases:

$$(m_k, h_k) \in J = \{(2,3), (3,2), (1,3), (3,1), (1,2), (2,1)\}.$$

It is easy to check that for every $(h_k, m_k) \in J$,

$$\left[T_{h_k}(1) T_{m_k}(1)\right]^6 = I.$$

Here, we have a periodicity, $S^{(k+12)} = S^{(k)} \left[T_{h_k}(1) T_{m_k}(1)\right]^6 = S^{(k)}$, which is impossible by the condition $f^{(k)}_r < f^{(k)}_{m_k} < f^{(k)}_{h_k}$. Hence, $f^*_1 = f^*_2 = f^*_3$ must hold. □

Theorems 3 and 4 are generalizations of Theorem 2 of Lagarias et al. [12], since they require only convexity and boundedness from below. Moreover, their proofs are much simpler.

6. Convergence of the Simplex Sequences

Here, we study the convergence of the simplex sequence $\left\{S^{(k)}\right\}$ generated by both Nelder–Mead methods. If $S^{(k)} \to S^{\infty} = \left[x^*_1, \ldots, x^*_{n+1}\right]$, then (provided that f is continuous)

$f_i^{(k)} \to f(x_i^*)$ for $i = 1, \dots, n+1$ as $k \to \infty$. The limit vertices x_i^* can be different (see, , e.g., Examples 1 and 2 of [20]).

Ideally, $S^{(k)}$ should converge to some limit of the form $S^\infty = [x^*, \dots, x^*] = x^* e^T$, where $x^* \in \mathbb{R}^n$ is a stationary point of f. MacKinnon's example [19] and Example 3 of [15] show that x^* is not always a stationary point.

If B_k converges to a rank one matrix of the form $B^\infty = we^T$ ($w \in \mathbb{R}^{n+1}$), then $S^{(k)} = S^{(0)} B_k \to S^{(0)} we^T = \widehat{x} e^T$ and diameter $\left(S^{(k)}\right) \to 0$. There are several examples (see, e.g., [14,15,20,21]), where the simplex sequence $\left\{S^{(k)}\right\}$ converges to a limit S^∞ with diameter $(S^\infty) > 0$ and rank (S^∞), rank$(B^\infty) \geq 2$.

The following necessary and sufficient result holds for both algorithms.

Lemma 10. *Assume that $S^{(0)}$ is non-degenerate (affinely independent). Then, $\left\{S^{(k)}\right\}$ (generated by Algorithm 1 or 2) converges to some $S^\infty \in \mathbb{R}^{n\times(n+1)}$ if and only if $\{B_k\}$ converges to some B^∞.*

Proof. In both cases $S^{(k)} = S^{(0)} \prod_{i=0}^k X_i = S^{(0)} B_k \quad (X_i \in \mathbb{T})$, where $\mathbb{T} = \widetilde{\mathcal{T}}$ (in case of Algorithm 1) or $\mathbb{T} = \mathcal{T}$ (in case of Algorithm 2). If $B_k \to B^\infty$, then $S^{(k)} = S^{(0)} B_k \to S^{(0)} B^\infty := S^\infty$. Since $e^T B_k = e^T$, we have

$$\begin{bmatrix} e^T \\ S^{(k)} \end{bmatrix} = \begin{bmatrix} e^T \\ S^{(0)} \end{bmatrix} B_k.$$

By assumption, the first matrix on the right is invertible and so

$$B_k = \begin{bmatrix} e^T \\ S^{(0)} \end{bmatrix}^{-1} \begin{bmatrix} e^T \\ S^{(k)} \end{bmatrix}.$$

If $S^{(k)} \to S^\infty$, then

$$B_k = \begin{bmatrix} e^T \\ S^{(0)} \end{bmatrix}^{-1} \begin{bmatrix} e^T \\ S^{(k)} \end{bmatrix} \to \begin{bmatrix} e^T \\ S^{(0)} \end{bmatrix}^{-1} \begin{bmatrix} e^T \\ S^\infty \end{bmatrix} = B^\infty.$$

□

From now on, we assume that $S^{(0)}$ is always non-degenerate (its columns are affinely independent). Lemma 10 is formally independent of f. However, the decision mechanism of the NM algorithm determines the next simplex and the resulting matrix product. Certain configurations clearly cannot occur. Such is the case of strictly convex functions, when no shrinking can occur or $T(1)P_2$ cannot be repeated in a sequence more than five times (for the case $n = 2$). In general, it is hard to tell what sequences (products) are possible for a given f.

Lemma 11. *If $B_k = \prod_{i=0}^k X_i \to B^\infty$ ($X_i \in \mathbb{T}, \mathbb{T} = \widetilde{\mathcal{T}}$ or $\mathbb{T} = \mathcal{T}$), then rank $(B^\infty) \leq n$.*

Proof. Assume that rank $(B^\infty) = n + 1$. Then, B^∞ is invertible. Since all of $X_i \in \mathbb{T}$ is invertible, $\lim_{k\to\infty} B_{k-1}^{-1} = (B^\infty)^{-1}$. As $X_k = B_{k-1}^{-1} B_k$,

$$\lim_{k\to\infty} X_k = \lim_{k\to\infty} \left(B_{k-1}^{-1}\right) \lim_{k\to\infty} (B_k) = (B^\infty)^{-1} B^\infty = I.$$

Hence, $X_k \to I$ is necessary for an invertible B^∞. Since $\mathbb{T} = \widetilde{\mathcal{T}}$ or $\mathbb{T} = \mathcal{T}$ is a finite set and its elements are different from I, this cannot happen and rank $(B^\infty) \leq n$. □

Sylvester's theorem on rank (see, e.g., Mirsky [30]) implies that

$$\text{rank}(S^\infty) \geq \text{rank}\left(S^{(0)}\right) + \text{rank}(B^\infty) - (n+1).$$

Since, by assumption, rank $\left(S^{(0)}\right) = n$, we have $n \geq \text{rank}\,(S^{\infty}) \geq \text{rank}\,(B^{\infty}) - 1$. In Example 4 of [15], rank $\left(S^{(0)}\right) = n$, rank $(B^{\infty}) = n$ and rank $(S^{\infty}) = n - 1$.

The following simple result characterizes the possible limits of infinite matrix products.

Lemma 12. *Assume that $B_k = \prod_{i=0}^{k} X_i \to B^{\infty}$ ($X_i \in \mathbb{T}$) and X_s occurs infinitely often in the product $\prod_{i=0}^{\infty} X_i$, then every nonzero row of B^{∞} is a left eigenvector of X_s belonging to the eigenvalue $\lambda = 1$.*

Proof. Since X_s appears infinitely many times in the product $\prod_{i=0}^{\infty} X_i$, there is a subsequence of $\left\{B_{i_j}\right\}$ with rightmost factor X_s, say

$$B_{i_1} X_s, B_{i_2} X_s, \ldots,$$

where the B_{i_j}s are products of X_is. Since $B_{i_j} \to B^{\infty}$, we have $B_{i_j} X_s \to B^{\infty} X_s = B^{\infty}$. □

Define the (left) 1-eigenspace of matrix A by $E_1(A) = \{x : xA = x\}$. The rows of B^{∞} belong to $E_1(X_s)$. If several X_s, say X_{s_i} ($i = 1, \ldots, m$), occur infinitely often in the product $\prod_{i=0}^{\infty} X_i$, then the rows of B^{∞} belong to $\cap_{i=1}^{m} E_1(X_{s_i})$.

Since $e^T \in E_1\left(T_i P^{(i)}\right)$ for all $T_i P^{(i)} \in \mathcal{T}$ and $e^T \in E_1\left(T^{(i)}\right)$ for all $T^{(i)} \in \widetilde{\mathcal{T}}$, we have $e^T \in \cap_{i=1}^{m} E_1\left(T_{s_i} P^{(s_i)}\right)$ and $e^T \in \cap_{i=1}^{m} E_1\left(T^{(s_i)}\right)$, respectively. If $\cap_{i=1}^{m} E_1\left(T_{s_i} P^{(s_i)}\right) = \left\{\lambda e^T : \lambda \in \mathbb{R}\right\}$ or $\cap_{i=1}^{m} E_1\left(T^{(s_i)}\right) = \left\{\lambda e^T : \lambda \in \mathbb{R}\right\}$, then B^{∞} has the form we^T for some $w \in \mathbb{R}^{n+1}$.

Note that for any $T \in \widetilde{\mathcal{T}}$ or $T \in \mathcal{T}$, $e^T T = e^T$. Hence, if $B_k \to B^{\infty}$, then $e^T B_k = e^T \to e^T B^{\infty}$, that is, $e^T = e^T B^{\infty}$. If $B^{\infty} = we^T$, then $e^T B^{\infty} = e^T w e^T = e^T$ implies that $e^T w = 1$.

We recall the following definitions and results (see, e.g., Hartfiel [31,32]).

A right infinite matrix product is an expression $A_1 A_2 \cdots A_k A_{k+1} \cdots$. A set $\mathcal{M}$ of $n \times n$ matrices has the right convergence property (RCP), if all possible right infinite products $\prod_{i=1}^{\infty} A_i$ ($A_i \in \mathcal{M}$) converge.

A set $\mathcal{M}$ of $n \times n$ matrices is product-bounded if there is a constant $\beta > 0$ such that $\|A_1 \cdots A_k\| \leq \beta$ for all k and all $A_1, \ldots, A_k \in \mathcal{M}$.

If $\mathcal{M}$ is an RCP set, then $\mathcal{M}$ is also product-bounded.

Lemma 13. *If $\mathcal{M}$ is an RCP set, $A_1, \ldots, A_k \in \mathcal{M}$ and λ is an eigenvalue of $A_1 A_2 \cdots A_k$, then $|\lambda| < 1$ or $\lambda = 1$, and this eigenvalue is simple. Hence, each matrix of $\mathcal{M}$ must satisfy this condition.*

The matrices $T(2)P_1$ and $T_j(2)$ have at least one eigenvalue greater than 1, and $\|T(2)P_1\| > 1$ and $\|T_j(2)\| > 1$ in any induced matrix norm. Hence, $\left\{[T(2)P_1]^k\right\}$ and $\left\{T_j(2)^k\right\}$ are unbounded. There are also examples when the Nelder–Mead algorithm produces unbounded simplex sequences $S^{(k)} = S^{(0)}[T(2)P_1]^k$ or $S^{(k)} = S^{(0)}[T(1)P_1]^k$ (see, e.g., [14,15,20]). It is clear that the whole matrix set $\widetilde{\mathcal{T}}$ or $\mathcal{T}$ is not an RCP set.

Hence, we must seek for subsets $\mathcal{M}$ of $\widetilde{\mathcal{T}}$ or $\mathcal{T}$, which are RCPs. However, Blondel and Tsitsiklis [33] proved that the product boundedness of a finite matrix set $\mathcal{M}$ is algorithmically undecidable and it remains undecidable even in the special case, when $\mathcal{M}$ consists of only two matrices. Since product boundedness is a weaker property than the RCP from which it follows, and although it is algorithmically undecidable, it seems difficult to decide the RCP property in general. This might answer question (e) of Wright.

However, it is relatively easy to identify one particular RCP set. For both algorithms, the corresponding shrinking matrices form an RCP set.

Lemma 14. *Both* $\mathcal{M}_1 = \left\{T_j^{shr} : 1 \le j \le n+1\right\}$ *and* $\mathcal{M}_2 = \{T_{shr}P : P \in \mathcal{P}_{n+1}\}$ *are RCP sets.*

Proof. By using induction on k, we can prove that for $k \ge 1$,

$$\prod_{i=1}^{k} T_{j_i}^{shr} = \prod_{i=1}^{k}\left(\frac{1}{2}I + \frac{1}{2}e_{j_i}e^T\right) = \frac{1}{2^k}I + \sum_{i=1}^{k}\frac{1}{2^i}e_{j_i}e^T$$

and

$$\prod_{i=1}^{k} T_{shr}P^{(j_i)} = \prod_{i=1}^{k}\left(\frac{1}{2}I + \frac{1}{2}e_1e^T\right)P^{(j_i)} = \frac{1}{2^k}\prod_{i=1}^{k}P^{(j_i)} + \sum_{i=1}^{k}\frac{1}{2^i}e_{\ell_i}e^T \quad (\ell_1 = 1).$$

For the last formula, note that if P is a permutation matrix, then $e_i^T P = e^T$ and $Pe_i = e_j$ for some j. In element-wise partial ordering, $T_j^{shr} \ge 0$ holds for $j = 1, \ldots, n+1$ and

$$0 \le \prod_{i=1}^{k} T_{j_i}^{shr} = \frac{1}{2^k}I + \sum_{i=1}^{k}\frac{1}{2^i}e_{j_i}e^T \le ee^T.$$

Here, $\frac{1}{2^k}I \to 0$ as $k \to \infty$, and the sequence $\left\{\sum_{i=1}^{k}\frac{1}{2^i}e_{j_i}e^T\right\}_{k=1}^{\infty}$ is monotone increasing in component-wise partial ordering. Hence, $\prod_{i=1}^{\infty} T_{j_i}^{shr}$ is convergent. We also have that $0 \le \prod_{i=1}^{k} T_{shr}P^{(j_i)} \le ee^T$, $\frac{1}{2^k}\prod_{i=1}^{k}P^{(j_i)} \to 0$ as $k \to \infty$, and $\left\{\sum_{i=1}^{k}\frac{1}{2^i}e_{j_i}e^T\right\}_{k=1}^{\infty}$ is monotone increasing in partial ordering. Hence, $\prod_{i=1}^{\infty} T_{shr}P^{(j_i)}$ is convergent. □

Remark 2. *Corollary 2 implies that* $E_1(T_{shr}P) = \left\{\lambda e^T : \lambda \in \mathbb{R}\right\}$ *for all* $P \in \mathcal{P}_{n+1}$*. If matrices* $T_{shr}P^{(s_i)}$ $(i = 1, \ldots, m)$ *occur infinitely often in the product* $\prod_{i=1}^{\infty} T_{shr}P^{(j_i)}$ *and* $\prod_{i=1}^{k} T_{shr}P^{(j_i)} \to B^{\infty}$ $(k \to \infty)$*, then the nonzero rows of* B^{∞} *belong to* $\cap_{i=1}^{m} E_1\left(T_{shr}P^{(j_i)}\right) = \left\{\lambda e^T : \lambda \in \mathbb{R}\right\}$*. Hence,* $\prod_{i=1}^{k} T_{shr}P^{(j_i)} \to we^T$ *for some* w*. Lemma 1 implies that* $E_1\left(T_j^{shr}\right) = \left\{\lambda e^T : \lambda \in \mathbb{R}\right\}$ *for all* $j = 1, \ldots, n+1$*. It also follows from the previous argument that* $\prod_{i=1}^{k} T_{j_i}^{shr} \to we^T$ *for some* w*.*

The $S^{(k)} \to \hat{x}e^T$ for the infinitely repeated shrinking process also follows from Cantor's intersection theorem.

Lemma 15. *Define the matrix*

$$F = \begin{bmatrix} 1 & -e^T \\ 0 & I_n \end{bmatrix} \quad \left(F^{-1} = \begin{bmatrix} 1 & e^T \\ 0 & I_n \end{bmatrix}\right). \tag{23}$$

Then, for all $T \in \mathbb{T}$ $(\mathbb{T} = \mathcal{T}$ *or* $\mathbb{T} = \widetilde{\mathcal{T}})$*, the matrix* $F^{-1}TF$ *has the common block lower triangular form*

$$F^{-1}TF = \begin{bmatrix} 1 & 0 \\ b(T) & C(T) \end{bmatrix}, \tag{24}$$

where $b(T) \in \mathbb{R}^n$ *and* $C(T) \in \mathbb{R}^{n \times n}$*.*

The result was proved for $T \in \mathcal{T}$ in [14,15]. The proof of case $\widetilde{\mathcal{T}}$ is similar. Since both $\widetilde{\mathcal{T}}$ and $\mathcal{T}$ are finite sets, there exist two constants $\gamma_b, \gamma_C > 0$ such that $\|b(T)\| \le \gamma_b$ and $\|C(T)\| \le \gamma_C$ for all $T \in \widetilde{\mathcal{T}}$ or $T \in \mathcal{T}$. For $T_j(\alpha) \in \widetilde{\mathcal{T}}$, the corresponding $C(T_j(\alpha))$ has $n-1$ eigenvalues of $\lambda = 1$ and $\|C(T_j(\alpha))\| \ge 1$. For $T(\alpha)P_j \in \mathcal{T}$ $(j > 2)$, $\|C(T(\alpha)P_j)\| \ge 1$. However, for $T(\alpha)P_j \in \mathcal{T}$ with $|\alpha| < 1$ and $j = 1, 2$, the corresponding $C(T(\alpha)P_j)$s have all their eigenvalues in the open unit disk (see Theorem 1). Hence, they can be elements of an RCP set.

If

$$A_i = \begin{bmatrix} 1 & 0 \\ b_i & C_i \end{bmatrix} \in \mathbb{R}^{(n+1)\times(n+1)} \quad (C_i \in \mathbb{R}^{n\times n},\, i \geq 1), \tag{25}$$

then

$$L_k = \prod_{i=1}^{k} A_i = \begin{bmatrix} 1 & 0 \\ \sum_{i=1}^{k}\left(\prod_{j=1}^{i-1} C_j\right) b_i & \prod_{i=1}^{k} C_i \end{bmatrix} = \begin{bmatrix} 1 & 0 \\ x_k & \prod_{j=1}^{k} C_i \end{bmatrix}. \tag{26}$$

If the infinite matrix product $\prod_{i=1}^{\infty} T_i$ ($T_i \in \mathcal{T}$ $(i \geq 1)$ or $T \in \widetilde{\mathcal{T}}$ $(i \geq 1)$) converges to a rank-one matrix of the form we^T, that is, if $\prod_{i=1}^{k} T_i \to we^T$ $(w^T = [w_1, \widehat{w}^T], \widehat{w} \in \mathbb{R}^n)$, then it follows that

$$\prod_{i=1}^{k} \begin{bmatrix} 1 & 0 \\ b(T_i) & C(T_i) \end{bmatrix} \to F^{-1} we^T F = \begin{bmatrix} e^T w & 0 \\ \widehat{w} & 0_{n\times n} \end{bmatrix}. \tag{27}$$

Lemma 16. *For the convergence of the simplex sequence to a common point, it is necessary and sufficient that both*

$$\prod_{i=1}^{k} C(T_i) \to 0 \quad (k \to \infty) \tag{28}$$

and

$$\sum_{i=1}^{k} \left(\prod_{j=1}^{i-1} C(T_j)\right) b(T_i) \to \widetilde{x} \quad (k \to \infty) \tag{29}$$

hold for some vector $\widetilde{x}$.

It is clear that the infinite products of $\mathcal{M}_1$ or $\mathcal{M}_2$ satisfy these conditions. We recall the following simple result.

Lemma 17 ([14–16])**.** *Assume that* $\left\|\prod_{j=1}^{k} C_j\right\| \leq c_k$, $\sum_{k=1}^{\infty} c_k$ *is convergent* ($< \infty$) *and that* $\|b_k\| \leq \gamma$ *for all k. Then,* $L_k = \prod_{j=1}^{k} A_j$ *converges and*

$$\lim_{k\to\infty} L_k = \begin{bmatrix} 1 & 0 \\ \widetilde{x} & 0 \end{bmatrix} \tag{30}$$

for some $\widetilde{x}$.

If $\|C_i\| \leq q < 1$ for all i, then the speed of convergence is linear (geometric). This follows from $\left\|\prod_{i=1}^{k} C_i\right\| \leq q^k$ and the inequality

$$\|\widetilde{x} - x_k\| \leq \gamma \sum_{i=k+1}^{m} \left\|\prod_{j=1}^{i-1} C_i\right\| \leq \gamma \frac{q^k}{1-q} \quad (m > k)$$

with $m \to \infty$.

We recall the following result (see Hartfiel [31], Corollary 6.4, Daubechies and Lagarias [34]).

Lemma 18. *Let* $\mathcal{M}$ *be a compact matrix set. Then, every infinite product, taken from* $\mathcal{M}$*, converges to 0 if there is a norm* $\|\cdot\|$ *such that* $\|A\| \leq q$, $q < 1$*, for all* $A \in \mathcal{M}$.

Hence, there must be a norm $\|\cdot\|$ such that $\|C_i\| \leq q < 1$, for all $T_{shr}P_j$ (or T_j^{shr}). Note that Lemma 18 is not constructive. However, for the convergence of $\prod_{i=1}^{\infty} T_i \to we^T$ (for all $T_i \in \mathcal{M}$), the condition $\|C(T_i)\| \leq q < 1$ $(T_i \in \mathcal{M})$ is necessary and sufficient.

For Algorithm 1, $\mathcal{M}_1$ is the only subset that satisfies this requirement. For Algorithm 2, define the set $\widehat{\mathcal{M}_2} = \left\{T(\alpha)P_j : \alpha = \pm\frac{1}{2},\ j = 1,2\right\} \cup \mathcal{M}_2$ and assume that

(A) *there exists a matrix norm* $\|\cdot\|_\vartheta$ *(induced by a vector norm* $\|\cdot\|_\vartheta$*) such that if* $T_iP^{(i)} \in \widehat{\mathcal{M}_2}$*, then* $\left\|C\left(T_iP^{(i)}\right)\right\|_\vartheta < 1$.

If (A) holds, then $\widehat{\mathcal{M}_2}$ is an RCP set and every infinite product $\prod_{i=1}^{\infty} T_iP^{(i)}$ ($T_iP^{(i)} \in \widehat{\mathcal{M}_2}$) has the form we^T for some w. It follows from Theorem 1 that for $T_iP^{(i)} \in \mathcal{T} \setminus \widehat{\mathcal{M}_2}$, $\left\|C\left(T_iP^{(i)}\right)\right\|_\vartheta \geq 1$.

The existence of such norm $\|\cdot\|_\vartheta$ of the form $\|A\|_\vartheta = \left\|S^{-1}AS\right\|_2$ was proved in [14] for $n = 1,2,3$ and in [15] for $n = 1,2,\ldots,8$. For the cases $n = 9,10$, the corresponding matrices S, found experimentally by direct search methods, are given in Appendix A. Since the eigenvalues of $T(\alpha)P_j$ ($|\alpha| < 1$, $j = 1,2$) that lay inside the unit disk converge to 1 for $n \to \infty$ (see Lemma 3 of [15]), it is hard to find a norm with property (A) for greater n values.

The obtained RCP sets $\mathcal{M}_1$ and $\widehat{\mathcal{M}_2}$ are quite narrow. However, we can keep the convergence of any infinite product from $\mathcal{M}_1$ or $\widehat{\mathcal{M}_2}$ by inserting an infinite number of matrices from $\mathbb{T}$ under proper conditions (for Algorithm 2, see [14,15]). For Algorithm 2, it was also observed [16] that several fixed-length matrix products have spectral norms less than 1. Upon these observations, the convergence set of both algorithms can be significantly expanded.

Assume again that $\mathbb{T} = \widetilde{\mathcal{T}}$ or $\mathbb{T} = \mathcal{T}$. Moreover, introduce the notation

$$\mathbb{T}^\ell = \left\{T_{i_1}\cdots T_{i_\ell} : T_{i_j} \in \mathbb{T},\ j = 1,2,\ldots,\ell\right\} \tag{31}$$

and consider the product $\prod_{j=1}^{m\ell} T_{i_j} = \prod_{j=1}^{m}\left(\prod_{t=1}^{\ell} T_{i_{(j-1)\ell+t}}\right)$, where

$$\widehat{T}_j = \prod_{t=1}^{\ell} T_{i_{(j-1)\ell+t}} \in \mathbb{T}^\ell \quad (j = 1,\ldots,m).$$

Let

$$F^{-1}T_{i_j}F = \begin{bmatrix} 1 & 0 \\ b\left(T_{i_j}\right) & C\left(T_{i_j}\right) \end{bmatrix} = \begin{bmatrix} 1 & 0 \\ b_{i_j} & C_{i_j} \end{bmatrix}.$$

Then,

$$F^{-1}\left(\prod_{j=1}^{m\ell} T_{i_j}\right)F = \begin{bmatrix} 1 & 0 \\ \sum_{j=1}^{m\ell}\left(\prod_{t=1}^{j-1} C_{i_t}\right)b_{i_j} & \prod_{j=1}^{m\ell} C_{i_j} \end{bmatrix} \tag{32}$$

and

$$\begin{aligned} F^{-1}\widehat{T}_jF &= \begin{bmatrix} 1 & 0 \\ \sum_{r=1}^{\ell}\left(\prod_{t=1}^{r-1} C_{i_{(j-1)\ell+t}}\right)b_{i_{(j-1)\ell+r}} & \prod_{\tau=1}^{\ell} C_{i_{(j-1)\ell+\tau}} \end{bmatrix} \\ &= \begin{bmatrix} 1 & 0 \\ \widehat{b}_j & \widehat{C}_j \end{bmatrix} = \begin{bmatrix} 1 & 0 \\ b\left(\widehat{T}_j\right) & C\left(\widehat{T}_j\right) \end{bmatrix}, \end{aligned}$$

where $\left\|\widehat{b}_j\right\| \leq \left(\sum_{r=1}^{\ell-1} \gamma_C^{r-1}\right)\gamma_b =: \gamma_{\widehat{b}}$ and $\left\|C\left(\widehat{T}_j\right)\right\| \leq \gamma_C^\ell =: \gamma_{\widehat{C}}$ are uniformly bounded.

Assume that $k = \ell m$ and consider the expression

$$F^{-1}\left(\prod_{j=1}^{m} \widehat{T}_j\right)F = \begin{bmatrix} 1 & 0 \\ \sum_{j=1}^{m}\left(\prod_{t=1}^{j-1} \widehat{C}_t\right)\widehat{b}_j & \prod_{j=1}^{m} \widehat{C}_j \end{bmatrix}. \tag{33}$$

which is clearly equal to (32).

For $0 < q < 1$, define the sets

$$\mathbb{T}_q^\ell = \left\{T_{i_1}\cdots T_{i_\ell} \in \mathbb{T}^\ell : \|C(T_{i_1}\cdots T_{i_\ell})\| \le q\right\}, \quad \mathbb{T}_Q^\ell = \mathbb{T}^\ell \setminus \mathbb{T}_q^\ell. \tag{34}$$

For any $\widehat{T}_j \in \mathbb{T}_q^\ell$, we have the estimates $\left\|\widehat{b}_j\right\| \le \widehat{\gamma}_{\widehat{b}}$ and $\left\|\widehat{C}_j\right\| \le q < 1$. Hence, by Lemma 17, the set $\mathbb{T}_q^\ell$ is an RCP set, and the limits of the infinite products $\prod_{j=1}^\infty \widehat{T}_j$ are of the form we^T.

Now, consider the product

$$\prod_{j=1}^{k} T_{i_j} = \prod_{j=1}^{m\ell+r} T_{i_j} = \left(\prod_{j=1}^{m\ell} T_{i_j}\right)\left(\prod_{\tau=m\ell+1}^{m\ell+r} T_{i_\tau}\right) = \left(\prod_{j=1}^{m} \widehat{T}_j\right)\left(\prod_{\tau=m\ell+1}^{m\ell+r} T_{i_\tau}\right)$$

where $\widehat{T}_j \in \mathbb{T}_q^\ell$ $(j = 1,\ldots,m)$, $T_{i_\tau} \in \mathbb{T}$ $(\tau = m\ell+1,\ldots m\ell+r)$ and $0 \le r < \ell$ are fixed $(\ell \ge 1)$. The corresponding reduced form is

$$F^{-1}\left(\prod_{j=1}^{k} T_{i_j}\right)F = \begin{bmatrix} 1 & 0 \\ x_k & \left(\prod_{j=1}^{m}\widehat{C}_j\right)\left(\prod_{\tau=m\ell+1}^{m\ell+r} C_{i_\tau}\right) \end{bmatrix}$$

Consider the product $\prod_{j=1}^m \widehat{T}_j$ $(\widehat{T}_j \in \mathbb{T}^\ell)$, where the number of factors $\widehat{T}_j$ that belong to $\mathbb{T}_q^\ell$ is $r_1(m)$. Clearly, $0 \le r_1(m) \le m$. There exists a $\kappa \in \mathbb{N}$ such that $\frac{1}{q^{\kappa-1}} \le \widehat{\gamma}_C \le \frac{1}{q^\kappa}$. Assume that there is a number $\mu \in (0,1]$ such that $(1+\kappa)r_1(m) - \kappa m \ge \mu m$. Then,

$$\begin{aligned}\left\|\left(\prod_{j=1}^{m}\widehat{C}_j\right)\left(\prod_{\tau=m\ell+1}^{m\ell+r} C_{i_\tau}\right)\right\| &\le q^{r_1(m)}\widehat{\gamma}_C^{m-r_1(m)}\gamma_C^r \\ &\le \gamma_C^{\ell-1} q^{r_1(m)-\kappa(m-r_1(m))} = \gamma_C^{\ell-1} q^{(1+\kappa)r_1(m)-\kappa m} \\ &\le \gamma_C^{\ell-1} q^{\mu m} \le \gamma_C^{\ell-1} q^{\mu\left(\frac{k}{\ell}-1\right)} = \frac{1}{q}\gamma_C^{\ell-1}\left(q^{\frac{\mu}{\ell}}\right)^k.\end{aligned}$$

Since $\widehat{q} = q^{\frac{\mu}{\ell}} < 1$ and the conditions of Lemma 17 hold, we proved the following result.

Theorem 5. *Assume that $n \ge 2$, $S^{(0)}$ is non-degenerate, $\ell \ge 1$ is fixed and $\mathbb{T}_q^\ell$ is not empty. Let $r_1\left(\left\lfloor\frac{k}{\ell}\right\rfloor\right)$ be the number of ℓ products that belong to $\mathbb{T}_q^\ell$ during the first k iterations of the Nelder–Mead method (Algorithm 2). Moreover, assume that for an integer $\kappa \in \mathbb{N}$, $\frac{1}{q^{\kappa-1}} \le \widehat{\gamma}_C \le \frac{1}{q^\kappa}$ and for some $\mu \in (0,1]$, $r_1\left(\left\lfloor\frac{k}{\ell}\right\rfloor\right) \ge \frac{\mu+\kappa}{1+\kappa}\left\lfloor\frac{k}{\ell}\right\rfloor$. Then, the Nelder–Mead method (Algorithm 2) converges in the sense that*

$$\lim_{k\to\infty} x_j^{(k)} = \widehat{x} \quad (j = 1,2,\ldots,n+1) \tag{35}$$

with a convergence speed proportional to $Q\left(\left(q^{\frac{\mu}{\ell}}\right)^k\right)$.

Note that assumption $r_1(m) \ge \frac{\mu+\kappa}{1+\kappa}m$, which is a density condition, yields an infinite number of factors from the set $\mathbb{T}_Q^\ell$ as $m \to \infty$. The larger the ratio $\left|\mathbb{T}_q^\ell\right| / \left|\mathbb{T}^\ell\right|$, the wider the convergence set. Theorem 5 is clearly true for any pair of subsets $\mathcal{W}^\ell$ and $\mathcal{W}_q^\ell$ such that $\emptyset \ne \mathcal{W}_q^\ell \subset \mathcal{W}^\ell$ and $\mathcal{W} \subseteq \mathbb{T}$.

For Algorithm 1 and some $q < 1$, $\widetilde{T}_q^1 = \mathcal{M}_1$, to which we cannot add more elements, since $\|C(T_j(\alpha))\| \ge 1$ in any induced matrix norm. For Algorithm 2, $T_q^1 = \widehat{\mathcal{M}}_2$ under Assumption (A).

Theorem 5 was first proved for Algorithm 2 for $\ell = 1$ in [14,15] and for $\ell > 1$ in [16] in a slightly different way and form. The conditions were verified for a spectral norm and $q = 0.99$ up to dimension $n = 6$.

The computational results of Section 7 show that for $\ell > 1$, $\mathcal{M}_1^\ell \subsetneqq \widetilde{\mathcal{T}}_q^\ell$ and $\widehat{\mathcal{M}}_2^\ell \subsetneqq \mathcal{T}_q^\ell$ (see also [16]). Hence, Theorem 5 indeed extends the convergence sets $\mathcal{M}_1$ and $\widehat{\mathcal{M}}_2$ (quite significantly).

Upon the basis of the earlier version of Theorem 5, a stochastic convergence result was proved for Algorithm 2 in [16]. Since Theorem 5 holds for both algorithms, this can be extended for Algorithm 1 as well. However, it will be published later for reasons discussed in Section 7.

7. Computational Results and Conclusions

Here, we consider the cardinality of set $\mathbb{T}_q^\ell$ for both algorithms with and without shrinking operations.

Define the quantities $\widetilde{r}(n,\ell) = \left|\widetilde{\mathcal{T}}_q^\ell\right| / \left|\widetilde{\mathcal{T}}^\ell\right|$ and $r(n,\ell) = \left|\mathcal{T}_q^\ell\right| / \left|\mathcal{T}^\ell\right|$. The larger the ratio $\widetilde{r}(n,\ell)$ or $r(n,\ell)$, the wider the convergence set. Table 1 and 2 contain the computed values $\widetilde{r}(n,\ell)$ or $r(n,\ell)$, respectively.

Table 1. Ratios for Algorithm 1.

			$\widetilde{\mathbf{r}}(\mathbf{n},\ell)$			
$\mathrm{n} \setminus \ell$	2	3	4	5	6	7
2	0.39556	0.52563	0.59953	0.65873	0.70331	0.7403
3	0.28	0.3755	0.45766	0.52766	0.58397	0.63187
4	0.28	0.34304	0.40802	0.46877	0.52489	0.57442

Table 2. Ratios for Algorithm 2.

			$\widetilde{\mathbf{r}}(\mathbf{n},\ell)$			
$\mathrm{n} \setminus \ell$	2	3	4	5	6	7
2	0.71111	0.83615	0.90204	0.94099	0.9641	0.97795
3	0.85185	0.93746	0.97389	0.98912	0.9956	0.99824

Note that $\widetilde{r}(n,\ell) \geq 1/5^\ell$ and $r(n,\ell) \geq \left[\frac{(n+1)!}{3n+3+(n+1)!}\right]^\ell$. The ratio $r(n,\ell)$ cannot achieve 1 (see the examples within Sections 1 and 6). For increasing n, the $(n+1)!$ shrinking matrices somehow distort the real situation, since it is unlikely that all possible shrinking steps occur for one particular function.

It is clear that Algorithm 2 has a wider convergence set than Algorithm 1.

If we discard the shrinking operations (the case of strictly convex functions), then Theorem 5 also holds for the pair of sets

$$\widetilde{\mathcal{T}}_w^\ell = \left\{T_{i_1} \cdots T_{i_\ell} : T_{i_j} \in \widetilde{\mathcal{T}} \setminus \mathcal{M}_1,\ j = 1, \ldots, n+1\right\},$$

$$\widetilde{\mathcal{T}}_w^\ell(q) = \left\{T_{i_1} \cdots T_{i_\ell} \in \widetilde{\mathcal{T}}_w^\ell : \|C(T_{i_1} \cdots T_{i_\ell})\| \leq q\right\}$$

and

$$\mathcal{T}_w^\ell = \left\{T_{i_1} \cdots T_{i_\ell} : T_{i_j} \in \mathcal{T} \setminus \mathcal{M}_2,\ j = 1, 2, \ldots, \ell\right\},$$

$$\mathcal{T}_w^\ell(q) = \left\{T_{i_1} \cdots T_{i_\ell} \in \mathcal{T}_w^\ell : \|C(T_{i_1} \cdots T_{i_\ell})\| \leq q\right\},$$

instead of $\mathbb{T}^\ell$ and $\mathbb{T}^\ell_q$, respectively. Define again the ratios $\tilde{r}_w(n,\ell) = \left|\tilde{\mathcal{T}}^\ell_w(q)\right| / \left|\tilde{\mathcal{T}}^\ell_w\right|$ and $r_w(n,\ell) = \left|\mathcal{T}^\ell_w(q)\right| / \left|\mathcal{T}^\ell_w\right|$. Note that for $n \le 10$ and some $0 < q < 1$, $r_w(n,\ell) \ge 4^\ell / (3n+3)^\ell$. Table 3 and 4 contain the computed values of $\tilde{r}_w(n,\ell)$ and $r_w(n,\ell)$, respectively.

Table 3. Ratios for Algorithm 1 without shrinking.

			$\tilde{\mathbf{r}}_\mathbf{w}(\mathbf{n},\boldsymbol{\ell})$			
$\mathbf{n} \setminus \boldsymbol{\ell}$	2	3	4	5	6	7
2	0.18056	0.24537	0.2963	0.32823	0.35566	0.37868
3	0	0.026367	0.062897	0.093401	0.11847	0.13871
4	0	0	0.00555	0.016575	0.029086	0.041108

Table 4. Ratios for Algorithm 2 without shrinking.

			$\tilde{\mathbf{r}}_\mathbf{w}(\mathbf{n},\boldsymbol{\ell})$			
$\mathbf{n} \setminus \boldsymbol{\ell}$	2	3	4	5	6	7
2	0.34568	0.46914	0.54687	0.61437	0.67157	0.71873
3	0	0.1169	0.24257	0.3278	0.39074	0.44305

The comparison of the last two tables shows again that Algorithm 1 has a smaller convergence set and a slower convergence rate than Algorithm 2.

We can observe that the larger the ℓ, the larger is the ratio in all four tables. Although the ratios cannot achieve 1, it is conjectured that they might be quite close to 1 for a large ℓ.

For both algorithms, we can observe the increase in convergence sets, if ℓ is increasing. For larger n values, this increase is slower, especially if no shrinking is allowed. However, this requires an enormous computational time to check, even for modest pairs of n and ℓ. The reason for this behavior is unknown and yet to be investigated. However, we can establish the following conclusions. In case of the convergence of the simplex sequence, Algorithm 2 is better than Algorithm 1. Hence, the ordering of Lagarias et al. [12] significantly improved the classical Nelder–Mead simplex method.

Theorem 5 indicates that the generated simplex sequence has a strong tendency to converge to one common limit point. This, together with Lemma 2 (or Lemma 3), may guarantee that the Nelder–Mead simplex algorithm generates good practical results.

In the case of the convergence of function values at vertices, the novel results are essentially the same for both methods. The extension of Theorems 2, 3 or 4 for $n > 2$ is an open question. For $n = 2$, all proofs exploit (implicitly or explicitly) an involuntory property, which does not hold for $n \ge 3$. How the convexity assumptions can be relaxed further is also an open question. If f is quasiconvex on $\mathbb{R}^2$ and bounded from below, then $f_2^* = f_3^*$. However, the other parts of the proof do not work.

Funding: This research received no external funding.

Institutional Review Board Statement: Not applicable.

Informed Consent Statement: Not applicable.

Data Availability Statement: Data is contained within the article or supplementary material.

Acknowledgments: The author is indebted to J. Abaffy and L. Szeidl for their help and comments. He is also indebted to the unknown referees whose observations and suggestions helped to improve this paper.

Conflicts of Interest: The author declares no conflicts of interest.

Appendix A

The matrices for the norm $\|A\|_\vartheta = \left\|S^{-1}AS\right\|_2$ and $n = 9, 10$.

The case where $n = 9$:

$$S = \begin{bmatrix} -1.2571 & -0.0128 & 0.5028 & 0.5047 & 2.3173 & -1.2682 & -1.4404 & -2.2437 & 1.1064 \\ -1.1260 & -1.8225 & -1.0287 & 0.0972 & -1.3367 & -1.2773 & -0.3227 & 0.7005 & 2.0421 \\ -1.7197 & -0.0434 & -0.1394 & 0.5115 & 0.1233 & -0.2385 & -0.9384 & 0.4162 & -2.5873 \\ 0.0424 & -0.0296 & 0.2238 & 1.3031 & 1.0884 & -0.0319 & 1.7164 & 2.1411 & 0.7368 \\ 1.1717 & 1.7668 & 0.2037 & 0.6359 & -1.3285 & -1.7570 & 0.1136 & -0.3241 & -1.0724 \\ -0.0138 & -0.9442 & 0.0342 & -0.8301 & 0.0499 & 0.2340 & 2.8213 & -1.0245 & -0.5072 \\ -0.0718 & 0.5593 & 2.5306 & -0.5053 & -1.5393 & 0.9299 & -0.4054 & 0.1010 & 0.0832 \\ 0.5396 & 0.2878 & -0.1743 & -2.5974 & 0.9844 & -0.1686 & 0.0242 & 1.0429 & -0.1559 \\ 2.1307 & -1.4421 & 0.4763 & 0.5795 & -0.0088 & 0.4682 & -1.0355 & 0.0462 & -0.5015 \end{bmatrix}$$

The case where $n = 10$:

$$S = \begin{bmatrix} 2.0568 & 4.2600 & -0.8876 & -0.1566 & -1.9386 & -0.0651 & 0.2271 & -1.8056 & 0.1294 & 0.0664 \\ 0.4312 & -0.0886 & 1.8330 & 4.6222 & -0.0374 & -0.0071 & -0.0483 & 0.8386 & 0.2741 & -0.0184 \\ 3.4415 & -1.8884 & 0.8095 & -1.5183 & 1.2164 & -1.3902 & -0.0710 & -0.0837 & -1.6783 & -0.0210 \\ -2.1716 & -0.0643 & -1.0052 & 0.0034 & -1.0715 & -0.8995 & 2.6609 & 0.0733 & -2.7365 & 0.5002 \\ -1.3924 & -0.2185 & 0.0169 & 0.3663 & -0.0497 & -0.4920 & -4.0342 & -0.1023 & -1.1653 & 1.0472 \\ -0.1115 & -0.9027 & 1.9524 & -2.1948 & -2.7482 & 0.0134 & 0.0722 & 1.3448 & 1.2371 & -0.3717 \\ -0.5131 & -0.0069 & 0.8486 & -0.0600 & 0.7613 & 3.9197 & 0.5919 & -0.5482 & -1.0459 & -0.4000 \\ 0.9491 & -1.8435 & -3.2427 & 0.4871 & -0.5311 & 0.8950 & -0.6546 & 0.7932 & 0.6149 & -0.3206 \\ -1.0854 & 0.3223 & -0.1177 & -0.5542 & 0.7593 & -1.2426 & 0.0307 & 0.0804 & 0.2414 & -3.5372 \\ -0.0960 & 1.6263 & -0.1063 & -0.4209 & 1.6908 & -0.0096 & 0.1650 & 2.4456 & 0.5468 & 1.1978 \end{bmatrix}$$

References

1. Nelder, J.A.; Mead, R. A simplex method for function minimization. *Comput. J.* **1965**, *7*, 308–313. [CrossRef]
2. Spendley, W.; Hext, G.; Himsworth, F. Sequential application of simplex designs in optimisation and evolutionary operation. *Technometrics* **1962**, *4*, 441–461. [CrossRef]
3. Audet, C.; Hare, W. *Derivative-free and Blackbox Optimization*; Springer: Berlin/Heidelberg, Germany, 2017. [CrossRef]
4. Conn, A.; Scheinberg, K.; Vicente, L. *Introduction to Derivative-Free Optimizations*; SIAM: Philadelphia, PA, USA, 2009. [CrossRef]
5. Kelley, C. *Iterative Methods for Optimization*; SIAM: Philadelphia, PA, USA, 1999. [CrossRef]
6. Kochenderfer, M.; Wheeler, T. *Algorithms for Optimization*; The MIT Press: Cambridge, MA, USA, 2019.
7. Walters, F.; Morgan, S.; Parker, L.; Deming, S. *Sequential Simplex Optimization*; CRC Press LLC: Boca Raton, FL, USA,1991.
8. Larson, J.; Menickelly, M.; Wild, S. Derivative-free optimization methods. *Acta Numer.* **2019**, *28*, 287–404. [CrossRef]
9. Rykov, A. *System Analysis. Models and Methods of Decision Making and Search Engine Optimization (in Russian)*; MISiS: Moscow, Russia, 2009.
10. Heeman, P. On Derivative-Free Optimisation Methods. Master's Thesis, University of Bergen, Bergen, Norway, 2023.
11. Rios, L.; Sahinidis, N. Derivative-free optimization: A review of algorithms and comparison of software implementations. *J. Glob. Optim.* **2013**, *56*, 1247–1293. [CrossRef]
12. Lagarias, J.; Reeds, J.; Wright, M.; Wright, P. Convergence properties of the Nelder-Mead simplex method in low dimensions. *SIAM J. Optimiz.* **1998**, *9*, 112–147. [CrossRef]
13. Kelley, C. Detection and remediation of stagnation in the Nelder-Mead algorithm using an sufficient decrease condition. *SIAM J. Optimiz.* **1999**, *10*, 43–55. [CrossRef]
14. Galántai, A. Convergence theorems for the Nelder-Mead method. *J. Comput. Appl. Mech.* **2020**, *15*, 115–133. [CrossRef]
15. Galántai, A. Convergence of the Nelder-Mead method. *Numer. Algorithms* **2022**, *90*, 1043–1072. [CrossRef]
16. Galántai, A. A stochastic convergence result for the Nelder-Mead simplex method. *Mathematics* **2023**, *11*, 1998. [CrossRef]
17. Wright, M. Direct search methods: Once scorned, now respectable. In *Numerical Analysis 1995 (Proceedings of the 1995 Dundee Biennial Conference in Numerical Analysis)*; Griffiths, D., Watson, G., Eds.; Addison-Wesley Longman: Harlow, UK, 1996; pp. 191–208.
18. Wright, M. Nelder, Mead, and the other simplex method. In *Documenta Mathematica, Extra Volume: Optimization Stories*; FIZ Karlsruhe GmbH: Eggenstein-Leopoldshafen, Germany, 2012; pp. 271–276.
19. McKinnon, K. Convergence of the Nelder-Mead simplex method to a nonstationary point. *SIAM J. Optimiz.* **1998**, *9*, 148–158. [CrossRef]
20. Galántai, A. Convergence of the Nelder-Mead method for convex functions. *Acta Polytech. Hung.* **2024**, *21*, 185–202. [CrossRef]
21. Galántai, A. A convergence analysis of the Nelder-Mead simplex method. *Acta Polytech. Hung.* **2021**, *18*, 93–105. [CrossRef]

22. Hendrix, E.; G.-Tóth, B. *Introduction to Nonlinear and Global Optimization*; Springer: Berlin/Heidelberg, Germany, 2010. [CrossRef]
23. Oldenburger, R. Infinite powers of matrices and characteristic roots. *Duke Math. J.* **1940**, *6*, 357–361. [CrossRef]
24. Meyer, C. *Matrix Analysis and Applied Linear Algebra*; SIAM: Philadelphia, PA, USA, 2000.
25. Horn, R.; Johnson, C. *Matrix Analysis*; Cambridge University Press: Cambridge, UK, 2013. [CrossRef]
26. Barnett, S. *Matrices: Methods and Applications*; Clarendon Press: Oxford, UK, 1990.
27. Kittaneh, F. Singular values of companion matrices and bounds of zeros of polynomials. *SIAM J. Matrix. Anal. Appl.* **1995**, *16*, 333–340. [CrossRef]
28. Montano, E.; Salas, M.; Soto, R. Positive matrices with prescribed singular values. *Proyecciones* **2008**, *27*, 289–305. [CrossRef]
29. Trench, W. Characterization and properties of matrices with k-involutory symmetries. *Linear Algebra Its Appl.* **2008**, *429*, 2278–2290. [CrossRef]
30. Mirsky, L. *An Introduction to Linear Algebra*; Dover Publications, Inc.: New York, NY, USA, 1990.
31. Hartfiel, D. *Nonhomogeneous Matrix Products*; World Scientific: Singapore, 2002. [CrossRef]
32. Beyn, W.J.; Elsner, L. Infinite products and paracontracting matrices. *Electron. J. Linear Al.* **1997**, *2*, 1–8. [CrossRef]
33. Blondel, V.; Tsitsiklis, J. The boundedness of all products of a pair of matrices is undecidable. *Syst. Control Lett.* **2000**, *41*, 135–140. [CrossRef]
34. Daubechies, I.; Lagarias, J. Sets of Matrices All Infinite Products of Which Converge. *Linear Algebra Its Appl.* **1992**, *161*, 227–263. [CrossRef]

algorithms

MDPI

Article

New Insights into Fuzzy Genetic Algorithms for Optimization Problems

Oleksandr Syzonov [1], Stefania Tomasiello [1,2,*] and Nicola Capuano [3]

[1] Institute of Computer Science, University of Tartu, 50090 Tartu, Estonia; sizonov2000@gmail.com
[2] Department of Industrial Engineering, University of Salerno, 84084 Fisciano, Italy
[3] Department of Information Engineering, Electrical Engineering and Applied Mathematics, University of Salerno, 84084 Fisciano, Italy; ncapuano@unisa.it
* Correspondence: stomasiello@unisa.it or stefania.tomasiello@ut.ee

Abstract: In this paper, we shed light on the use of two types of fuzzy genetic algorithms, which stand out from the literature due to the innovative ideas behind them. One is the Gendered Fuzzy Genetic Algorithm, where the crossover mechanism is regulated by the gender and the age of the population to generate offspring through proper fuzzy rules. The other one is the Elegant Fuzzy Genetic Algorithm, where the priority of the parent genome is updated based on the child's fitness. Both algorithms present a significant computational burden. To speed up the computation, we propose to adopt a nearest-neighbor caching strategy. We first performed several experiments, using some well-known benchmark functions, and tried different types of membership functions and logical connectives. Afterward, some additional benchmarks were retrieved from the literature for a fair comparison against published results, which were obtained by means of former variants of fuzzy genetic algorithms. A real-world application problem, which was retrieved from the literature and dealt with rice production, was also tackled. All the numerical results show the potential of the proposed strategy.

Keywords: nearest neighbor; caching; fuzzy rules; fitness; priority

1. Introduction

Fuzzy genetic algorithms (FGAs) are kinds of genetic algorithms (GAs) that employ fuzzy logic (FL) at some point in the process of mutation, crossover, selection, or updating the hyperparameters. The fuzzy reasoning has been exploited by several authors for the complex problem of dynamic control of GA parameters. There is ample literature devoted to the application of FGAs, e.g., in security [1,2], engineering [3,4], web applications [5], and manufacturing [6]. Despite the numerous articles of this type, there are not so many that introduce variants of FGAs. In [7], the authors proposed FL-controlled GAs, where mutation and crossover probabilities were dynamically adjusted by FL controllers. The application problem they considered was simulations of economic dispatch and emission dispatch. In [8], the authors presented dynamically controlled GA simulations, with population size and mutation and crossover rates adjusted using the FL controller. Inputs to the controller were genotypic diversity and phenotypic diversity, and outputs of the controller were population size change, mutation rate change, and crossover rate change. At any rate, in [9], it was stated that using a population diversity metric in an FGA can significantly improve premature convergence. Input variables of the fuzzy controller were population diversity, average population diversity divided by the maximum population diversity, and the number of unchanged greatest fitness values. As in some of the previously reviewed works, authors used mutation and crossover probabilities as output variables. An improved version of the FGAs in [8] was presented in [10], modifying some fuzzy inference rules and introducing a scaling factor for normalizing the input variables. In [11], a GA with varying population size (GAVaPS) was enhanced with a fuzzy logic controller by adapting the crossover probability according to a fuzzy rule base. The resulting method was called fuzzy-based lifetime extended GA (LTexGA). A hybrid fuzzy genetic algorithm (HFGA) was developed in [12]. In this FGA variant, the average fitness value and the best fitness value of each generation were adopted for the dynamical tuning of the crossover and mutation probabilities. The HFGA was used to tackle the problem of an optimal grouping of tank crews, based on training time, qualifications, and specialization of soldiers. The authors of [13] tried to improve the adaptive method of [10]. In the proposed improved fuzzy genetic algorithm (IFGA), the average fitness value and standard deviation changes between two consecutive generations were selected as the input variables, using two adaptive scaling factors to normalize the input variables. An interesting variant of FGA was introduced in [14] a few years later. The authors proposed a gendered fuzzy genetic algorithm (GFGA), i.e., GA with gendered selection based on fuzzy rules and age/lifetime formulas. GFGA was successfully used to generate the weights of a neural network (NN). More recently, the elegant fuzzy genetic algorithm (EFGA) was proposed [15] with a min-heap data structure for prioritizing genomes for the optimization of the initial weights of complex-valued NNs. In the EFGA, mutation rate, crossover rate, and reproductive group size were updated using fuzzy rules. In [3], the authors introduced a bilinear fuzzy fitness function in a standard GA to design optimal diagrid structures for tall buildings under engineering design constraints. In [2], the population selection was based on fuzzy clustering, with no explicit mention of fuzzy rules.

Here, we focus on FGAs using fuzzy rules. GAs remain largely unexplained to the end-user in terms of their search properties. An attempt to make GAs explainable was made by using surrogate models, which gave a partial understanding of the solutions limited to a single generation [16]. Needing to be trained, such a model comes with an additional computational cost. Fuzzy rules for controlling the GA's search properties bring an enormous advantage: they can add explainability.

At any rate, in all the cited articles, the computational effort has not been properly discussed. Here, we consider the most recent fuzzy-rule-based variants of GAs, i.e., EFGA, along with GFGA, with the latter conceived in a very different way with respect to the other FGAs. We compare their performance on a set of benchmarks, noticing that there is no comparative analysis of the two in the literature. In fact, these kinds of FGAs were originally introduced as part of a learning algorithm of some type of NN. Here, the aim is not to replicate such applications but to explore the use and possible enhancement of the EFGA and GFGA for more general optimization problems. We discuss a possible strategy to speed up the fuzzy inference in these FGA variants. Such a strategy is inspired by nearest-neighbor caching (NNC), which is well known in the context of web applications [17] and wireless networks [18], but, to the best of our knowledge, was never applied to any specific case.

Some preliminary experiments on some well-known benchmarks were performed. The outcome of such experiments is that the results obtained by using the NNC are sufficiently accurate without any losses in performance and give a speed-up of over 90%, thus making it a viable alternative. For a fair comparison against former variants of FGAs, some case studies were retrieved from the literature, which include Dejong's functions [19] and the traveling salesman problem (TSP) [11], thus confirming the good performance of the approach. Additionally, a real-world application problem related to rice production was investigated following the model and the GA results discussed in [20].

The contribution of the paper can be summarized as follows:

- An NNC strategy is discussed as a way to significantly speed up the fuzzy inference in two promising FGAs, from the explainability perspective, i.e., GFGA and EFGA, without any performance losses;
- Variants of the original GFGA and EFGA, in terms of fuzzy logic settings, are explored;
- A comparison against traditional and state-of-the-art techniques, which are both fuzzy and non-fuzzy, is presented over benchmark cases and a real-world application problem.

The paper is structured as follows. The next section is devoted to recalling some notions on fuzzy sets and the two considered FGA variants, i.e., EFGA and GFGA. In Section 3, the NNC strategy is described. Section 4 deals with the numerical experiments and discussion. Finally, some conclusions close the paper.

2. Fuzzy Genetic Algorithms: Dealing with Individuals and Fitness

In this section, we briefly present the considered FGAs, along with a few preliminaries.

2.1. Preliminaries

2.1.1. Fuzzy Sets

A fuzzy set generalizes the classical set, allowing not only for the classical set's 0–1 membership degrees but also a continuum of values in the range [0, 1]. A fuzzy set A is uniquely defined over a universe of discourse U by a membership function (MF) μ, $\mu_A : U \to [0,1]$. A fuzzy set is normal when its highest membership degree equals 1.

The most common MF is the triangular function. Another popular option is the trapezoidal MF:

$$\mu(x) = \begin{cases} 0 & \text{if } x \leq a \\ \frac{x-a}{b-a} & \text{if } a \leq x \leq b \\ 1 & \text{if } b \leq x \leq c \\ \frac{d-x}{d-c} & \text{if } c \leq x \leq d \\ 0 & \text{if } d \leq x \end{cases} \quad (1)$$

which models the linguistic expression "approximately b to c" or "close to [b, c]". Dealing with a central interval (i.e., [b, c]), instead of a point as for the triangular MF, this type of MF allows to take into account a higher level of uncertainty.

It is worth recalling the pi-shaped MF:

$$\mu(x) = \begin{cases} 0, & \text{if } x \leq a \\ 2(\frac{x-a}{b-a})^2, & \text{if } a \leq x \leq \frac{a+b}{2} \\ 1 - 2(\frac{x-b}{b-a})^2, & \text{if } \frac{a+b}{2} < x \leq b \\ 1, & \text{if } b < x \leq c \\ 1 - 2(\frac{x-c}{d-c})^2, & \text{if } c < x \leq \frac{c+d}{2} \\ 2(\frac{x-d}{d-c})^2, & \text{if } \frac{c+d}{2} < x \leq d \\ 0, & \text{if } x \geq d \end{cases} \quad (2)$$

which represents a smoother alternative to the trapezoidal MF.

The connectives from the classical logic AND, OR are generalized in fuzzy logic via t-norms and t-conorms, respectively. A typical t-norm is the minimum operator, and a typical t-conorm is the maximum operator.

A fuzzy partition can be seen as a series of m partially overlapping normal fuzzy sets with $m \geq 2$. Such fuzzy sets represent the terms of a linguistic variable. For instance, the linguistic variable temperature may have the terms {cold, cool, warm}, each one expressed by a fuzzy set, i.e., a proper membership value. As in our natural language, the terms represent the attributes of the linguistic variable. Any linguistic variable can assume a certain value, i.e., membership degree, which has relevance in fuzzy IF-THEN rules. Figure 1 recalls the generic N terms of a linguistic variable (here denoted as term_1,. . . ,term_N), determined by syntactic rules to form eventually fuzzy rules. Examples of fuzzy rules will be provided in the next sections.

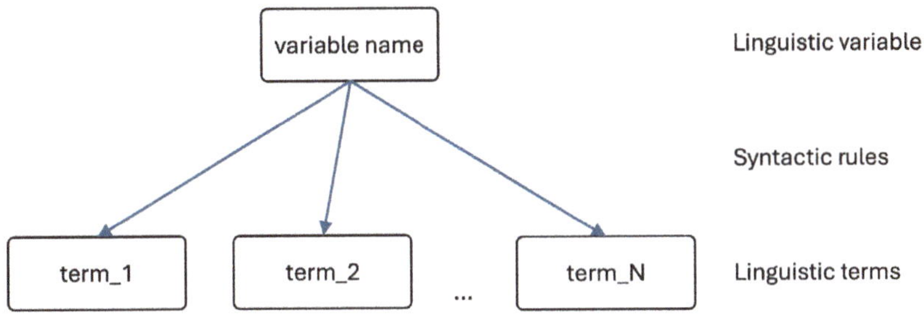

Figure 1. Linguistic variable, terms, and syntactic rules.

2.1.2. Genetic Algorithms

Natural evolution is the source of inspiration for GAs, and typical evolutionary terminology can be found through the algorithm. One might think of the optimization problem as the environment and the proposed solution as a single entity inside it. Individual fitness is correlated with the quality of the candidate solution as determined by the objective function. The population is first randomly initialized by the algorithm, which then iterates through generations. Every individual in a generation is assessed for fitness, and new individuals (offspring) are produced depending on that evaluation. By applying mutations or merging parents (crossover), a new population is created.

2.2. EFGA

In EFGA, as in traditional GAs, an initial set of individuals is randomly generated. The approach relies on a min-heap data structure keyed at fitness in order to reduce the number of fitness function calls. The min-heap data structure is a binary tree, where each node is associated with a pair of numbers, i.e., the index assigned to the chromosome as a unique identifier and the fitness value. A simple example of min-heap is shown in Figure 2 where the integer in any node represents the above-mentioned index and the other value is the fitness value. The solutions that show minimal fitness are put on top levels of the tree, and they are more likely to be selected. The root presents the best solution. The algorithm is stopped when either the difference between the best and worst solution is below a certain threshold value (e.g., 0.003) or when the number of iterations exceeds the maximum iterations. Additionally, a large initial population is created to increase the likelihood of being close to local minima early on. However, large populations also have a lot of low-quality solutions. The authors of the original approach refer to the Pareto front by allowing only the top 20% of the individuals to reproduce.

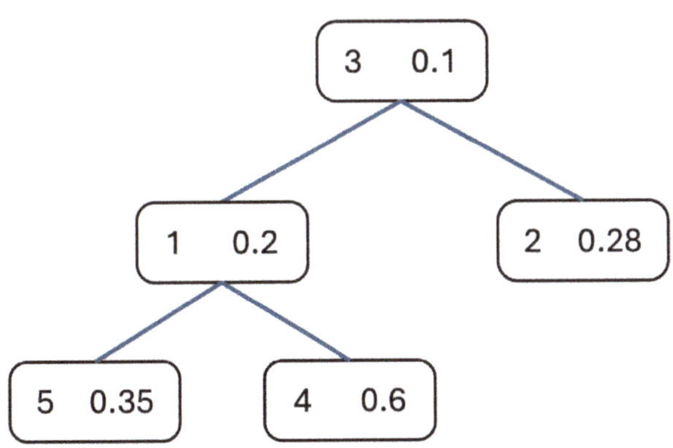

Figure 2. A simple example of a min-heap scheme.

In EFGA, fuzzy rules are used to update the simulation parameters. Average fitness, best fitness, and average fitness change are used as inputs. The first two variables are associated with the term set *{Excellent, Acceptable, Poor}*, and the last one with the term set *{Low, Medium, High}*. Crossover rate and mutation rate may take the linguistic values *{Low, Medium, High}*, while reproductive group size may take *{Low, Medium, Large}*. Fuzzy rules are also used for priority updates. After evaluating a child's genome fitness, the parent genome priority is updated. Linguistic values for the child's genome fitness are *{Excellent, Acceptable, Poor}*, the term set for the priority is *{Very Low, Low, Medium, High, Very High}*. Let $N_{priority}$ and $N_{fitness}$ be the number of terms of priority and fitness, respectively. We notice that $N_{priority} = 2N_{fitness} - 1$.

Remark 1. *We observe that the term sets for fitness and priority, respectively, could include more attributes as follows:*

- *{Great, Good, Fair, Average, Bad} and {Very Low, Low, Moderate, Medium, High, Very High, Extremely High, Exceptionally High, Extraordinary}, i.e.,* $N_{fitness} = 5$;
- *{Exceptional, Great, Good, Average, Mediocre, Poor, Terrible} and {Very Low, Low, Moderately Low, Below Average, Average, Above Average, Moderate, High, Very High, Extremely High, Exceptionally High, Extraordinary, Exceptional}, i.e.,* $N_{fitness} = 7$;

having increased the linguistic attributes of priority proportionally.

An example of a fitness linguistic variable, with five terms expressed as trapezoidal MFs, is shown in Figure 3. In the original method, triangular MFs were adopted, but here we investigate trapezoidal MFs, in addition to pi-shaped MFs, as they allow for a higher degree of uncertainty, as explained in Section 2.1.1.

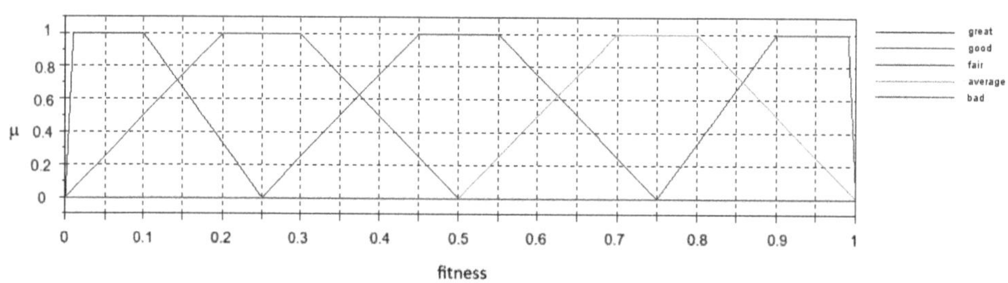

Figure 3. Fitness linguistic variable with trapezoidal MFs ($N_{fitness} = 5$).

The rules are of the form

```
IF f1 IS term_i AND f2 IS term_j THEN priority is term_(2N-i-j-2)
```

where $f1$ and $f2$ are the fitness of the parent genomes.

The goal is to move parent genomes with offspring having better fitness higher in the priority queue and parent genomes with worse offspring lower in the priority queue.

Regarding diversity, some formulas are meant to preserve it, as explained in the following.

Let $\rho_1^{(k)}, \rho_2^{(k)}$ be two chromosomes in the kth generation. A crossover is performed by randomly selecting two positions $\overline{p}_1, \overline{p}_2 \in (0, 1]$, with $\overline{p}_1 < \overline{p}_2$.

The crossover operation is performed according to the following:

$$\rho_1^{(k+1)}(\overline{p}_1, \overline{p}_2) = \gamma\rho_1^{(k)}(\overline{p}_1, \overline{p}_2) + (1-\gamma)\rho_2^{(k)}(\overline{p}_1, \overline{p}_2) \quad (3)$$

$$\rho_2^{(k+1)}(\overline{p}_1, \overline{p}_2) = (1-\gamma)\rho_1^{(k)}(\overline{p}_1, \overline{p}_2) + \gamma\rho_2^{(k)}(\overline{p}_1, \overline{p}_2), \quad (4)$$

where $\rho_i^{(j)}(\overline{p}_1, \overline{p}_2)$, with $i = 1, 2$, and $j = k, k+1$, represents a subchromosome, from position $\overline{p}_1$ to $\overline{p}_2$, and $\gamma \in [-1, 1]$. As the crossover is performed on the subchromosome, the latter is (randomly) altered while some parent's genes are retained. This helps with diversity.

2.3. GFGA

The original method splits the whole population into two groups with different genders. While in standard GAs, a crossover between any two individuals is possible, in the GFGA, this is possible only for opposite-gender individuals, as it works in nature. Moreover, selection rules are entirely different for male and female genomes. Some male genomes are randomly selected and matched with female genomes based on population diversity and male genomes' age via fuzzy rules. Each genome has an age, which is simply the number of iterations that have passed since its initialization. Selection is applied based on the maximum age of the individual; when its age is greater than the maximum age, the genome is removed from the population.

The linguistic variable *age* has four terms: *{Infant, Teenager, Adult, Elderly}*. The term set for the population diversity is *{Very Low, Low, Medium, High}*. At the selection stage, males are selected randomly, and the age of female partners is estimated based on the age of male partners and population diversity using fuzzy rules. Then, a partner with the age closest to the desired one is selected, and crossover is performed.

Remark 2. *The term set for age and partner age could be the following (where $N_{age} = N_{partner-age}$):*

- *{Infant, Child, Teenager, Adult, Elderly};*
- *{Newborn, Infant, Toddler, Child, Teenager, Adult, Elderly};*

including more linguistic attributes.

Similarly, the term set for diversity could be the following:

- *{Very Low, Low, Medium, High, Very High};*
- *{Extremely Low, Very Low, Low, Medium, High, Very High, Extremely High};*

Notice that for the highest age term, the second-highest age term for potential partners is prioritized if there is the highest diversity. With the lowest two diversity terms, there is the lowest age term for potential partners. Hence, the general rule can be written as follows:

```
IF age IS term_i AND diversity IS term_j  THEN partner_age IS
term_{2N-(i+j)+2}}
```

except for the lowest diversity term or the highest age term and second-lowest diversity term:

```
IF (age IS term_i AND diversity is term_1) OR (age IS term_N
AND diversity IS term_2) THEN partner_age IS term_1.
```

Let us assume $N_{age} = N_{partner\_age} = N_{diversity} = 5$, for example. Then the following rule is obtained:

```
IF age IS adult AND diversity IS high  THEN partner_age IS adult
```

since the index of the age and diversity terms is 4, resulting in an index for the partner age, which equals 4.

For $N_{age} = N_{partner\_age} = N_{diversity} = 7$, a similar rule is obtained:

```
IF age IS adult AND diversity IS very high THEN partner_age IS adult
```

since the index of the age and diversity terms is 5, resulting in an index for the partner age, which equals 6.

For low and very low diversity, the following rule holds with $N_{age} = N_{partner\_age} = N_{diversity} = 5$.

```
IF (age IS adult AND diversity is very low) OR (age IS elderly
AND diversity IS low) THEN partner_age IS infant.
```

Let m and M be the minimum and the maximum age in the population, and let n and P denote the population size and half of the population, respectively. The population *diversity* linguistic variable is defined as follows:

$$D_{pop}(P) = \begin{cases} very \quad high & P \leq L + t \\ high & m + t < P \leq m + t + 1 \\ medium & m + t + 1 < P \leq m + t + 2 \\ low & m + t + 2 < P \leq m + t + 3 \\ very \quad low & P > m + t + 3 \end{cases} \tag{5}$$

where $t = \left[\lambda \frac{m+M}{n}\right]$, $\lambda = \left[\frac{n}{10}\right]$, with $[.]$ denoting the rounding operation returning the nearest integer number.

Figure 4 shows the *diversity* variable (5) with trapezoidal MFs.

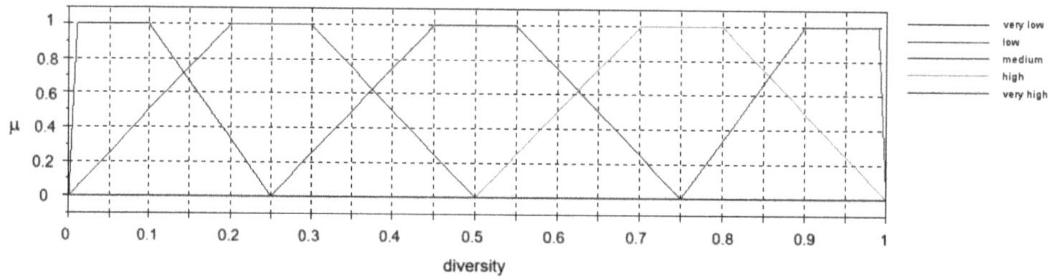

Figure 4. Diversity linguistic variable with trapezoidal MFs ($N_{diversity} = 5$).

3. Nearest-Neighbor Caching

Here, we refer to caching as a technique to avoid repetitive computations thanks to some values stored in memory. Such values are retrieved when a function call with the same or similar inputs is needed. This may significantly reduce the runtime. Formally, there exists a metric space (X, d) over the presented vector $\mathbf{x}$, and a threshold radius R. When a new vector $\overline{\mathbf{x}}$ (query) arrives, and if there is a stored vector $\mathbf{x}$ in the cache with $d(\mathbf{x}, \overline{\mathbf{x}}) \leq R$, the request can be satisfied by the cache without any new computations. Let k be the size of the cache. It is expected that k increases as R tends to zero [21].

Operationally, the vector space is stored in a matrix. Distances from a query vector to the matrix rows are computed. A certain number of closest neighbors are selected, and their distances are stored. After finding k nearest neighbors, a weighted average of them is computed as an approximation of a certain function value:

$$\overline{g_i} = \sum_{i=1}^{k} w_i g_i, \tag{6}$$

where g_i is a value of a function at the ith nearest point and $w_i \in [0, 1]$ is a weight such that $\sum_{i=1}^{k} w_i = 1$.

The weights of the nearest neighbors w_i can be computed as normalized L_2 distances to the nearest neighbors d_i:

$$w_i = \frac{d_i}{\sum_{j=1}^{N} d_j}. \tag{7}$$

For the remainder of this work, EFGA-NNC and GFGA-NNC will denote the implementation of the respective methods with NNC.

4. Experiments and Discussion

4.1. Initial Experiments on Popular Benchmarks

Experiments were run on the same initial population (ensured with the random seed) through 500 generations. The population size was 100, as in [15]. For any considered benchmark function, there were 20 runs, considering two possible dimensions, i.e., $D \in \{5, 1000\}$. The minimum type t-norm and the product type t-norm were considered. Similarly, the maximum type and the sum type t-conorm were employed. Regarding the membership functions (MFs), two options were explored, i.e., pi-shaped and trapezoid. The aim here was to explore variants of the original EFGA and GFGA, where only the triangular MFs and the minimum t-norm were used.

For EFGA, it was chosen $N_{fitness} = \{3, 5, 7\}$. The same values for N_{age} are in GFGA. A larger number of terms would not be beneficial for interpretability.

Four typical benchmarks for algorithms in optimization tasks were considered:

- The Schwefel function

$$f_1(\mathbf{x}) = 418.9829d - \sum_{i=1}^{D} x_i \sin(\sqrt{|x_i|}), \quad x_i \in [-500, 500], \tag{8}$$

- The Rastrigin function

$$f_2(\mathbf{x}) = 10D + \sum_{i=1}^{D} \left(x_i^2 - 10\cos(2\pi x_i)\right), \quad x_i \in [-5.12, 5.12], \tag{9}$$

- The Griewank function

$$f_3(\mathbf{x}) = \sum_{i=1}^{D} \frac{x_i^2}{4000} - \prod_{i=1}^{D} \cos(\frac{x_i}{\sqrt{i}}) + 1, \quad x_i \in [-40, 40], \tag{10}$$

- The Ackley function

$$f_4(\mathbf{x}) = -a\, e^{\left(-b\sqrt{\frac{1}{D}\sum_{i=1}^{D} x_i^2}\right)} - e^{\left(\frac{1}{D}\sum_{i=1}^{D}\cos(c\, x_i)\right)} + a + e, \quad x_i \in [-32.768, 32.768]. \tag{11}$$

The choice of the benchmark functions and the dimension has been motivated by the presence of relevant results in the literature, as will be discussed in the following.

The different MFs did not affect the results significantly: a difference of less than 5% on average was observed. Regarding t-norm or t-conorm types, we observed that the best results were obtained by using the combination of min t-norm and max t-conorm, and this combination was kept for the rest of the experiments, including those discussed in the next sections. We observed that the results between the two types of implementation (i.e., the original one and that with NNC) had no significant differences, i.e., not greater than 3%. Similarly, assuming $N_{fitness} > 5$ for EFGA-NNC and $N_{age} > 5$ for GFGA-NNC did not produce any significant changes in the results and even worsened them. Regarding the MFs, we observed a slightly better performance by using pi-shaped MFs. Hence, we fixed it for all the remaining experiments. Instead, the NNC strategy allowed to significantly shorten the runtime. The same experiments with the same fitness function and parameters were run several times, as described before, and mean runtime and runtime standard deviation were compared. The ratio s between the runtime with and without NNC in the case of EFGA is $s = 0.077 \pm 0.022$ and $s = 0.014 \pm 0.007$ for GFGA. Comparing the runtime of EFGA-NNC and GFGA-NNC to that of the standard GA, the latter is still more competitive, with 30–50% shorter runtimes on average, where the worst case is the high-dimensional one (i.e., D = 1000). This computational burden may be seen as the cost of explainability [16].

The results using the different techniques are depicted in Figure 5a,b. Both for $D = 5$ and $D = 1000$, there is no significant difference between the results by EFGA-NNC and

GAs, except for the case of the Shwefel function. In that case, the GA outperforms the other approaches, especially for $D = 5$. At any rate, the advantage over the standard GA is boosting explainability. Regarding the GFGA-NNC, it performs better for problems with higher dimensions, being comparable to the other approaches. The PSO outperforms the other approaches for the Rastrigin and Griewank functions. There is no significant difference among all the approaches in the case of the Ackley function, both for $D = 5$ and $D = 1000$. In all the cases, except for the Ackley function, with any dimension, the highest standard deviation was observed for the GFGA-NNC. For the EFGA-NNC, it was found that the standard deviation was generally comparable to that related to the GA or even less in a couple of cases (both for D = 5 and D = 1000), suggesting better stability of the EFGA-NNC with respect to the GFGA-NNC.

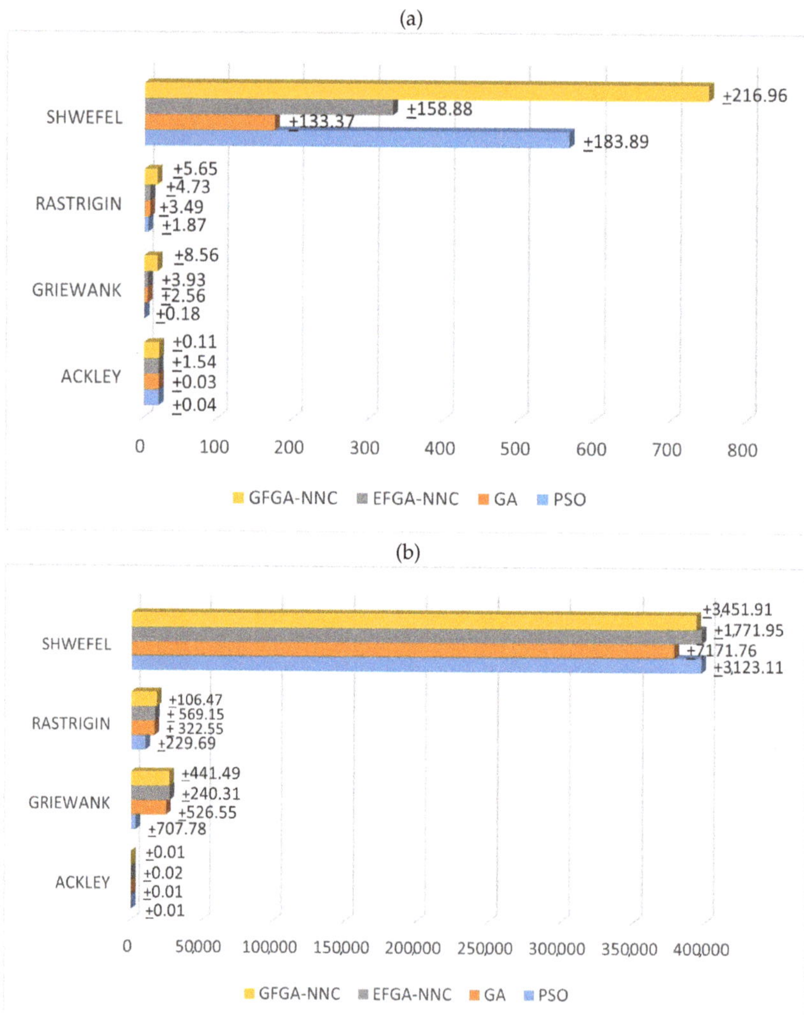

Figure 5. First set of benchmark functions. Average fitness value for the following: (**a**) D = 5; (**b**) D = 1000.

The f_2 function, with $D = 5$, was also considered in [22] and the results by a fuzzy logic controlled genetic algorithm (FCGA) [7], an improved fuzzy genetic algorithm (IFGA) [13] and another kind of FGA [23] were compared. The latter outperformed the other methods, providing an average fitness value of around 100 after 14 generations, while IFGA provided the closest value to the optimum in the least average number of generations, i.e., 1205. The authors used a population of 20 individuals. With the same population size and after 14 generations, we obtained 85 and 68 by GFGA-NNC and EFGA-NNC, respectively. In Figure 6a,b, one can see the effect of the type of MF, t-norm, and t-conorm on the result by EFGA-NNC, where $m = N_{fitness}$, for the sake of simplicity. One can observe that the best result was obtained by using the min t-norm and max t-conorm (max-min on the plot). Even though there is not much difference due to the type of MF, it is possible to see that the best result was obtained with $m = 5$ and pi-shaped MF. Similar behavior was observed over the other benchmarks.

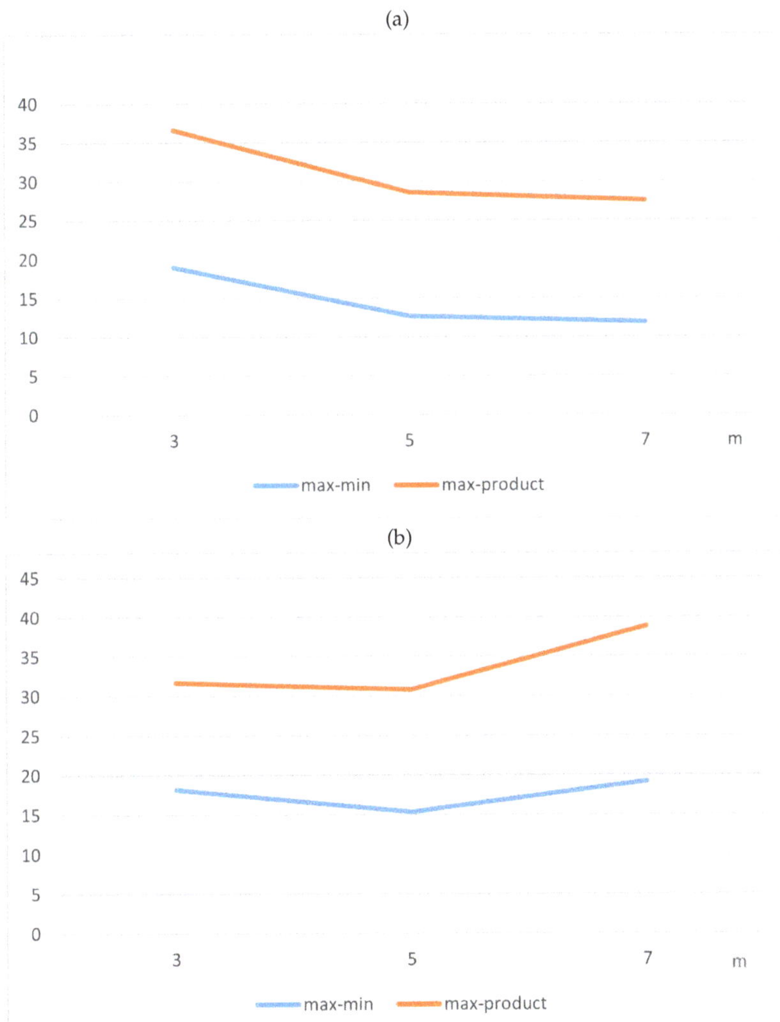

Figure 6. Rastrigin function (D = 5). Average fitness value after 500 generations with the following: (**a**) Pi-shaped MF; (**b**) Trapezoidal MF.

The benchmark functions f_1, f_2, and f_4 with the same domains and $D = 1000$ have been considered in [24], where the same number of runs was also adopted (i.e., 20). The authors of [24] used a standard GA as a baseline but with a population size of 50 and a number of iterations of 1000. They performed the experiments with their approach, i.e., the adaptive dimensionality reduction genetic algorithm (ADRGA), the hybrid genetic and simulated annealing algorithm (HGSA), and the adaptive genetic algorithm (AGA). We consider their published results for a fair comparison, reporting our results after 1000 generations. In Figure 7, the mean and standard deviation (sd) for the above-mentioned benchmark functions are shown. For the Ackley function, the results by EFGA-NNC and GFGA-NNC are quite close to that by ADRGA, also considering that the standard deviation for the latter is 20 times greater than that of EFGA and GFGA. A similar situation can be observed for the Shwefel function, with a standard deviation for the ADRGA, which is almost 5 and 4 times greater than that by EFGA-NNC and GFGA-NNC, respectively. For the Rastrigin function, the ADRGA outperforms. Unfortunately, in [24], there is no information about the computational cost of ADRGA or runtime.

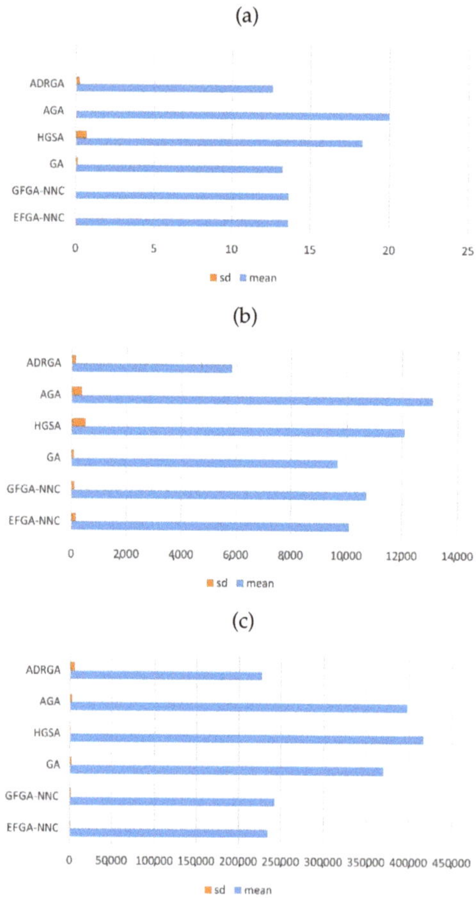

Figure 7. Comparison against state-of-the-art approaches in [24] ($D = 1000$; 1000 generations) for the following: (**a**) The Ackley function; (**b**) The Rastrigin function; (**c**) The Shwefel function.

4.2. Additional Benchmarks from the Literature

Here, we consider two functions retrieved from [19], i.e.,

$$g_1(x_1, x_2, x_3) = \sum_{i=1}^{3} x_i^2, \quad x_i \in [-5.12, 5.12], \tag{12}$$

with minimum $g_1(0,0,0) = 0$,

$$g_2(x_1, x_2) = 100(x_2 - x_1^2)^2 + (1 - x_1)^2, \quad x_i \in [-2.048, 2.048]. \tag{13}$$

with minimum $g_2(1,1) = 0$.

By the approach in [19], the global minimum for g_1 is reached after 100 generations with a final population size of 120 (though the accuracy is not reported in the paper). By the same approach, the global optimum was approached with accuracy $O(10^{-2})$ after 40 generations with a final population size of 60 for g_2. With a population size of 120 generations for both cases, we obtained an accuracy of $O(10^{-4})$ for g_1 and $O(10^{-3})$ for g_2 by EFGA-NNC. The results by GFGA-NNC are worse, particularly for g_1 with an accuracy of $O(10^{-3})$. All this can be observed in Figure 8. It can be seen that the convergence process is faster for EFGA-NNC, especially for the g_1 function, while the difference is not that significant in the case of the g_2 function. It must be recalled that, here, we refer to the best average results by EFGA-NNC and GFGA-NNC, obtained by using pi-shaped MFs, $N_{fitness} = 5$, $N_{age} = N_{diversity} = 5$.

4.3. Traveling Salesman Problem

The traveling salesman problem is another typical benchmark. Given a list of cities and the distances between each pair of cities, the TSP aims to find the shortest possible path such that all cities are visited just once. For a p-city TSP, the distances x_{ij} between the cities c_i and c_j are collected into a $p \times p$ matrix, $\mathbf{X} = \{x_{ij}\}$, known as the cost matrix. The number of its entries represents the number of variables of the problem, of course. The TSP is an NP-hard problem. Here we refer to [11] for a comparison against published results. Following [11], three problems were taken from TSPLIB95 [25]. They are all symmetric TSPs with Euclidean distance, namely Eil51, Eil76, Rat99, i.e., $p = 51$, $p = 76$ and $p = 99$, respectively. As in [11], we fixed a population size of 100, with 200 generations, through 30 runs. The age linguistic variable in [11] has three terms, i.e., young, middle-aged, and old. In GFGA-NNC, we tried both term sets presented in Section 2.2, even though the first term set with five terms provided the best results. Similarly, for the other term sets, even for EFGA-NNC, increasing from 5 to 7, the number of terms did not produce any significant improvement. For both approaches, the pi-shaped MF was adopted. The best results, i.e., the shortest path length by each approach, are shown in Figure 9. Although neither EFGA-NNC nor GFGA-NNC could outperform the baseline, they represent an improvement with respect to LTexGA [11]. The distance from the baseline grows as the problem cardinality increases. In all cases, EFGA-NNC performed better than GFGA-NNC and LTexGA, keeping the shortest distance from the baseline. EFGA-NNC offers a result very close to the baseline, i.e., 477.34 against 426, especially for Eil51, while it is 525.46 by LTexGA and 601.34 by GFGA.

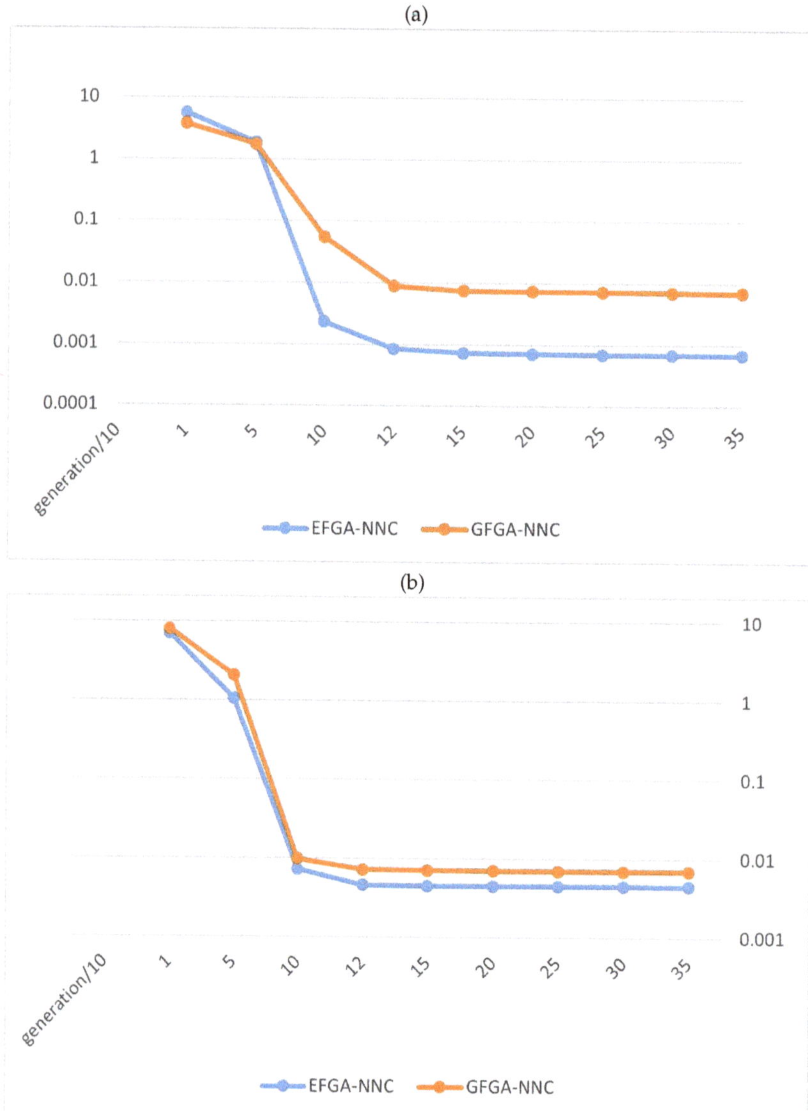

Figure 8. Average fitness value over the generations for the following: (**a**) The g_1 function; (**b**) The g_2 function.

Let r denote the ratio between the solution obtained by the proposed approaches and the baseline. Figure 10 shows how r varies with p for all the approaches. It can be observed that increasing the problem cardinality does not significantly affect the performance of EFGA-NNC and GFGA-NNC, while LTexGA seems to be influenced by it.

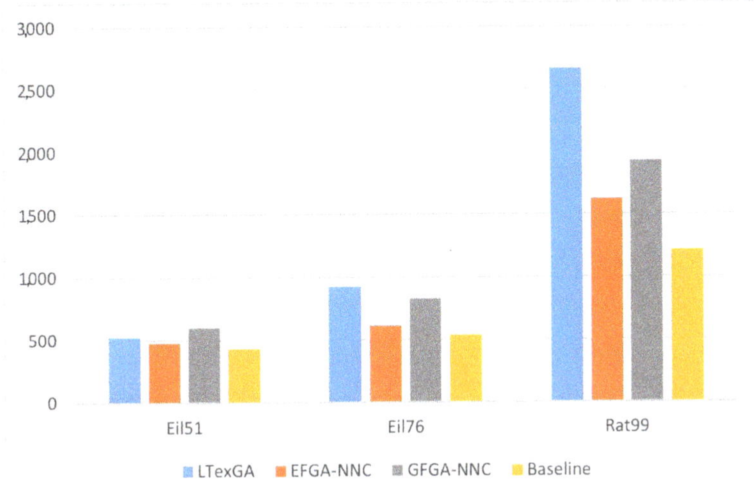

Figure 9. TSP: shortest path length by each approach.

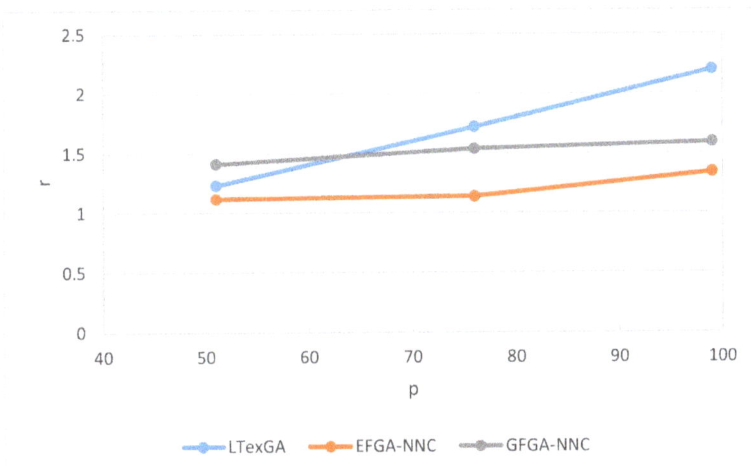

Figure 10. r vs. p.

4.4. A Real-World Problem: The Case of Rice Production

Rice is a staple crop in Asia, with China and India as the major producers in the world [26]. While rice is not a staple crop in most of Europe, there are countries, such as Italy, that have developed specific traditions around rice cultivation and cuisine. Italy is the main rice producer in Europe [26], with many well-known rice-based recipes. One important stage in the production of rice from paddy (rice with husk) is milling, which is the process of removing the husk and a portion of the bran using techniques like drying, winnowing, and whiting. This process is energy-demanding, requiring proper optimization. Here, we rely on a model discussed in [20], dealing with a case study from Iran, which is solved by GA. The optimization of converting paddy to white rice foresees three goals, i.e., maximize the output energy, minimize the global warming potential (GWP), and maximize the net profit. The GWP contribution represents the cost of CO_2 emission. It takes into account climate change damages, and it is one of the most important outputs of the Life Cycle Assessment (LCA) analysis, especially when dealing with crop production and supply chains [27–29]. The multi-objective optimization (MOO) problem above can be tackled as a single-objective optimization (SOO) problem (e.g., see [30]). This means that the benefits to be maximized are converted into costs to be minimized. The SOO problem can be tackled by summing up all the single contributions, and the resulting function to be minimized can be written as follows:

$$F(x_1, \ldots, x_6) = \sum_{i=1}^{3} \sum_{j=1}^{6} \alpha_{ij} x_j = \mathbf{A}\mathbf{x}, \qquad x_j \in X_j, \tag{14}$$

where x_1 is the energy equivalent of human labor, x_2 is the energy equivalent of machinery, x_3 is the energy equivalent of natural gas, x_4 is the energy equivalent of nylon, and x_5 is the energy equivalent of transportation. These independent variables can be arranged into a vector **x**. The respective domains (in MJ TIP^{-1}) are the following: $X_1 = [44, 67]$, $X_2 = [92, 346]$, $X_3 = [675, 2005]$, $X_4 = [45784, 81580]$, $X_5 = [934, 2240]$, $X_6 = [44, 133]$. The matrix **A** of the coefficients α_{ij} is the following:

$$\mathbf{A} = \begin{pmatrix} -0.99 & -2.70 & -1.35 & -2.02 & 0.18 & -0.19 \\ -0.28 & -0.01 & -0.30 & 3.15 & 2.21 & 0.21 \\ -2.64 & -0.44 & -1.58 & 2.85 & 0.42 & 0.65 \end{pmatrix} \tag{15}$$

All the numerical values above have been retrieved from [20]. In [20], the best solution was found after 26 generations. No information about the runtime was reported in [20]. The best solutions by EFGA-NNC and GFGA-NNC were retrieved after 43 generations. They are all depicted in Figure 11. The difference between the solutions by EFGA-NNC and the baseline [20] is around 10% on average. Same for GFGA-NNC.

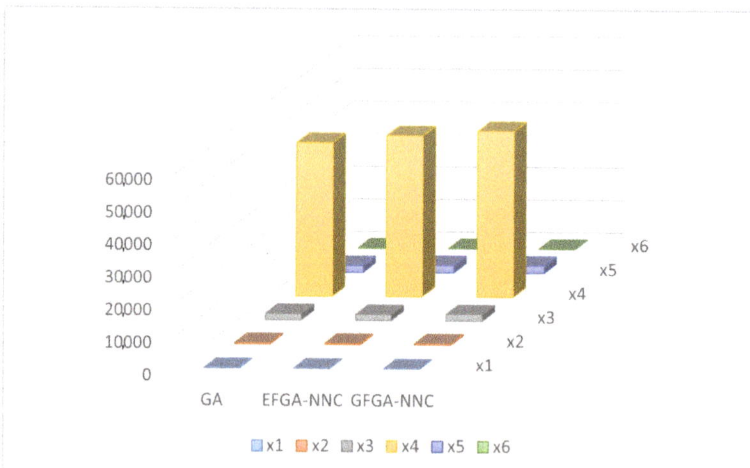

Figure 11. The rice production problem: best solutions.

5. Conclusions

In this paper, we focused on two kinds of FGAs, i.e., GFGA and EFGA, which have greatly contributed to the explainability of GAs but present a significant computational burden. To address this issue, we proposed employing a nearest-neighbor caching strategy. We performed several experiments on popular benchmarks, trying different membership functions and logical connectives. Algorithms based on the proposed strategy were compared to the original ones, and different hyperparameters were tried. The results showed significant time savings by means of the proposed NNC strategy, which is the ratio between the runtime with and without NCC $\frac{9}{200}$ on average. First, some classical benchmarks were considered, and each FGA type was compared to standard GA and PSO. Except for one case, EFGA with NNC and GA had almost the same performance, while a similar situation was observed for GFGA with NNC for problems with higher dimensions. Secondly, some cases from the literature were considered for a fair comparison against former variants of FGAs. Finally, a real-world application problem was tackled by comparing the results against the ones in the literature. All the results show, in general, the good performance of EFGA with NNC.

The advantage of such a technique is explainability. Even though the NNC allowed for a significant reduction in the runtime, as mentioned before, the standard GA still presents a shorter runtime. This is not surprising since, as discussed in the existing literature, explainability usually brings an additional computational cost. In future work, we plan to investigate how to efficiently deploy the technique for the training of neural networks in order to have an explainable learning process.

Author Contributions: Conceptualization, O.S. and S.T.; methodology, O.S. and S.T.; software, O.S. and N.C.; validation, O.S., S.T. and N.C.; investigation, O.S.; resources, O.S. and S.T.; data curation, O.S., S.T. and N.C.; writing—original draft preparation, S.T.; writing—review and editing, S.T.; supervision, S.T. All authors have read and agreed to the published version of the manuscript.

Funding: This research received no external funding.

Informed Consent Statement: Not applicable.

Data Availability Statement: No new data were created or analyzed in this study. Data sharing is not applicable to this article.

Acknowledgments: Stefania Tomasiello acknowledges support through the project *Big data and machine learning applications: developing a research direction—Increasing the knowledge intensity of Ida-Viru entrepreneurship* co-funded by the European Union.

Conflicts of Interest: The authors declare no conflicts of interest.

References

1. Lv, S.; Chen, H. Security Protection Technology in Multi-Attribute Data Transmission Based on Fuzzy Genetic Algorithm. *Wirel. Pers. Commun.* **2022**, *127*, 897–917. [CrossRef]
2. Rajasekar, V.; Predić, B.; Saracevic, M.; Elhoseny, M.; Karabasevic, D.; Stanujkic, D.; Jayapaul, P. Enhanced multimodal biometric recognition approach for smart cities based on an optimized fuzzy genetic algorithm. *Sci. Rep.* **2022**, *12*, 622. [CrossRef] [PubMed]
3. Ashtari, P.; Karami, R.; Farahmand-Tabar, S. Optimum geometrical pattern and design of real-size diagrid structures using accelerated fuzzy-genetic algorithm with bilinear membership function. *Appl. Soft Comput.* **2021**, *110*, 107646. [CrossRef]
4. Jahan, A.; Mollazadeh, M.; Akbarpour, A.; Khatibinia, M. Health monitoring of pressurized pipelines by finite element method using meta-heuristic algorithms along with error sensitivity assessment. *Struct. Eng. Mech.* **2023**, *87*, 211–219.
5. Song, P.; Chen, C.; Zhang, L. Evaluation Model of Click Rate of Electronic Commerce Advertising Based on Fuzzy Genetic Algorithm. *Mob. Netw. Appl.* **2022**, *27*, 936–945. [CrossRef]
6. Wang, S.; Hui, J.; Zhu, B.; Liu, Y. Adaptive Genetic Algorithm Based on Fuzzy Reasoning for the Multilevel Capacitated Lot-Sizing Problem with Energy Consumption in Synchronizer Production. *Sustainability* **2022**, *14*, 5072. [CrossRef]
7. Song, Y.H.; Wang, G.S.; Wang, P.Y.; Johns, A.T. Environmental/economic dispatch using fuzzy logic controlled genetic algorithms. *Gener. Transm. Distrib.* **1997**, *144*, 377. [CrossRef]
8. Subbu, R.; Sanderson, A.C.; Bonissone, P.P. Fuzzy logic controlled genetic algorithms versus tuned genetic algorithms: An agile manufacturing application. In Proceedings of the 1998 IEEE International Symposium on Intelligent Control (ISIC) Held Jointly with IEEE International Symposium on Computational Intelligence in Robotics and Automation (CIRA), Piscataway, NJ, USA, 17 September 1998; pp. 434–440.
9. Wang, K. A new fuzzy genetic algorithm based on population diversity. In Proceedings of the 2001 IEEE International Symposium on Computational Intelligence in Robotics and Automation (Cat. No.01EX515), Banff, AB, Canada, 29 July–1 August 2001; pp. 108–112.
10. Yun, Y.; Gen, M. Performance analysis of adaptive genetic algorithms with fuzzy logic and heuristics. *Fuzzy Optim. Decis. Mak.* **2003**, *2*, 161–175. [CrossRef]
11. Last, M.; Eyal, S. A fuzzy-based lifetime extension of genetic algorithms. *Fuzzy Sets Syst.* **2005**, *149*, 131–147. [CrossRef]
12. Liu, H.; Xu, Z.; Abraham, A. Hybrid fuzzy-genetic algorithm approach for crew grouping. In Proceedings of the 5th International Conference on Intelligent Systems Design and Applications (ISDA05), Wroclaw, Poland, 8–10 September 2005; pp. 332–337.

13. Li, Q.; Tong, X.; Xie, S.; Liu, G. An improved adaptive algorithm for controlling the probabilities of crossover and mutation based on a fuzzy control strategy. In Proceedings of the 2006 Sixth International Conference on Hybrid Intelligent Systems (HIS06), Rio de Janeiro, Brazil, 13–15 December 2006; pp. 50–56.
14. Jafari, S.A.; Mashohor, S.; Varnamkhasti, M.J. Committee neural networks with fuzzy genetic algorithm. *J. Pet. Sci. Eng.* **2011**, *76*, 217–223. [CrossRef]
15. Venkatanareshbabu, K.; Nisheel, S.; Sakthivel, R.; Muralitharan, K. Novel elegant fuzzy genetic algorithms in classification problems. *Soft Comput.* **2018**, *23*, 5583–5603. [CrossRef]
16. Singh, M.; Brownlee, A.E.I.; Cairns, D. Towards Explainable Metaheuristic: Mining Surrogate Fitness Models for Importance of Variables. In Proceedings of the GECCO22, Boston, MA, USA, 9–13 July 2022; pp. 1785–1793.
17. Pandey, S.; Broder, A.; Chierichetti, F.; Josifovski, V.; Kumar, R.; Vassilvitskii, S. Nearest-Neighbor Caching for Content-Match Applications. In Proceedings of the International World Wide Web Conference (WWW 2009-ACM), Madrid, Spain, 20–24 April 2009; pp. 441–451.
18. Niesen, U.; Shah, D.; Wornell, G.W. Caching in Wireless Networks. *IEEE Trans. Inf. Theory* **2012**, *58*, 6524–6540. [CrossRef]
19. Bagheri, E.; Deldari, H. Dejong Function Optimization by means of a Parallel Approach to Fuzzified Genetic Algorithm. In Proceedings of the 11th IEEE Symposium on Computers and Communications (ISCC'06), Cagliari, Italy, 26–29 June 2006; pp. 675–680.
20. Nabavi-Pelesaraei, A.; Rafiee, S.; Mohtasebi, S.S.; Hosseinzadeh-Bandbafha, H.; Chau, K.W. Assessment of optimized pattern in milling factories of rice production based on energy, environmental and economic objectives. *Energy* **2019**, *169*, e1259–e1273. [CrossRef]
21. Chierichetti, F.; Kumar, R.; Vassilvitskii, S. Similarity Caching. In Proceedings of the 28th ACM SIGMOD-SIGACT-SIGART Symposium on Principles of Database Systems (PODS-09), San Diego, CA, USA, 9–12 June 2009; pp. 127–135.
22. Li, Q.; Yin, Y.; Wang, Z.; Liu, G. Comparative Studies of Fuzzy Genetic Algorithms. In Proceedings of the Advances in Neural Networks—ISNN 2007: 4th International Symposium on Neural Networks, ISNN 2007, Nanjing, China, 3–7 June 2007.
23. Li, Q.; Zheng, D.; Tang, Y.; Chen, Z. A New Kind of Fuzzy Genetic Algorithm. *J. Univ. Sci. Technol.* **2001**, *1*, 85–89.
24. Kuang, T.; Hu, Z.; Xu, M. A Genetic Optimization Algorithm Based on Adaptive Dimensionality Reduction. Available online: https://onlinelibrary.wiley.com/doi/10.1155/2020/8598543 (accessed on 5 November 2024).
25. Available online: http://comopt.ifi.uni-heidelberg.de/software/TSPLIB95/ (accessed on 5 November 2024).
26. Faostat-Crops and Livestock Products. Available online: www.fao.org/faostat/en/#data/QCL (accessed on 5 November 2024).
27. Tomasiello, S.; Uzair, M.; Liu, Y.; Loit, E. Data-driven approaches for sustainable agri-food: Coping with sustainability and interpretability. *J. Ambient. Intell. Humaniz. Comput.* **2023**, *14*, 16867–16878 [.CrossRef]
28. D'Arienzo, M.P.; Raritá, L. Management of Supply Chains for the Wine Production. In Proceedings of the International Conference on Numerical Analysis and Applied Mathematics 2019, ICNAAM 2019, Rhodes, Greece, 23–28 September 2019.
29. Raritá, L. *A Genetic Algorithm to Optimize Dynamics of Supply Chains*; AIRO Springer Series; Springer: Berlin/Heidelberg, Germany, 2022; Volume 8, pp. 107–115.
30. Naderi, R.; Nikabadi, M.S.; Tabriz, A.A.; Pishvaee, M.S. Supply chain sustainability improvement using exergy analysis. *Comput. Ind. Eng.* **2021**, *154*, 107142. [CrossRef]

algorithms

MDPI

Article

Issues on a 2–Dimensional Quadratic Sub–Problem and Its Applications in Nonlinear Programming: Trust–Region Methods (TRMs) and Linesearch Based Methods (LBMs)

Giovanni Fasano [1,*,†], Christian Piermarini [2,†] and Massimo Roma [2,†]

1 Department of Management, Venice School of Management, Ca' Foscari University, S. Giobbe, Cannaregio 873, 30121 Venice, Italy
2 Department of Computer, Control and Management Engineering, Sapienza University of Rome, Via Ariosto 25, 00185 Rome, Italy; piermarini@diag.uniroma1.it (C.P.); massimo.roma@diag.uniroma1.it (M.R.)
* Correspondence: fasano@unive.it
† These authors contributed equally to this work.

Abstract: This paper analyses the solution of a specific quadratic sub-problem, along with its possible applications, within both constrained and unconstrained Nonlinear Programming frameworks. We give evidence that this sub–problem may appear in a number of Linesearch Based Methods (LBM) schemes, and to some extent it reveals a close analogy with the solution of trust–region sub–problems. Namely, we refer to a two-dimensional structured quadratic problem, where five linear inequality constraints are included. Finally, we detail how to compute an exact global solution of our two-dimensional quadratic sub-problem, exploiting first order Karush-Khun-Tucker (KKT) conditions.

Keywords: nonlinear programming; quadratic linearly constrained optimization; KKT conditions

1. Introduction

There are plenty of real problems where the minimization of a twice continuously differentiable functional $f : \mathbb{R}^n \to \mathbb{R}$ is sought, (possibly) subject to several linear and nonlinear constraints. Among authoritative textbooks, where such problems are widely detailed, we can surely find [1–3]. Such general problems typically require the solution to a sequence of simple sub-problems with the following pattern:

$$\begin{array}{ll} \min\limits_{x} & \varphi_k(x) \\ \text{s.t.} & x \in D_k, \end{array} \qquad D_k : \left\{ \begin{array}{l} A_k x + u_k = 0 \\ B_k x + v_k \leq 0, \end{array} \right. \tag{1}$$

where $A_k \in \mathbb{R}^{m_k \times n}$, $B_k \in \mathbb{R}^{p_k \times n}$, $u_k \in \mathbb{R}^{m_k}$, $v_k \in \mathbb{R}^{p_k}$, $m_k, p_k \geq 1$ and $k \geq 1$. Furthermore, $\varphi_k(x)$ represents a *model* of the smooth function $f(x)$ at the current iterate x_k and the feasible set D_k represents a *linearization* of the constraints.

As is well known, affine and quadratic polynomials based on Taylor's expansion are often adopted to represent the models $\{\varphi_k(x)\}$, but valid alternatives also include least squares approximations, Radial Basis Functions, metamodels based on Splines, B-Splines, Kriging, etc. [4,5]. We remark that the advantage of solving the sequence of sub-problems (1) in place of the original nonlinearly constrained problem, within a suitable convergence framework, essentially relies on their simplicity. In particular, in this paper our focus is on investigating the role and the properties of the next problem (2), that represents a special sub-case of the more general problem (1). More specifically, we consider the case where in (1) the feasible set D_k includes only a *finite number of inequalities*, and the function $\varphi_k(x)$ is a quadratic functional, i.e., we focus on the sub-problem and drop the dependency on k

$$\begin{aligned} \min_{\alpha,\beta} \quad & \varphi(x) = \frac{1}{2}x^T Qx + b^T x + c \\ & x = \bar{x} + \alpha d + \beta z \\ & a_1 \le \alpha \le b_1 \\ & a_2 \le \beta \le b_2 \\ & \epsilon_1 \alpha + \epsilon_2 \beta \le \epsilon_3 \end{aligned} \tag{2}$$

where $Q \in \mathbb{R}^{n \times n}$, $b, \bar{x} \in \mathbb{R}^n$, $c \in \mathbb{R}$, $d, z \in \mathbb{R}^n$ are given n-real search directions, and $a_1 \le b_1$, $a_2 \le b_2$. Despite the apparent specific structure of (2), a number of real applications may benefit from its solution, as partly described in Section 4 (see also [6] for a general perspective and [7] for a more recent similar viewpoint within neural network frameworks).

As an example of versatility for the structure of (2), both in TRMs and LBMs, we will shortly consider how it may be possibly successfully embedded within the framework of Truncated Newton's methods (TNMs—see Table 1).

Table 1. A standard framework for linesearch-based TNMs for large-scale problems. The alternative of possibly using negative curvature directions allows for convergence to stationary limit points which fulfill second-order necessary optimality conditions.

Set $x_0 \in \mathbb{R}^n$
Set $\eta_k \in [0, 1)$ for any k, with $\{\eta_k\} \to 0$
OUTER ITERATIONS
for $k = 0, 1, \ldots$
    Compute $b \approx \nabla f(x_k)$ and $Q \approx \nabla^2 f(x_k)$; if $\|b\|$ is small then STOP
        INNER ITERATIONS
        - Compute d_k, which approximately solves Newton's equation $Qd + b = 0$, i.e., it satisfies the *truncation rule* $\|Qd_k + b\| \le \eta_k \|b\|$
        - Possibly compute a bounded negative curvature direction z_k at x_k

    Use a criterion to either combine d_k and z_k, or choose between d_k and z_k
    If the directions d_k and z_k were combined, set $v_k(\alpha) = \omega_1(\alpha)d_k + \omega_2(\alpha)z_k$, and use a *curvilinear linesearch* procedure to select $\alpha \leftarrow \alpha_k$. Otherwise, set $v_k(\alpha) = \alpha\bar{d}$ with $\bar{d} \in \{d_k, z_k\}$, and use an *Armijo-type procedure* to select $\alpha \leftarrow \alpha_k$
    Update $x_{k+1} = x_k + v_k(\alpha_k)$
endfor

Where (see also [8–14])

- $d \in \mathbb{R}^n$ represents an approximate Newton-type direction, at the current feasible point $\bar{x} \in \mathbb{R}^n$;
- $z \in \mathbb{R}^n$ represents a negative curvature direction for the nonlinear function $f(x)$, at the current feasible point $\bar{x} \in \mathbb{R}^n$;
- $Q \in \mathbb{R}^{n \times n}$ represents the exact/approximate Hessian matrix of $f(x)$ at $\bar{x}$;
- $b \in \mathbb{R}^n$ represents the exact/approximate Gradient vector of $f(x)$ at $\bar{x}$;
- α and β are steplengths along the directions $d, z \in \mathbb{R}^n$ (i.e., following the taxonomy of Table 1, we have $\alpha \leftarrow \omega_1(\alpha)$ and $\beta \leftarrow \omega_2(\alpha)$), with $-\infty < a_1 \le b_1 < +\infty$ and $-\infty < a_2 \le b_2 < +\infty$. The constraint $\varepsilon_1 \alpha + \varepsilon_2 \beta \le \varepsilon_3$ potentially plays a multi-purpose role, modeling for instance the *gradient-related* property for the search direction $\alpha d + \beta z \in \mathbb{R}^n$ at $\bar{x}$, i.e.,

$$(\alpha d + \beta z)^T \nabla f(\bar{x}) \le -\bar{c} \|\nabla f(\bar{x})\|^h, \tag{3}$$

being

$$\begin{cases} \varepsilon_1 = d^T \nabla f(\bar{x}), \\ \varepsilon_2 = z^T \nabla f(\bar{x}), \\ \varepsilon_3 = -\bar{c} \|\nabla f(\bar{x})\|^h, \qquad \bar{c}, h > 0. \end{cases}$$

The availability of an (exact) global solution for (2) may also suggest some alternatives to Table 1, either selecting a TRM or a LBM framework, or combining the two approaches. In particular, the scheme in Table 2 represents an immediate *acceleration scheme* for linesearch-based TNMs with respect to Table 1, in case the global convergence of $\{x_k\}$ to stationary limit points is simply sought. Note that selecting negative values for a_1, a_2 and positive ones for b_1, b_2 allows us to possibly perform the following:

- reverse the directions d_k and z_k;
- use (2) in the light of simulating a *dogleg-like* [3] procedure for TRMs, also in LBM nonconvex frameworks.

In Table 2, the global convergence to stationary limit points is easily preserved by using similar results adopted for Table 1.

Table 2. A standard framework for linesearch-based TNMs for large-scale problems which exploits the sub-problem (2). Differences with respect to Table 1 are quite evident.

Set $x_0 \in \mathbb{R}^n$
Set $\eta_k \in [0, 1)$ for any k, with $\{\eta_k\} \to 0$
OUTER ITERATIONS
for $k = 0, 1, \ldots$
  Compute $b \approx \nabla f(x_k)$ and $Q \approx \nabla^2 f(x_k)$; if $\|b\|$ is small then STOP
    INNER ITERATIONS
    - Compute d_k, which approximately solves Newton's equation $Qd + b = 0$, i.e., it satisfies the *truncation rule* $\|Qd_k + b\| \le \eta_k \|b\|$
    - Set $z_k = -b$

  Compute α^* and β^* by solving (2); then, update the trust-region parameters a_1, a_2, b_1, b_2
  Set $v_k = \alpha^* d_k + \beta^* z_k$, and use an *Armijo-type procedure* to select the steplength α_k along the direction v_k
  Update $x_{k+1} = x_k + \alpha_k v_k$
endfor

As a further alternative case for considering the exact global solution of (2), with respect to Tables 1 and 2, we have the scheme in Table 3, where we suitably combine the strategies used in TRMs and LBMs to ensure global convergence (in particular, TRMs require the fulfillment of a sufficient reduction of the model in order to force a sufficient decrease in the objective function, so that they do not need any linesearch procedure, possibly implying a reduced computational burden with respect to LBMs. Conversely, LBMs easily compute an effective search direction but they need to perform a linesearch procedure, because they do not include any -direct- function reduction mechanism based on the local quadratic model). In particular, if the test $Ared_k / Pred_k > \rho$ is fulfilled, there is no need to perform a linesearch procedure, since the global convergence for $\{x_k\}$ is preserved by the trust-region framework. We also remark that, in Table 3, the computation of both $\varphi(x_k)$ and $\varphi(x_k + v_k)$ is required, regardless of the outcomes of the test $Ared_k / Pred_k \not> \rho$, since in any case these quantities must be computed.

Table 3. A framework for combining trust-region and linesearch approaches within TNMs for large-scale problems, exploiting the sub-problem again (2). Differences with respect to Tables 1 and 2 are quite evident.

Set $x_0 \in \mathbb{R}^n$
Set $\eta_k \in [0,1)$ for any k, with $\{\eta_k\} \to 0$. Set $\rho > 0$
OUTER ITERATIONS
for $k = 0, 1, \ldots$
  Compute $b \approx \nabla f(x_k)$ and $Q \approx \nabla^2 f(x_k)$; if $\|b\|$ is small then STOP
    INNER ITERATIONS
      - Compute d_k, which approximately solves Newton's equation $Qd + b = 0$, i.e., it satisfies the *truncation rule* $\|Qd_k + b\| \le \eta_k \|b\|$
      - Set $z_k = -b$

  Compute α^* and β^* by solving (2); then, set $v_k = \alpha^* d_k + \beta^* z_k$,
  $Ared_k = f(x_k) - f(x_k + v_k)$, $Pred_k = \varphi(x_k) - \varphi(x_k + v_k)$
  If $Ared_k / Pred_k \not> \rho$, use an *Armijo-type procedure* to select the steplength α_k along v_k; otherwise, skip the linesearch procedure
  Update the trust-region parameters a_1, a_2, b_1, b_2
  Update $x_{k+1} = x_k + \alpha_k v_k$
endfor

Finally, there is a chance to further exploit the scheme (2) in a TNM framework based on the linesearch procedure, in order to ensure global convergence properties for the sequence $\{x_k\}$ to stationary limit points satisfying the *second-order necessary optimality conditions* (namely, those stationary points where the Hessian matrix is positive and semidefinite). The resulting scheme is proposed in Table 4 and potentially does not require additional comments. The above examples give an overview of the possible basic contexts where the solution of the sub-problem (2) is sought. Hence, to some extent, specifically exploiting issues on its solution may yield a tool for practitioners working in Nonlinear Programming frameworks. We remark that, both in Sections 3 and 6, the reader may find additional guidelines for possible alternatives and extensions to the use of global solutions of (2).

Table 4. A framework of linesearch-based approaches within TNMs for large-scale problems: solving the sub-problem (2) successfully allows for the convergence of the sequence $\{x_k\}$ to limit points satisfying second-order necessary optimality conditions. Differences with respect to Tables 1–3 are quite evident.

Set $x_0 \in \mathbb{R}^n$
Set $\eta_k \in [0,1)$ for any k, with $\{\eta_k\} \to 0$
OUTER ITERATIONS
for $k = 0, 1, \ldots$
  Compute $b \approx \nabla f(x_k)$ and $Q \approx \nabla^2 f(x_k)$; if $\|b\|$ is small then STOP
    INNER ITERATIONS
      - Compute d_k, which approximately solves Newton's equation $Qd + b = 0$, i.e., it satisfies the *truncation rule* $\|Qd_k + b\| \le \eta_k \|b\|$

  Compute a suitable negative curvature direction z_k for $f(x)$ at x_k
  Compute α^* and β^* by solving (2); then, set $v_k = \alpha^* d_k + \beta^* z_k$. Update the trust-region parameters a_1, a_2, b_1, b_2
  Use an *Armijo-type procedure* to select the steplength α_k along v_k
  Update $x_{k+1} = x_k + \alpha_k v_k$
endfor

We also highlight that our perspective both differs from SQP (Sequential Quadratic Programming) methods—see, for example, the seminal paper [15], and approaches from the literature where LSMs and TRMs have been combined. Indeed, in the basic structure of SQPs (see also [16,17]), inner and outer iterations are performed. At each outer iteration, the pair given by primal-dual variables is computed, and a problem similar to (2) is addressed. On the contrary, we do not intend to propose a (novel) framework of global convergence for Nonlinear Programming, but rather we suggest the generality of the scheme (2) within a number of cases from the literature. Furthermore, we specifically focus on the exact solution of the quadratic sub-problem (2) as well.

On the other hand, in the seminal paper [18] and in the more recent ones [19–21], linesearch and trust-region techniques are integrated in a unified framework. Conversely, our point of view merely intends to bridge the gap between them.

The structure of the present paper is as follows. In Section 2, we describe the conditions ensuring the feasibility of our problem. In Section 3, we reveal the basic motivations for our analysis and outcomes. Section 4 reports relevant remarks, highlighting how general our proposal can be. Section 5 includes the Karush-Kuhn-Tucker conditions associated with problem (2), along with precise guidelines to find a global minimum for it. Finally, Section 6 provides some conclusions and suggestions for future work.

As regards the symbols adopted in the paper, $\|x\|_1$, $\|x\|_2$ and $\|x\|_\infty$ are, respectively, used to indicate the 1-norm, the 2-norm, and the ∞-norm of the vector $x \in \mathbb{R}^n$ or the real $n \times m$ matrix x. Given the n-real vectors x and y, we indicate their standard inner product with $x^T y$. Given the matrix $A \in \mathbb{R}^{m \times n}$, we then indicate by A^+ its Moore-Penrose pseudoinverse matrix, i.e., the unique matrix, such that $AA^+A = A$, $A^+AA^+ = A^+$, $(AA^+)^T = AA^+$, $(A^+A)^T = A^+A$. With $A \succeq 0$ ($A \succ 0$), we indicate a positive semidefinite (positive definite) matrix A.

2. Feasibility Issues for Our Quadratic Problem

Here, we consider some feasibility issues for the linear inequality constrained quadratic problem (2). Clearly, (2) just includes the two real unknowns α and β. Moreover, as regards the existence of solutions for (2), we have the following result.

Lemma 1 (Feasibility). *Let the problem (2) be given and assume that the real values a_1, b_1, a_2, b_2 are finite, with $a_1 \le b_1$ and $a_2 \le b_2$. Then, (2) admits solutions if and only if at least one of the following conditions holds:*

Cond. I: $\epsilon_1 = \epsilon_2 = 0$ *and* $\epsilon_3 \ge 0$.

Cond. II: $\epsilon_1 = 0$ *and* $\epsilon_2 \neq 0$; *moreover,*

- *if* $\epsilon_2 > 0$, *then* $a_2 \le \epsilon_3/\epsilon_2$
- *if* $\epsilon_2 < 0$, *then* $b_2 \ge \epsilon_3/\epsilon_2$.

Cond. III: $\epsilon_1 \neq 0$ *and* $\epsilon_2 = 0$; *moreover,*

- *if* $\epsilon_1 > 0$, *then* $a_1 \le \epsilon_3/\epsilon_1$
- *if* $\epsilon_1 < 0$, *then* $b_1 \ge \epsilon_3/\epsilon_1$.

Cond. IV: $\epsilon_1 \neq 0, \epsilon_2 \neq 0, -\epsilon_1/\epsilon_2 < 0$, *moreover,*

- *if* $\varepsilon_1 > 0$ *and* $\varepsilon_2 > 0$, *then* $a_2 \le -(\epsilon_1/\epsilon_2)a_1 + (\epsilon_3/\epsilon_2)$
- *if* $\varepsilon_1 < 0$ *and* $\varepsilon_2 < 0$, *then* $b_2 \ge -(\epsilon_1/\epsilon_2)b_1 + (\epsilon_3/\epsilon_2)$.

Cond. V: $\epsilon_1 \neq 0, \epsilon_2 \neq 0, -\epsilon_1/\epsilon_2 > 0$, *moreover,*

- *if* $\varepsilon_1 < 0$ *and* $\varepsilon_2 > 0$, *then* $a_2 \le -(\epsilon_1/\epsilon_2)b_1 + (\epsilon_3/\epsilon_2)$
- *if* $\varepsilon_1 > 0$ *and* $\varepsilon_2 < 0$, *then* $b_2 \ge -(\epsilon_1/\epsilon_2)a_1 + (\epsilon_3/\epsilon_2)$.

Proof of Lemma 1. For the sake of simplicity, we refer to Figure 1. The objective function in (2) is continuous, so that the existence of solutions follows from the compactness and nonemptiness of the feasible region. In this regard, the compactness is a consequence of assuming a_1, b_1, a_2, b_2 finite. Furthermore, it is not difficult to realize that the feasible set of (2) is nonempty as long as at least one among the five conditions, Cond. I–Cond. V, is fulfilled, where the dashed-dotted line in Figure 1 represents the line associated with the last inequality constraint in (2). In particular, Cond. IV refers to the corner points **A** and **B** of Figure 1, while Cond. V refers to the vertices **C** and **D**. □

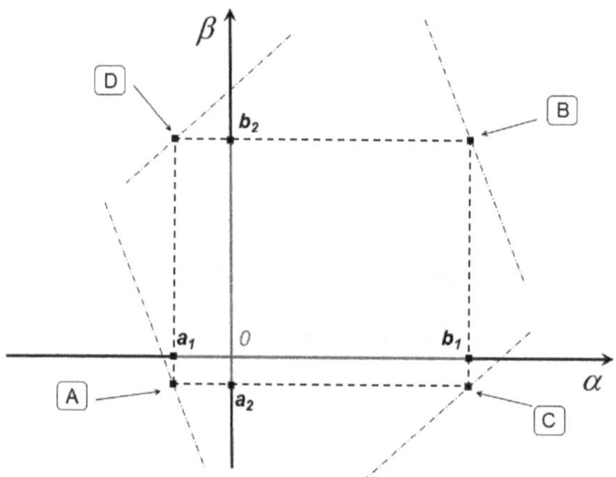

Figure 1. A graphical representation of the feasible set in (2). The dashed-dotted lines represent all the extreme choices for the last inequality constraint in (2).

3. On the Use of Quadratic Sub-Problems Within TRMs and LBMs, in Large-Scale Optimization

Here, we give details about a possible motivation for our proposal, in order to reduce the gap between two renowned classes of optimization methods, namely TRMs and LBMs. We are indeed persuaded that such a viewpoint may suggest a number of possible enhancements, to improve both the last classes of methods.

In this regard, observe that a TRM for large-scale problems is an iterative procedure that generates the sequence of n-real iterates $\{x_k\}$, and seeks at any step k for the solution of the trust-region sub-problem

$$\begin{aligned} \min_s \ & q_k(s) = f(x_k) + \nabla f(x_k)^T s + \frac{1}{2} s^T Q_k s \\ & \|s\|_2 \leq \Delta_k, \end{aligned} \tag{4}$$

where x_k is the current iterate, Q_k represents the *exact/approximate* Hessian matrix $\nabla^2 f(x_k)$, and $\Delta_k > 0$ represents the radius of the trust-region, i.e., the compact subset where the model $q_k(s)$ needs to be validated (for an exhaustive description of TRMs for Nonlinear Programming, the reader can refer to [2]). A number of possible variants of (4) can be introduced when n is large, including iterative updating strategies for both Q_k and Δ_k, and a number of approximate/sophisticated/refined schemes for its solution are available in the literature.

A distinguishing feature of TRMs, with respect to LBMs, is that at iteration k the methods in the first class attempt to determine the stepsize α_k and the search direction d_k at once, so that $x_{k+1} = x_k + s_k \equiv x_k + \alpha_k d_k$, where s_k indeed approximately/exactly solves (4). Conversely, in LBMs, the computations of α_k and d_k are independent, as detailed later on in this paper. In particular, (see also [3]) the effective computation of s_k in TRMs properly attempts to comply with the following issues:

- s_k can be computed by either an exact (small- and medium-scale problems) or an approximate (large-scale problems) procedure;
- In order to prove the global convergence of the sequence $\{x_k\}$ to stationary limit points satisfying either first- or second-order necessary optimality conditions, s_k is

required to provide a *sufficient reduction* of the quadratic model $q_k(s)$, i.e., the difference $q_k(0) - q_k(s_k)$ is asked to satisfy a condition like $(c > 0)$

$$q_k(0) - q_k(s_k) \geq c\|\nabla f(x_k)\|_2 \min\left\{\Delta_k, \frac{\|\nabla f(x_k)\|_2}{1 + \|Q\|_2}\right\};$$

- s_k can be computed by an approximate procedure, e.g., by adopting a Cauchy step or using the Steihaug conjugate gradient (see [22,23]), regardless of Q_k signature. Then, the approximate solution of (4) is merely sought on a *linear manifold* of a dimension of one or at most two, rather than on the entire subset $B \equiv \{s \in \mathbb{R}^n \ : \ \|s\|_2 \leq \Delta_k\}$;
- Depending on a number of additional assumptions, TRMs can prove to be globally convergent to either a simple stationary limit point, or to a point which satisfies second-order necessary optimality conditions [2];
- Finding the exact/accurate solution of the sub-problem (4) is in general quite a cumbersome task in large-scale problems, representing a difficult goal that is often (when possible) skipped.

On the other hand, to some extent, LBMs represent the counterpart of TRMs. Indeed, to yield the next iterate $x_{k+1} = x_k + \alpha_k d_k$, they perform the computation of the steplength α_k and the direction d_k as separate tasks. Furthermore, unlike for TRMs, the novel iterate x_{k+1} in LBMs can be also obtained by adopting the more general update

$$x_{k+1} = x_k + \alpha_k d_k + \beta_k z_k, \tag{5}$$

with d_k and z_k now being two search directions summarizing different information on the function $f(x)$, and α_k and β_k being stepsizes. In particular:

- when $z_k \equiv 0$ (or $\beta_k \equiv 0$ for any k), then d_k represents a Newton-type direction, being typically computed by approximately solving Newton's equation $\nabla^2 f(x_k)d = -\nabla f(x_k)$ at the current iterate x_k. Then, an Armijo-type linesearch procedure is applied along d_k to compute α_k, provided that d_k is gradient-related (see e.g., [3]) at x_k;
- when $z_k \neq 0$, then d_k represents a Newton-type direction again, while z_k is typically a *negative curvature direction* for $f(x)$ at x_k, which approximates an eigenvector associated with the least negative eigenvalue of $\nabla^2 f(x_k)$. The vector z_k plays an essential role, when LBMs' convergence to stationary points satisfying the second-order necessary optimality conditions needs to be proved. In the last case, the computation of the steplengths α_k and β_k is often carried out at once, (as in *curvilinear* linesearch procedures—see [24]), or the steplength computation is carried out by pursuing independent tasks (see, for example, [25]). We highlight that in (5), when both $d_k \neq 0$ and $z_k \neq 0$, we may experience difficulties related to properly scaling the two search directions.

As a general class of efficient algorithms within LBMs for large-scale problems, we find Truncated Newton methods (TNMs) coupled with a linesearch procedure (see Table 1). Similarly to general TRMs, they are evidently based on possibly computing d_k and z_k after exploiting the second-order Taylor's expansion of $f(x)$ at x_k. However, a couple of quite disappointing issues arise when applying linesearch-based TNMs, namely:

- Unlike trust-region based TNMs, at iterate x_k, the search of a stationary point for a quadratic polynomial model of $f(x)$ (i.e., Newton's equation) is performed on $\mathbb{R}^n$, so that the quadratic expansion is not trusted on a more reliable compact subset (trust-region) of $\mathbb{R}^n$. Thus, the search direction d_k might show poor performance when the iterates in the sequence $\{x_k\}$ are far from a stationary limit point x^*. More specifically, note that in case $\nabla^2 f(x_k) \succeq 0$; then, solving Newton's equation and the trust-region sub-problem

$$\min_d \ q_k(d) = f(x_k) + \nabla f(x_k)^T d + \frac{1}{2} d^T \nabla^2 f(x_k) d$$
$$\|d\|_2 \leq \gamma_k,$$

for any $\gamma_k \geq \|[\nabla^2 f(x_k)]^+ \nabla f(x_k)\|_2$ yields the same solutions. Conversely, when $\nabla^2 f(x_k)$ is indefinite, then Newton's equation provides a saddle point for $q_k(d)$, that might be interpreted as a solution to a trust-region sub-problem (the interested reader may consider the paper [26] for some extensions). Furthermore, from this perspective, we remark that in LBMs, solving (2) where $d_k = -\nabla f(x_k)$, $z_k \equiv 0$, and $\varepsilon_1 = \varepsilon_2 = \varepsilon_3 = 0$, is to a to large extent equivalent to computing the Cauchy step when solving (4). Indeed, in the last case, the trust-region constraint in (4), in principle, can be equivalently replaced by the compact feasible set (box constraints) in (2), after setting $\varepsilon_1 = \varepsilon_2 = \varepsilon_3 = 0$. On the other hand, in case $\nabla^2 f(x_k) \succeq 0$, and setting (2) $d_k = -\nabla f(x_k)$ and $z_k = -[\nabla^2 f(x_k)]^+ \nabla f(x_k)$, along with $\varepsilon_1 = \varepsilon_2 = \varepsilon_3 = 0$, then, with similar reasoning, the solution to (2) closely resembles the application of the *dogleg method* when solving (4). Finally, since the coefficients a_1, a_2 in (2) may have negative values, we may potentially reverse the directions d_k and z_k when solving (2). Thus, following the idea behind (3), the scheme (2) suggests that, in case $\nabla^2 f(x_k)$ is also indefinite, (2) easily generalizes the proposals in [27]. In fact, following (3), we are able to exactly compute a global minimum (α^*, β^*) for (2), regardless of the signature of Q, so that the resulting direction $\alpha^* d_k + \beta^* z_k$ is gradient-related at x_k.

- As in (5), the search directions d_k and z_k might be suitably combined in a curvilinear framework (see, for example, [24]). However, to our knowledge, the selection of α_k and β_k in the literature is seldom performed with a joint procedure to separately assess α_k and β_k, i.e., α_k and β_k are rarely chosen as independent parameters. Hence, in the literature of linesearch-based TNMs, the linesearch procedure that starts from x_k and yields x_{k+1} explores a one-dimensional manifold (regular curve), rather than considering $x_k + \alpha d_k + \beta z_k$ as a two-dimensional manifold with independent real coefficients α and β.

In this regard, using (2) within LBMs tends to partially compensate for the drawbacks in the last two items, in light of the great success that TRMs have achieved in the last decade. In particular, using (2) within linesearch-based TNMs, our aim is that of developing a simple tool which could possibly carry out the following:

1. Combines at iterate x_k two independently computed vectors, namely $d_k, z_k \in \mathbb{R}^n$, by exactly computing a *global minimum* (we recall that, conversely, a global solution of the trust-region sub-problem (4) is often only approximately computed) for the two-dimensional constrained problem (2), being $\bar{x} \leftarrow x_k$, $d \leftarrow d_k$, $z \leftarrow z_k$;
2. Adaptively updates the parameters a_1, a_2, b_1, b_2 in (2), when the iterate x_k changes, following the rationale behind the update of Δ_k in (4), and retaining the strong convergence properties of TRMs. This fact is of remarkable interest, since in (2) the information associated with the search directions d_k and z_k is suitably trusted in a *compact subset* of $\mathbb{R}^n$ (namely, the box constraints $a_1 \leq \alpha \leq b_1, a_2 \leq \beta \leq b_2$);
3. *Exactly computes a cheap global minimum* (α^*, β^*) for (2), so that the vector $\alpha^* d_k + \beta^* s_k$ is then provided to a standard linesearch procedure such as the Armijo rule, to ensure that the global convergence of the sequence $\{x_k\}$ to stationary (limit) points is preserved;
4. Allows for the convergence of subsequences of the iterates $\{x_k\}$ to stationary limit points, where either first- or second-order necessary optimality conditions are fulfilled;
5. *Preserves generality* within a wide range of optimization frameworks, as reported in the next Section 4;
6. *Combines* the effects of d_k and z_k, skipping all the drawbacks related to a possible different scaling between these directions. We recall that since d_k and z_k are generated through the application of different methods, then the comparison of their performances may be biased by the latter generating methods.

The TNMs sketched in Tables 2–4 are only examples of proposals in light of the previous comments.

4. How General Is the Model (2) in Nonlinear Programming Frameworks?

This section is devoted to reporting a number of real constrained optimization schemes from Nonlinear Programming, whose formulation is encompassed in (2). We can see that for some of the following schemes (see Figure 2), it is possible that more than one reformulation can be considered in the framework (2).

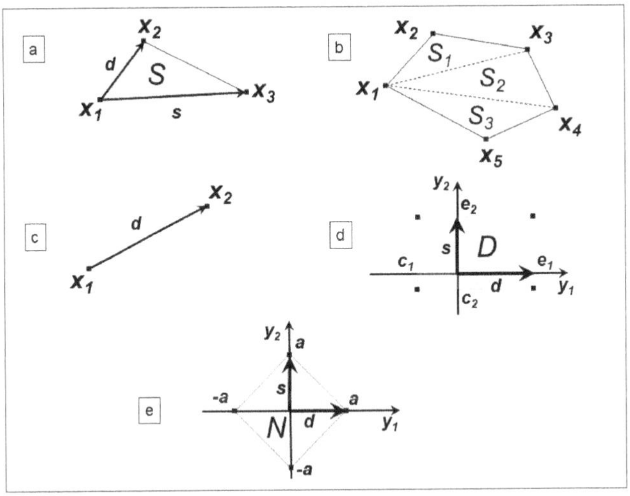

Figure 2. Examples where the structure of the feasible set in (2) is helpful: case (**a**) is treated in Section 4.1, case (**b**) is treated in Section 4.2, case (**c**) is treated in Section 4.3, case (**d**) is treated in Section 4.4 and case (**e**) is treated in Section 4.5.

4.1. Minimization over a Bounded Simplex

We consider the problem of minimizing a quadratic functional over the simplex $S \subset \mathbb{R}^n$, such that

$$\min_x \frac{1}{2}x^T Qx + b^T x + c \\ x \in S, \tag{6}$$

where $S = \{x \in \mathbb{R}^n : x = \sum_{i=1}^3 \lambda_i x_i, \sum_{i=1}^3 \lambda_i = 1, \lambda_i \geq 0, i = 1,2,3\}$. Figure 2a reports an example of a simplex. In this regard, by simply setting in (2)

- $d = x_2 - x_1, \ z = x_3 - x_1,$
- $\bar{x} = x_1,$
- $a_1 = 0, \ b_1 = 1, \ a_2 = 0, \ b_2 = 1,$
- $\epsilon_1 = \epsilon_2 = \epsilon_3 = 1,$

the problem (6) is a special case of the problem (2).

4.2. Minimization over a Bounded Polygon

We consider the problem of minimizing a quadratic functional over a polygon $P \subset \mathbb{R}^2$, described by a finite number m of vertices (observe that the points in the polygon P must belong to a hyperplane $\pi \subset \mathbb{R}^n$, with $\pi : \omega^T x + \omega_0 = 0, \omega = (\omega_1, \ldots, \omega_n)^T \in \mathbb{R}^n, \omega_0 \in \mathbb{R}$, so that $\omega^T \bar{x} + \omega_0 = 0$ for any $\bar{x} \in P$.), i.e.,

$$\min_x \frac{1}{2}x^T Qx + b^T x + c \\ x \in P, \tag{7}$$

where $P = \{x \in \mathbb{R}^2 \, : \, x = \sum_{i=1}^{m} \lambda_i x_i, \, \sum_{i=1}^{m} \lambda_i = 1, \, \lambda_i \geq 0, \, i = 1, \dots, m\}$. Figure 2b reports an example of a polygon with $m = 5$. In this regard, the problem (7) can be split into to solution of the $(m-2)$ sub-problems

$$\begin{aligned} \min_x \ & \frac{1}{2} x^T Q x + b^T x + c \\ & x \in S_i, \qquad i = 1, \dots, m-2, \end{aligned} \tag{8}$$

where

$$S_i = \left\{ x \in \pi \subset \mathbb{R}^2 \, : \, x = \sum_{j \in \{1, i+1, i+2\}} \lambda_j x_j, \, \sum_{j \in \{1, i+1, i+2\}} \lambda_j = 1, \, \lambda_j \geq 0, \ \ j \in \{1, i+1, i+2\} \right\},$$

which are of the form (6). Thus, solving the problem (7) corresponds to solve a sequence of $(m-2)$ instances of the problem (2).

4.3. Minimization over a Bounded Segment

We consider the problem of minimizing a quadratic functional over a segment $L \subset \mathbb{R}^n$, i.e.,

$$\begin{aligned} \min_x \ & \frac{1}{2} x^T Q x + b^T x + c \\ & x \in L, \end{aligned} \tag{9}$$

where $L = \{x \in \mathbb{R}^n \, : \, x = \lambda x_1 + (1-\lambda) x_2, \, \lambda \in [0,1]\}$. Figure 2c reports an example of a segment. In this regard, by simply setting in (2)

- $d = x_2 - x_1, \ z \equiv 0,$
- $\bar{x} = x_1,$
- $a_1 = 0, \ b_1 = 1, \ a_2 = 0, \ b_2 = 0,$
- $\epsilon_1 = \epsilon_2 = \epsilon_3 = 0,$

the problem (9) is a special case of the problem (2).

4.4. Minimization over a Bounded Box in $\mathbb{R}^2$

We consider the problem of minimizing a quadratic functional over a box domain $D \subset \mathbb{R}^2$, i.e.,

$$\begin{aligned} \min_x \ & \frac{1}{2} x^T Q x + b^T x + c \\ & x \in D, \end{aligned} \tag{10}$$

where $D = \{x \in \mathbb{R}^2 \, : \, c_i \leq x_i \leq e_i, \, i = 1, 2\}$. Figure 2d reports an example of a box domain. In this regard, by simply setting in (2)

- $d = \begin{pmatrix} e_1 - c_1 \\ 0 \end{pmatrix}, \ z = \begin{pmatrix} 0 \\ e_2 - c_2 \end{pmatrix}$
- $\bar{x} = \begin{pmatrix} c_1 \\ c_2 \end{pmatrix}$
- $a_1 = 0, \ b_1 = 1, \ a_2 = 0, \ b_2 = 1,$
- $\epsilon_1 = \epsilon_2 = \epsilon_3 = 0,$

the problem (10) is a special case of the problem (2). As an alternative to the previous setting, we might also consider treating this case with a setting in (2), given by

- $d = \begin{pmatrix} \frac{e_1 - c_1}{2} \\ 0 \end{pmatrix}, \ z = \begin{pmatrix} 0 \\ \frac{e_2 - c_2}{2} \end{pmatrix}$
- $\bar{x} = \begin{pmatrix} \frac{e_1 + c_1}{2} \\ \frac{e_2 + c_2}{2} \end{pmatrix}$
- $a_1 = -1, \ b_1 = 1, \ a_2 = -1, \ b_2 = 1,$
- $\epsilon_1 = \epsilon_2 = \epsilon_3 = 0.$

4.5. Minimization Including a 1-Norm Inequality Constraint in $\mathbb{R}^2$

We consider the problem of minimizing a quadratic functional subject to the 1-norm inequality constraint $x \in N$, with $N \subset \mathbb{R}^2$, i.e.,

$$\begin{aligned} \min_x \ & \frac{1}{2}x^T Q x + b^T x + c \\ & x \in N, \end{aligned} \tag{11}$$

which is $N = \{x \in \mathbb{R}^2 \ : \ \|x\|_1 \leq a\}$. Figure 2e reports an example of such a constraint. In this regard, it suffices to recast (11) as in (8), where

- $m = 6$
- $\bar{x} = x_1 = 0$
- $x_2 = \begin{pmatrix} 0 \\ 1 \end{pmatrix}, x_3 = \begin{pmatrix} 1 \\ 0 \end{pmatrix}, x_4 = \begin{pmatrix} 0 \\ -1 \end{pmatrix}, x_5 = \begin{pmatrix} -1 \\ 0 \end{pmatrix}, x_6 = x_2,$

so that four instances of the problem (2) need to be solved.

4.6. Minimization Including an ∞-Norm Inequality Constraint in $\mathbb{R}^2$

We consider the problem of minimizing a quadratic functional subject to the ∞-norm inequality constraint $x \in E$, with $E \subset \mathbb{R}^2$, i.e.,

$$\begin{aligned} \min_x \ & \frac{1}{2}x^T Q x + b^T x + c \\ & x \in E, \end{aligned} \tag{12}$$

which is $E = \{x \in \mathbb{R}^2 \ : \ \|x\|_\infty \leq a\}$. In this regard, we obtain similar results with respect to Section 4.5. Indeed, by simply setting in (2)

- $d = \begin{pmatrix} a \\ 0 \end{pmatrix}, \ z = \begin{pmatrix} 0 \\ a \end{pmatrix}$
- $\bar{x} = 0$
- $a_1 = -1, \ b_1 = 1, \ a_2 = -1, \ b_2 = 1,$
- $\epsilon_1 = \epsilon_2 = \epsilon_3 = 0,$

the problem (12) is a special case of the problem (2).

4.7. Minimization Including a 2-Norm Inequality Constraint in $\mathbb{R}^2$

We consider the problem of minimizing a quadratic functional in $\mathbb{R}^2$ subject to the 2-norm inequality constraint $\|x\|_2 \leq \gamma$, with $\gamma \geq 0$, i.e.,

$$\begin{aligned} \min_x \ & \frac{1}{2}x^T Q x + b^T x + c \\ & \|x\|_2 \leq \gamma, \qquad x \in \mathbb{R}^2. \end{aligned} \tag{13}$$

In this regard, it suffices to observe that the solution of (2) provides both a

- <u>LOWER bound</u> to the solution of (13), as long as we set (following Section 4.6)
  - $d = \begin{pmatrix} \gamma \\ 0 \end{pmatrix}, \ z = \begin{pmatrix} 0 \\ \gamma \end{pmatrix}$
  - $\bar{x} = 0$
  - $a_1 = -1, \ b_1 = 1, \ a_2 = -1, \ b_2 = 1,$
  - $\epsilon_1 = \epsilon_2 = \epsilon_3 = 0,$
- <u>UPPER bound</u>: to the solution of (13), as long as we follow the indications in Section 4.5, i.e., we recast and solve (11) as in (8), where
  - $m = 6$
  - $\bar{x} = x_1 = 0$

- $x_2 = \begin{pmatrix} 0 \\ \gamma \end{pmatrix}, x_3 = \begin{pmatrix} \gamma \\ 0 \end{pmatrix}, x_4 = \begin{pmatrix} 0 \\ -\gamma \end{pmatrix}, x_5 = \begin{pmatrix} -\gamma \\ 0 \end{pmatrix}, x_6 = x_2,$

  so that four instances of the problem (2) need to be solved.

Note that the dogleg-like methods for the approximate solution of the trust-region problem (4), in the convex case, equivalently solve the sub-problem (13) with just a couple of unknowns.

5. KKT Conditions and the Fast Solution of Problem (2)

Replacing the expression of the vector x in (2) within the objective function, we easily obtain the equivalent problem

$$\begin{array}{ll} \min\limits_{\alpha,\beta} & \varphi(\alpha,\beta) \\ \mathcal{P}: & \left\{ \begin{array}{l} a_1 \le \alpha \le b_1 \\ a_2 \le \beta \le b_2 \\ \epsilon_1\alpha + \epsilon_2\beta \le \epsilon_3, \end{array} \right. \end{array} \tag{14}$$

where

$$\left\{ \begin{array}{l} \varphi(\alpha,\beta) = \frac{1}{2}\begin{pmatrix} \alpha \\ \beta \end{pmatrix}^T \begin{pmatrix} t & u \\ u & w \end{pmatrix} \begin{pmatrix} \alpha \\ \beta \end{pmatrix} + \begin{pmatrix} y \\ h \end{pmatrix}^T \begin{pmatrix} \alpha \\ \beta \end{pmatrix} + q \\ t = d^T Q d \\ u = d^T Q z = z^T Q d \\ w = z^T Q z \\ y = (Q\bar{x} + b)^T d \\ h = (Q\bar{x} + b)^T z \\ q = \left(\frac{1}{2}Q\bar{x} + b\right)^T \bar{x} + c. \end{array} \right.$$

Observe that transforming (2) into (14) only requires the computation of two additional matrix-vector products (i.e., Qd and Qz), along with six inner products. The problem (14) is a constrained quadratic problem, such that first-order Fritz-John optimality conditions do not require additional constraint qualifications (since all the constraints are linear). Thus, after considering its Lagrangian function

$$\begin{aligned} \mathcal{L}(\alpha,\beta,\mu_1,\mu_2,\mu_3,\mu_4,\mu_5) \quad = \\ \varphi(\alpha,\beta) - \mu_1(\alpha - a_1) + \mu_2(\alpha - b_1) - \mu_3(\beta - a_2) + \mu_4(\beta - b_2) + \mu_5(\epsilon_1\alpha + \epsilon_2\beta - \epsilon_3) \end{aligned}$$

we have the next set of equalities/inequalities representing the associated KKT conditions:

$$\left\{ \begin{array}{l} \begin{pmatrix} t & u \\ u & w \end{pmatrix} \begin{pmatrix} \alpha^* \\ \beta^* \end{pmatrix} + \begin{pmatrix} y \\ h \end{pmatrix} + \begin{pmatrix} -\mu_1^* + \mu_2^* + \epsilon_1\mu_5^* \\ -\mu_3^* + \mu_4^* + \epsilon_2\mu_5^* \end{pmatrix} = \begin{pmatrix} 0 \\ 0 \end{pmatrix} \\ \begin{pmatrix} \alpha^* \\ \beta^* \end{pmatrix} \in \mathcal{P} \\ \mu_1^*[\alpha^* - a_1] = 0 \\ \mu_2^*[\alpha^* - b_1] = 0 \\ \mu_3^*[\beta^* - a_2] = 0 \\ \mu_4^*[\beta^* - b_2] = 0 \\ \mu_5^*[\epsilon_1\alpha^* + \epsilon_2\beta^* - \epsilon_3] = 0 \\ \mu_i^* \ge 0, \qquad i = 1,\ldots,5. \end{array} \right. \tag{15}$$

The remaining part of the present section will be devoted to analyze all the possible solutions of (15), with the aim of possibly computing a global minimum for (2). In this regard, exploiting the solutions of (15) evidently undergoes a reduction, allowing us to analyze the cases (I)–(XII) in Figure 3.

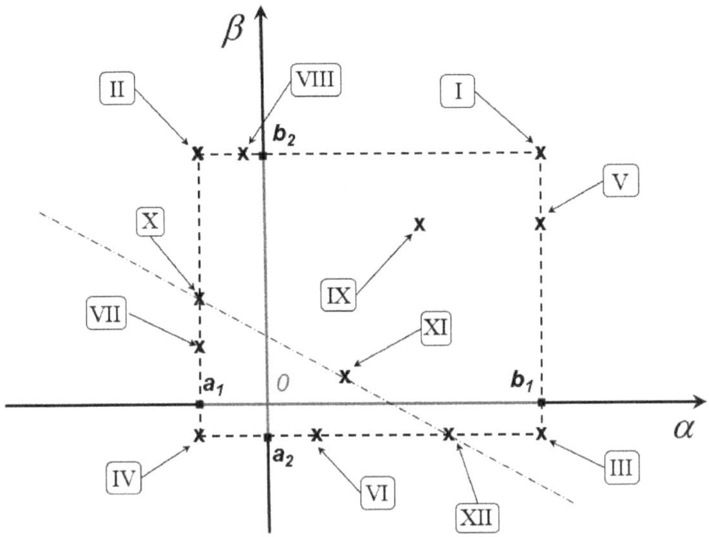

Figure 3. Overview of possible solutions (I)–(XII) for KKT conditions in (15).

Observing that in (15) the multipliers μ_i^* and $i = 1, \ldots, 5$ must fulfill nonnegativity conditions, it is not difficult to realize that computing all the KKT points satisfying (15) can turn out to be a burdensome task, including a number of sub-cases depending on the possible combinations of signs for the parameters $a_1, b_1, a_2, b_2, \epsilon_1, \epsilon_2$ and ϵ_3. Conversely, a global minimizer for (2) can be equivalently exploited by analyzing all the possible solutions of (15) uniquely in terms of α^* and β^*, without requiring the computation of the multipliers as well. Hence, we limit our analysis to consider the computation of α^* and β^* in the cases (I)–(XII) of Figure 3, where

- Cases (I), (II), (III), (IV) are associated with possible solutions in the vertices of the box constraints;
- Cases (V), (VI), (VII), (VIII) are associated with possible solutions on the edges of the box constraints;
- Case (IX) represents a possible feasible unconstrained minimizer for the objective function in (2);
- Cases (X), (XI), (XII) are associated with possible solutions, making the last inequality constraint in (14) active.

Then, in Lemma 2, we will provide a simple theoretical result which justifies our simplification, with respect to computing all the KKT points. In this regard, we preliminarily set $i = 1$ and consider the next cases from Figure 3, with $\{y_i\}$ being the sequence of *tentative* solution points of (14):

- Case (I): We set $\bar{\alpha} = b_1, \bar{\beta} = b_2$. If $\boxed{\epsilon_1\bar{\alpha} + \epsilon_2\bar{\beta} \leq \epsilon_3}$, then set

$$P_1 = \begin{pmatrix} b_1 \\ b_2 \end{pmatrix}, \qquad \varphi_i = \varphi(b_1, b_2), \qquad y_i = P_1, \qquad i = i + 1; \tag{16}$$

- Case (II): We set $\bar{\alpha} = a_1, \bar{\beta} = b_2$. If $\boxed{\epsilon_1\bar{\alpha} + \epsilon_2\bar{\beta} \leq \epsilon_3}$, then set

$$P_2 = \begin{pmatrix} a_1 \\ b_2 \end{pmatrix}, \qquad \varphi_i = \varphi(a_1, b_2), \qquad y_i = P_2, \qquad i = i + 1; \tag{17}$$

- Case (III): We set $\bar{\alpha} = b_1$, $\bar{\beta} = a_2$. If $\boxed{\epsilon_1\bar{\alpha} + \epsilon_2\bar{\beta} \le \epsilon_3}$, then set

$$P_3 = \begin{pmatrix} b_1 \\ a_2 \end{pmatrix}, \qquad \varphi_i = \varphi(b_1, a_2), \qquad y_i = P_3, \qquad i = i + 1; \tag{18}$$

- Case (IV): We set $\bar{\alpha} = a_1$, $\bar{\beta} = a_2$. If $\boxed{\epsilon_1\bar{\alpha} + \epsilon_2\bar{\beta} \le \epsilon_3}$, then set

$$P_4 = \begin{pmatrix} a_1 \\ a_2 \end{pmatrix}, \qquad \varphi_i = \varphi(a_1, a_2), \qquad y_i = P_4, \qquad i = i + 1; \tag{19}$$

- Case (V): We set $\bar{\alpha} = b_1$ and possibly compute the solution $\bar{\beta} = -(ub_1 + h)/w$ of the equation

$$\frac{d\varphi(b_1, \beta)}{d\beta} = w\beta + ub_1 + h = 0,$$

so that:

  - if $\boxed{w \neq 0 \text{ AND } \epsilon_1\bar{\alpha} + \epsilon_2\bar{\beta} \le \epsilon_3}$, then set

$$P_5 = \begin{pmatrix} \bar{\alpha} \\ \bar{\beta} \end{pmatrix}, \qquad \varphi_i = \varphi(\bar{\alpha}, \bar{\beta}), \qquad y_i = P_5, \qquad i = i + 1; \tag{20}$$

  - if $\boxed{w = 0 \text{ AND } ub_1 + h \neq 0}$, then there is no solution for Case (V);
  - if $\boxed{w = 0 \text{ AND } ub_1 + h = 0}$, then set $\bar{\beta} \in [a_2, b_2]$ as any value satisfying $\epsilon_1\bar{\alpha} + \epsilon_2\bar{\beta} \le \epsilon_3$, and compute P_5 as in (20);

- Case (VI): We set $\bar{\beta} = a_2$ and possibly compute the solution $\bar{\alpha} = -(ua_2 + y)/t$ of the equation

$$\frac{d\varphi(\alpha, a_2)}{d\alpha} = t\alpha + ua_2 + y = 0,$$

so that:

  - if $\boxed{t \neq 0 \text{ AND } \epsilon_1\bar{\alpha} + \epsilon_2\bar{\beta} \le \epsilon_3}$, then set

$$P_6 = \begin{pmatrix} \bar{\alpha} \\ \bar{\beta} \end{pmatrix}, \qquad \varphi_i = \varphi(\bar{\alpha}, \bar{\beta}), \qquad y_i = P_6, \qquad i = i + 1; \tag{21}$$

  - if $\boxed{t = 0 \text{ AND } ua_2 + y \neq 0}$, then there is no solution for Case (VI);
  - if $\boxed{t = 0 \text{ AND } ua_2 + y = 0}$, then set $\bar{\alpha} \in [a_1, b_1]$ as any value satisfying $\epsilon_1\bar{\alpha} + \epsilon_2\bar{\beta} \le \epsilon_3$, and compute P_6 as in (21);

- Case (VII): We set $\bar{\alpha} = a_1$ and possibly compute the solution $\bar{\beta} = -(ua_1 + z)/w$ of the equation

$$\frac{d\varphi(a_1, \beta)}{d\beta} = w\beta + ua_1 + h = 0,$$

so that:

  - if $\boxed{w \neq 0 \text{ AND } \epsilon_1\bar{\alpha} + \epsilon_2\bar{\beta} \le \epsilon_3}$, then set

$$P_7 = \begin{pmatrix} \bar{\alpha} \\ \bar{\beta} \end{pmatrix}, \qquad \varphi_i = \varphi(\bar{\alpha}, \bar{\beta}), \qquad y_i = P_7, \qquad i = i + 1; \tag{22}$$

  - if $\boxed{w = 0 \text{ AND } ua_1 + h \neq 0}$, then there is no solution for Case (VII);
  - if $\boxed{w = 0 \text{ AND } ua_1 + h = 0}$, then set $\bar{\beta} \in [a_2, b_2]$ as any value satisfying $\epsilon_1\bar{\alpha} + \epsilon_2\bar{\beta} \le \epsilon_3$, and compute P_7 as in (22);

- Case (VIII): We set $\bar{\beta} = b_2$ and possibly compute the solution $\bar{\alpha} = -(ub_2 + y)/t$ of the equation
$$\frac{d\varphi(\alpha, b_2)}{d\alpha} = t\alpha + ub_2 + y = 0,$$
so that:
  - if $\boxed{t \neq 0 \text{ AND } \epsilon_1\bar{\alpha} + \epsilon_2\bar{\beta} \leq \epsilon_3}$, then set
$$P_8 = \begin{pmatrix} \bar{\alpha} \\ \bar{\beta} \end{pmatrix}, \qquad \varphi_i = \varphi(\bar{\alpha}, \bar{\beta}), \qquad y_i = P_8, \qquad i = i + 1; \tag{23}$$
  - if $\boxed{t = 0 \text{ AND } ub_2 + y \neq 0}$, then there is no solution for Case (VIII);
  - if $\boxed{t = 0 \text{ AND } ub_2 + y = 0}$, then set $\bar{\alpha} \in [a_1, b_1]$ as any value satisfying $\epsilon_1\bar{\alpha} + \epsilon_2\bar{\beta} \leq \epsilon_3$, and compute P_8 as in (23);
- Case (IX): If $\boxed{tw - u^2 \neq 0}$, we compute the solution
$$\begin{pmatrix} \bar{\alpha} \\ \bar{\beta} \end{pmatrix} = -\begin{pmatrix} t & u \\ u & w \end{pmatrix}^{-1} \begin{pmatrix} y \\ h \end{pmatrix}$$
of the linear system
$$\begin{cases} \varphi_\alpha(\alpha, \beta) = t\alpha + u\beta + y = 0 \\ \varphi_\beta(\alpha, \beta) = u\alpha + w\beta + h = 0; \end{cases}$$
otherwise, in case $\boxed{tw - u^2 = 0 \text{ AND } ((th - uy \neq 0) \text{ OR } (uh - wy \neq 0))}$, then there is no solution for Case (IX);

otherwise, in case $\boxed{tw - u^2 = 0 \text{ AND } (th - uy = 0) \text{ AND } (uh - wy = 0)}$, then we have three sub-cases:

1. $\boxed{t > 0}$: then, recalling that we are in the sub-case where equations $\varphi_\alpha(\alpha, \beta) = 0$ and $\varphi_\beta(\alpha, \beta) = 0$ yield the same information, we exploit equation $\varphi_\alpha(\alpha, \beta) = 0$ and we set $\alpha = -(u\beta + y)/t$. Thus, from the bounds and the last inequality in (14), we obtain
$$\begin{cases} a_2 \leq \beta \leq b_2 \\ (\varepsilon_2 t - \varepsilon_1 u)\beta \leq \varepsilon_3 t + \varepsilon_1 y \\ a_1 t + y \leq -u\beta \leq b_1 t + y \end{cases}$$
which yield the next three cases:
   - $\boxed{\varepsilon_2 t - \varepsilon_1 u > 0}$: admitting other three sub-cases, namely
     * $\boxed{u > 0}$, so that we set
$$\beta_1 = \max\left\{a_2, -\frac{b_1 t + y}{u}\right\} \leq \beta \leq \min\left\{b_2, \frac{\varepsilon_3 t + \varepsilon_1 y}{\varepsilon_2 t - \varepsilon_1 u}, -\frac{a_1 t + y}{u}\right\} = \beta_2$$
     * $\boxed{u = 0}$, so that we set
$$\beta_1 = a_2 \leq \beta \leq \min\left\{b_2, \frac{\varepsilon_3 t + \varepsilon_1 y}{\varepsilon_2 t}\right\} = \beta_2$$

* $\boxed{u < 0}$, so that we set

$$\beta_1 = \max\left\{a_2, -\frac{a_1 t + y}{u}\right\} \le \beta \le \min\left\{b_2, \frac{\varepsilon_3 t + \varepsilon_1 y}{\varepsilon_2 t - \varepsilon_1 u}, -\frac{b_1 t + y}{u}\right\} = \beta_2$$

- $\boxed{\varepsilon_2 t - \varepsilon_1 u = 0}$: admitting no solution for Case (IX) as long as the condition $\boxed{\varepsilon_3 t + \varepsilon_1 y < 0}$ holds. Conversely, in case $\boxed{\varepsilon_3 t + \varepsilon_1 y \ge 0}$, we have the three cases:
  * $\boxed{u > 0}$, so that we set

$$\beta_1 = \max\left\{a_2, -\frac{b_1 t + y}{u}\right\} \le \beta \le \min\left\{b_2, -\frac{a_1 t + y}{u}\right\} = \beta_2$$

  * $\boxed{u = 0}$, so that we set

$$\beta_1 = a_2 \le \beta \le b_2 = \beta_2$$

  * $\boxed{u < 0}$, so that we set

$$\beta_1 = \max\left\{a_2, -\frac{a_1 t + y}{u}\right\} \le \beta \le \min\left\{b_2, -\frac{b_1 t + y}{u}\right\} = \beta_2$$

- $\boxed{\varepsilon_2 t - \varepsilon_1 u < 0}$: corresponding to the three cases:
  * $\boxed{u > 0}$, so that we set

$$\beta_1 = \max\left\{a_2, -\frac{b_1 t + y}{u}, \frac{\varepsilon_3 t + \varepsilon_1 y}{\varepsilon_2 t - \varepsilon_1 u}\right\} \le \beta \le \min\left\{b_2, -\frac{a_1 t + y}{u}\right\} = \beta_2$$

  * $\boxed{u = 0}$, so that we set

$$\beta_1 = \max\left\{a_2, \frac{\varepsilon_3 t + \varepsilon_1 y}{\varepsilon_2 t - \varepsilon_1 u}\right\} \le \beta \le b_2 = \beta_2$$

  * $\boxed{u < 0}$, so that we set

$$\beta_1 = \max\left\{a_2, -\frac{a_1 t + y}{u}, \frac{\varepsilon_3 t + \varepsilon_1 y}{\varepsilon_2 t - \varepsilon_1 u}\right\} \le \beta \le \min\left\{b_2, -\frac{b_1 t + y}{u}\right\} = \beta_2$$

2. $\boxed{t = 0}$: then, recalling that we are in the sub-case where equations $\varphi_\alpha(\alpha, \beta) = 0$ and $\varphi_\beta(\alpha, \beta) = 0$ yield the same information, with $tw - u^2 = 0$, we exploit equation $\varphi_\alpha(\alpha, \beta) = 0$ with $t = u = 0$. Therefore, we have

$$\begin{cases} a_2 \le \beta \le b_2 \\ y = 0 \\ a_1 \le \alpha \le b_1 \end{cases}$$

which yield the next two cases:

- $\boxed{y = 0}$ This case implies that the objective function is constant (i.e., $\varphi(\alpha, \beta) = q$), so that we set

$$\beta_1 = a_2 \le \beta \le b_2 = \beta_2$$

- $\boxed{y \ne 0}$: admitting no solution for Case (IX)

3. $\boxed{t < 0}$: then, recalling that we are again in the sub-case where equations $\varphi_\alpha(\alpha, \beta) = 0$ and $\varphi_\beta(\alpha, \beta) = 0$ yield the same information, we exploit equation $\varphi_\alpha(\alpha, \beta) = 0$ and we set $\alpha = -(u\beta + y)/t$. Thus, from the bounds and the last inequality in (14), we obtain

$$\begin{cases} a_2 \le \beta \le b_2 \\ (\varepsilon_2 t - \varepsilon_1 u)\beta \ge \varepsilon_3 t + \varepsilon_1 y \\ b_1 t + y \le -u\beta \le a_1 t + y \end{cases}$$

which yield the next three cases:

- $\boxed{\varepsilon_2 t - \varepsilon_1 u > 0}$: admitting other three cases, namely
    * $\boxed{u > 0}$, so that we set

$$\beta_1 = \max\left\{a_2, -\frac{a_1 t + y}{u}, \frac{\varepsilon_3 t + \varepsilon_1 y}{\varepsilon_2 t - \varepsilon_1 u}\right\} \le \beta \le \min\left\{b_2, -\frac{b_1 t + y}{u}\right\} = \beta_2$$

    * $\boxed{u = 0}$, so that $b_1 t + y \le -u\beta \le a_1 t + y$ is always fulfilled and we set

$$\beta_1 = \max\left\{a_2, \frac{\varepsilon_3 t + \varepsilon_1 y}{\varepsilon_2 t}\right\} \le \beta \le b_2 = \beta_2$$

    * $\boxed{u < 0}$, so that we set

$$\beta_1 = \max\left\{a_2, -\frac{b_1 t + y}{u}, \frac{\varepsilon_3 t + \varepsilon_1 y}{\varepsilon_2 t - \varepsilon_1 u}\right\} \le \beta \le \min\left\{b_2, -\frac{a_1 t + y}{u}\right\} = \beta_2$$

- $\boxed{\varepsilon_2 t - \varepsilon_1 u = 0}$: admitting no solution for Case (IX) as long as the condition $\boxed{\varepsilon_3 t + \varepsilon_1 y > 0}$ holds. Conversely, in case $\boxed{\varepsilon_3 t + \varepsilon_1 y \le 0}$ we have the three cases:
    * $\boxed{u > 0}$, so that we set

$$\beta_1 = \max\left\{a_2, -\frac{a_1 t + y}{u}\right\} \le \beta \le \min\left\{b_2, -\frac{b_1 t + y}{u}\right\} = \beta_2$$

    * $\boxed{u = 0}$, so that we set

$$\beta_1 = a_2 \le \beta \le b_2 = \beta_2$$

    * $\boxed{u < 0}$, so that we set

$$\beta_1 = \max\left\{a_2, -\frac{b_1 t + y}{u}\right\} \le \beta \le \min\left\{b_2, -\frac{a_1 t + y}{u}\right\} = \beta_2$$

- $\boxed{\varepsilon_2 t + \varepsilon_1 u < 0}$: corresponding to the three cases
    * $\boxed{u > 0}$, so that we set

$$\beta_1 = \max\left\{a_2, -\frac{a_1 t + y}{u}\right\} \le \beta \le \min\left\{b_2, \frac{\varepsilon_3 t + \varepsilon_1 y}{\varepsilon_2 t - \varepsilon_1 u}, -\frac{b_1 t + y}{u}\right\} = \beta_2$$

    * $\boxed{u = 0}$, so that we set

$$\beta_1 = a_2 \le \beta \le \min\left\{b_2, \frac{\varepsilon_3 t + \varepsilon_1 y}{\varepsilon_2 t - \varepsilon_1 u}\right\} = \beta_2$$

* $\boxed{u < 0}$, so that we set

$$\beta_1 = \max\left\{a_2, -\frac{b_1 t + y}{u}\right\} \leq \beta \leq \min\left\{b_2, \frac{\varepsilon_3 t + \varepsilon_1 y}{\varepsilon_2 t - \varepsilon_1 u}, -\frac{a_1 t + y}{u}\right\} = \beta_2.$$

Thus, overall, for Case (IX), if $\beta_1 \leq \beta_2$, we set

$$\bar{\beta} = (\beta_1 + \beta_2)/2, \qquad \bar{\alpha} = \begin{cases} -(u\bar{\beta} + y)/t & t \neq 0 \\ (a_1 + b_1)/2 & t = 0, \end{cases}$$

along with

$$P_9 = \begin{pmatrix} \bar{\alpha} \\ \bar{\beta} \end{pmatrix}, \qquad \varphi_i = \varphi(\bar{\alpha}, \bar{\beta}), \qquad y_i = P_9, \qquad i = i + 1, \tag{24}$$

otherwise, if $\beta_1 > \beta_2$, there is no solution for Case (IX);

- Case (X): We set $\bar{\alpha} = a_1$ with $\varepsilon_1 a_1 + \varepsilon_2 \beta = \varepsilon_3$, and we distinguish among three cases:
  - if $\boxed{\varepsilon_2 = 0 \text{ AND } \varepsilon_3 = \varepsilon_1 a_1}$, then set $\beta_1 = a_2 \leq \beta \leq b_2 = \beta_2$;
  - if $\boxed{\varepsilon_2 = 0 \text{ AND } \varepsilon_3 \neq \varepsilon_1 a_1}$, then there is no solution for Case (X);
  - if $\boxed{\varepsilon_2 \neq 0}$, then set

$$\beta_1 = \max\left\{a_2, \frac{\varepsilon_3 - \varepsilon_1 a_1}{\varepsilon_2}\right\} \leq \beta \leq \min\left\{b_2, \frac{\varepsilon_3 - \varepsilon_1 a_1}{\varepsilon_2}\right\} = \beta_2.$$

Set $\bar{\beta} = (\beta_1 + \beta_2)/2$ with

$$P_{10} = \begin{pmatrix} a_1 \\ \bar{\beta} \end{pmatrix}, \qquad \varphi_i = \varphi(a_1, \bar{\beta}), \qquad y_i = P_{10}, \qquad i = i + 1; \tag{25}$$

- Case (XI): We distinguish among the next four cases:
  - if $\boxed{\varepsilon_1 = \varepsilon_2 = 0 \text{ AND } \varepsilon_3 \geq 0}$, then set $\bar{\alpha} = (a_1 + b_1)/2$, $\beta_1 = a_2 \leq \beta \leq \beta_2 = b_2$; otherwise, there is no solution for Case (XI);
  - if $\boxed{\varepsilon_1 > 0}$, then $\alpha = (-\varepsilon_2 \beta + \varepsilon_3)/\varepsilon_1$ and we analyze three sub-cases:
    1. If $\boxed{\varepsilon_2 > 0}$, then set

$$\beta_1 = \max\left\{a_2, \frac{\varepsilon_1 b_1 - \varepsilon_3}{-\varepsilon_2}\right\} \leq \beta \leq \min\left\{b_2, \frac{\varepsilon_1 a_1 - \varepsilon_3}{-\varepsilon_2}\right\} = \beta_2;$$

    2. If $\boxed{\varepsilon_2 = 0}$, then set $\beta_1 = a_2 \leq \beta \leq b_2 = \beta_2$;
    3. If $\boxed{\varepsilon_2 < 0}$, then set

$$\beta_1 = \max\left\{a_2, \frac{\varepsilon_1 a_1 - \varepsilon_3}{-\varepsilon_2}\right\} \leq \beta \leq \min\left\{b_2, \frac{\varepsilon_1 b_1 - \varepsilon_3}{-\varepsilon_2}\right\} = \beta_2;$$

  - if $\boxed{\varepsilon_1 = 0 \text{ AND } \varepsilon_2 \neq 0}$, then set $\bar{\beta} = \varepsilon_3/\varepsilon_2$, $\bar{\alpha} = (a_1 + b_1)/2$; if ($\bar{\beta} < a_2$ OR $\bar{\beta} > b_2$), then there is no solution for Case (XI);
  - if $\boxed{\varepsilon_1 < 0}$, then $\alpha = (-\varepsilon_2 \beta + \varepsilon_3)/\varepsilon_1$ and we analyze three sub-cases:
    1. If $\boxed{\varepsilon_2 > 0}$, then set

$$\beta_1 = \max\left\{a_2, \frac{\varepsilon_1 a_1 - \varepsilon_3}{-\varepsilon_2}\right\} \leq \beta \leq \min\left\{b_2, \frac{\varepsilon_1 b_1 - \varepsilon_3}{-\varepsilon_2}\right\} = \beta_2;$$

    2. If $\boxed{\varepsilon_2 = 0}$, then set $\beta_1 = a_2 \leq \beta \leq b_2 = \beta_2$;

3. If $\boxed{\varepsilon_2 < 0}$, then set

$$\beta_1 = \max\left\{a_2, \frac{\varepsilon_1 b_1 - \varepsilon_3}{-\varepsilon_2}\right\} \le \beta \le \min\left\{b_2, \frac{\varepsilon_1 a_1 - \varepsilon_3}{-\varepsilon_2}\right\} = \beta_2.$$

Set $\bar{\beta} = (\beta_1 + \beta_2)/2$ and $\bar{\alpha} = (-\varepsilon_2\bar{\beta} + \varepsilon_3)/\varepsilon_1$; if $a_1 \le \bar{\alpha} \le b_1$, then set

$$P_{11} = \begin{pmatrix} \bar{\alpha} \\ \bar{\beta} \end{pmatrix}, \qquad \varphi_i = \varphi(\bar{\alpha}, \bar{\beta}), \qquad y_i = P_{11}, \qquad i = i+1, \tag{26}$$

otherwise, there is no solution for Case (XI);

- Case (XII): We set $\bar{\beta} = a_2$ with $\varepsilon_1\alpha + \varepsilon_2 a_2 = \varepsilon_3$, and we distinguish among three cases:
  - if $\boxed{\varepsilon_1 = 0 \text{ AND } \varepsilon_3 = \varepsilon_2 a_2}$, then set $\bar{\alpha} = (a_1 + b_1)/2$;
  - if $\boxed{\varepsilon_1 = 0 \text{ AND } \varepsilon_3 \neq \varepsilon_2 a_2}$, then there is no solution for Case (XII);
  - if $\boxed{\varepsilon_1 \neq 0}$, then set

$$\alpha_1 = \max\left\{a_1, \frac{\varepsilon_3 - \varepsilon_2 a_2}{\varepsilon_1}\right\} \le \alpha \le \min\left\{b_1, \frac{\varepsilon_3 - \varepsilon_2 a_2}{\varepsilon_1}\right\} = \alpha_2.$$

Set $\bar{\alpha} = (\alpha_1 + \alpha_2)/2$ with

$$P_{12} = \begin{pmatrix} \bar{\alpha} \\ a_2 \end{pmatrix}, \qquad \varphi_i = \varphi(\bar{\alpha}, a_2), \qquad y_i = P_{12}, \qquad i = i+1. \tag{27}$$

The next lemma justifies the role of the last analysis for the computation of possible solutions of (14).

Lemma 2. *Given the problem (14), and let the assumptions of Lemma 1 hold. Consider the sequence of the m points $\{y_i\}$ and the sequence of the m values $\{\varphi_i\}$, from (16)–(27), which are relabelled so that for any index $i \ge 2$, we have*

$$\varphi_{i-1} \le \varphi_i \le \varphi_{i+1}.$$

Then, if

$$\hat{\ss} \in \underset{1 \le i \le m}{\operatorname{argmin}}\{\varphi_i\}$$

then, the point $y_{\hat{\ss}}$ is a global minimum for (14).

Proof of Lemma 2. The existence of a global minimum y^* and the corresponding value $\varphi(y^*)$ for (14) is ensured by Lemma 1. Moreover, each global minimum of (14) naturally fulfills KKT conditions, so that each global minimum must belong to the sequence $\{y_i\}$. Now, assume by contradiction that there exists a point $\tilde{y} \in \{y_i\}$, with $\tilde{y} \in \operatorname{argmin}_{1\le i\le m}\{\varphi_i\}$, but $\tilde{y}$ is *not* a global minimum. This yields the contradictory fact that $\varphi(y^*) > \varphi(\tilde{y})$. □

6. Conclusions and Future Work

We have considered a very relevant issue within Nonlinear Programming, namely the solution of a specific constrained quadratic problem, whose exact global solution can be easily computed after analyzing the first-order KKT conditions associated with it. We also highlighted that our proposal may, to a large extent, suggest guidelines for the research of novel LBMs, by drawing inspiration from TRMs. This last observation represents a promising tool, in order to provide algorithms which guarantee global convergence to stationary limit points, satisfying either first- or second-order necessary optimality conditions. In particular, we can summarize the following promising lines of research, for large-scale problems which iteratively generate the sequences of points

$$\begin{cases} x_{k+1} = x_k + \alpha_k d_k \\ x_{k+1} = x_k + \alpha_k d_k + \beta_k z_k \end{cases} \qquad \longleftarrow \qquad \text{for LBMs}$$

$$x_{k+1} = x_k + s_k \qquad \longleftarrow \qquad \text{for TRMs}$$

which are d_k, z_k, and s_k search directions at the current iterate x_k:

- Developing novel iterative LBMs (e.g., linesearch-based TNMs), where the search direction d_k (e.g., a Newton-type direction) is possibly combined with another direction z_k (e.g., the steepest descent at x_k, a negative curvature direction at x_k, etc.) through the use of (14). Then, comparing the efficiency of the novel methods with more standard linesearch-based approaches from the literature could give indications of the reliability of the ideas in this paper;
- Developing novel hybrid methods where the rationale behind alternating trust-region or linesearch-based techniques is exploited. In particular, the iterative scheme $x_{k+1} = x_k + \alpha_k d_k + \beta_k z_k$ (respectively, $x_{k+1} = x_k + \alpha_k d_k$) might be considered, where the search directions d_k and z_k, along with the steplengths α_k and β_k (respectively, d_k and α_k), are alternatively computed by solving
    1. A trust-region sub-problem like (4), so that a sufficient reduction in the quadratic model is ensured;
    2. A sub-problem like (14), so that the solution $\alpha^* d_k + \beta^* z_k$ is a promising gradient-related direction to be used within a linesearch procedure.

  In order to preserve the global convergence to stationary points satisfying either first- or second-order necessary optimality conditions;
- Specifically, comparing the use of *dogleg methods* (within TRMs) vs. the application of (14) coupled with a linesearch technique. This issue is tricky, since dogleg methods are applied to trust-region sub-problems like (4), including a general quadratic constraint (i.e., the trust-region constraint), while in (14) all the constraints are linear, so that the exact global solution of (14) is easily computed. Moreover, the last issue might shed light also on the opportunity (possibly) of privileging an efficient linesearch procedure applied to a (coarsely computed) gradient-related search direction, in place of a precise computation of the search direction in LBMs, using an inexpensive linesearch procedure. In other words, it is at present questionable if coupling a *coarse* computation of the vectors d_k and z_k with an *accurate* linesearch procedure would be more preferable than coupling the *accurately* computed vectors d_k and z_k with a *cheaper* linesearch procedure;
- Introducing nonmonotone stabilization techniques (see e.g., [28]) combining nonmonotonicity with any of the above ideas, for both TRMs and LBMs.

Author Contributions: G.F., C.P. and M.R. have equally contributed to this paper. All authors have read and agreed to the published version of the manuscript.

Funding: This research received no external funding.

Informed Consent Statement: Informed consent was obtained from all subjects involved in the study.

Data Availability Statement: Data are contained within the article.

Acknowledgments: Giovanni Fasano and Massimo Roma thank *INδAM* (Istituto Nazionale di Alta Matematica) for the support they received.

Conflicts of Interest: The authors declare no conflicts of interest. The funders had no role in the design of the study; in the collection, analyses, or interpretation of data; in the writing of the manuscript; or in the decision to publish the results.

References

1. Ben-Tal, A.; Nemirovski, A. Lectures on Modern Convex Optimization: Analysis, Algorithms, and Engineering Applications. In *MPS-SIAM Series on Optimization*; SIAM: Philadelphia, PA, USA, 2001.
2. Conn, A.R.; Gould, N.I.M.; Toint, P.L. Trust-region methods. In *MPS-SIAM Series on Optimization*; SIAM: Philadelphia, PA, USA, 2000.
3. Nocedal, J.; Wright, S.J. *Numerical Optimization*, 2nd ed.; Springer: New York, NY, USA, 2006.
4. Micchelli, C.A. Interpolation of scattered data: Distance matrices and conditionally positive definite functions. *Constr. Approx.* **1986**, *2*, 11–22. [CrossRef]
5. Myers, D.E. Kriging, cokriging, radial basis functions and the role of positive definiteness. *Comput. Math. Appl.* **1992**, *24*, 139–148. [CrossRef]
6. Conn, A.R.; Gould, N.I.M.; Sartenaer, A.; Toint, P.L. On Iterated-Subspace Minimization Methods for Nonlinear Optimization. In *Proceedings on Linear and Nonlinear Conjugate Gradient-Related Methods*; Adams, L., Nazareth, L., Eds.; SIAM: Philadelphia, PA, USA, 1996; pp. 50–78.
7. Shea, B.; Schmidt, M. Why line search when you can plane search? SO-friendly neural networks allow per-iteration optimization of learning and momentum rates for every layer. *arXiv* **2024**, arXiv:2406.17954.
8. Caliciotti, A.; Fasano, G.; Nash, S.; Roma, M. An adaptive truncation criterion for Newton-Krylov methods in large scale nonconvex optimization. *Oper. Res. Lett.* **2018**, *46*, 7–12. [CrossRef]
9. McCormick, G.P. A modification of Armijo's step-size rule for negative curvature. *Math. Program.* **1977**, *13*, 111–115. [CrossRef]
10. Moré, J.J.; Sorensen, D.C. On the use of directions of negative curvature in a modified Newton method. *Math. Program.* **1979**, *16*, 1–20. [CrossRef]
11. Nash, S.G. A survey of truncated-Newton methods. *J. Comput. Appl. Math.* **2000**, *124*, 45–59. [CrossRef]
12. Fasano, G.; Roma, M. Iterative computation of negative curvature directions in large scale optimization. *Comput. Optim. Appl.* **2007**, *38*, 81–104. [CrossRef]
13. De Leone, R.; Fasano, G.; Roma, M.; Sergeyev, Y.D. Iterative Grossone-Based Computation of Negative Curvature Directions in Large-Scale Optimization. *J. Optim. Theory Appl.* **2020**, *186*, 554–589. [CrossRef]
14. Curtis, F.E.; Robinson, D.P. Exploiting negative curvature in deterministic and stochastic optimization. *Math. Program.* **2019**, *176*, 69–94. [CrossRef]
15. Gill, P.E.; Wong, E. Sequential Quadratic Programming Methods. In *Mixed Integer Nonlinear Programming*; Lee, J., Leyffer, S., Eds.; The IMA Volumes in Mathematics and Its Applications; Springer: New York, NY, USA, 2012; Volume 154.
16. Fletcher, R.; Gould, N.I.; Leyffer, S.; Toint, P.; Wächter, A. Global convergence of a trust-region SQP-filter algorithm for general nonlinear programming. *SIAM J. Optim.* **2002**, *13*, 635–659. [CrossRef]
17. Wang, J.; Petra, C.G. A Sequential Quadratic Programming Algorithm for Nonsmooth Problems with Upper-Objective. *SIAM J. Optim.* **2023**, *33*, 2379–2405. [CrossRef]
18. Nocedal, J.; Yuan, Y. Combining trust-region and line-search techniques. In *Advances in Nonlinear Programming*; Yuan, Y., Ed.; Kluwer: Boston, MA, USA, 1998; pp. 157–175.
19. Tong, X.; Zhou, S. Combining Trust Region and Line Search Methods for Equality Constrained Optimization. *Numer. Funct. Anal. Optim.* **2006**, *24*, 143–162. [CrossRef]
20. Waltz, R.A.; Morales, J.L.; Nocedal, J.; Orban, D. An interior algorithm for nonlinear optimization that combines line search and trust region steps. *Math. Program.* **2006**, *107*, 391–408. [CrossRef]
21. Pei, Y.; Zhu, D. A trust-region algorithm combining line search filter technique for nonlinear constrained optimization. *Int. J. Comput. Math.* **2014**, *91*, 1817–1839. [CrossRef]
22. Dembo, R.S.; Eisenstat, S.C.; Steihaug, T. Inexact Newton methods. *SIAM J. Numer. Anal.* **1982**, *19*, 400–408. [CrossRef]
23. Steihaug, T. The Conjugate Gradient method and Trust Regions in large scale optimization. *SIAM J. Numer. Anal.* **1983**, *20*, 626–637. [CrossRef]
24. Lucidi, S.; Rochetich, F.; Roma, M. Curvilinear stabilization techniques for truncated Newton methods in large scale unconstrained optimization. *SIAM J. Optim.* **1998**, *8*, 916–939. [CrossRef]
25. Gould, N.I.M.; Lucidi, S.; Roma, M.; Toint, P.L. Solving the trust-region subproblem using the Lanczos method. *SIAM J. Optim.* **1999**, *9*, 504–525. [CrossRef]
26. De Leone, R.; Fasano, G.; Sergeyev, Y.D. Planar methods and Grossone for the Conjugate Gradient breakdown in Nonlinear Programming. *Comput. Optim. Appl.* **2018**, *71*, 73–93. [CrossRef]
27. Grippo, L.; Lampariello, F.; Lucidi, S. A truncated Newton method with nonmonotone linesearch for unconstrained optimization. *J. Optim. Theory Appl.* **1989**, *60*, 401–419. [CrossRef]
28. Grippo, L.; Lampariello, F.; Lucidi, S. A class of nonmonotone stabilization methods in unconstrained optimization. *Numer. Math.* **1991**, *59*, 779–805. [CrossRef]

algorithms

MDPI

Article

A Transfer Matrix Norm-Based Framework for Multimaterial Topology Optimization: Methodology and MATLAB Implementation

Giacomo Galuppini [1] and Paolo Venini [2,*]

[1] Department of Electrical, Computer and Biomedical Engineering, University of Pavia, Via Ferrata 5, 27100 Pavia, Italy; giacomo.galuppini@unipv.it

[2] Department of Civil Engineering and Architecture, University of Pavia, Via Ferrata 3, 27100 Pavia, Italy

* Correspondence: paolo.venini@unipv.it

Abstract: This paper introduces a novel method for the topology optimization of multimaterial structures. The core innovation lies in minimizing an appropriate norm of the transfer matrix that links external loads to system outputs. Several formulations of structural compliance are considered within a framework that accommodates both single and multiple inputs and outputs. The multimaterial interpolation scheme follows the approach proposed by Yi and collaborators, which is adapted to the general topology optimization framework proposed by Venini and collaborators. While the proposed method is inherently capable of addressing the dynamic response of multimaterial structures, this study focuses exclusively on static topology optimization. The extension to dynamic topology optimization is deferred to a fortcoming study, which requires a multimaterial interpolation of inertial properties that is currently in an advanced development stage. Finally, a series of numerical case studies illustrate the key features of the proposed approach.

Keywords: multimaterial topology optimization; transfer matrix norms; MATLAB

1. Introduction

Topology optimization (TO) is a powerful computational design methodology that seeks optimal material distribution within a given domain, subject to physics-based constraints and objectives [1,2]. Traditionally applied to single-material structures, TO has enabled the synthesis of highly efficient, lightweight, and innovative components across engineering disciplines. However, as manufacturing technologies such as additive manufacturing and multi-material printing have advanced [3], there is a growing need to extend TO methodologies to accommodate multiple material phases. This demand is driven by the potential of multimaterial structures to offer superior performance by strategically combining the desirable properties of different materials—e.g., stiffness, energy absorption, heat conduction, or cost-efficiency—in a single optimized layout.

Multimaterial Topology Optimization (MMTO) introduces several new challenges that distinguish it from its single-material counterpart. First, the interpolation of material properties becomes substantially more complex. Unlike the conventional SIMP (Solid Isotropic Material with Penalization) scheme [2] which interpolates between solid and void, MMTO must handle transitions across multiple discrete materials without producing physically meaningless mixtures or gray zones. Techniques such as the Discrete Material

Optimization (DMO) [4], peak functions [5], and ordered SIMP interpolations [3] have been developed to address these challenges.

More recent efforts have focused on improving the manufacturability and computational efficiency of MMTO. Zheng et al. [6] proposed a mapping-based interpolation function implemented in a compact MATLAB code, extending the educational 88-line TO code [7] to handle 2D and 3D multimaterial structures. Their approach yields clear 0–1 solutions and ensures discrete material regions suitable for fabrication. Similarly, Dinh et al. [8] introduced a single-variable-based interpolation function with continuous derivatives to enhance optimization convergence and reduce design dimensionality.

A significant issue in MMTO is phase mixing at material boundaries, which complicates both physical interpretation and manufacturing. Asadpoure et al. [9] tackled this by developing a multi-phase topology optimization framework with independent design variables for each phase. Their approach employs Heaviside projections with phase-specific length-scale control, aggregated constraints to avoid overfilling, and consistent penalization for phase separation. This method enables large-scale optimization and mitigates the tendency for undesired material interpenetration.

Beyond mechanical compliance, MMTO has also been extended to address more complex physical behaviors such as plasticity and transient thermal conduction. Jia et al. [10] introduced a multiobjective MMTO framework for elastoplastic composite structures that simultaneously enhances stiffness, delays yielding, and maximizes energy dissipation. Their use of return mapping algorithms and history-dependent adjoint sensitivities marks an important advance in integrating realistic material behavior into TO. On the thermal side, Ogawa et al. [4] applied the DMO method to MMTO for unsteady heat conduction, emphasizing its robustness against material ordering and its capability to optimize thermal pathways in time-dependent problems.

Several common threads emerge across these studies. First, interpolation strategies remain a central concern: whether using multi-variable SIMP [11], DMO [12], or peak/ordered functions [3,5], each method seeks to balance fidelity, computational cost, and manufacturability. As far as other optimization methods are concerned, Huang and Xie (2023) [13], enhanced the BESO (Bi-directional Evolutionary Structural Optimization) method by incorporating buckling constraints, effectively addressing pseudo buckling modes in structural topology optimization; Huang (2020) developed an element-based topology optimization algorithm for a smooth design [14] that was further investigated in Huang and Li (2022), [15], that introduced a three-field FPTO (Floating Projection Topology Optimization) method employing linear material interpolation, facilitating the generation of clear 0/1 designs without the need for penalization schemes. Huang et al. (2020) [16], developed the SEMDOT (Smooth-Edged Material Distribution for Optimizing Topology) algorithm, which effectively produces optimized topologies with smooth and clear boundaries, outperforming traditional methods in certain applications. Second, controlling the interface between material phases—whether through filtering, projection, or constraint aggregation—proves essential in achieving physically meaningful designs. Third, scalability and code accessibility, as emphasized by Zheng et al. [6] and Sanders [17], are key to enabling broader adoption of MMTO by researchers and practitioners.

Despite these advances, MMTO remains a vibrant and rapidly evolving field. Open challenges include the integration of nonlinear material behaviors, uncertainty quantification, multi-physics coupling, and automated manufacturability constraints. Moreover, emerging fabrication technologies continue to push the boundaries of what is realizable, demanding increasingly sophisticated optimization formulations.

Given this background, the present paper proposes an innovative formulation for the topology optimization of Multi-Input (multi-loaded) Multi-Output (multi-objective) structures in a multi-material setting. While the framework naturally includes classical compliance optimization as a special case, it is sufficiently general to accommodate alternative compliance definitions and other objective functions tailored to specific applications. The central idea behind the proposed approach is the explicit formulation of the input-output transfer matrix, denoted as $\boldsymbol{G}$, which maps applied loads to the structural response.

In this framework, the structural response vector—often referred to as the output vector—can be any linear combination of the state vector components. Notably, structural compliance, which is the primary objective in most topology optimization studies, can inherently be expressed as a linear combination of displacements at the locations where loads are applied. The applied loads themselves serve as coefficients of this combination, as demonstrated later in this paper. Consequently, the optimization problem is formulated to minimize a suitable measure of the "amplitude" of the transfer matrix $\boldsymbol{G}$, which, from an engineering perspective, corresponds to reducing the impact of applied loads on the structure.

A key component of this methodology is the selection of an appropriate metric to quantify the amplitude of the transfer matrix. The Singular Value Decomposition (SVD) [18,19] emerges as a natural tool for this purpose. Not only does SVD facilitate the computation of matrix norms that serve as amplitude measures, but it also provides valuable insights into system behavior. Furthermore, since these norms are uniquely defined in terms of the singular values of $\boldsymbol{G}$, the computation of sensitivities is significantly simplified—reducing the problem to evaluating the gradients of singular values with respect to design parameters. The formulation presented in this paper extends to MMTO the one presented in [20] that is restricted to standard topology optimization in a SIMP framework. As for the 3D idealization, the formulation introduced in [21] shall be extended to the multimaterial case in a forthcoming contribution.

The remainder of this paper is structured as follows. Section 2 outlines the fundamental principles of the adopted multi-material interpolation scheme, which builds upon the methodology proposed in [22]. The section provides a detailed account of the interpolation strategy, along with a comprehensive description of the two-stage filtering process employed to mitigate checkerboarding and prevent the emergence of grey regions within the design domain.

Section 3 introduces the key properties of Singular Value Decomposition (SVD), emphasizing its relevance in engineering applications. A central insight is that the input and output singular vectors are linked through a gain defined by the corresponding singular value, making this framework particularly effective for both static and dynamic topology optimization problems, including the classical minimum compliance formulation. The section also provides an explicit derivation of the transfer matrix for the widely used case of displacement-based finite element discretization. Furthermore, it formulates the general topology optimization problem and presents a semi-analytic approach for computing the sensitivities of both the objective function and the constraints.

Section 4 focuses on selecting the components of the output vector, which define specific optimization objectives, with an emphasis on compliance and its generalized variants. By integrating compliance and its generalized forms into the output vector, the optimization process can systematically address multiple performance criteria, leading to structures that are both efficient and robust under various loading conditions.

Section 5 discusses the MATLAB implementation of the MMTO algorithm proposed in this work, and made available to the reader.

The effectiveness of the proposed methodology is demonstrated in Section 6, which presents numerical results and comparative analyses with standard topology optimization techniques (where applicable). The focus is placed on deriving general insights into the relative performance and robustness of various norm-based optimization criteria.

Finally, Section 7 offers concluding remarks and outlines potential directions for future research.

2. Multi-Material Interpolation

The objective of this section is twofold. First, the adopted multi-material interpolation scheme is introduced, accompanied by selected implementation details. Second, an explicit semi-analytical expression is derived for evaluating the gradient of the stiffness matrix. This formulation will be shown to play a critical role in the derivation of the sensitivity of the objective function, defined as a suitable norm of the input/output transfer matrix.

2.1. Design Variables, Intermediate Filtered Variables and Projected Variables

We adopt the multi-material interpolation scheme originally proposed in [22] and subsequently examined in [6], where the emphasis was placed on its MATLAB implementation. The design variables are grouped in an $NE \times NM$ matrix:

$$x = \begin{bmatrix} x_1^1 & x_2^1 & \dots & x_{NM}^1 \\ x_1^2 & x_2^2 & \dots & x_{NM}^2 \\ \vdots & \vdots & \ddots & \vdots \\ x_1^{NE} & x_2^{NE} & \dots & x_{NM}^{NE} \end{bmatrix} \quad (1)$$

where NE represents the number of finite elements, and NM denotes the number of materials. To avoid mesh-dependency and checkerboard issues, a standard density filter is applied that reads:

$$\tilde{x}_i^e = \frac{\sum_{j \in N_e} H_{ej} x_i^j}{\sum_{j \in N_e} H_{ej}} \quad (2)$$

where N_e is a proper set of elements surrounding element e, and H_{ej} is a weight factor. A projection function is further used to avoid grey patterns in the final structure at convergence. One then writes:

$$\overline{\tilde{x}_i^e} = \frac{\tanh(\beta\eta) + \tanh(\beta(\tilde{x}_i^e - \eta))}{\tanh(\beta\eta) + \tanh(\beta(1-\eta))} \quad (3)$$

where β controls the slope, and η is the transition parameter.

2.2. Interpolation Function for Multi-Materials

The Young's modulus of each element is interpolated between all the materials as:

$$E(\overline{\tilde{x}_1^e}, \overline{\tilde{x}_2^e}, \dots, \overline{\tilde{x}_{NM}^e}) = \sum_{i=1}^{NM} \left(\frac{\|\overline{\tilde{x}^e}\|_p}{\|\overline{\tilde{x}^e}\|_1 + \delta} \overline{\tilde{x}_i^e} \right)^n (E_i - E_{void}) + E_{void} \quad (4)$$

where (all the components of $\overline{\tilde{x}^e}$ are non-negative and therefore absolute values are omitted):

- $\|\overline{\tilde{x}^e}\|_p = \left[(\overline{\tilde{x}_1^e})^p + (\overline{\tilde{x}_2^e})^p + \dots + (\overline{\tilde{x}_{NM}^e})^p \right]^{1/p}$ is the p-norm of vector $\overline{\tilde{x}^e}$;
- $\|\overline{\tilde{x}^e}\|_1 = \overline{\tilde{x}_1^e} + \overline{\tilde{x}_2^e} + \dots + \overline{\tilde{x}_{NM}^e}$ is the 1-norm of vector $\overline{\tilde{x}^e}$;
- n is a standard SIMP penalization factor, often set equal to 3;
- δ is a small positive constant that avoids singularities to show up.

For the purpose of sensitivity analysis, it is appropriate to rewrite Equation (4) in the form:

$$E(\tilde{x}_1^e, \tilde{x}_2^e, \ldots, \tilde{x}_{NM}^e) = \sum_{i=1}^{NM} (\psi_i)^n (E_i - E_{void}) + E_{void}, \tag{5}$$

where the i–th interpolation function ψ_i reads:

$$\psi_i(\tilde{x}_1^e, \tilde{x}_2^e, \ldots, \tilde{x}_{NM}^e) = \frac{\|\tilde{x}^e\|_p}{\|\tilde{x}^e\|_1 + \delta} \tilde{x}_i^e. \tag{6}$$

An interpolation of the type shown in Equation (4) may introduce non-differentiability issues, as the p-norm is generally non-differentiable when p is an odd integer. However, this limitation does not apply in the present context, since all components of the vectors $\overline{\tilde{x}_i^e}$ are constrained to be non-negative. Nonetheless, the adoption of general optimization algorithms that do not rely on gradient information remains a short-term objective and will be addressed in a forthcoming contribution.

2.3. Sensitivity Analysis

Let $\mathcal{F}(\overline{\tilde{x}}^e)$ and $\mathcal{V}(\overline{\tilde{x}}^e)$ denote, for the time being, the objective function and a generic constraint, respectively. Throughout the remainder of this paper, these will correspond to a suitable norm of the transfer matrix and a standard volume constraint. With regard to the components of the gradient vectors, the application of the chain rule yields the following expressions:

$$\frac{\partial \mathcal{F}}{\partial x_i^j} = \frac{\partial \mathcal{F}}{\partial \overline{\tilde{x}}_i^e} \frac{\partial \overline{\tilde{x}}_i^e}{\partial \tilde{x}_i^e} \frac{\partial \tilde{x}_i^e}{\partial x_i^j}, \quad \frac{\partial \mathcal{V}}{\partial x_i^j} = \frac{\partial \mathcal{V}}{\partial \overline{\tilde{x}}_i^e} \frac{\partial \overline{\tilde{x}}_i^e}{\partial \tilde{x}_i^e} \frac{\partial \tilde{x}_i^e}{\partial x_i^j}, \tag{7}$$

Upon differentiating Equation (3), one gets the derivative of the projected density with respect to the filtered density, that reads:

$$\frac{\partial \overline{\tilde{x}}^e}{\partial \tilde{x}^e} = \frac{\beta(1 - \tanh^2(\beta(\tilde{x}^e - \eta)))}{\tanh(\beta\eta) + \tanh(\beta(1 - \eta))} \tag{8}$$

Upon recalling Equation (2), the derivative of the filtered density with respect to the original design variable may be written as:

$$\frac{\partial \tilde{x}_i^e}{\partial x_i^j} = \frac{H_{ej}}{\sum_{j \in N_e} H_{ej}} \tag{9}$$

The Derivative of the Stiffness Matrix

Let $\underline{\underline{K}}_0$ be the stiffness matrix of the standard quadrilateral element for unitary Young modulus. Thanks to Equation (5), the element stiffness matrix may be written as:

$$\underline{\underline{K}}(\overline{\tilde{x}_1^e}, \overline{\tilde{x}_2^e}, \ldots, \overline{\tilde{x}_{NM}^e}) = \left[\sum_{i=1}^{NM} (\psi_i)^n (E_i - E_{void}) + E_{void} \right] \underline{\underline{K}}_0 \tag{10}$$

The gradient of the stiffness matrix therefore reads:

$$\frac{\partial \underline{\underline{K}}}{\partial \overline{\tilde{x}}_h^e} = \left[\sum_{i=1}^{NM} \frac{\partial (\psi_i)^n}{\partial \overline{\tilde{x}}_h^e} (E_i - E_{void}) \right] \underline{\underline{K}}_0, \tag{11}$$

where:

$$\frac{\partial(\psi_i)^n}{\partial\overline{\tilde{x}_j^e}} = n(\psi_i)^{n-1}\frac{\partial\psi_i}{\partial\overline{\tilde{x}_j^e}}. \tag{12}$$

The derivative of the interpolation function ψ with respect to the generic projected variable $\overline{\tilde{x}_i^e}$ admits two distinct expressions, depending on whether $h = i$ or $h = j \neq i$:

$$\frac{\partial(\psi_i)}{\partial\overline{\tilde{x}_i^e}} = \frac{||\overline{\tilde{x}^e}||_p}{||\overline{\tilde{x}^e}||_1 + \delta} + \left(\frac{\frac{\partial||\overline{\tilde{x}^e}||_p}{\partial\overline{\tilde{x}_i^e}}}{||\overline{\tilde{x}^e}||_1 + \delta} - \frac{\frac{\partial||\overline{\tilde{x}^e}||_1}{\partial\overline{\tilde{x}_i^e}}||\overline{\tilde{x}^e}||_p}{(||\overline{\tilde{x}^e}||_1 + \delta)^2}\right)\overline{\tilde{x}_i^e} \tag{13}$$

and

$$\frac{\partial(\psi_i)}{\partial\overline{\tilde{x}_j^e}} = \left(\frac{\frac{\partial||\overline{\tilde{x}^e}||_p}{\partial\overline{\tilde{x}_j^e}}}{||\overline{\tilde{x}^e}||_1 + \delta} - \frac{\frac{\partial||\overline{\tilde{x}^e}||_1}{\partial\overline{\tilde{x}_j^e}}||\overline{\tilde{x}^e}||_p}{(||\overline{\tilde{x}^e}||_1 + \delta)^2}\right)\overline{\tilde{x}_i^e}, \tag{14}$$

where the derivatives of the p-norm and the 1-norm respectively writes as follows:

$$\frac{\partial||\overline{\tilde{x}^e}||_p}{\partial\overline{\tilde{x}_i^e}} = \left(\overline{\tilde{x}^e}_1^p + \overline{\tilde{x}^e}_2^p + \ldots + \overline{\tilde{x}^e}_{NM}^p\right)^{\frac{1}{p}-1}\overline{\tilde{x}_i^e}^{p-1} \tag{15}$$

$$\frac{\partial||\overline{\tilde{x}^e}||_1}{\partial\overline{\tilde{x}_i^e}} = 1 \tag{16}$$

Ultimately, Equation (11) provides the key component required to compute the gradient of the cost function, as demonstrated in the following.

3. The Multi-Material Topology Optimization Problem

The reader is referred to [20] for further details within a standard single-material SIMP framework.

3.1. Matrix Norms via Singular Values

With reference to Figure 1, the (parameter-dependent) matrix $\boldsymbol{G}(\boldsymbol{p})$ mapping the $r \times 1$ input vector $\boldsymbol{f}$ to the $n_o \times 1$ output vector $\boldsymbol{z}$ is given by

$$\boldsymbol{G}(\boldsymbol{p}) = \boldsymbol{C}\boldsymbol{K}(\boldsymbol{p})^{-1}\boldsymbol{B} \tag{17}$$

where $\boldsymbol{B}$ is an $n \times r$ matrix that distributes the loads to the appropriate degrees of freedom, $\boldsymbol{C}$ is an $n_o \times n$ output matrix that extracts the relevant components of the output vector $\boldsymbol{z}$, and $\boldsymbol{K}(\boldsymbol{p})$ is the (parameter-dependent) system stiffness matrix, and $\boldsymbol{p}$ is a vector of design parameters. For any fixed $\boldsymbol{p}$, the corresponding transfer matrix $\boldsymbol{G}$ admits a Singular Value Decomposition (SVD) [19], which can be written as:

$$\boldsymbol{G} = \boldsymbol{U}\boldsymbol{\Sigma}\boldsymbol{V}^H, \tag{18}$$

where:

- $\boldsymbol{U}$ and $\boldsymbol{V}$ are unitary matrices of dimensions $l \times l$ and $m \times m$, respectively.
- $\boldsymbol{\Sigma}$ is an $l \times m$ diagonal matrix with the form:

$$\boldsymbol{\Sigma} = \begin{cases} \begin{bmatrix} \boldsymbol{\Sigma}_1 \\ \mathbf{0} \end{bmatrix}, & \text{if } l \geq m, \\ \begin{bmatrix} \boldsymbol{\Sigma}_1 & \mathbf{0} \end{bmatrix}, & \text{if } l \leq m, \end{cases}$$

where:

$$\boldsymbol{\Sigma}_1 = \text{diag}\{\sigma_1, \sigma_2, \ldots, \sigma_r\}, \tag{19}$$

with $r = \min(l, m)$ denoting the rank of $\boldsymbol{G}$. The singular values are sorted in descending order,

$$\overline{\sigma} \equiv \sigma_1 \geq \sigma_2 \geq \cdots \geq \sigma_r \equiv \underline{\sigma} > 0, \tag{20}$$

as indicated in Equation (20), in order to define the matrix norms introduced below, which will be used in the sequel:

- 2-Norm:

$$||\boldsymbol{G}||_2 = \sigma_1, \tag{21}$$

- Nuclear Norm:

$$||\boldsymbol{G}||_N = \sum_{k=1}^{r} \sigma_k. \tag{22}$$

- Frobenius Norm:

$$||\boldsymbol{G}||_F = \sqrt{\sum_{k=1}^{r} \sigma_k^2}. \tag{23}$$

The matrix norms introduced above lay the groundwork for the multi-material topology optimization problem formulated in the following section.

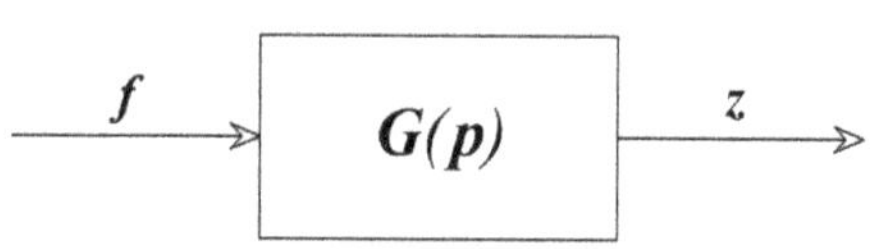

Figure 1. System Representation in Standard Input/Output Form.

3.2. *The Topology Optimization Problem*

For the sake of simplicity, the design variable matrix introduced in Equation (1) will henceforth be reformatted into a column vector, denoted by $\boldsymbol{p}$, defined as

$$\boldsymbol{p} = [\boldsymbol{p}_1, \boldsymbol{p}_2, \ldots, \boldsymbol{p}_{NM}]^T, \tag{24}$$

where each block $\boldsymbol{p}_i$ corresponds to the design variables associated with the i-th material phase. The typical $\boldsymbol{G}$-based optimal design problem may then be formulated as follows:

$$\begin{cases} \min\limits_{\boldsymbol{p}} \mathcal{F}(\boldsymbol{p}) & = & ||\boldsymbol{G}(\boldsymbol{p})|| \\ \text{s.t.} & & \boldsymbol{G}(\boldsymbol{p}) = \boldsymbol{C}\boldsymbol{K}^{-1}(\boldsymbol{p})\boldsymbol{B} \\ & & V_i(\boldsymbol{p}_i) \leq V_{\text{max}i}, \quad i = 1, \ldots, NM \\ & & \boldsymbol{p}_{\text{min}} \leq \boldsymbol{p} \leq \boldsymbol{p}_{\text{max}} \end{cases}, \tag{25}$$

where:

- $(25)_1$ defines the objective function $\mathcal{F}(\boldsymbol{p})$ to be minimized;
- $(25)_2$ provides the explicit definition of the input/output transfer matrix, highlighting its dependence on the design variable vector $\boldsymbol{p}$;
- $(25)_3$ imposes a standard volume constraint on each individual material. In the present formulation, all materials are subject to the same volume constraint, although non-uniform constraints may also be considered;
- the inequalities on the design variables in $(25)_4$ are to be interpreted component-wise. As is customary, each design variable is bounded between zero (void) and one (full material).

The optimization problem in (25) is solved numerically using the Method of Moving Asymptotes (MMA) [23]. To this end, the user must supply the current values of the objective function and the constraints, together with their sensitivities with respect to the design variables. When the objective function to be minimized is a matrix norm that depends on the singular values of the transfer matrix, a semi-analytical computation of the sensitivities is required. Specifically, using the chain rule, one must compute the derivatives of both the transfer matrix and its singular values with respect to each design variable, denoted generically by p_k. Further details on this procedure are provided below.

3.3. Gradient of the Input/Output Transfer Matrix

Under the assumption that neither $\boldsymbol{B}$ nor $\boldsymbol{C}$ depend on the design variable vector $\boldsymbol{p}$, and recalling that the derivative of a matrix inverse is given by:

$$\frac{\partial \boldsymbol{K}^{-1}}{\partial p_k} = -\boldsymbol{K}^{-1}\frac{\partial \boldsymbol{K}}{\partial p_k}\boldsymbol{K}^{-1}, \tag{26}$$

the gradient of the input/output transfer matrix $\boldsymbol{G}(\boldsymbol{p})$ with respect to the design variable p_k becomes:

$$\frac{\partial \boldsymbol{G}(\boldsymbol{p})}{\partial p_k} = \boldsymbol{C}\frac{\partial \boldsymbol{K}^{-1}}{\partial p_k}\boldsymbol{B} = \boldsymbol{C}\boldsymbol{K}^{-1}\frac{\partial \boldsymbol{K}}{\partial p_k}\boldsymbol{K}^{-1}\boldsymbol{B}, \tag{27}$$

where the gradient of the stiffness matrix, $\partial \boldsymbol{K}/\partial p_k$, is computed using Equation (11).

3.4. Gradient of the Singular Values

As for the gradient of the generic h-th singular value σ_h, one can write (see also [19]):

$$\frac{\partial \sigma_h}{\partial p_k} = \mathcal{R}\left(\boldsymbol{u}_h^H \frac{\partial \boldsymbol{G}(\boldsymbol{p})}{\partial p_k}\boldsymbol{v}_h\right), \tag{28}$$

where $\mathcal{R}(\cdot)$ denotes the real part of a (possibly) complex number, and $\boldsymbol{u}_h$ and $\boldsymbol{v}_h$ are the left and right singular vectors corresponding to σ_h.

3.5. Gradient of the Norm of the Input/Output Transfer Matrix G

Since the spectral ($|\cdot|_2$), nuclear ($|\cdot|_N$), and Frobenius ($|\cdot|_F$) norms of $\boldsymbol{G}$ are uniquely determined by its singular values, Equation (28), combined with a proper application of the chain rule, enables the computation of the sensitivity of any norm appearing in (25)1. In general, the sensitivity of each norm with respect to the design variable p_k reads:

$$\frac{\partial ||\boldsymbol{G}||}{\partial p_k} = \sum_{h=1}^{r} \frac{\partial ||\boldsymbol{G}||}{\partial \sigma_h}\frac{\partial \sigma_h}{\partial p_k}, \tag{29}$$

where the second factor on the right-hand side is provided by Equation (28), and the first factor depends on the chosen norm:

$$\begin{aligned} \frac{\partial ||G||_2}{\partial \sigma_h} &= \delta_{1h}, \quad \text{(Kronecker delta)} \\ \frac{\partial ||G||_N}{\partial \sigma_h} &= 1 \\ \frac{\partial ||G||_F}{\partial \sigma_h} &= \frac{\sigma_h}{\sqrt{\sigma_1^2 + \ldots + \sigma_r^2}}. \end{aligned} \tag{30}$$

It is worth noting that these expressions assume all singular values are distinct and strictly positive. In cases where singular values are repeated (i.e., multiplicities greater than one) or approach zero, the singular value decomposition may become non-differentiable or numerically unstable. In such situations, regularization techniques or subdifferential approaches may be employed to ensure a well-defined and robust sensitivity analysis [24].

4. Definition of System Output and Optimization Goal

Referring again to Figure 1, the output vector z defines the optimization objective and depends linearly on the state vector x, which collects all the generalized displacements resulting from the finite element discretization. This relationship is expressed as:

$$z = Cx. \tag{31}$$

A notable advantage of employing a norm minimization strategy is its ability to address multiple objectives—represented by the components of the output vector z—simultaneously through a standard scalar minimization procedure. This approach effectively transforms a multi-objective optimization problem into a single-objective one. Such scalarization techniques are widely recognized for their efficacy in simplifying complex optimization tasks while preserving the integrity of the original multi-objective problem [25].

Regarding the components of the output vector z, a few feasible options are outlined below.

4.1. Any Component of the State Vector x

In structural analysis, it is often necessary to ensure that specific displacements do not exceed predefined thresholds. When focusing on a particular displacement—denoted as the i-th component, x_i, of the state vector x—this component can be incorporated into the output vector by augmenting the output matrix C with an appropriate row. Specifically, to include x_i as the h-th component of z, the h-th row of C should be defined as:

$$C_{hj} = \begin{cases} 1 & h = j \\ 0 & h \neq j \end{cases} \tag{32}$$

4.2. Any Component of the Strain and Stress Tensors

As long as the structure is geometrically and constitutively linear, there exist discrete linear versions of the compatibility and constitutive equations such that any component of the strain and stress tensors may be respectively written as:

$$\begin{aligned} \varepsilon &= \tilde{B}x \\ \sigma &= D\tilde{B}x \end{aligned} \tag{33}$$

Therefore, $\tilde{\boldsymbol{B}}$ and $\boldsymbol{D}\tilde{\boldsymbol{B}}$ are line vectors that should be appended to the output matrix $\boldsymbol{C}$ to include ε and σ in the output vector.

4.3. The Structural Compliance

The proposed approach allows for the minimization of a wide variety of compliance, in a general multi-load setting. Reference is made to [20] for a detailed description of the approach.

Let $\boldsymbol{F} = [F_1, F_2, \dots F_N]$ be the $N \times 1$ column vector of the external forces and $\boldsymbol{U} = [u_1, u_2, \dots u_N]$ the displacement vector that is dual to $\boldsymbol{F}$ in virtual work sense. The following should be noted:

- any $u_i \in \boldsymbol{U}$ coincides with a proper component of the state vector $\boldsymbol{x}$;
- to any load $F_i \in \boldsymbol{F}$, a compliance is associated that is defined as $\mathcal{C}_i = F_i u_i$;
- the overall compliance is defined as $\mathcal{C} = \boldsymbol{F} \cdot \boldsymbol{U} = \sum_{i=1}^{N} F_i u_i$.

That said, the following cases are investigated:

4.3.1. SISO Case

The single input is represented by the force vector $\boldsymbol{F}$ whose components act simultaneously, and the single output is the overall scalar compliance: $\mathcal{C} = \boldsymbol{F} \cdot \boldsymbol{U}$.

4.3.2. MISO Case

Each component of the force vector $\boldsymbol{F}$ represents an independent input, meaning that N independent structural analyses are performed. In each analysis, a single load component is applied in isolation. The scalar output is the total compliance, defined as $\mathcal{C} = \sum_{i=1}^{N} F_i u_i$, and is computed *a posteriori* by summing the N individual contributions $\mathcal{C}_i = F_i u_i$ associated with each load case.

4.3.3. SIMO Case

The single input is represented by the force vector $\boldsymbol{F}$, whose components act simultaneously in a single load case. The output vector $\boldsymbol{z}$ contains N components, each corresponding to an individual compliance term, defined as $\mathcal{C}_i = F_i u_i$, for $i = 1, \dots, N$.

4.3.4. MIMO Case

This is the most general case. Each individual component F_i of the input vector $\boldsymbol{F}$ is assumed to act independently, resulting in N distinct load cases. On the output side, each corresponding compliance term $\mathcal{C}_i = F_i u_i$ forms a component of the output vector $\boldsymbol{z}$, thereby enabling a fully decoupled multi-input, multi-output (MIMO) representation.

5. Implementation Details Using MATLAB

The MMTO algorithm discussed in this work is implemented using MATLAB 2023b, and the software code is freely accessible online (see Code Availability Statement). Specifically, the code is organized as follows:

- `main_stat.m` is the main file of the project, containing all user-defined parameters of MMTO algorithm and case study. Once their are properly setup, `main_stat.m` calls three further scripts, namely `initialize_mma.m`, which contains the parameters of the MMA algorithm, `start_mma.m`, which solves the MMTO problem using the MMA algorithm, and `plot_densities.m`, which takes care of producing the figures displayed in this paper.
- The `casestudy` folder contains the code required to define the setup of the case study, including the definition of loads (`defineLoads.m`), the definition of supports

(defineSupports.m), and the computation of the dynamic system matrices B and C (prepareDynamicSystemSS_BandC.m).

- The costandconstraints folder contains the code required to define the objective function and the constraints of the MMTO problem. Specifically, the function compliance_stat_mma.m calls two dedicated subroutines to evaluate the objective function and its derivatives (computeCostWithSensitivities.m), and evaluate constraint violations and their derivatives (computeConstraintsWithSensitivities.m) at the current guess for the MMTO problem.
- The filteringprojectioninterpolation folder contains the code implementing the filtering, projection and interpolation functions for Multi-Materials, as discussed in Section 2.
- The matrixnorms folder contains the code required to carry out the computation of tensor norms (pnormmat.m and pnormmatmenouno.m) and their derivatives (pnormmatsens.m), as discussed in Section 3.
- The stiffnesscomputations folder contains the code required to compute the stiffness matrix of the system, based on the desired finite element discretization, relying on the subroutines prepareStiffnessMatIdxs.m, prepareFEM.m and computeStiffnessMatrix.m.
- The MMA folder contains the code of the MMA algorithm, as made available by [23].
- The utils folder contains the computeDependentParams.m function, which computes additional parameters of the MMTO algorithm, based on the user-defined parameters set in main_stat.m.

Thanks to its modularity, the proposed code structure allows the new user to successfully run the code by focusing on the main_stat.m file only. If required, a number of variations of the case studies discussed in this work can be straightforwardly obtained by implementing minor modifications to the subroutines contained in the casestudy folder. Furthermore, the code can be straightforwardly adjusted for use with any MATLAB optimization algorithm, by defining suitable objective and constraint functions based on computeCostWithSensitivities.m, and computeConstraintsWithSensitivities.m.

6. Numerical Results

Two representative numerical problems are investigated next:

1. a simply supported cantilevered beam;
2. an L-shaped appended structure.

Both structures are composed of three distinct materials and are subjected to two external loads, allowing the application of all the idealized input/output configurations discussed earlier—namely, SISO, MISO, SIMO, and MIMO.

6.1. Simply Supported Cantilevered Beam

6.1.1. General Data

Figure 2 illustrates the geometry and loading conditions of the problem under investigation. Taking advantage of structural symmetry, a half-model—as shown in Figure 2—is adopted for the numerical simulations. The computations are carried out using the following parameters:

- Poisson ratio: $\nu = 0.3$;
- Number of elements in vertical (Y) direction: 2^6;
- Number of elements in horizontal (X) direction: 3×2^6;

- Side length: $L = 2^6$ (by this choice a uniform mesh of square elements of unit size is generated);
- $P = Q = 1$;
- Volume fraction = 0.5;
- SIMP penalization factor: $n = 3$;
- p-norm parameter = 6;
- $\delta = 10^{-9}$.

As for the multi-material characterization, the number of materials is set to $NM = 3$, with corresponding Young's moduli $E_1 = 1$, $E_2 = 2$, and $E_3 = 3$. It is worth noting that this configuration is based on the system previously studied in [20], with the sole difference being the extension to a three-material setting in the present work.

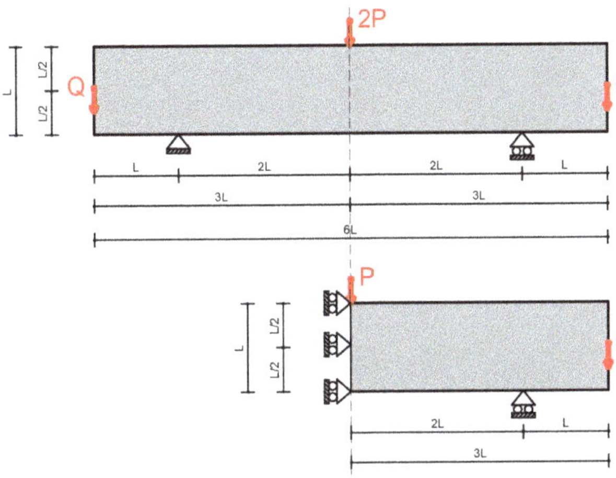

Figure 2. Numerical example 1: full and half structure using symmetry.

6.1.2. SISO System

The optimal topology at convergence obtained using the SISO model is presented in Figure 3. In the following, blue, green, and yellow are used to represent the materials with $E_1 = 1$, $E_2 = 2$, $E_3 = 3$, respectively. Interestingly, the single-material solution obtained using the approach presented in [20]—see Figure 4—appears to consist of (nearly) uniformly sized elements. In contrast, the components of the multi-material structure exhibits varying dimensions, with stronger materials tending to form thinner features. This behavior confirms the capability of the proposed approach to effectively exploit the available materials: since the objective is to minimize compliance, the algorithm allocates larger volumes to weaker materials in order to enhance the overall structural stiffness.

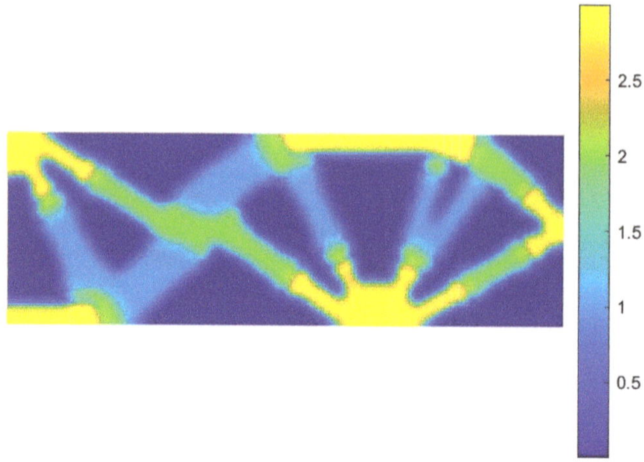

Figure 3. Cantilevered structure: (SISO)-Optimal topology at convergence.

Figure 4. Cantilevered structure: (SISO)-Optimal topology at convergence using standard SIMP.

6.1.3. Stability of the Results with Respect to Mesh Resolution: A Trade-Off with Computational Cost

In the context of topology optimization, the stability of results with respect to mesh resolution presents a twofold challenge:

- On one hand, the numerical method may be sensitive to the level of discretization, implying that accurate and reliable solutions can only be achieved with sufficiently fine meshes.
- On the other hand, finer meshes enable a richer representation of the design space, allowing for a broader range of topologies to be captured as the continuous, infinite-dimensional domain is projected onto its discrete, finite-dimensional counterpart.

The two aspects above are investigated next solving three times the very same SISO problem of Section 6.1.2. The meshes of the three runs are as follows

1. Number of elements in Y direction = 16; number of elements in X direction = 48;
2. Number of elements in Y direction = 32; number of elements in X direction = 96;
3. Number of elements in Y direction = 64; number of elements in X direction = 192;

The optimal topologies along with the convergence curves for the three cases are respectively shown in Figures 5–7. The CPU time needed to achieve convergence for the three runs were respectively as follows:

$$t_1 = 539 \text{ s}\,, t_2 = 4382 \text{ s}\,, t_3 = 23453 \text{ s} \tag{34}$$

6.1.4. MIMO System-Norm 2

Figure 8 displays the optimal topology obtained in the MIMO setting using the spectral 2-norm. The design confirms the key feature observed in the SISO case: the weaker the material, the wider the corresponding structural component. A comparison with the standard single-material SIMP solution shown in Figure 9 reveals an additional characteristic. In the SIMP-based topology, the top and bottom chords constitute the most substantial elements of the truss, with their cross-sectional dimensions decreasing from midspan toward the supports. This behavior is not replicated in the multi-material design, where the top and bottom chords are composed of the stiffest material, resulting in nearly uniform and comparatively slender members throughout. Figure 10 depicts the convergence curve obtained solving the optimization.

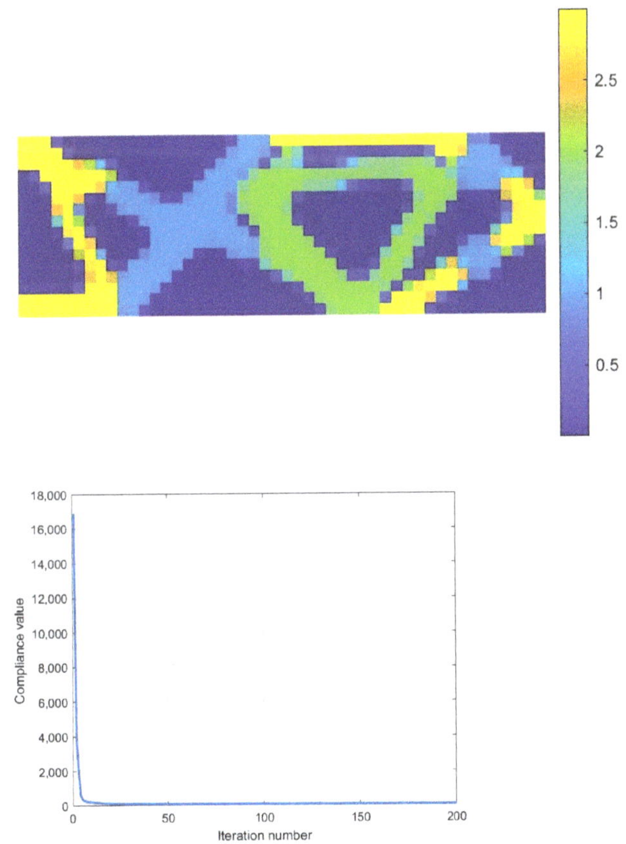

Figure 5. Stability assessment wrt the mesh size-Number of elements in Y direction = 16. (**up**) Optimal Topology, (**bottom**) Convergence curve.

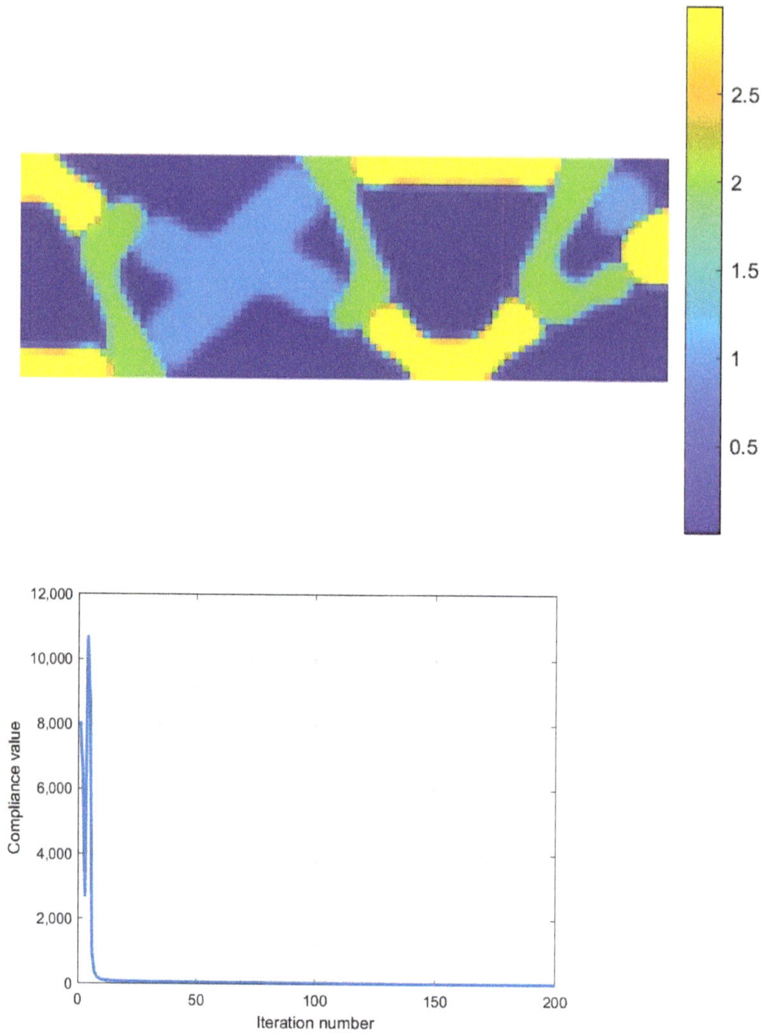

Figure 6. Stability assessment wrt the mesh size-Number of elements in Y direction = 32. (**up**) Optimal Topology, (**bottom**) Convergence curve.

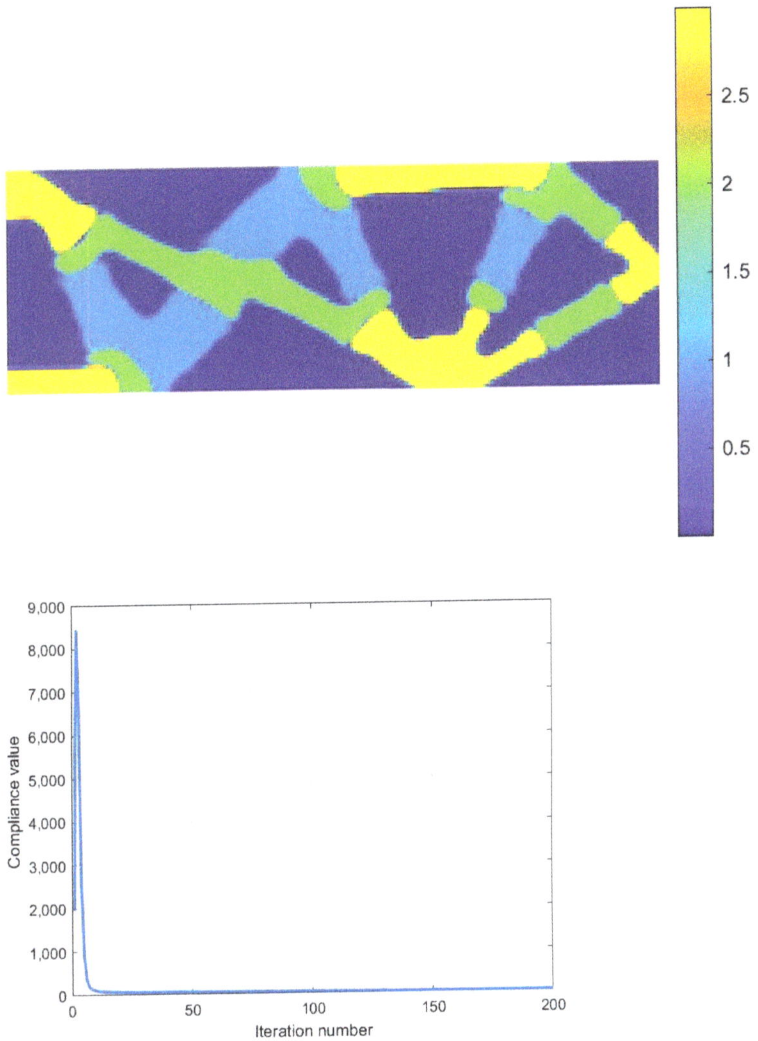

Figure 7. Stability assessment wrt the mesh size-Number of elements in Y direction = 64. (**up**) Optimal Topology, (**bottom**) Convergence curve.

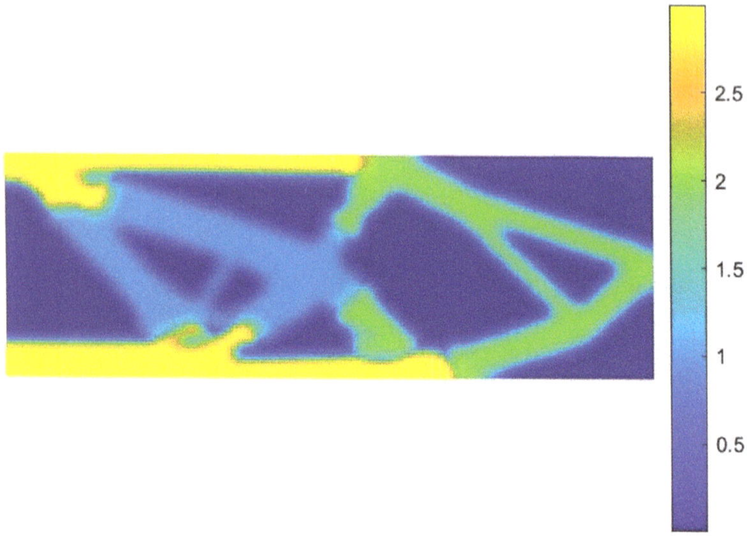

Figure 8. Cantilevered structure: (MIMO-2 Norm)-Optimal topology at convergence.

Figure 9. Cantilevered structure: (MIMO-2 Norm)-Optimal topology at convergence using standard SIMP.

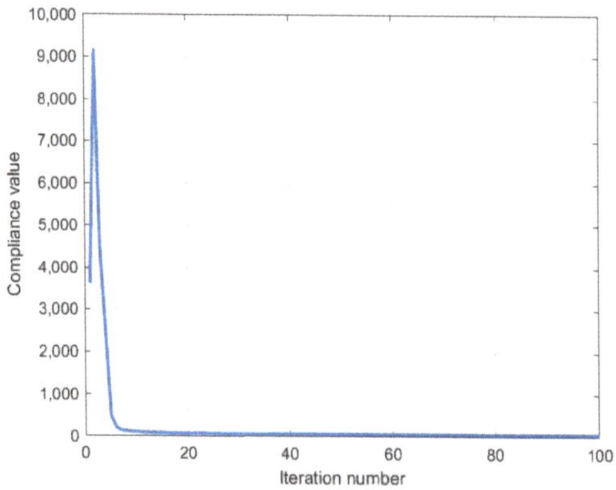

Figure 10. Cantilevered structure: (MIMO-2 Norm)-Convergence curve.

6.1.5. Effect of the SIMP Penalization Factor n on the Optimal Topology

The same MIMO example introduced in Section 6.1.4 is now employed to investigate the sensitivity of the resulting optimal topology to variations in the SIMP penalization factor n. As established in [26], the choice $n = 3$ is particularly significant, as it renders the SIMP interpolation thermodynamically consistent and compliant with theoretical bounds. Nevertheless, from a numerical standpoint, it is insightful to examine how the optimal topology evolves as a function of n. To this end, a parametric study is conducted in which optimal designs corresponding to $n = 3, 4, 5$ are computed and compared. Optimal topologies are shown in Figures 11–13, for the choices $n = 3, 4, 5$, respectively. The superiority of the choice $n = 3$ is demonstrated by the values of the compliance that are reported in Table 1.

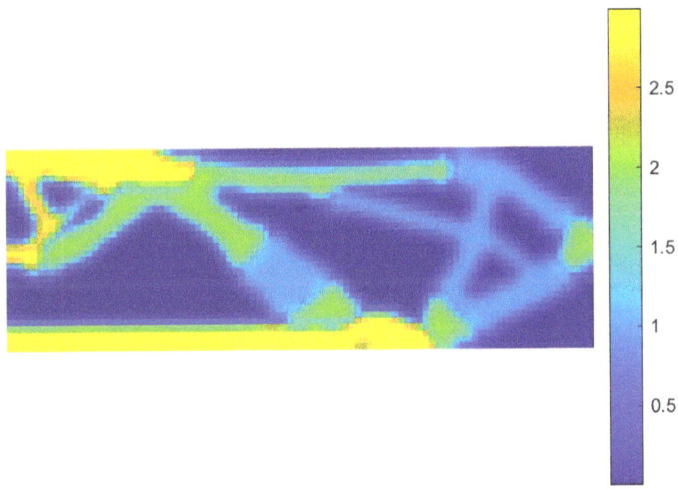

Figure 11. Cantilevered structure: (MIMO, 2-norm solution)-Optima topology at convergence using standard SIMP-$n = 3$.

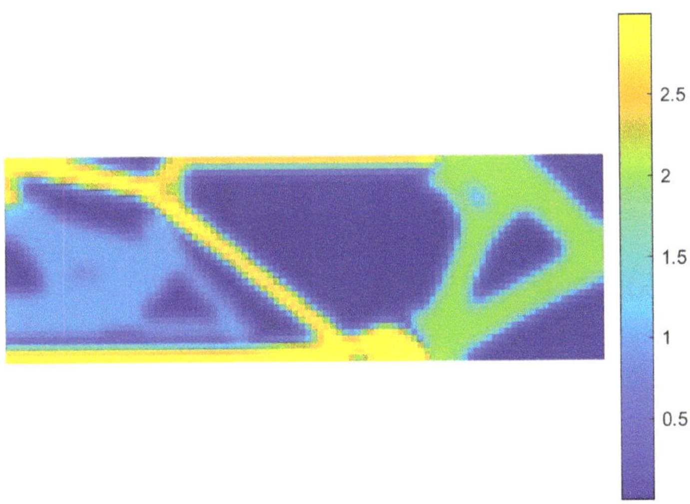

Figure 12. Cantilevered structure: (MIMO, 2-norm solution)-Optima topology at convergence using standard SIMP-$n = 4$.

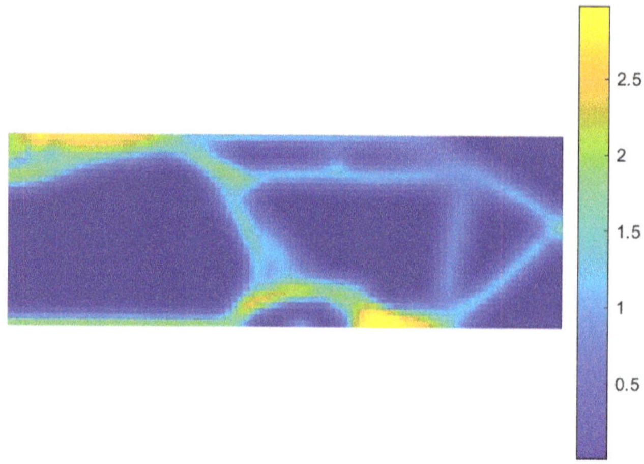

Figure 13. Cantilevered structure: (MIMO, 2-norm solution)-Optima topology at convergence using standard SIMP-$n = 5$.

Table 1. Compliance values vs. SIMP exponent n.

	$n = 3$	$n = 4$	$n = 5$
Compliance	58.14	78.64	2451.72

6.1.6. MIMO System-Nuclear Norm

Figure 14 shows the optimal topology obtained in the MIMO setting using the nuclear norm, while Figure 15 illustrates the corresponding convergence history. The convergence path is fast and monotonic; the initial increase in the objective function is attributed to a temporary violation of the volume constraints, which the solver successfully addresses starting from the third iteration.

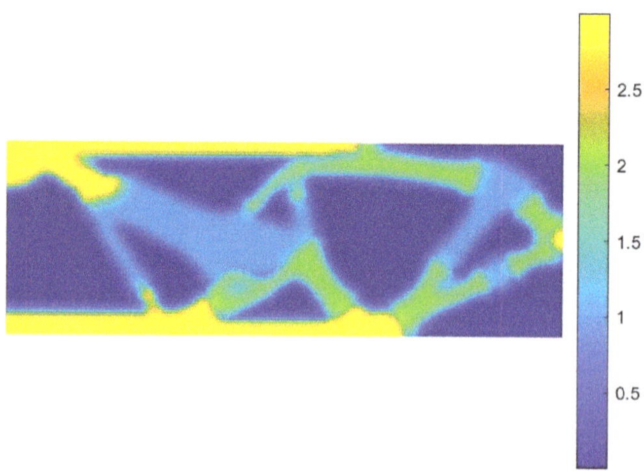

Figure 14. Cantilevered structure: (MIMO-Nuclear Norm)-Optimal topology at convergence.

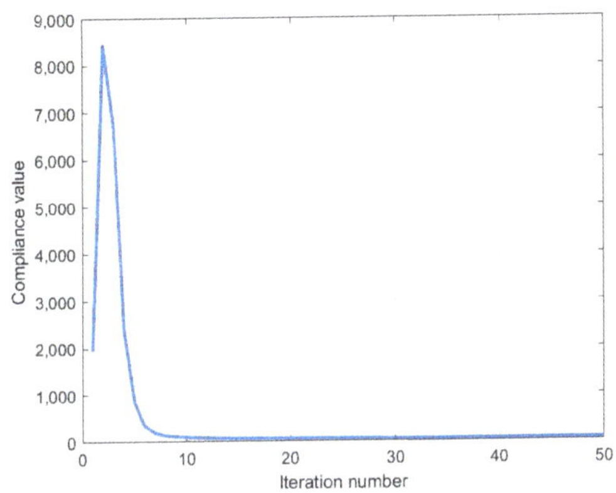

Figure 15. Cantilevered structure: (MIMO-Nuclear norm)-Convergence curve.

6.1.7. MIMO System-Frobenius Norm

Figure 16 presents the optimal topology obtained in the MIMO setting using the Frobenius norm, while Figure 17 depicts the associated convergence history. Similar observations to those made for the nuclear norm-based solution apply here. This similarity is expected, as both norms depend on all non-zero singular values (two in the present case), resulting in comparable optimal topologies and convergence behaviors. In contrast, the 2-norm depends solely on the largest singular value, which explains the significantly different topology obtained in that case.

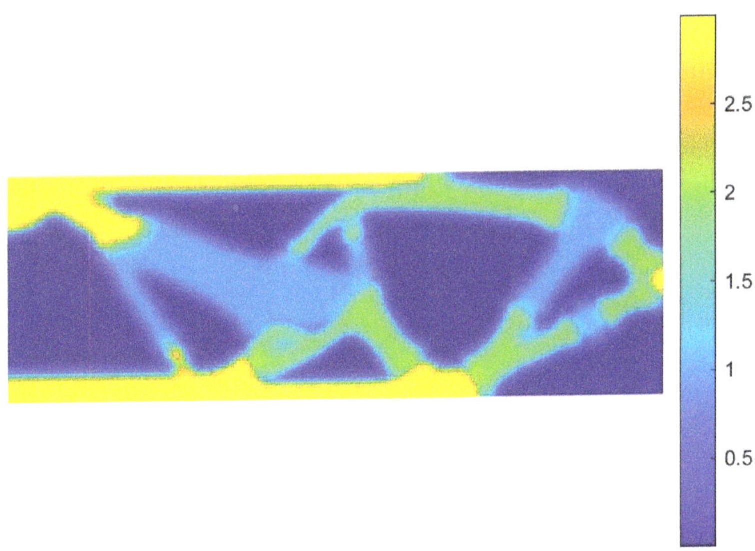

Figure 16. Cantilevered structure: (MIMO-Frobenius Norm)-Optimal topology at convergence.

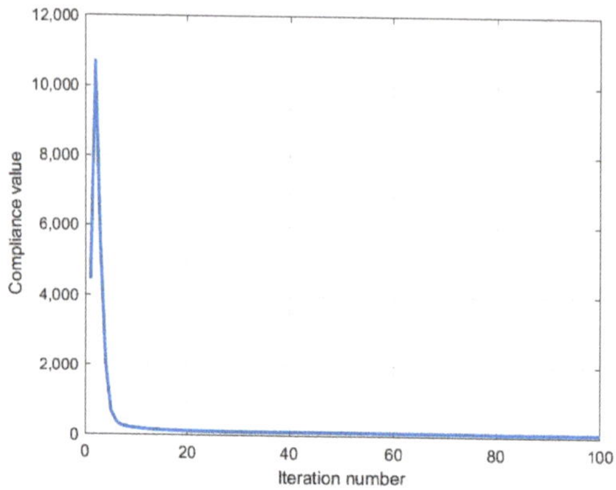

Figure 17. Cantilevered structure: (MIMO-Frobenius norm)-Convergence curve.

6.2. L-Shaped Structure

Figure 18 illustrates the geometry and loading conditions of the L-shaped problem under investigation. The following numerical parameters are used:

- Poisson ratio: $\nu = 0.3$;
- Number of elements in vertical (Y) direction: 2×2^5;
- Number of elements in horizontal (X) direction: 3×2^5;
- Side length: $L = 2^6$;
- $P = Q = 1$;
- Volume fraction = 0.5;
- SIMP penalization factor: $n = 3$;
- p-norm parameter = 6;
- $\delta = 10^{-9}$.

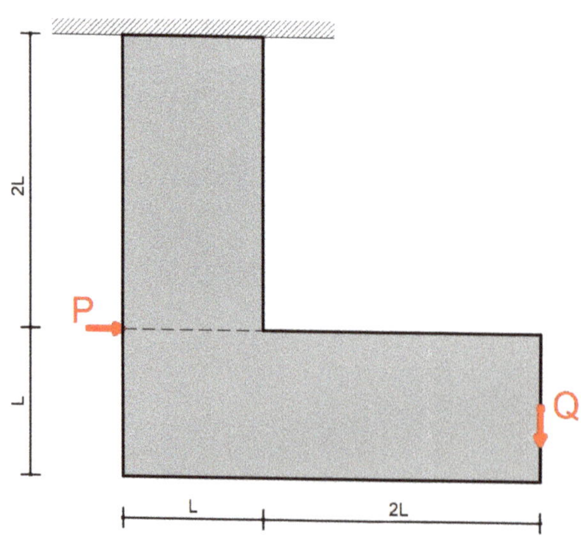

Figure 18. Numerical example 2: L-shaped structure.

6.2.1. SISO System

The optimal topology at convergence obtained using the SISO model is shown in Figure 19. In the visual representation, blue, green, and yellow correspond to the materials with $E_1 = 1$, $E_2 = 2$, and $E_3 = 3$, respectively. As previously observed, the components of the multi-material structure exhibit varying thicknesses, with stiffer materials typically forming thinner, more localized features.

To assess the performance of the optimization algorithm, Figure 20 presents the evolution of the objective function over the course of the optimization. Although a total of 50 iterations were executed, convergence was effectively achieved after approximately 25 iterations, demonstrating the efficiency of the Method of Moving Asymptotes in this context.

A note is in order regarding the influence of the horizontal load P on the optimal topology. In the standard benchmark formulation of the L-shaped problem, the horizontal load is typically absent. As a result, the vertical leg of the structure is often reduced to a pair of vertical load-bearing chords positioned at the left and right edges.

The introduction of the horizontal load P, however, significantly alters the structural response. In addition to the need to sustain axial forces (tension and compression), the left vertical leg must now exhibit substantial bending rigidity to effectively counteract the moment induced by P. This additional requirement gives rise to a more complex and articulated topology in that region, aimed at minimizing the compliance associated with the horizontal loading component.

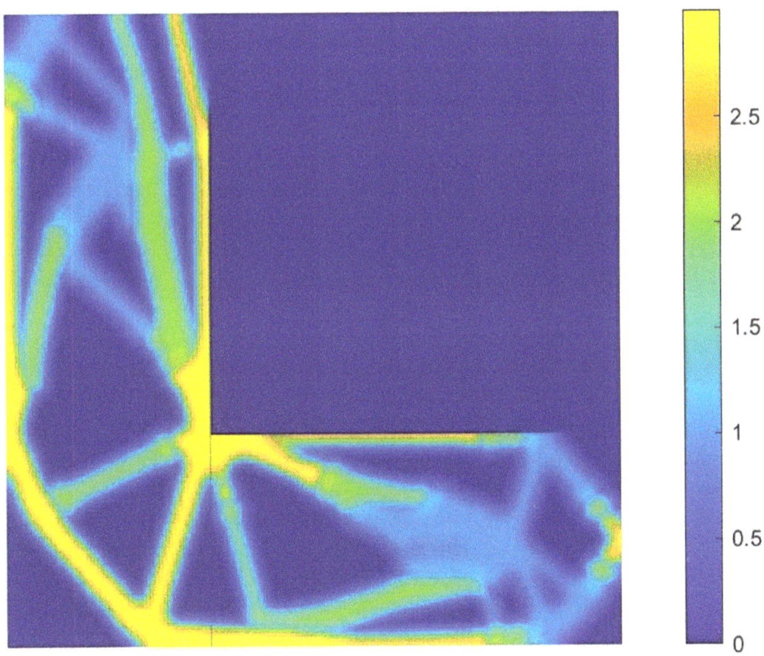

Figure 19. L-shaped structure: (SISO) Optimal topology at convergence.

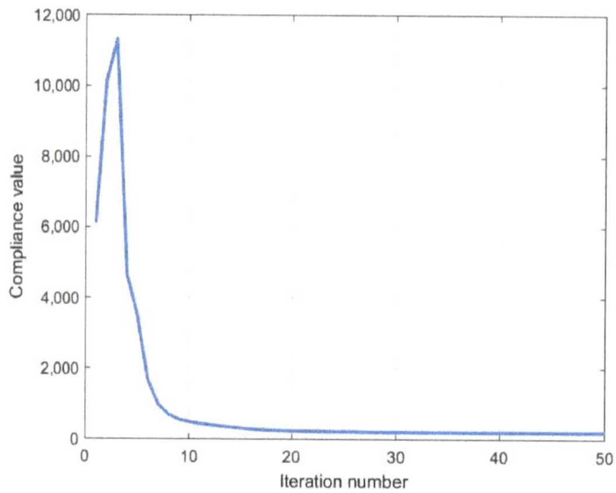

Figure 20. L-shaped structure: (SISO) Convergence curve.

6.2.2. MIMO System-2 Norm

Figure 21 presents the optimal topology obtained by minimizing the 2-norm of the transfer matrix within a MIMO framework. The most notable feature of the resulting design is the emergence of a plate-like resisting member within the horizontal leg, where the majority of the weakest material phase (depicted in blue) is concentrated.

This behavior is closely tied to the nature of the 2-norm, which corresponds to the largest singular value of the transfer matrix and thus emphasizes the most critical input/output direction in terms of energy transfer or compliance [27]. Consequently, the optimization process concentrates design efforts on stiffening the dominant deformation mode, allocating stronger materials to regions of peak stress—namely, the vertical load-bearing chords and the reentrant corner, which is known to be subjected to significant stress concentrations (and therefore strains, and displacements). To conclude, Figure 22 depicts the convergence curve obtained during the optimization.

6.2.3. MIMO System-Nuclear Norm

Figure 23 presents the optimal topology obtained by minimizing the Frobenius norm of the transfer matrix within a MIMO (Multiple-Input Multiple-Output) framework. A distinctive feature of this topology is the presence of two pairs of "twin" tendons located on the left and right sides of the vertical leg.

The outer tendons (highlighted in yellow) reduce the compliance associated with the vertical load Q, and their layout aligns with conventional design principles. In contrast, the inner tendons (depicted in green) facilitate the bending mechanism required to mitigate the compliance contribution from the horizontal load P. As previously observed, the green tendons appears thicker than the yellow ones due to their fabrication from a mechanically weaker material, which necessitates a larger cross-sectional area to ensure the desired structural performance.

The convergence behavior, depicted in Figure 24, confirms the same trend noted in the previously discussed cases.

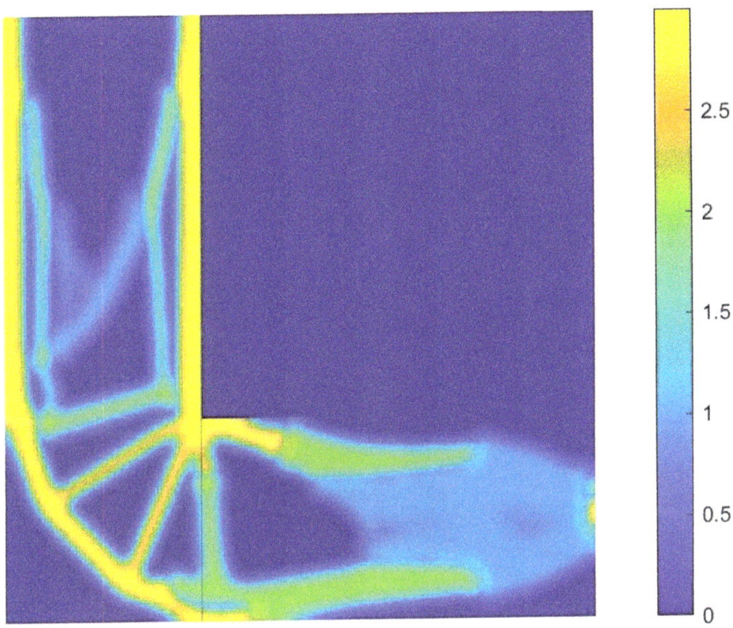

Figure 21. L-shaped structure: (MIMO-2 Norm) Optimal topology at convergence.

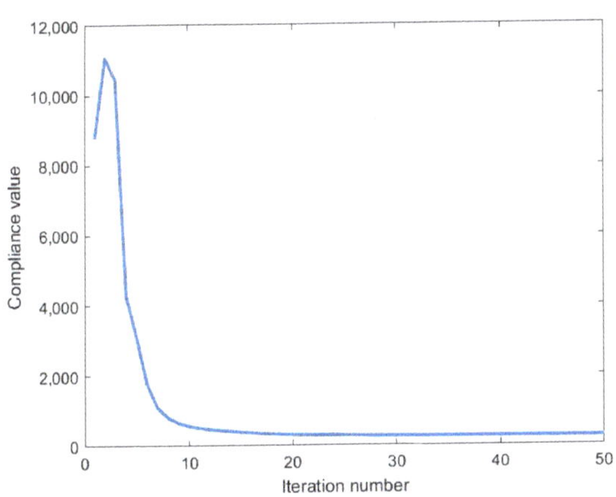

Figure 22. L-shaped structure: (MIMO-2 Norm) Convergence curve.

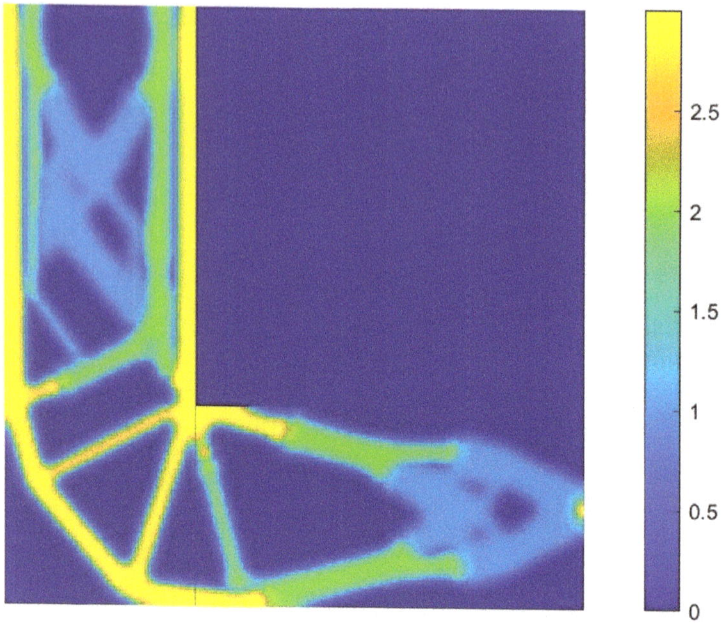

Figure 23. L-shaped structure: (MIMO-Nuclear Norm) Optimal topology at convergence.

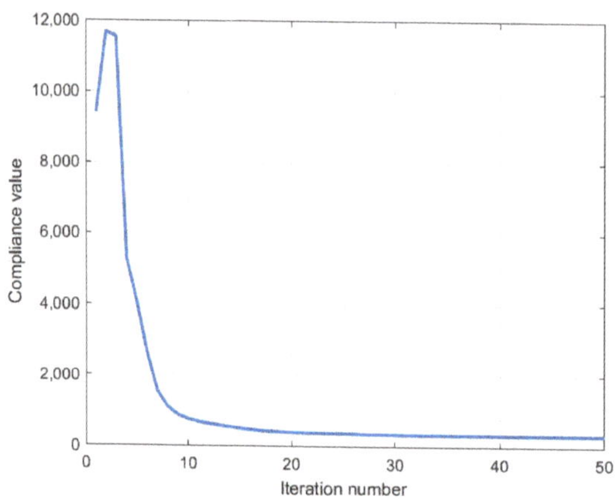

Figure 24. L-shaped structure: (MIMO-Nuclear Norm) Convergence curve.

6.2.4. MIMO System-Frobenius Norm

Figure 25 illustrates the optimal topology resulting from the minimization of the Frobenius norm of the transfer matrix within a MIMO (Multiple-Input Multiple-Output) framework. The most notable feature of this topology is the presence of two "twin" tendons located on the right side of the vertical leg.

The outer tendon (highlighted in yellow) primarily serves to reduce the compliance associated with the vertical load Q, and its configuration is consistent with classical designs. In contrast, the inner tendon (shown in green) contributes to the bending mechanism required to reduce the compliance component related to the horizontal load P. As previously noted, the green tendon appears thicker than the yellow one due to its composition from a mechanically weaker material, which necessitates a larger cross-section to achieve the desired structural response.

The convergence curve is shown in Figure 26 that confirms the same pattern already commented on for the other cases.

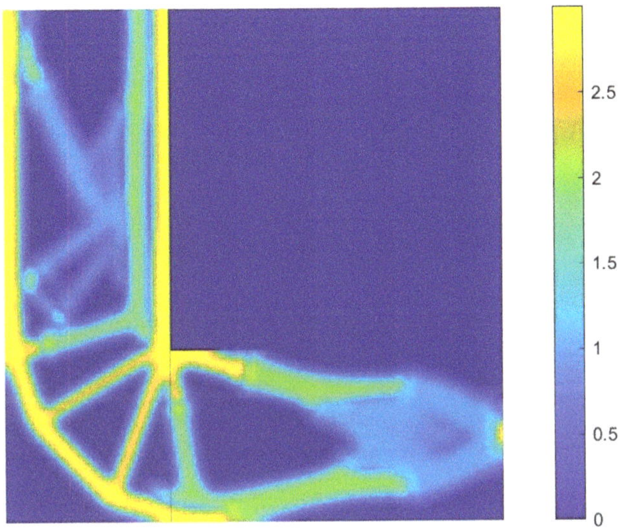

Figure 25. L-shaped structure: (MIMO-Frobenius Norm) Optimal topology at convergence.

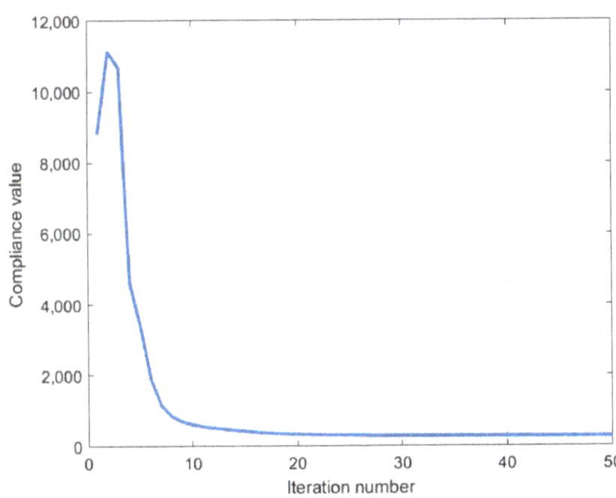

Figure 26. L-shaped structure: (MIMO-Frobenius Norm) Convergence curve.

6.3. 3D Problem

The proposed topology optimization formulation is general enough to be applicable to any linear elastic problem. 3D structures do not represent an exception and can be addressed and solved using our approach, provided that a suitable finite-element discretization is adopted to model the structure to be designed. The standard eight-noded elements are therefore used to model the 3D problem of Figure 27 that shows the full problem and the half-domain that has been actually discretized using symmetries. Regarding the parameters of the numerical problem, the following values are used:

- Number of elements in the transverse (Y) direction: 30;
- Number of elements in the longitudinal (X) direction: 90;
- Number of elements in the vertical, top-down (Z) direction: 3;
- SIMP penalization factor: 3;
- p-norm parameter: 6;
- Filter distance: $r_{min} = 3$;
- Initial value of the filter parameter: $\beta = 1$;
- $\delta = 10^{-9}$;
- Number of materials: 3;
- Young moduli: $E_1 = 5$, $E_2 = 3$, $E_3 = 1$;
- Poisson ratio: $\nu = 0.3$.

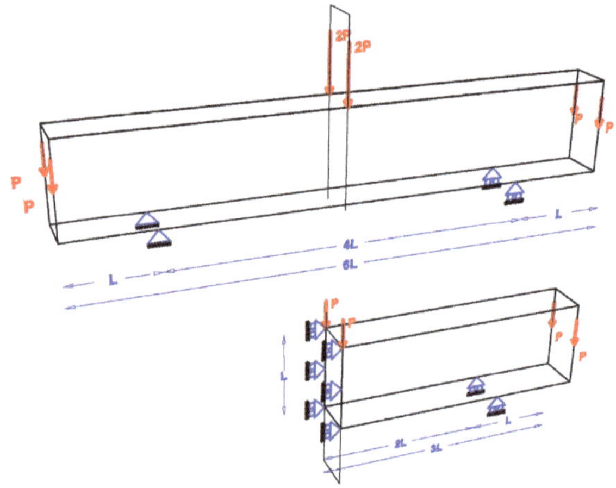

Figure 27. 3d Problem-Geometry, loads and constraints.

Only a SISO scheme is considered such that all loads are supposed to act simultaneously and concur to define a unique scalar compliance. The optimal topology at convergence is shown in Figure 28 where red, blue, and green are used for material 1 (stronger), 2 (intermediate) and 3 (weaker), respectively. Figure 29 shows the variation on the objective function, i.e., the overall compliance, with the iterations.

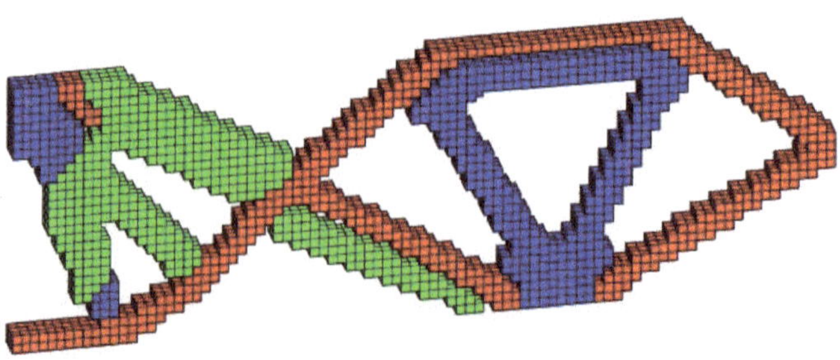

Figure 28. 3d Problem-Optimal topology at convergence (red = material 1, blue = material 2, green = material 3).

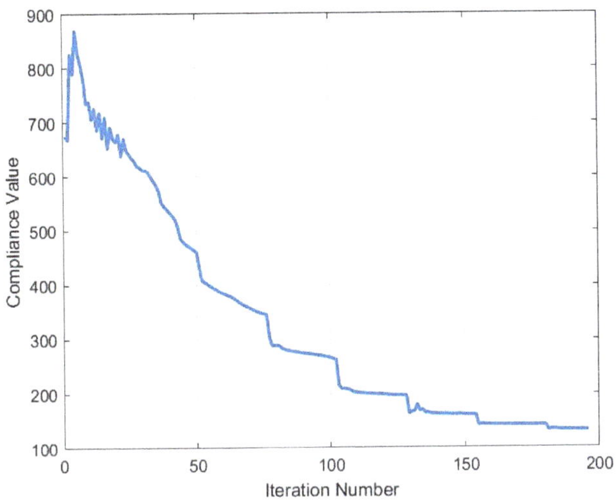

Figure 29. 3d Problem-Convergence history.

7. Conclusions

This paper introduced a novel framework for MMTO based on the explicit use of the input/output transfer matrix G, which maps external loads to selected structural responses. By minimizing a suitable matrix norm of G, the proposed approach enables the systematic design of high performance structures under general multiload, multiobjective settings, while seamlessly extending the classical compliance minimization problem as a special case.

The methodology incorporates an advanced multimaterial interpolation scheme that ensures well-defined discrete material distributions. Coupled with a robust two-stage filtering strategy, this formulation effectively mitigates numerical instabilities such as checkerboarding and gray regions. The use of Singular Value Decomposition (SVD) plays a central role, not only in defining the optimization objective but also in facilitating a semianalytical sensitivity analysis, which is crucial for the computational efficiency and robustness of the optimization process.

The numerical results obtained on 2D and 3D benchmark problems—including domains of simply supported and L-shaped—demonstrate the flexibility and effectiveness of the proposed framework. The method yields intuitive material layouts that reflect optimal load transfer paths and leverage the distinct stiffness properties of available materials. Moreover, the comparative analysis of different matrix norms (2-norm, nuclear norm, Frobenius norm) highlights the versatility of the approach in capturing various aspects of structural performance.

Ongoing extensions include dynamic response, thermal problems and compliant mechanism design.

Author Contributions: Conceptualization, P.V.; methodology, P.V.; software, P.V. and G.G.; validation, P.V.; formal analysis, P.V.; investigation, P.V.; resources, P.V.; writing—original draft preparation, P.V. and G.G.; writing—review and editing, P.V. and G.G.; visualization, P.V. and G.G.; supervision, P.V.; project administration, P.V. All authors have read and agreed to the published version of the manuscript.

Funding: This research received no external funding.

Data Availability Statement: This study did not involve any data. The software code related to this work is available online at https://github.com/GiacomoGaluppini?tab=repositories, accessed on 1 April 2025.

Conflicts of Interest: The authors declare no conflict of interest.

References

1. Bendsøe, M.; Kikuchi, N. Generating optimal topologies in structural design using a homogenization method. *Comput. Methods Appl. Mech. Eng.* **1988**, *71*, 197–224. [CrossRef]
2. Sigmund, O. A 99 line topology optimization code written in Matlab. *Struct. Multidiscip. Optim.* **2001**, *21*, 120–127. [CrossRef]
3. Zuo, K.; Saitou, K. Ordered SIMP interpolation for multi-material topology optimization. *Struct. Multidiscip. Optim.* **2017**, *55*, 477–491.
4. Ogawa, S.; Yonekura, K.; Suzuki, K. Multimaterial topology optimization of unsteady heat conduction problems based on discrete material optimization. *Int. J. Heat Mass Transf.* **2024**, *225*, 125353. [CrossRef]
5. Yin, L.; Ananthasuresh, G. Topology optimization of compliant mechanisms with multiple materials using a peak function material interpolation scheme. *Struct. Multidiscip. Optim.* **2006**, *33*, 199–214. [CrossRef]
6. Zheng, R.; Yi, B.; Peng, X.; Yoon, G.H. An Efficient Code for the Multi-Material Topology Optimization of 2D/3D Continuum Structures Written in Matlab. *Appl. Sci.* **2024**, *14*, 657. [CrossRef]
7. Andreassen, E.; Clausen, A.; Schevenels, M.; Lazarov, B.S.; Sigmund, O. Efficient topology optimization in MATLAB using 88 lines of code. *Struct. Multidiscip. Optim.* **2011**, *43*, 1–16. [CrossRef]
8. Dinh, T.; Hedayatrasa, S.; Bormann, F.; Bosman, M.; Paepegem, W.V. A smooth single-variable-based interpolation function for multi-material topology optimization. *Eng. Comput.* **2024**, *40*, 2841–2855. [CrossRef]
9. Asadpoure, A.; Rahman, M.; Nejat, S.; Javidannia, L.; Valdevit, L.; Guest, J.; Tootkaboni, M. Topology optimization with multi-phase length-scale control. *Int. J. Mech. Sci.* **2025**, *291–292*, 110086. [CrossRef]
10. Jia, Y.; Li, W.; Zhang, X. Multimaterial topology optimization of elastoplastic composite structures. *J. Mech. Phys. Solids* **2025**, *196*, 106018. [CrossRef]
11. Hvejsel, C.; Lund, E. Material interpolation schemes in topology optimization. *Int. J. Numer. Methods Eng.* **2011**, *85*, 575–599.
12. Bruyneel, M. A general and efficient multi-material topology optimization approach using the shape function with penalization. *Struct. Multidiscip. Optim.* **2010**, *42*, 529–553.
13. Huang, X.D.; Xie, Y.M. Bi-directional evolutionary structural optimization with buckling constraints. *Struct. Multidiscip. Optim.* **2023**, *67*, 123–135.
14. Huang, X. Smooth topological design of structures using the floating projection. *Eng. Struct.* **2020**, *208*, 110330. [CrossRef]
15. Huang, X.D.; Li, W. Three-field floating projection topology optimization of continuum structures. *Comput. Methods Appl. Mech. Eng.* **2022**, *398*, 115259. [CrossRef]
16. Fu, Y.F.; Rolfe, B.; Chiu, N.S.L.; Wang, Y.; Huang, X.; Ghabraie, K. Smooth-edged material distribution for optimizing topology algorithm. *Adv. Eng. Softw.* **2020**, *149*, 102921. [CrossRef]
17. Sanders, E.D.; Pereira, A.; Aguiló, M.A.; Paulino, G.H. PolyMat: An efficient Matlab code for multi-material topology optimization. *Struct. Multidiscip. Optim.* **2018**, *58*, 2727–2759. [CrossRef]
18. Brunton, S.; Kutz, J. *Data-Driven Science and Engineering—Machine Learning, Dynamical Systems and Control*; Cambrige University Press: Cambridge, MA, USA, 2019.
19. Strang, G. *Linear Algebra and Learning from Data*; Cambridge Press: Wellesley, MA, USA, 2019.
20. Venini, P.; Pingaro, M. Static and dynamic topology optimization: An innovative unifying approach. *Struct. Multidiscip. Optim.* **2023**, *66*, 85. [CrossRef]
21. Venini, P. A new rational approach to multi-input multi-output 3D topology optimization. *Comput. Struct.* **2024**, *298*, 107362. [CrossRef]
22. Yi, B.; Yoon, G.; Zheng, R.; Liu, L.; Li, D.; Peng, X. A unified material interpolation for topology optimization of multi-materials. *Comput. Struct.* **2023**, *282*, 107041. [CrossRef]
23. Svanberg, K. The method of moving asymptotes—A new method for structural optimization. *Int. J. Numer. Methods Eng.* **1987**, *24*, 359–373. [CrossRef]
24. Lewis, A.S.; Overton, M.L. Eigenvalue optimization. *Acta Numer.* **1996**, *5*, 149–190. [CrossRef]
25. Ehrgott, M. *Multicriteria Optimization*; Springer Science & Business Media: Berlin/Heidelberg, Germany, 2005.

26. Bendsøe, M.P.; Sigmund, O. Material interpolation schemes in topology optimization. *Arch. Appl. Mech.* **1999**, *69*, 635–654. [CrossRef]
27. Trefethen, L.N.; Bau, B. *Numerical Linear Algebra*; SIAM: Philadelphia, PA, USA, 1997.

MDPI AG
Grosspeteranlage 5
4052 Basel
Switzerland
Tel.: +41 61 683 77 34

Algorithms Editorial Office
E-mail: algorithms@mdpi.com
www.mdpi.com/journal/algorithms

www.ingramcontent.com/pod-product-compliance
Lightning Source LLC
Chambersburg PA
CBHW061551220926
51850CB00005B/305
9783725873586